ENVIRONMENT COMMITTEE

Second Report

COASTAL ZONE PROTECTION AND PLANNING

Volume II

Minutes of Evidence and Appendices

Ordered by The House of Commons *to be printed*
12 *March* 1992

LONDON : HMSO

£33·60 net

The cost of printing and publishing this volume is estimated by HMSO at £56,361·00.

MINUTES OF EVIDENCE

TAKEN BEFORE THE ENVIRONMENT COMMITTEE

MONDAY 4 NOVEMBER 1991

Members present:

Sir Hugh Rossi, in the Chair.

Mr Ralph Howell   Mr Anthony Steen
Mr Robert B Jones   Mr Hugo Summerson

Memorandum by the Department of the Environment, the Ministry of Agriculture, Fisheries and Food, the Home, Scottish, Welsh and Northern Ireland Offices, the Departments of Energy, Transport and Education and Science, the Treasury and the Crown Estate Commissioners.

COASTAL POLICY

Contents

I INTRODUCTION

1. The Committee has requested of the Department of the Environment a comprehensive Memorandum setting out the Government's policies on coastal zone protection and planning. This Memorandum has been prepared by the Departments listed above.

2. Part II of this Memorandum describes the overall pattern of Government policies affecting the coast. The following parts deal with individual policies and systems in more detail:

Part III   Control over development
Part IV   Protection of wildlife and the landscape
Part V   Direct regulatory control of polluting activities
Part VI   Coastal defence
Part VII   Coastal fisheries.

Parts VIII and IX summarise current developments in policy and indicate the Government's intentions for further work.

3. There are important links between these policies and systems. For example, planning permission may be refused because of the impact of the proposed development on a Site of Special Scientific Interest established under wildlife protection provisions; or on the grounds that the development would cause pollution unacceptable in the context of the proposed location, even though it may meet more general pollution control standards. Coastal defence policy takes account of the need to protect the landscape and wildlife; and onshore coast protection works are themselves subject to the development control system, including environmental impact assessment.

II. THE COASTAL ZONE

4. Nowhere in the United Kingdom is more than 135 km from the sea. The United Kingdom's 15,000 km of coast is much larger, in relation to land area, than that of most other countries. Thus coastal issues are of a particular significance.

5. The conventional definition of the " coast " is that stretch of land bordering the sea, but it is accepted that any geographic definition of the coast may include both tracts of land and sea where terrestrial and marine environments interact. The breadth of the coastal environment therefore varies depending on the issue under consideration. Flora and fauna can be affected by windborne salt for miles inland, whereas the onshore effects of marine pollution tend to be restricted to a narrow zone of land immediately adjoining the sea.

6. The range of issues within the coastal zone is wide, including many of the issues relevant to either the marine or land environment, or to the interaction between them. The Government's policies are similarly wide-ranging, covering control over development, landscape and wildlife conservation, environmental protection, coastal defence and coastal fisheries. Government Departments—notably the Ministry of Defence and Scottish Office—are major holders of land at the coast, and the Crown Estate owns the sea bed and half of the foreshore (ie the area between high and low water mark). This Memorandum covers all of the more significant issues; however no Memorandum could possibly encompass every matter affecting the coastal zone.

7. The livelihood of many people depends on the coastal zone; many more use it for recreation. In the coastal zone as elsewhere the Government aims to secure the conditions for a thriving economy, consistent with good stewardship of the environment. The policies of each Department therefore seek to balance economic and other interests with environmental and safety factors on a sustainable basis.

Responsibility for coastal policies

8. In England, general policy responsibility for environmental matters in the coastal zone, on both sea and land, lies with the Department of the Environment (DOE), whilst the Ministry of Agriculture, Fisheries and Food (MAFF) is responsible for coastal defence policy, fisheries and the control of dumping of waste at sea. In Scotland, Wales and Northern Ireland the majority of DOE's and MAFF's functions are undertaken by the territorial Departments. the Departments of Transport, of Energy and of Education and Science have roles which span the United Kingdom, arising from their responsibilities for shipping, oil and gas and sport and recreation respectively. The Home Office is responsible for byelaws in England and Wales.

9. Local authorities' powers under the Planning Acts, their responsibilities for coastal defence, their extensive statutory environmental regulation functions and their byelaw making powers give them a prominent role in the landward part of the coastal zone. Authorities can also have an influence over matters where other regulatory agencies have the main powers, especially through their rights to be consulted by bodies such as the National Rivers Authority (NRA), Her Majesty's Inspectorate of Pollution (HMIP), English Nature, the Countryside Commission and their Scottish and Welsh equivalents.

10. The Government has issued a consultation paper setting out proposals for the creation of a new Environment Agency for England and Wales which would bring together HMIP, related functions of the NRA, and responsibility for waste regulation. The role of the NRA in relation to the new body will need careful consideration. The consultation paper therefore examines possible options for establishing a strong regulatory pollution control body whilst ensuring that there is no weakening of the protection and effective management of the water environment.

11. The Secretary of State for Scotland announced on 23 September proposals to establish a Scottish Environment Protection Agency. Although the detailed functions of the Agency will be the subject of consultation it is envisaged that it will inherit the functions of Her Majesty's Industrial Pollution Inspectorate (HMIPI), the river purification authorities and the air quality and waste regulation responsibilities of district councils. It is not thought that coast protection, currently a responsibility of regional and islands councils, will fall to the agency but the marine functions of river purification authorities in estuaries and near-coastal waters would transfer. As consultation and legislation are required the Agency will not come into existence until 1994 at the earliest. The timing of these institutional changes will be considered alongside the proposed local government reform in Scotland.

12. The Secretary of State for Northern Ireland will be considering whether it is necessary to propose any change in the present institutional arrangements for the care of the environment in the Province.

Scotland

13. Scotland's coastline differs considerably in extent and character from that of the rest of the United Kingdom. At about 10,200 km it is over twice as long as the coastline of the rest of the United Kingdom. Much of the West Coast consists of fjord type lochs; and 95 per cent of the United Kingdom's islands are in Scotland.

14. Whilst much of the legislative and policy framework on coastal matters is common throughout the United Kingdom, there are some significant differences in institutional arrangements in Scotland. The Scottish Office has wide responsibilities, including its own Environment and Agriculture and Fisheries Departments, generally covering a similar range of issues to their English counterparts. There are distinct non-departmental public bodies (NDPBs) and some statutes specific to Scotland.

15. There are also differences of degree and kind in the issues affecting the Scottish coast. For example, there is public concern in Scotland, as elsewhere in the United Kingdom, about pollution of the sea and beaches. By contrast an issue which has received much more attention in Scotland than in England and Wales is the environmental impact and location of marine fish farms.

16. The intensity of recreational and other use of the English coastline, especially on the south coast, is not generally experienced in Scotland, and conflicts between a multiplicity of activities are rare. There have been pressures of major industrial developments, particularly in the 1970s for North Sea oil and gas related developments and currently including large scale coastal quarrying, which give rise to conflict with natural heritage considerations including international obligations. But the intrinsic character and unspoilt quality of so much of Scotland's extensive coastline make it a conservation resource of United Kingdom and international value. The challenge facing the regulatory authorities in Scotland is how to safeguard this resource without prejudicing local and national economic interests.

17. Coast protection in Scotland does not raise major issues despite its long coastline. Schemes are few in number and small in scale, and are generally uncontroversial. Similarly, flood prevention in Scotland raises no major problems in coastal policy context. Coastal flooding tends to be localised and infrequent, and authorities have sufficient powers to carry out protective works.

Wales

18. The Welsh coastline is some 1,200 km long, nearly two thirds of which is steeply cliffed. The remainder includes significant areas of coastal lowlands and extensive sand dune systems. There are several notable offshore islands.

19. Much of the coastline is subject to specific conservation designations. For instance two National Parks and three coastal Areas of Outstanding Natural Beauty have extensive lengths of coastline. 445 km of Heritage Coast have been defined. There is one marine nature reserve and the designation of a second is under consideration. There are also numerous SSSIs and national nature reserves.

20. However, the diversity of the coastline leads to contrasting land uses over relatively limited stretches. Many parts of the Welsh coastline are significant for tourism and recreation, forming an important element in the Welsh economy. There is also demand for land for urban development or energy or communication needs.

21. The legislative and policy framework on coastal matters is essentially the same as in England and the institutional arrangements are briefly described in paragraph 8 above.

Northern Ireland

22. In Northern Ireland many of the functions undertaken in Great Britain by local authorities are carried out by Government Departments, including the Department of the Environment for Northern Ireland (DOE (NI)).

23. The Northern Ireland coastline has not suffered from industrial degradation, and much of it is significant in scenic terms. Pressure for recreational activity is increasing. There is a relatively small fish farming industry.

III. CONTROL OF DEVELOPMENT

24. The coastal zone provides one of the most striking instances of the success of the United Kingdom's planning and development control system. An example is the comparison of the development that took place along the south coast of England before and after planning controls took effect. Britain retains long stretches of undeveloped coast, well protected by the planning system. Much of this has been acquired by the National Trust under its Enterprise Neptune programme, and the strength of our planning system has helped to keep the price of such land low and thus made that programme more cost-effective.

25. Where development is appropriate, the planning system takes all material considerations into account and promotes solutions which respect the environment. These arrangements have helped to avoid overdevelopment and inappropriate development of the coastline which has occurred in some other countries.

26. Sections III A and III B set out the regimes controlling development and activity above and below the low water mark respectively. Section III C, which describes the role of private legislation and Orders, and section III D, which deals with recreational craft, apply both above and below the low water mark. The European Community obligation under Directive 85/337/EEC to carry out environmental assessment of certain sorts of project before they can proceed applies both above and below the low water mark. The Directive requires the consideration of an environmental statement, which must be published and sent to public bodies with environmental responsibilities for comment. The means by which that Directive is implemented for each type of development are described in the appropriate paragraphs below.

III A. **Development on land**

Development Plans

27. Above the mean low water mark, local planning authorities have powers to control the development and use of land under the Town and Country Planning Acts. The Planning and Compensation Act 1991 requires local authorities in England and Wales to prepare comprehensive development plans: county structure plans and district wide local plans in shire areas and unitary development plans in the metropolitan areas and London. There are special arrangements for National Parks. The 1991 Act enhances the status of such plans. In Scotland, structure plans are prepared by regional and islands councils, and approved by the Secretary of State. The preparation of local plans is a matter for district councils except where there are single tier planning authorities. In Northern Ireland development plans are prepared by the DOE (NI) under the Planning (Northern Ireland) Order 1991.

28. Coastal planning authorities' development plans will include as appropriate particular policies for the coastal zone, and take national and regional planning guidance into account. For reasons of simplification the 1991 Act provides that the development plan system in England and Wales should not in future include subject plans (except in the special cases of minerals planning and waste disposal). Local authorities' policies and proposals relating specifically to the coast must therefore be included in general local and structure plans. However authorities may co-operate to prepare supplementary guidance where issues are best addressed on a basis extending across their boundaries. Some groups of planning authorities—for example, on the south coast—with common coastal or estuarial interests already act closely together. The risk of conflict between the plans of neighbouring authorities may also be minimised by appropriate regional guidance and, for districts in the same county, by policies in structure plans.

29. The Government is aware of a number of individual local authorities—for instance, Sefton Metropolitan Borough Council and Hampshire County Council—who have taken initiatives in coastal zone management using their existing powers and voluntary arrangements. The Government is following these initiatives with interest. Research commissioned by the Department (see paragraph 33) is presently reviewing their effects.

Development Control

30. Development plans provide the general framework to guide development, and are a key element in deciding applications for planning permission. Development control ensures that all material considerations, including such matters as landscape protection, nature conservation and the defence of the man-made heritage, are taken into account. Each planning decision involves a judgement between such considerations and the desirability of development. The appropriate location of development, and the imposition of conditions regulating its effects, are important elements in the regime of environmental protection. Developments in certain classes must undergo environmental assessment before they may be granted planning permission.

Government Guidance—England and Wales

31. In the past, physical restraints to development such as coastal erosion, flooding and landslides have not been considered in many development plans or in some development control decisions. The Government has begun to rectify this situation by publishing Planning Policy Guidance Note 14, " Development on Unstable Land ". Both in preparing development plans and in carrying out development control, local planning

authorities have to take account of national and regional policy guidance. This includes the long-standing DOE Circular 12/72 (Welsh Office Circular 36/72) on the planning of the undeveloped coast, which includes advice on the creation of " Heritage Coasts ", and DOE Circular 17/82 (Welsh Office Circular 15/82 and MAFF circular LDW 1/82) on the control of development in areas subject to the risk of flooding. Many other areas of national guidance for general application are also relevant to development in coastal areas.

32. There is however no current up-to-date planning guidance directed specifically to coastal issues. In view of the particular sensitivity and importance of the coast, and the pressures for development, the Government recognises the need for such guidance and is preparing a Planning Policy Guidance note on coastal planning, which it aims to publish in draft for public consultation by the end of the year. Further advice on control of development in coastal areas which could be subject to flooding will be included in an updated version of DOE Circular 17/82 (MAFF circular LDW/1/82 and WO Circular 15/82) " Development in flood risk areas: Liaison between planning authorities and Water Authorities " (now the NRA). This updated advice will take account of the possible consequences of rising sea levels in the longer term.

33. A DOE research project is now reviewing all relevant published material on coastal planning and management, covering powers, policies and actions for the planning, management and conservation of the coastal zone. This research is part of a larger project on earth science information requirements for supporting coastal planning and management. Interim results will be available at the end of October, and will be taken into account in the preparation of the Planning Policy Guidance note on coastal planning policy described above. A copy of the research report will be sent to the Committee.

34. The Welsh Office is presently reviewing Strategic Planning Guidance for Wales, in consultation with Welsh local authorities. Coastal planning is likely to be amongst the subjects dealt with by that review. Authorities have been asked to submit their contributions to the review by early in 1992.

Government guidance—Scotland

35. In Scotland, Coastal Planning Guidelines were issued in 1974, and restated in the National Planning Guidelines in 1981. These guidelines, which were designed originally to manage pressure on the coast arising from North Sea oil and gas developments, identified Preferred Development Zones and Preferred Conservation Zones. Although they appeared simple and direct in character, they were underpinned by and derived from a thorough detailed study of coastal activities, and fully recognised the many statutory designations such as SSSIs and National Nature Reserves.

36. This guidance stood the test of time by being subsequently incorporated into development plans. However it is now somewhat dated, and the Environment White Paper committed the Government to review all Scottish planning policy guidance, including that relating to the coast. The Scottish Office Environment Department have therefore recently indicated to Scottish Planning Authorities that coastal planning will be one of the subjects to be addressed in the new comprehensive National Planning Policy Guidance series now in preparation. This guidance will take account of current pressures for development on the Scottish coast, such as recreation and leisure projects and large scale quarries, and define priorities for conservation.

Government Guidance—Northern Ireland

37. In Northern Ireland the DOE (NI) is the planning authority. In dealing with planning applications in coastal areas it is aware of the importance of these areas in the general environment in Northern Ireland. It is also reviewing its rural planning policy and where this affects coastal areas will be ensuring that adequate protection is retained.

Designations

38. There are no planning powers or statutory designations which apply specifically to the coast or the coastal zone. However the planning system offers great flexibility to use planning, landscape and heritage instruments to enable local authorities to protect and enhance coastal areas. National Park and Area of Outstanding Natural Beauty designations and the definition of Heritage Coasts (in Scotland, National Scenic Area, Area of Great Landscape Value and in future Natural Heritage Area) add force to this process in appropriate areas. Where valuable parts of the built heritage fall within the coastal zone, local authorities can give special protection by designating conservation areas.

Drainage

39. Drainage activities on sea and estuary coasts in Great Britain, including salt marshes, are subject to normal planning controls and environmental assessment requirements. Drainage for agricultural purposes may benefit from permitted development rights, which can in certain circumstances be withdrawn. Drainage works in the course of flood defence are described in Part VI below.

Geographical extent

40. The planning system operates only within local authorities' jurisdiction. That jurisdiction generally ends at the low water mark although in some cases it is more extensive, for instance where boundaries are drawn between the outermost mean low water marks of an estuary.

III B. Control of development and use below the low water mark

41. The great diversity of development on land requires that there is a single, comprehensive system to control development and use of land. Such a system is possible because the land is capable of division into discrete packets in identifiable ownership.

42. Below the low water mark different areas may be more or less subject to particular uses, but division along rigid boundaries is impractical. The Crown Estate Commissioners are the landlords of the sea bed, but the sea itself is owned by no one. Regulation is therefore generally targeted on particular activities rather than on defined areas. This is practicable because there is a much smaller variety of development below the low water mark, and less direct conflict between uses competing for available space. Furthermore, there are aspects of activity below the low water mark which have no analogy on land, for instance the historic right of free navigation. Controls ensuring safety of navigation therefore apply to all activities.

43. Consideration of coastal development proposals is inevitably complex, given the wide range of issues which must be taken into account. But this does mean that businesses may require consents from about five and sometimes as many as eight different organisations for works in tidal waters. Ministers have therefore agreed, as part of the Deregulation Initiative, a review of the scope for rationalising this system of consents. Where controls cannot be removed, the Government is considering how procedures can be simplified.

44. The need to protect the environment is taken into account in the various approvals systems. In particular, there is provision for environmental assessment of development proposals likely to have a significant effect on the environment. Paragraphs 46 to 67 below describe each regime in turn.

Role of the Crown Estate Commissioners

45. The sea bed lying between Mean Low Water Mark (or Spring Low Water in Scotland), and the limit of territorial waters, together with about half the foreshore of the United Kingdom (that part of the shore between low and high water marks), form part of the Crown Estate. Almost any activity on the foreshore owned by the Crown Estate and on the sea bed requires approval and consent from the Commissioners acting as landlords. But the two main activities which have the greatest environmental sensitivity are marine aggregate extraction (mainly in England and Wales) and marine fish farming (mainly in Scotland).

Marine aggregate extraction

46. Minerals from the seabed account for about 16 per cent of the supply of sand and gravel for the construction industry in England and Wales, as well as playing a vital role in coast protection (see paragraph 115). It is Government policy to encourage the use of marine aggregates wherever this is possible without introducing a risk of coast erosion or unacceptable damage to sea fisheries and the marine environment.

47. Marine aggregate extraction is controlled through the non-statutory Government View procedure. The Commissioners consult informally with Fisheries Departments, the Department of Transport, the Department of Energy, Coast Protection authorities and heritage and conservation bodies. All applications for production licences are also advertised in the local papers and the fishing trade press. The responses to consultation are passed to the Department of the Environment (or Scottish or Welsh Office), who are responsible for the formulation of the Government View. Before finalising the View, the Department consults with other Government Departments whose interests are affected by the proposals. The Crown Estate does not issue a licence if there is a negative Government View, and it accepts in every case all conditions included in a favourable View by incorporating them in its licence. The conditions of a licence are contractual terms enforceable as such by the Commissioners. The Government View Procedure was revised in 1989 and incorporates environmental assessment requirements.

Marine fish farming

48. Marine fish farming—mainly for salmon—has expanded rapidly around the West coast and islands of Scotland and there are now 600 sites leased by the Crown Estate Commissioners. Shellfish farming is more widely dispersed around the coasts of the United Kingdom. The Commissioners have progressively developed a broad based consultative framework for considering environmental issues, in association with the Secretary of State for Scotland. This involves seeking reaction from the relevant statutory and other bodies to all new site applications and embraces the procedures involved in the discharge of the Commissioners' responsibilities as competent authority for environmental impact assessment under EC rules. Thus the Commissioners have a wide range of viewpoints available to them before taking a decision and, in particularly contentious cases, an Advisory Committee (with an independent Chairman and Deputy Chairman) on which relevant statutory bodies are represented, consider the issues. Separate control arrangements have been developed in Shetland under powers in the Zetland County Council Act 1974; similar arrangements apply in parts of Orkney (see paragraph 68).

49. The rapid development of the industry in the 1980s and the role of the Crown Estate's site leases as both a generator of income and a regulatory system has been subject to criticism. The dual role has been recently and fully explored in a report (HC 141 1989–90) by the House of Commons Select Committee on Agriculture. There has been no evidence of serious environmental damage resulting from the manner in which the industry has been controlled; and it is doubtful that any different control regime would have substantially affected the environmental outcome.

50. In its response to the Committee's report (HC 116 1990–91) the Government agreed with the Committee's advocacy of an evolutionary approach to control and committed itself to developing revised guidance on the location and siting of marine fish farms. Fish-farming is subject to a range of other regulatory controls such as those operated by Fisheries Departments in respect of fish diseases and river purification authorities in Scotland (the NRA in England and Wales) in respect of the discharge of effluent, with particular attention to chemical loadings. The landward components of fish farms are subject to planning control. There is a substantial programme of research funded by Government and by the industry itself on matters of husbandry, including environmental effects.

51. In Northern Ireland marine fish farms are licensed by the Department of Agriculture for Northern Ireland under the Fisheries Act (Northern Ireland) 1966. This includes the exclusive right to cultivate marine fish and shellfish in specified waters. While the applicant requires the consent of the Crown Estate Commissioners, all applications for licences under the Fisheries Act are subject to a wide consultation procedure undertaken by the Department of Agriculture for Northern Ireland which includes advertisements in local papers and direct contact with known interested parties including environmental groups. There are also provisions requiring the holding of public local inquiries if there are substantial objections to an application for a licence. A fish culture licence is not issued by the Department of Agriculture for Northern Ireland unless DOE (NI) gives its consent under the Water Act (Northern Ireland) 1972. The DOE (NI) also controls the discharge of effluent from fish farms under the 1972 Act.

Transport infrastructure

52. Procedures for approving road, railway and airport projects in tidal waters are in practice much the same as those applying on land. Various categories of these developments require environmental assessment. These requirements may be discharged through consideration by the local planning authority or through Private Bill procedures (see paragraph 69). The Department of Transport itself undertakes assessments for trunk roads in England, and it (or the DOE (NI)) arranges them for some works in ports and harbours. Telecommunication and power cables, some temporary works, individual moorings and small pontoons do not require an Assessment.

Development affecting navigation

53. Proposed works in tidal waters, and in designated areas on the Continental Shelf, which would obstruct or endanger navigation require the consent or approval of the Secretary of State for Transport under the Coast Protection Act 1949. The system of consents is used to control hazards to navigation around the coasts and in shipping lanes, either by refusing consent or more normally by use of conditions on, for instance, lighting requirements. The Secretary of State's powers are confined to navigational considerations and do not allow a more general view on the merits of a proposal.

54. Applicants for consent under the 1949 Act may be required to advertise and the Department is considering making this requirement more general. Those likely to be affected must be informed. Other parties, such as shipping and fishing interests, are invited to comment on the navigational aspects of proposals. Public inquiries are not generally required for consents under the 1949 Act, but if an inquiry is required for some other aspect of a proposal, it can be extended to encompass the consent.

Shipwreck sites

55. There are many thousands of sites of shipwrecks in United Kingdom coastal waters. Some of these are important historic or military remains, and it is the Government's policy to protect them from inappropriate interference by designating them under the Protection of Wrecks Act 1973 or the Protection of Military Remains Act 1986, as appropriate.

56. It is an offence under the 1986 Act to tamper with aircraft or vessels, and associated human remains, at a protected place. Damaging, or diving on, the site of a designated historic or dangerous wreck without a licence from the Secretary of State is an offence under the 1973 Act. Licences are issued to allow archaeological investigations of historic wreck sites. These are subject to conditions about the way the activity is carried out and the supervision and expertise of those involved. On 1 April 1991, responsibility for designating historic wreck sites and licensing activity on them was transferred from the Department of Transport to the Departments with responsibility for archaeology on land in each part of the United Kingdom. The Ministry of Defence is responsible for military remains.

57. The Department of Transport is responsible for the designation and protection of dangerous wrecks, and for the system of Receiver of Wreck which operates under the Merchant Shipping Act 1894 and is designed to protect the rights of owners, mariners and salvors. Since 1 April 1991 the commission payable to Receivers for recovered wreck has been waived.

Ports

58. There are some 300 harbours around the coast of Great Britain and nearly all with any significant degree of commercial or recreational use operate under statutory powers, which are set out in local Acts or Orders made under the Harbours Act 1964. The local circumstances of the harbours are very varied; and so is the nature and function of the harbour authorities themselves. Some deal exclusively with regulatory or conservancy functions to do with the navigation of the waters within the harbour limits; others have a mix of activities including commercial operations, for instance the handling and storage of goods passing through the harbour.

59. Harbour authorities need statutory powers to construct and maintain harbour works below the high water mark. Where their existing powers are insufficient they apply for Harbour Orders which are made by the appropriate Minister (in most cases the Secretary of State for Transport). If there are objections to a proposed Order a public inquiry is almost always held. A licence is also required under the Food and Environment Protection Act 1985 (from MAFF in England and Wales and the Scottish Office Agriculture and Fisheries Department in Scotland) for construction work below the mean high water mark.

60. Applications for Harbour Orders involving any significant works must be accompanied by an environmental assessment. In reaching a decision on the application for an Order the Secretary of State must take into account the assessment together with the comments of relevant bodies who have been consulted about it. Harbour works below low water mark which are not authorised by a Harbour Order and which do not require planning permission were brought within the environmental assessment regime under Regulations made in 1989.

61. The Department is discussing delegation to Harbour Authorities of the consents described in paragraph 53 above. This would not require primary legislation. A right of objection or appeal to the Secretary of State would remain. Harbour Authorities already have to consider applications for works licences, which deal with issues which overlap with the Secretary of State's consents, and duplication of effort would be significantly reduced. Other possibilities for delegation are also being considered.

Oil and gas extraction and pipelines

62. Under the Petroleum (Production) Act 1934 a licence is required to explore for or extract oil or gas on land or on the United Kingdom continental shelf. Some 40 existing licences are on or near the coast in Great Britain.

63. Licences for oil and gas exploration and production below the low water mark may carry conditions, restrictions, or inhibitions on activity. These may require licensees to consult or notify interested parties of exploration or development plans; they may include seasonal restrictions on activities, or designate particular parts of a licensed area as either no-go areas or, for example, areas which cannot be drilled. The making of such conditions follows full consultation of other Government Departments and agencies.

64. The resulting licence conditions, restrictions and inhibitions relate specifically to the licensed area concerned. The Government has no general regulatory power to apply the conditions agreed for new licences retrospectively to those already in existence. However the Secretary of State has a further, limited opportunity to consider additional environmental conditions when programmes for the development of a field or the abandonment of structures and pipelines offshore are being agreed. Development consent conditions are agreed under the Model Clauses attached to each licence. Conditions related to the abandonment of offshore structures and pipelines are attached to the abandonment programmes agreed under the Petroleum Act 1987. In both cases the Secretary of State has the power to impose programmes or conditions where these are not agreed.

65. Current conditions and restrictions in force on various licences reflect environmental, fishing, defence and transport interests. Applicants for licences carry out analyses of environmental impact which may also include the potential impact on leisure activities and amenities.

Renewable energy sources

66. A variety of renewable enery sources might operate within the coastal zone, including tidal, wind and wave power schemes and use of seaweed or other marine plants as a fuel. All these technologies are still at the research, development and demonstration stage, and their economic viability for commercial scale use has yet to be established. The Government is supporting studies on tidal energy barrages on the rivers Mersey, Severn, Wye and Conwy in order to reduce the uncertainties on costs, performance and regional and environmental issues to the point where it will be possible to take decisions on whether or not to plan for construction. The Government would promulgate appropriate consent procedures before any such development was begun.

67. The Department of the Environment and the Department of Energy are presently preparing planning policy advice on how local planning authorities should handle applications for renewable energy developments, with particular reference to their treatment in coastal areas.

Scottish Islands Councils

68. The three Scottish Islands Councils are unique in being unitary authorities and also responsible for control of pollution in coastal waters, although the consent of the Secretary of State for Scotland is required for any discharges made by the islands councils themselves. In Shetland and Orkney, powers are further advanced by local Acts based on ports and harbours legislation which provide the local authority with powers to licence "works" within defined harbour areas; in the case of Shetland these extend to the surrounding territorial waters. These powers were based on the need to control oil and gas related activities, but have subsequently provided a means to control other activities such as marine fish farming. The authority is able to grant licences subject to conditions, including on local investment. The system operates well, probably due to both the inherent nature of the legislation and its operation by unitary authorities which also embrace ports and pollution responsibilities and which have provided support for activities which have been taken into control. The legislative regime incorporates a third party right of appeal.

III C. Private Bill procedure

69. Certain types of development project both above and below the low water mark have at present to be approved by the Private Bill procedure. From the 1991–92 Parliamentary Session Private Bills for the authorisation of works must be accompanied by an environmental statement unless the Secretary of State determines that this is unnecessary. The Government proposes to introduce legislation whereby these types of development will be authorised by Ministerial Order, rather than by private Act. This will apply to those works such as tidal barrages and artificial islands which interfere with rights of navigation in English and Welsh territorial waters. The Order making procedure will incorporate environmental assessment where this is required under EC law.

III D. Recreational craft

70. Coastal areas are used by large numbers of people for a wide range of recreational activities, ranging from swimming, sailing, rowing, diving and angling to power boating, water skiing and use of water scooters. Co-operation between neighbouring local authorities and between authorities and sports clubs is encouraged, in order to create coherent plans for recreational use of coastal areas which reconcile the interests of the many varieties of participant. Activities can be managed so as not to create conflict by co-ordinated planning of the use of coastal resources, by provision of sites with suitable supporting facilities where there is extensive use, and by education of participants about respect for other users and for the environment. The Department of Education and Science is currently carrying out a review of safety in water sports, which is likely to promote action on the education of participants in order to improve safety.

71. The Department of Transport's controls on commercial shipping are being progressively extended to all craft for trade and hire (for example pleasure and charter boats and sail training craft). In the main, however, small recreational craft are not regulated in the same way. Safety issues raised by these craft are different. They do not carry paid crew, paying passengers or potentially dangerous cargo. The main problems are congestion in popular areas and poor seamanship.

72. There are estimated to be some 150,000 sea-going recreational craft in use excluding surfboards, water scooters and similar craft. The main safety problem is not the craft themselves but the way in which they are used. Craft are sometimes taken out in conditions which exceed the capabilities either of the craft or those on board, or are used—as is sometimes the case with water scooters—without sufficient consideration for others. A significant portion of HM Coastguard's resources have been devoted to education of users, working closely with the Royal Yachting Association and other user associations.

73. Local authorities have powers to control the use of recreational craft up to 1000 metres beyond the low water mark by making byelaws under the Public Health Acts 1936 and 1961 (for England and Wales) and the Civic Government (Scotland) Act 1982, subject to confirmation by the relevant Minister. Byelaws can set aside areas for bathing, regulate the navigation of vessels used for pleasure purposes within such areas, and regulate the use of pleasure boats to prevent danger, obstruction or annoyance to bathers in the sea. They may be used to regulate the speed and navigation of all pleasure boats, including power boats, water skiers and water scooters, and to require silencers to be fitted to pleasure boats.

IV. Protection of Wildlife and Landscape

74. The United Kingdom has the finest and most diverse coastline in Europe. The range of coastal landscapes include the fjords of west Scotland, the cliff lines of south west Wales and Cornwall, the unique chalk geology of Dorset and many estuaries, bays and enclosed harbours. Not only is the coast a distinctive feature of our landscape, but coastal habitats are home for many internationally important species of plants and wildlife. The importance of the coastline has been reflected in national systems to protect its landscape

and wildlife which go back over 40 years. These have succeeded in affording greater protection to the most valuable sections of our coastline than in any other EC country. Above the low water mark, protection of coastal landscape and wildlife is achieved largely through the same mechanisms as apply to other parts of the country.

75. Landscape and wildlife conservation are the responsibility of the Department of the Environment and the territorial Departments. Policy is largely implemented through NDPBs; in England, the Countryside Commission for the landscape and English Nature for wildlife. In Wales, the Environmetal Protection Act 1990 brought these responsibilities together in one agency, the Countryside Council for Wales (CCW). The Natural Heritage (Scotland) Act 1991 will bring together from 1 April 1992 the Countryside Commission for Scotland and the Nature Conservancy Council for Scotland (NCCS) into a new body, Scottish Natural Heritage (SNH). In Northern Ireland, the DOE (NI) is responsible for nature and countryside conservation.

IV A. Landscape conservation

76. The conservation of the coastal landscape is an important element in the more general system of landscape protection, which in England and Wales is largely founded on those areas designated under the National Parks and Access to the Countryside Act 1949. Five of the 10 National Parks in England and Wales contain stretches of coastline as do almost half of the 39 Areas of Outstanding Natural Beauty (AONBs) so far designated. In both types of area a stricter than normal planning regime operates which gives high priority to the conservation of the landscape and thereby to the protection of some of the finest stretches of our undeveloped coastline.

77. The Countryside Commission is responsible for encouraging public access to the countryside. It does this through the designation of long distance footpaths and bridleways and by encouraging local authorities to promote the development of more local routes. A long distance route around the South West peninsula some 829 km long was opened in 1978. This route affords public access to some of the most magnificent stretches of coastline in England. Other long distance routes provide access to the coastline in Norfolk, Cleveland, and along the South Downs.

78. In addition, 1970 the Countryside Commission introduced the concept of Heritage Coasts. These are not statutorily designated, but the Commission has identified some 1460 km—or one third of the coastline of England and Wales—as of sufficient quality in terms of its scenic beauty and opportunities for enjoyment by the public as to be worthy of Heritage Coast status. The Department encourages local authorities to define Heritage Coasts and to include appropriate policies for their planning in development plans. The great majority of Heritage Coasts are afforded further protection by their inclusion within National Parks or AONBs. Heritage Coasts have promoted a focus on the management needs of the finest stretches of our coastline. Local authorities have been encouraged and grant aided to develop Heritage Coast services, including appointment of Heritage Coast officers and development of management plans. In January 1991 the Commission published an updated policy document on Heritage Coasts which set out proposals for taking the concept forward by strengthening management measures and ensuring adequate resources are available. The Government intends to respond to this statement shortly. The Heritage Coast programme has been closely associated with the development of the National Trust's " Enterprise Neptune " initiative, under which the Trust has taken many of the finest stretches of coastline into protective ownership. The Trust has raised some £15 million so far and secured the protection of some 770 km of coastline, of which 540 km are Heritage Coasts.

79. In 1991 the Countryside Commission in England launched, on behalf of the Government, a new initiative, Countryside Stewardship, which is aimed at restoring and enhancing traditional landscapes. Farmers and land managers may enter into contracts to undertake environmental works designed to recreate distinctive elements in the landscape. Countryside Stewardship is avowedly experimental but some £13 million has been made available for the first three years. The scheme is discretionary and targeted towards a series of prescriptions considered of special significance. One of these is aimed at restoring traditional landscapes along the coastal strip. Public access is an important feature of Stewardship, and has been given special emphasis in the coastal prescription. A similar scheme is being considered for Wales.

80. The only national landscape designation in Scotland is the National Scenic Area; 31 of these have a coastal element. In addition regional authorities have identified designations such as Areas of Great Landscape Value in their development plans. The Natural Heritage (Scotland) Act 1991 created a new designation, the Natural Heritage Area, designed to safeguard and enhance areas combining high nature conservation and landscape qualities. The Government is currently considering the result of a consultation exercise on this new designation, and will expect SNH to make recommendations after consultation with relevant interests for the implementation of this designation in specific areas. At present it is not envisaged that this designation will extend beyond the low water mark. The National Trust for Scotland owns stretches of the Scottish coast.

81. In Northern Ireland the only landscape designation is AONB; 73 per cent of the coastline (471 km) is so designated.

IV B. **Wildlife conservation**

82. Conservation of wildlife is primarily achieved through the designation by the conservation agencies (English Nature, the NCCS and the CCW) of Sites of Special Scientific Interest (SSSIs), and in Northern Ireland by the designation of the equivalent Areas of Special Scientific Interest (ASSIs) by the DOE (NI). A significant number of the total of over 5,600 SSSIs contain a coastal element. Within SSSIs owners and occupiers are prohibited from carrying out specified potentially damaging operations unless the consent of the relevant conservation agency is obtained or a management agreement entered into. There is also a strict planning regime. Local authorities are required to consult the relevant conservation agency and take the conservation interest of these sites into account, including the international interest where appropriate, when developments which might affect them are proposed. The new integrated designation in Scotland, the Natural Heritage Area (see paragraph 80) will cover nature conservation as well as landscape issues.

83. The conservation agencies establish National Nature Reserves (NNRs) and ensure that the primary objective of their management is the enhancement of nature conservation. They can also make byelaws for their protection. Local authorities throughout the country have analogous powers to establish Local Nature Reserves (LNRs). Most NNRs and LNRs are in private ownership, but voluntary organisations such as the Royal Society for the Protection of Birds and the County Wildlife Trusts have also played a valuable part by acquiring and managing their own reserves, many of which are on or near the coast. The Crown Estate Commissioners co-operate with conservancy groups in granting management leases on the foreshore.

84. Section 3 of the Wildlife and Countryside Act 1981 carries forward the provisions of the Protection of Birds Act 1954 in providing for the establishment, with the agreement of owners, of Areas of Special Protection (bird sanctuaries). Such areas may under certain circumstances extend below the low water mark. Section 36 of the 1981 Act also introduced Marine Nature Reserves which may be established for conservation or research purposes and be regulated by byelaws. Very similar provision is made in the Nature Conservation and Amenity Lands (Northern Ireland) Order 1985. MNRs have been designated at Lundy in 1986 and Skomer in 1990. There are currently no formally designated Marine Nature Reserves in Scotland, although a voluntary marine nature reserve operates at St Abb's Head. Proposals for an MNR at Loch Sween are under review in the light of initial local consultation.

85. In 1986 the then Nature Conservancy Council (NCC) identified 14 marine areas off the Scottish coast which it termed Marine Consultation Areas (MCAs). These have no statutory authority but identify areas considered by NCC to be of particular distinction in respect to the quality and sensitivity of the marine environment. A further 15 areas have since been identified and a document containing the citations for each of the 29 areas was published by NCC in 1990. The selection of MCAs allows bodies which consult with NCC to be aware of marine conservation issues within particular areas. The main use of marine consultation areas by NCC has been in support of its response to consultations undertaken by the Crown Estate on proposed marine fish farm leases (see paragraph 48). The Environment White Paper "This Common Inheritance" (Cm 1200) included commitments to review existing legislation of relevance to marine conservation and to consider extending the concept of MCAs to England and Wales. The report (Cm 1655) on the first year of implementation of "This Common Inheritance" announced that the Government will issue a consultation document by January 1992 containing a first list of proposed MCAs in England and Wales, and draft Government guidelines addressed to regulatory agencies on the operation of an MCA scheme. The consultation scheme would initially operate in areas where pressures on conservation are strong, and will be monitored to see how well it addresses the problems facing coastal areas.

86. The Crown Estate identified separately in 1989 in its guidelines on the location and siting of marine fish farms a list of areas which it termed Very Sensitive Areas for marine fish farming. The revised guidance being prepared by the Scottish Office (see paragraph 50) will update these Very Sensitive Areas in the light of advice from the Government's statutory advisers on wildlife and conservation in Scotland and other consultees, together with recent research on the likely impact of fish farms on the marine environment.

87. The EC Directive on the Conservation of Wild Birds requires the designation of Special Protection Areas (SPAs). Forty of these have been designated so far, including 27 on the coast, and the conservation agencies have identified a possible further 182. SPAs largely receive their protection through their parallel status as SSSIs or ASSIs. Wetlands of international importance are also required to be designated under the Ramsar Convention, to which SSSI status again affords the major protection. Forty-five Ramsar sites have so far been listed in the United Kingdom, of which 21 are on the coast.

88. The requirement for designation of SPAs extends to the limit of territorial waters. The Ramsar Convention extends to marine water up to six metres in depth. However the powers to notify SSSIs and ASSIs relate to the boundaries of local authority jurisdiction. The Government will examine whether new powers are necessary to achieve nature conservation objectives in the light of the proposed EC Habitats Directive.

V. CONTROL OF POLLUTION

89. Pollution of coastal waters can arise both from the landward side and from sea based activities. The main inputs of polluting material are via rivers and coastal discharges, although emissions to the atmosphere can also have an adverse effect on marine waters. For the future, in England and Wales the overriding requirement of pollution control in respect of coastal waters will be the need to meet the water quality objectives which must be established for all coastal waters under the Water Act 1989.

90. The United Kingdom is fully committed to meeting the requirements of the EC Bathing Water Directive as soon as is practicable through a £1·4 billion bathing water compliance programme. Over 450 United Kingdom bathing waters have been identified under this Directive: 77 per cent met the standards in 1990 compared with 51 per cent in 1986.

International Agreements

91. The Government has entered into a number of international agreements to protect the quality of the marine environment. The North Sea Conferences of 1984, 1987 and 1990 have produced wide ranging agreements to reduce inputs of pollutants, to regulate offshore activities, to limit pollution from ships, and to protect certain marine species. Government policy is to apply these agreements to all coastal waters. The Government participates actively in the London Dumping Convention and the Paris and Oslo Commissions to adopt measures to protect the marine environment.

92. The EC has also adopted Directives to control inputs of dangerous substances (Directive 76/464 and associated daughter directives), and to set water quality standards for particular purposes—for instance, Directive 76/160/EEC on bathing waters, and Directive 79/923/EEC on shellfish waters.

V A. **Pollution on and from land**

93. In England and Wales, the Water Act 1989 and the Environmental Protection Act 1990 provide the statutory basis for control of the main land based polluting activities.

94. Pollution from certain prescribed processes is subject to integrated pollution control by HMIP in England and Wales, and jointly by HMIPI and the river purification authorities in Scotland. This regime requires that potential polluting processes must apply best available techniques not entailing excessive costs (" BATNEEC ") in order to curb the creation and discharge of wastes. It applies in particular to processes which discharge substances most harmful to the environment, including the marine environment—those on the United Kingdom's " Red List ". HMIP is required to consult the NRA before granting discharge consents and to take account of their requirements for achieving water quality objectives.

95. All other discharges to water are controlled by the NRA in England and Wales. In particular the NRA has responsibility for implementing Government policy to require treatment for all substantial discharges of sewage. This policy is now enshrined in the EC Urban Waste Water Directive. It is estimated that over £2 billion will be spent by 2005 on improvements to sewage discharges in coastal waters in addition to the £1·4 billion previously committed to improve the quality of bathing waters.

96. Under the 1990 Act, to the landward side of mean low water mark, local authorities are responsible for regulation of waste disposal and discharges to air and land, and control of nuisances such as noise. In Great Britain local authorities, the Crown Estate Commissioners and Government Departments, amongst others, are under a duty to ensure that areas in their control to which the public have access are kept free from litter and refuse (including dog faeces) so far as practicable. Beaches above the level of the mean high water tide must therefore be cleaned to the standards set out in a Code of Practice. Preparations for similar measures in respect of Northern Ireland have begun.

97. Arrangements in Scotland are similar, with the river purification authorities (the seven River Purification Boards and the three Islands Councils) exercising water pollution control powers under the Control of Pollution Act 1974. In Scotland the Government has provided for a steeply rising programme of capital expenditure by regional and island councils as water sewerage, coast protection and flood prevention authorities. During the three years 1991–94 provision for the programme will amount to £624 million and by 1993–94 annual capital expenditure will be double the 1988–89 level. Further, last year the Government introduced a new sewerage improvement grant to assist Scottish sewerage authorities to reduce pollution or otherwise benefit the environment. These measures will enable faster progress to be made in attaining United Kingdom and EC standards.

98. Under the Water Act (Northern Ireland) 1972, DOE (NI) is responsible for the cleanliness of inland and coastal waters; this Act provides for a system of statutory consents for discharges to water.

99. Local authorities have responsibilities for regulation of waste disposal and some emissions to air though these activities are less likely to have an adverse effect on the marine environment.

100. The Government has proposed that a new Environment Agency should undertake several of these pollution control responsibilities (see paragraph 10).

V B. Pollution from sea based activities

101. Separate controls apply to sea based activities which could pollute the marine environment. Disposal of waste at sea is controlled by MAFF, the Scottish Office Agriculture and Fisheries Department and DOE (NI) under powers in the Food and Environment Protection Act 1985. The range of wastes that may be disposed of at sea and the concentration of contaminants in dumped waste are being steadily reduced. In particular the disposal of industrial waste should end in 1992, disposal of sewage sludge will end in 1998, and of practicable stone waste from collieries will no longer be dumped after 1997. Incineration at sea has been banned since 1990. In licensing disposal at sea the Act requires account to be taken of the need to protect the marine environment, the living resources which it supports and public health, and to prevent interference with legitimate uses of the sea.

Pollution from oil and gas rigs

102. Regulations made under the Petroleum (Production) Act 1934 (see paragraph 62 above) include model licence clauses covering avoidance of harmful methods of working. These require licencees to take all practicable steps to control the flow and prevent the escape or waste of petroleum, and to report any escape to the Minister and Her Majesty's Coastguard. Oil Spill Contingency Plans are required before any well is drilled, and before production platforms start operation. Discharges of water and drill cuttings contaminated with oil based drilling mud are regulated under the Prevention of Pollution Act 1971, which provides that it is an offence to permit oil to be discharged into the sea. The Secretary of State is able to grant exemptions, and to attach conditions setting out the maximum permissible amounts of oil that may be discharged. These conditions take into account international agreements on reducing pollution in the North Sea.

Pollution from vessels

103. Prevention of pollution from vessels is necessarily achieved through international co-operation. Regulations cover oils and oily residues, noxious and dangerous cargoes and ship-generated garbage. Regulations for the prevention of pollution by sewage from ships have yet to come into force, and new requirements are being developed aimed at limiting air pollution from ships. The principal measures adopted to prevent pollution include constraints on tanker operation and design, minimum standards for shipboard equipment and requirements for survey and certification to ensure compliance. Over 30 per cent of ships visiting United Kingdom ports are inspected to check that they comply with these requirements.

104. Improving standards of safety of navigation is one of the best ways of reducing accidental pollution, as reducing the chances of an accident automatically reduces the chances of pollution. Encouraging vessels to keep well away from shore is not necessarily the best way of achieving this, if the alternative is using more dangerous waters. Account has to be taken of the international right of innocent passage through the seas around our coasts, including territorial waters.

105. Measures taken include the setting up of Areas To Be Avoided (ATBAs), which are designed to protect particularly ecologically sensitive areas by keeping certain catagories of vessels out of them completely: they are only practicable when there is a safe all-weather alternative. Two ATBAs have been established in United Kingdom waters, one off the coast of Pembrokeshire and the other around the west coast of Shetland. Where ATBAs are impracticable, the Department of Transport establishes advisory arrangements. For example, the Department advises tanker masters to use the Deep Water Route to the west of the Hebrides rather than going through the Minch, unless weather conditions are such that the outer route is particularly dangerous.

106. One particularly successful measure has been the introduction of the radar-based and computer-controlled Channel Navigation Information System to improve the safety of navigation through the Straits of Dover, one of the world's most congested seaways. Since it began operation in 1972, the accident rate has been reduced by over 85 per cent, despite substantial increases in traffic. There has been no oil pollution incident in the area covered since 1978. However, the system depends on availability of sites for radar scanners and some other equipment in adjoining areas of high landscape quality.

Dealing with spillages

107. Major spillages of oil and other pollution incidents arising from vessels at sea are dealt with by the Marine Pollution Control Unit (MPCU) of the Department of Transport. MPCU aircraft carry out regular patrols to detect and deter illegal discharges. Additional patrols are mounted in response to specific reports of slicks. Ships and aircraft are required to report sightings of oil in British territorial waters. Whenever a vessel is identified as a possible source of a slick, rigorous follow-up action is taken and arrangements made for the ship to be inspected when it next enters a Western European port.

108. MPCU monitors all reports of oil spillages, and takes counter-pollution action when oil from a shipping casualty threatens United Kingdom interests, particularly its beaches, fisheries, seabirds and ecologically sensitive areas. The Unit maintains a national contingency plan with the resources to support it, including aircraft and vessels, MPCU is also responsible for dealing with pollution arising from spillages of chemicals or other hazardous substances from ships and has under contract a team of mariners experienced in the operation of chemical tankers. Responsibility for keeping shorelines clean rests with local authorities.

109. English Nature, NCCS and CCW advise MPCU on the effects of oil pollution incidents on wildlife. To help with contingency planning arrangements, MPCU commissioned the then NCC, with support from the British Petroleum Company, to produce an atlas of the coastline of Great Britain showing sensitive wildlife areas vulnerable to oil pollution. The 77 colour maps, which have now been distributed to coastal county and regional authorities as well as to Government Departments, show the broad distribution of wildlife habitats along the coast, concentrations of seabirds, marine mammals and other fauna and areas designated for their nature conservation importance, including SSSIs and nature reserves.

110. The costs of cleaning up pollution arising from tanker casualties are recovered in full from the owners, United Kingdom or foreign, via international compensation regimes. Spillages from offshore installations are usually dealt with by operators, with MPCU advising when necessary.

Scientific support for pollution prevention

111. The Agriculture Departments maintain marine laboratories which provide advice to all relevant Government Departments, the Crown Estate Commissioners, and the NRA about the potential impact on the marine environment of various activities, as a basis for regulatory action.

112. Research carried out by the Natural Environment Research Council, the NRA, the Agriculture Departments, the conservation agencies, universities and others examines the natural processes of the marine environment and the effects of human activities on the life of the seas. The Co-ordinating Committee on Marine Science and Technology has estimated that United Kingdom publicly funded research on the marine environment is currently around £53 million per annum.

VI. COAST DEFENCE

113. Much of Britain faces a constant threat of flooding by rivers and the sea, and long stretches of coastline particularly in the East of England are under attack from the combined actions of waves and weather. The danger that water poses to life, property and the environment is always with us and many civil engineering works designed to keep the sea at bay trace their origins back through the Middle Ages to the Romans. British expertise is recognised worldwide. However, although we can draw on long expertise, it is important that we can adapt to changing circumstances and benefit from improved understanding of coastal processes made possible by research and development.

VI A. England and Wales

114. Responsibility for coast defence policy in England and Wales lies with MAFF and the Welsh Office. The term coast defence embraces coast protection (defined as the protection of land from erosion or encroachment by the sea) and the sea defence element of flood defence (defined as the prevention of flooding of land by rivers or the sea).

Coast protection

115. The Coast Protection Act 1949 empowers maritime district councils to carry out works which have been approved by the appropriate Minister to protect against coastal erosion and encroachment by the sea. The Act requires councils to consult about all works apart from maintenance, repair and emergency works before submitting proposals to MAFF. Consultation includes the NRA, neighbouring Maritime District Councils, the County Council, harbour, conservancy and navigation authorities and fisheries committees and MAFF (for licence under the Food and Environment Protection Act 1985). MAFF consult other Government Departments, the Crown Estate Commissioners, the Countryside Commission and English Nature or the Countryside Council for Wales. Schemes must be advertised. Objections to coast protection schemes on the grounds that they will be detrimental to protection or will interfere with the exercise of statutory functions of others, if not withdrawn are resolved by a local inquiry or hearing. Similar arrangements for consultation and consideration of objections apply in Scotland.

Flood defence

116. The Land Drainage Act 1976 and the Water Act 1989 provide for the construction, improvement and maintenance of defences against inland and coastal flooding in England and Wales to be undertaken by the NRA through its ten Regional Flood Defence Committees (whose Chairmen are Ministerial appointees); by local authorities; and by some 250 Internal Drainage Boards (IDBs) which administer areas with special drainage needs in, for example, the Fens.

Co-ordination

117. Authorities are responsible for identifying coastal defence works within the Government's priorities and for carrying them forward. The NRA is required to exercise a general supervision over all matters relating to flood defence and must be consulted by maritime district councils about coast protection works. These arrangements take full advantage of local knowledge and have worked well over time. In addition, informal coastal groups have been formed which bring together bodies with coastal defence responsibilities (including local authorities and the NRA) with the aim of encouraging co-ordination and exchange of information. MAFF and the Welsh Office have recently initiated a coastal defence forum through which groups can liaise on technical matters of common interest, and which provides a new mechanism for encouraging strategic planning of coastal defences on a coastline basis.

VI B. Scotland

118. In Scotland, coast protection and flood prevention are governed by the Coast Protection Act 1949 and the Flood Prevention (Scotland) Act 1961 respectively. The Acts provide discretionary powers to islands and regional councils, as coast protection and flood prevention authorities, to protect the coast against erosion and encroachment from the sea and non-agricultural land from flooding. To this end coast protection schemes and flood prevention schemes are prepared by local authorities and have to be approved by the Secretary of State for Scotland. There are only eight mainland coast protection authorities in Scotland and, as a result, they cover significant lengths of coastlines; as regional councils they also have structure plan functions and are represented on the river purification boards which are responsible for flood warning. This minimises any conflict of interest at the boundaries between authorities, and provides every opportunity for a coherent regional policy to be developed for the coastal zone above the low water mark. Authorities are able to plan on a regional basis for flood prevention as for coast protection. However, authorities tend to give coast protection and flood prevention schemes low priority because of their discretionary nature. Some authorities as a matter of policy will only provide schemes to protect public property; the protection of private property is in their view a matter for the owner. Overall, arrangements in Scotland for coast protection and flood prevention appear to operate satisfactorily.

VI C. Northern Ireland

119. The Northern Ireland Departments of Agriculture, Environment and Economic Development each have responsibilities for coastal protection measures, divided up according to the " Bateman formula " which is based, broadly, on the nature of the interest which is under threat.

VI D. Environmental considerations

120. MAFF has developed a strategy for tackling the consequences of global warming which has been endorsed by the House of Lords Select Committee on Science and Technology. The strategy includes the refurbishment of defences to enable them to withstand the currently expected risk and to encourage designs that will allow for adjustments when predictions are clearer. Research and development is being progressed; predictions are being monitored and standards are being reviewed in association with authorities responsible for promoting schemes.

121. Safeguards are in place to ensure that coastal defences do not damage the environment but protect and enhance it. These arrangements draw the full range of environmental interests into the early planning stages of schemes, so that design decisions take account of those interests. Major changes have occurred during the past decade in the way in which coastal defences are engineered. There has been a swing away from the construction of sea walls towards more natural forms of defence. It is now recognised that the beach is an essential feature of coastal defence because of its ability to absorb energy in storms and to rehabilitate itself. Grant aid is available where the renewal of a beach forms part of a scheme.

122. The Conservation Guidelines for Drainage Authorities, jointly prepared by MAFF, DOE and the Welsh Office, assist authorities in carrying out their environmental responsibilities and cover consultation with environmental bodies about proposed works. The guidelines are currently being updated.

123. New inland and coastal defence works on land are subject to normal development control procedures, including environmental assessment requirements (see paragraph 30); below the low water mark they require the consent of the Crown Estate as landlord. Although improvement works do not require planning permission, they are subject to environmental assessment procedures, and an environmental statement must be prepared where they are likely to have significant environmental effects.

124. As an additional safeguard for the environment, no scheme is approved for grant aid unless English Nature or the CCW have no objection to the proposals. The Minister of Agriculture has also initiated periodic reviews by a group comprising representatives of interested bodies to ensure that policy reflects environmental concerns.

125. Arrangements are similar in Scotland where authorities have to obtain any necessary planning permission and to comply with requirements for environmental assessments.

VI E. Finance

126. The relevant Minister is empowered under the Coast Protection Act 1949 and the Land Drainage Act 1976 Acts to pay grant in respect of coast defence works. The relevant Departments monitor and evaluate schemes and authorities' future needs and plans in order to estimate the level of Government support necessary to take priority works forward, to approve schemes and to allocate grant. Priority for grant is given to flood warning and urban inland and coastal defence work; a higher level of grant aid is given to those authorities with the greatest need and the least resources, and to coastal defences.

127. Government funding for inland and coastal defence in England has been increased on four occasions since 1986–87. The increases arise from assessments of the state of the nation's flood defences which have demonstrated the need for significant increases in investment in order to reduce the risk of flooding during winter storms. These increases taken together mean the grant for NRA flood defence schemes will rise from £21·6 million in 1990–91 to £40·4 million (representing a total capital investment of £92 million) by 1993–94; for local authority flood defence and coast protection schemes the grant provision will increase over the same period from £18·3 million to £21·2 million (representing a capital spend of about £38 million); and for IDBs grant rises from £1·7 million to £1·8 million. A large part of the increased NRA provision will support coast defence work, notably along the east coast where defences constructed following the severe flooding of 1953 are nearing the end of their useful life. The extra grant for coast protection is primarily for the repair of defences damaged in the winter storms of 1989–90.

128. Expenditure on coast protection in Wales has risen markedly in recent years, and is expected to reach £6·5 million during the current financial year, including £4·5 million in grant aid. Part of this expenditure results from the storms of the winter of 1989–90 but most was planned or in progress before then. There is also grant provision of £2 million for NRA expenditure on flood defence in Wales, which reaches some £3 million per annum. In addition, local authorities and IDBs have undertaken a small number of flood defence schemes at a cost of up to £1 million in recent years.

129. The Secretary of State for Scotland is empowered under the 1949 and 1961 Acts to pay grants, at varying rates, to authorities on eligible schemes. Grants are not made for ordinary repair and maintenance work. Coast protection and flood prevention work comes within the regional and islands' councils water and sewerage programmes, provision for which has been substantially increased over recent years. In 1989–90 expenditure was £668,000 on coast protection and £912,000 on flood prevention: grant expenditure is also relatively small.

VI F. Research and warnings

130. MAFF funds the national flood and coastal defence research and development programme. Particular emphasis is given to studies of river and coastal processes, the implications of global warming and environmental impact. In managing the programme, the Department draws on the advice of an expert committee; this advice is currently being updated. Designs for flood and coastal defences take advantage of the latest research and development results, which have in particular drawn attention to the important role that a beach can play in defence against sea flooding and coastal erosion.

131. Warnings of storm surges are provided by the Storm Tide Warning Service funded mainly by MAFF. Recent improvements have been made in the Service to take account of the latest results of research and development, computer developments and the installation of additional and updated tide gauges. More use is also being made of information on the incidence and impact of waves; high priority is given to further study of this matter. MAFF oversees the arrangements and works closely with the NRA, which is responsible in England and Wales for the issue of flood warnings and co-ordination of local arrangements for dealing with flooding. In Scotland, the Agriculture Act 1970 enables river purification authorities to provide and operate flood warning systems for tidal waters.

VII. COASTAL FISHERIES

132. Byelaws controlling fishing in English and Welsh coastal waters are made by Sea Fisheries Committees. Membership includes Ministerial appointees acquainted with fishing, representatives of local authorities and the NRA. The 12 Committees in England and Wales have powers under the Sea Fisheries Regulations Act 1966 to regulate the taking of sea fish and shellfish in the interests of fisheries management. Such regulations include limiting the length of vessels, specifying the nature of gear which can be used, minimum landing sizes, areas of restricted fishing and regulation of shellfisheries. Their present jurisdiction is out to three miles from the coast, but Ministers have agreed to extend any Committee's jurisdiction by Order to six miles on application. The NRA also has the power to make byelaws to protect stocks of migratory fish. Both sea fisheries committee and NRA byelaws require the approval of the appropriate Minister.

133. In Northern Ireland there are no Sea Fisheries Committees but equivalent powers to regulate fisheries in coastal waters are exercised by the Department of Agriculture for Northern Ireland under the Fisheries Act (Northern Ireland) 1966. Fishing for salmon in the coastal waters of Northern Ireland is controlled by the Foyle Fisheries Commission and the Fisheries Conservancy Board under the Foyle Fisheries Act (Northern Ireland) 1952 and the Fisheries Act (Northern Ireland) 1966.

134. The management and conservation of inshore fisheries resources in Scotland are governed by the Inshore Fishing (Scotland) Act 1984. The Secretary of State for Scotland is empowered, after consultation with the fishing industry, to make Orders regulating fishing activities in Scottish inshore waters up to six miles from the coast. The current regime came into force in 1989 following extensive consultations with the industry. It includes all-year and seasonal restrictions on the use of mobile fishing gear in areas which are sensitive from a fisheries conservation perspective; restrictions on suction dredging for shellfish; seasonal restrictions on lobster and creel fishing in certain areas; and restrictions on the length of vessels fishing in the Firths of Forth and Clyde. The Secretary of State for Scotland may also establish licensing restrictions for inshore fisheries. There are, for example, restrictive licensing conditons for sandeel fisheries in Shetland and the Western Isles, under which the Shetland sandeel fishery has been closed for stock conservation reasons since June 1990. In Scotland, there is no equivalent of the English and Welsh Sea Fisheries Committees' local management role. The Cameron Committee on Scottish Inshore Fisheries concluded that, in Scotland, local management would result in considerable outlay of time and cost without any corresponding benefit to the modern, mobile catching industry.

135. Conservation of fish stocks in both coastal and offshore waters, to ensure that they are maintaind as a renewable resource, is achieved through the EC's Common Fisheries Policy. The policy's rules, set out in EC regulations, limit the proportion of a stock which may be caught, the nature of gear to be used in all EC waters, and the size of vessels operating within 12 miles of the coast. They also prescribe closed or limited fishing areas. Stricter national measures can be, and are, taken to strengthen those in the Common Fisheries Policy. This is particularly important for stocks where a national measure alone can substantially benefit their conservation. For example, there has for some years been provision in United Kingdom legislation for minimum landing sizes for crabs. Such minima are only now being proposed in the EC as a whole. Further measures introduced by the United Kingdom Government in the last two years include nursery areas for the protection of bass in certain inshore zones, and a requirement to use a square mesh panel in certain nets to facilitate the escape of juvenile fish.

136. The placing in the coastal waters of Great Britain of molluscan shellfish, whether of domestic or imported stock, must be licensed under the Sea Fisheries (Shellfish) Act 1967. This Act also provides for the granting of Several Orders, creating the exclusive right to cultivate shellfish in specified waters. There are 24 such Orders operating currently in Great Britain. Proposals must be advertised in draft, and there are full provisions for public inquiries before applicants may be given the right to cultivate certain species exclusively in designed zones. Other controls which apply include the licensing of releases on non-native fish and shellfish into the wild under the Wildlife and Countryside Act 1981. Consultation arrangements involving the conservation agencies are undertaken before non-native species are introduced.

VIII. CURRENT ISSUES

137. The foregoing paragraphs set out the full range of significant Government policies operating in the coastal zone. Given that the coastal zone includes sea, land and the inter-tidal area it is hardly surprising that the total number of agencies operating within it is large. It is frequently suggested that this multiplicity of agencies leads to confusion or to conflicting objectives in relation to environmental protection. The Government has not so far seen convincing evidence to support this claim. While it is always prepared to consider changes in structures to promote efficiency or to simplify access for users it does not believe that a single agency for all coastal matters covered in this memorandum is either feasible or desirable.

138. There is inevitably conflict between development of any kind and preservation of the existing environment, but in the Government's view appropriate mechanisms are in place to reconcile most such conflicts. In relation to all important issues there are either specific environmental regulations or duties in place or arrangements have been set up to ensure that environmental considerations are taken fully into account before development decisions are taken. This is not to say that all past decisions did take the environment sufficiently into account by today's standards.

139. Rather than concentrate on organisational change or bureaucratic simplification the Government considers that attention should be devoted to issues which arise from specific pressures and requirements. As explained above and as set out in Cm 1655 the Government is:

(a) preparing a Planning Policy Guidance Note on coastal planning;

(b) considering extension of the Marine Consultation Area concept to England and Wales, taking account of experience so far in Scotland;

(c) undertaking a review of existing legislation which affects marine conservation with a view to identifying gaps or aspects which need clarification;

(d) conducting a periodic review of coast protection and flood defence policies;

(e) monitoring the development by individual coastal local authorities or groups of authorities of voluntary coastal zone management plans: a research project designed to assist the preparation of the coastal zone Planning Policy Guidance Note will examine these;

(f) considering proposals from the Countryside Commission to revise and bring up to date guidance on Heritage Coasts;

(g) considering the scope for rationalisation of the consents required for development in coastal waters, as part of the Deregulation Initiative;

(h) preparing revised advice on the siting of marine fish farms in Scotland; and

(i) considering how the need to extend effective protection to wildlife sites below the low water mark designated under the EC Birds Directive and the draft EC Habitats Directive (if adopted) can best be achieved in the marine environment.

IX. CONCLUSION

140. The Government intends to report progress on coastal issues next year. Meanwhile it intends to continue to work with other agencies to preserve and enhance the beauty and variety of the United Kingdom coastline while recognising its importance as an area of economic activity.

Department of the Environment
4 *October* 1991

Examination of Witnesses

Mr PAUL McQUAIL, Grade 2, Planning, Rural Affairs and Built Heritage, Mr NEIL SUMMERTON, Grade 3, Director of Water and Mr BRIAN WILSON, Director A, Chief Planning Adviser, Department of the Environment and Dr JIM PARK, Grade 5, Head of Flood Defence Division, Ministry of Agriculture, Fisheries and Food, examined.

Chairman

1. May I welcome you to this first session of our inquiry into coastal zone protection and planning and apologise that we have detained you a little while. As it is the first occasion on which we are discussing this matter there were some preliminary ideas we wished to clear in our minds first of all. Mr McQuail, you are leading the team from the Department of the Environment. Would you be so good as to introduce yourself and your colleagues?

(*Mr McQuail*) Thank you, Chairman. We are very pleased to be here. I am Paul McQuail. I am the Deputy Secretary with responsibility, among other matters, for countryside and wildlife issues, for planning issues and one or two other things not directly relevant. On my right is Mr Neil Summerton who is the Director in charge of the Water Directorate. On my left is Mr Brian Wilson who is Chief Planning Adviser to the Department. On Mr Summerton's right is Dr Park who is the Head of the Flood Defence Division from the Ministry of Agriculture. As the memorandum was submitted on behalf of several Government Departments we are not all represented but the Ministry of Agriculture are close colleagues and associates.

2. Is there any other preliminary statement you wish to make on the matter before we proceed with our questions on the memorandum?

(*Mr McQuail*) No, that is all.

3. The first question I would ask is this. In official statements that have been made, certainly in the Environment White Paper, coasts do not seem to be covered in very much detail. Is there much of a priority attached to coastal policy in so far as Government are concerned?

(*Mr McQuail*) It has certainly become very salient in the last few years and that is an illustration of the way in which these matters move forward. The White Paper which you mentioned was of course a very wide-ranging document and it represented, as I know successive Secretaries of State have remarked, first thoughts and not the last word. Even though the coast and the coastal zone were not mentioned as distinct topics, many aspects of the White Paper bear fundamentally on the issues that are the subject of your inquiry and of the memorandum the Government have put in. Just to pick a few headings: planning, pollution control, conservation of wild life, litter, indeed global warming itself. The degree of coverage in the White Paper is not in itself—I can assure you on behalf of the Government on this point—an index of priority, certainly not a reliable one. It is not a question of an absolute answer to what priority the Government give but they and we are certainly very conscious of the importance of the coast across the range of our policies and activities. The memorandum itself sets out, for the first time bringing together these matters in such a complete form, a full statement of Government policies and activities. I only want to add two points about it. One is that many of the policies, though here brought together for the first time, are not new; many of them go back many years, in some cases centuries. Secondly, it is a very fast moving world. The policies that are noted there are subject to a constant process of review and adaptation as new facts come to light, as the world itself changes and as public perceptions themselves alter. Towards the end of the memorandum, in paragraph 139, there is a specific list of nine or ten areas where review or change is currently in hand and paragraph 140 ends the report

Chairman *Contd]*

by promising a progress report on coastal issues next year. In preparing that I am sure the Government will very much look forward to having the benefit of the Select Committee's conclusions.

4. I was very interested in your comment that this memorandum which you have submitted to the Committee is the first document ever to bring together all the various threads of Government policy as they relate to our coasts. That seems to confirm a recurring criticism that we seem to be getting in the written representations that we are receiving that not only is there no priority in so far as coastal policy is concerned but there is no national policy and there is a constant lament that there seems to be very little central direction, guidance or advice in these matters. How do you respond to that criticism?

(Mr McQuail) The points you make there are in part about organisation and in part relate to a criticism that we here are familiar with that there ought to be a single agency to pull all these things together. The memorandum itself says that the Government has not been persuaded that there should be a single agency. It develops an alternative approach, explicitly I think but certainly implicitly, in laying out a map in prose of the various activities and the various responsibilities of the many agencies of Government who, in our view, are inevitably involved in coastal matters on both sides of the shoreline. It is the Government's view that a single agency is not an appropriate answer; at any rate it has not been persuaded of that so far. Perhaps I may make two or three points about that rather than dealing *in extenso* with the whole field of inquiry because your question indeed raises the whole field of your inquiry.

5. My question really raised the matter of there being a lack of a national policy. You have now very skilfully shifted that to the argument as to whether or not there should be a single administration, which I will come to in a moment.

(Mr McQuail) Fine. The two points are pretty closely related. What the Government's position is, as set out in the paper, is that there should be a range of policies dealing case by case, aspect by aspect and problem by problem with the issues surrounding the coast, that there should be co-ordination between those policies but that the case has not been made and indeed that it is unsustainable for a single national policy or strategy.

6. On the question of administration, I have mentioned that one criticism we have received is the lack of a national policy which people could use as a guideline. The other criticism on the administrative side is that the structures that exist are complex, unclear, overlapping and spread across too many Government Departments and other organisations. So not only is there no policy but there is no one place to which people can turn for help and guidance. How do you respond to that other side of the coin?

(Mr McQuail) I respond to it by referring to the signatories on the front of the memorandum where we have eight Government Departments. I have seen estimates of the numbers of Government agencies involved and they seem to be agreed at being around 30; it depends who you count. The answer basically is that whatever further degree of co-ordination were

brought to bear it would be unrealistic to believe a single agency would be an appropriate answer. The interests represented by those agencies are not simply that they are arms of Government, they represent the reality of the forces that meet at the coastline. So a single agency, by internalising the conflict, would not in itself solve the problem: it would alter the terms of the argument somewhat.

7. Could I put it another way? Instead of talking about a single agency, could I ask whether it would be possible to reduce the number of cooks that you have in this particular kitchen cooking this particular broth? As well as all the Government Departments that are listed at the head of your memorandum—and virtually every other Cabinet Minister seems to be involved in this in some way or another—we have all the local authorities up and down the country who have any bit of coastline, we have HMIP, we have the NRA, we have English Nature and the Countryside Commission. I can understand why people are a little bothered with so many cooks. Do we need so many?

(Mr McQuail) Yes, I am sure it would be possible to reduce the number. Some of these possibilities are actively under consideration and the consultative document about the possible new environment agency deals with one aspect of that. The list of matters under review in paragraph 139 includes a review of the scope of rationalisation of consents required for development in coastal waters. Another one is a review of existing legislation affecting marine conservation with a view to identifying gaps or aspects needing clarification. It is however, a constant process—in our view appropriately—of a marginal adjustment as needs change.

8. I know it is always difficult to take people's empires away from them when there is a certain amount of resistance and any rationalisation process causes you problems. What is your view of the NRA's role in relation to sea defence? How will its responsibilities be organised when the proposed environment protection agency comes into being?

(Dr Park) Would it be helpful if I said just a little bit about how the NRA's role fits in with that of other agencies that are involved?

9. I know that MAFF and the NRA have one or two areas of disputation at the moment as to how the environment protection agency will affect them. I do not know whether you wanted to embark upon this; that will be the subject of another inquiry that we are going to have.

(Dr Park) No, I would rather just deal with the question of NRA's responsibilities for sea defence, that of other authorities who are responsible for coastal defences and the role of central Government. It might be helpful if I just set that out in summary. The coastal defence policy in central Government is split between the Ministry of Agriculture and the Welsh Office on understandable geographical grounds. In legislation we have two particular types of activity for protecting the coastline: one is coast protection, which is about preventing the loss of land through erosion and the other is to prevent the flooding of land, which we call sea defence. For the

Chairman *Contd]*

purposes of this memoranda we have incorporated them under a general term "coastal defence". The priorities that we are working to at the moment are set by those two central Government Departments. The priority is really to achieve cost effective protection of people and property against flooding and erosion and by provision of flood warnings. It is very important that we remember the role of flood warnings in protecting people and property. Coastal defence has a particular priority and it generally attracts a higher rate of grant from the central Government Departments. In devising protection in the form of coastal defences then we have to take account of global warming. That is probably a topic we will discuss later. It is important in thinking about protection to take account of the best understanding that is available and hence we have a R&D programme which is looking very much at the processes that take place on the coastline. As that understanding develops and as we know more about the environmental implications then the designs of defences are beginning to change. So we will be able to come up with what I think in common parlance have been called soft engineering solutions and that really is a consequence of a better understanding of natural processes. As far as the operating authorities are concerned, I think we have set out in the memorandum, and I am sure the National Rivers Authority have set it out in their memorandum, the detail of their statutory responsibilities. Basically they have a requirement to exercise general supervision over all matters relating to flood defence. That means that they have to look at all aspects of flooding risks and they have to be the source of advice to other authorities who are involved in providing flood defences. They have a requirement to carry out surveys which is to look at the standard of defence which is offered and the state of repair so that is the way you can begin to identify where further works need to be done in the future. As far as coast protection is concerned the maritime district councils are the operating authorities who have the power to carry out coast protection works which have been approved by the Ministers, either in the Ministry of Agriculture or the Welsh Office. Those councils must consult the National Rivers Authority and neighbouring councils about their works. A lot has been said about the strategic planning of defences and much has been going on to encourage that strategic planning both in temporal and spatial terms. The R&D programme is a very important part of that in terms of developing understanding, the threat of effects arising from global warming has also tended to encourage thoughts along that sort of line. One very important step towards achieving this has been the setting up of coastal groups to cover various stretches of the coastline and now, as far as we are aware, there are coastal groups established to cover what I would call the man-managed part of the coastline; I refer to man-managed coastal defence. Recently the two Departments have set up a coastal defence forum which draws together members from those coastal groups in order to assist them in their work and also particularly to encourage strategic planning. It might be better if I stopped there.

10. Is this approach that you have just been outlining to us, the regional grouping that you have mentioned, the same as what has been represented to us as coastal zone management? That is a term that has been cropping up. As far as I understand, what is meant by that is regional coastal strategies which adopt a holistic approach to coastal protection and planning. Is that being achieved by the groupings you have mentioned or is something different being urged upon us?

(Mr McQuail) I think it is a contribution to that, possibly not going as far as some would wish.

(Dr Park) The coastal groups that I referred to are not just regional because sometimes they are local, they are at a lower level in the sense that they cover a shorter stretch of the coastline. The key thing about them is that they draw together local authorities in the broadest terms with responsibilities and interests in the coastline. The particular emphasis for most of them in the early days has been to do with coastal defence but coastal planning and coastal zone planning are one of the parts of the function to some of the more forward-looking coastal groups. As far as we are concerned in the Ministry of Agriculture our particular interest has been in coastal defences. In establishing the forum that we have called together we have called it the coastal defence forum so that it is very clear that that is the issue that is being dealt with in that forum.

Mr Jones

11. You alluded to the place where the responsibilities of the Welsh Office and the Ministry of Agriculture met, namely the border, the estuaries in the north and south of Wales. Do they pursue the same policies on both sides of the border? Are any of these coastal groups to which you refer bodies that cover both sides of such an estuary?

(Dr Park) You are asking me to remember the details of some of the groups. There are specific groups which cover estuaries and straddle the interests across estuaries and therefore are there to co-ordinate the interests between local authorities' interests on either side of the estuary.

12. Is that true even where it is Wales and England?

(Dr Park) It is certainly true in North Wales; the Tidal Dee User Group. I am not quite sure about the situation in the Severn estuary.

13. Perhaps we could have some information on that in a follow-up to this.

(Dr Park) Yes; certainly.

Chairman

14. What I was trying to clear up was this. We have already established that on the one hand there seems to be this universal concern over fragmentation and overlapping of responsibilities on the one hand and on the other the Departments' insistence that a single strategic authority is not going to solve these particular problems. Is there somewhere in the middle a solution or even a bridge between these concepts? That is why I have referred specifically to the proposals that have been put

Chairman Contd]

forward for regional coastal strategies. It would certainly be less than a single authority but would it be better than what we have at the moment?

(Mr McQuail) It might well be. I mean to say, if I did not quite, that what the Government wants to encourage is an incremental approach. The way in which these groupings have developed, many of them starting from the point of view of coastal defence, has been in response to perceived demand and to the perception that it is impossible to deal local authority area by local authority area. On the whole the impetus for them has come from estuarial waters. Both the Department of the Environment and the Ministry of Agriculture encourage the contributions that these groups, with a membership made up locally according to local needs, can make to other planning systems. In the case of the Department of the Environment, for example, what the groupings have to say about development control about estuaries will be a powerful influence on structure plans, local plans and indeed working back up the chain into the regional guidance that the Department is now committed to producing for the whole country. It is a question of developing in some cases from a local initiative outwards.

15. Is this going to be integration on an *ad hoc* basis which seems to flow from what you are saying, or would it be possible to lay down some kind of concept for the taking together of a useful region, in the same way as our water policies eventually reduce themselves to a river basin concept. We had about 10 regions with responsibility for water based upon a river basin. That seems to have worked extremely well. Does the coastline resolve itself logically into groupings in that way?

(Mr McQuail) We do not believe that it does quite to the same extent and not to form a part of a universal system. We believe that there are greater differences at the coastline between land and sea than that model implies. That is not to say that the experience of bodies who are concerned with the coastline in their regional locality does not have important things to say to all those systems that meet at the coastline and indeed in some case overlap the coastline.

Mr Steen

16. Can I ask about the National Rivers Authority's contribution? Am I right—and if I am not please put me right—that the overall budget of the National Rivers Authority is about £420 million or thereabouts? I understand or seem to recollect that the amount they spend on coastal defence or in this field of flood control is something like two or three per cent of their total expenditure. Am I right in that?

(Mr Summerton) That is not correct.

17. Could you put me right?

(Mr Summerton) The position is that expenditure on flood defence, both capital expenditure on defences and work related to the maintenance of defences and so on is about 50 per cent of the expenditure of the National Rivers Authority and something like half their full-time staff, over 3,000 people, is deployed on flood defence activities.

18. Does that include coastal defence as well?

(Mr Summerton) That includes both flood defence on river lines and also coastal defence.

19. Do you have a breakdown of the actual coastal defence work? That perhaps may be the very small bit.

(Dr Park) We do not have a breakdown with us but I am sure if it is not part of NRA's evidence to you we can make sure you can have it.

Mr Howell

20. Am I correct in thinking that your answers so far indicate that your proposals in paragraph 139 reinforce the existing *ad hoc* approach to coastal policy?

(Mr McQuail) Yes, though we and the Government would prefer to call the policy approach an incremental and a creative one not, in a dismissive way, *ad hoc*. We do believe that the arrangements and the policies that are set out in the memorandum represent an up-to-date response to the issues as they are now perceived and identified. Having said that, the proposals are of course designed to reinforce and improve the existing approaches to policy. What is in paragraph 139 happens to be a current agenda for action and review. It is quite a substantial agenda and it is constantly changing. As a matter of fact we believe there would be some such agenda on the table for action whatever institutional or policy state of development we happen to be in. The whole attitude both of the public and the Government to the coast and the needs of coastal zone management is as constantly shifting as the sands. We need to keep up-to-date with it. This is our idea of what is needed at the moment to keep up-to-date.

Mr Steen

21. Could I ask you one thing about the need? Coastal management and flood defence is obviously a question of regions and parts. Is it your view that there is actually a worsening situation in terms of the dangers to more areas of Britain or are we talking about very small amounts of mileage of the coastline if you take Great Britain as a whole?

(Mr Summerton) I wanted to add a point which related to the generality of the discussion so far which relates to the issues of a strategic plan, national plan, agencies and institutional arrangements. The point is this. It seems to me the national Government is always faced with a tension as to how it organises itself in relation to particular areas. Naturally the national Government is organised on the basis of functions, of functional areas of activity, whatever they are. It often becomes difficult in my judgement when you start asking the Government to look at types of area on a national basis so that you can, as it were, draw around the coastline the coastal zone all around the country, and say that the Government should lay down a national plan for such an area, as distinct from developing policies in its various functional responsibilities which bear on that area— and probably on other areas as well—and which need to be made, obviously, to fit together in a sensible way. If you ask yourself the question, what should a

Mr Steen *Contd]*

national coastal policy contain, and certainly if you ask the question, what should a national coastal plan contain, you begin to see some of the difficulties which might be entailed. You have to begin to focus down on smaller geographical areas when you begin to talk in terms of territorial plans as distinct from functional policies relating to activities. It seems to me that that is a problem with which the national Government is always faced. It has faced it in relation to inner city policy, for example, over the last 20 years. I just request that the Committee perhaps bear that in mind in thinking about this issue of whether there should be a national policy for the coastline as such and certainly in relation to national agencies. Of course sometimes we do organise certain activities on a national agency basis; sometimes they are dealt with on a local territorial basis.

Mr Jones

22. Some other countries have an integrated coastal planning or protection approach do they not?

(Mr Summerton) I believe that some other countries do. The question that I raise is what is the precise content of such a thing? It is easy enough to express generalities like: the coastline should be protected from development, and things like that. But that is not a policy which it seems to me one would wish sensibly to apply to *all* areas of the coastline.

23. Has the Department paid attention to what is going on in other countries in order that we may learn lessons, positive and negative, from what is happening?

(Mr Summerton) As a general point I would say that certainly as far as my area of responsibility, which is water, is concerned, yes, we are taking an interest, naturally, in what goes on in other countries. I do not know whether others here would like to speak specifically to the issue of coastal planning.

(Mr McQuail) We are indeed making inquiries. Paragraph 33 of the memorandum refers to the review of all relevant published material and that is not confined to these shores on coastal planning and management and so on. The paragraph says that " Interim results will be available at the end of October ". I am afraid they turned out not to be available at the end of October but we will certainly let the Committee have access to those as soon as we can.

Mr Steen

24. Are you talking about the planning and policy guidance?

(Mr McQuail) No. This is a DOE research project which is reviewing experience elsewhere on the basis of the best that is known and thought about it.

Mr Howell

25. On a narrow issue of coast defence, would you agree that something does need to be done to stop a situation, as exists in my constituency, where there are two stretches of coastline governed by one

Ministry and two by another? It does seem to be quite foolish in this situation.

(Mr McQuail) I am not sure that I know the detail. I do not know whether it means anything to Dr Park.

(Dr Park) Mr Howell and I debated the issue with the Minister of State.

26. We have debated this issue many times. What I am trying to establish is whether there is any move to bring the defence of stretches of the coastline within one authority.

(Dr Park) I think the particular instance you are referring to is a distinction between a sea defence responsibility of the National Rivers Authority which abuts a coastal protection responsibility of a maritime District Council in north Norfolk. The important point to remember is that both pieces of legislation which give the policy authority to Government Departments are the responsibility of the Ministry of Agriculture in England. So there is one Government Department that takes that responsibility. I think the particular instance you refer to is a very useful way of bringing into the discussion the major study that was done on the east coast, the east coast sea defence management study, which was carried out by consultants on behalf of the National Rivers Authority, funded through grant aid from the Ministry of Agriculture, which for the first time ever, certainly nationally and possibly for the first time ever internationally, has looked at coastal processes, at the movements of materials, at the factors affecting the movement of materials, at the sources of materials, at what is happening to the coastline in terms of erosion of the foreshore, and has brought that together in a major database. That database is now available to all the authorities operating on that stretch of coastline from the Humber through to the Thames. We will be continuing funding the maintenance of that database and we will be making sure that the authorities who are responsible in that area draw on that database for designing defences. One very interesting point about this is that it demonstrates the currency of the technical understanding. For one of those particular defences you referred to there has been a change in approach in the last two years and has caused concern to you as the constituency MP know this. That change of approach has come about because this study has demonstrated that the original understanding of those processes was incorrect and that the new understanding, based on this new knowledge, has led to a different design. That is an important point that we have to recognise; that we are at the limits of technical understanding in designing many of these defences. The R & D programme that I referred to earlier is a very important part of making sure those designs are going to stand up for the 30 to 70 years that we expect a defence to last.

27. I am grateful for that explanation. What I was really trying to get at was the division between the cliff area and the below sea level area. It still seems to me to be sensible and logical that this should be brought under the administration of one Ministry.

(Dr Park) I can confirm that it is the responsibility

Mr Howell *Contd]*

of the Ministry of Agriculture in England and the Welsh Office in Wales.

28. The entire area?

(Dr Park) Yes. The administration of coast protection and sea defence comes under the Ministry of Agriculture in England.

Mr Steen

29. Could you complete the answer to the question I was asking which was really whether there is a growing problem of the miles of coastline affected by the excursions and incursions of the sea? Is this a growing problem? Is this something which is increasing in terms of miles of the coastline?

(Dr Park) Could I apologise to start with: the figures I have are in kilometres. Do you mind? I cannot convert them to miles in my head. The England and Wales coastline is approximately 4,500 kilometres. Roughly one half of that requires no man-made defence against either erosion or flooding; being positive, roughly half requires man-made defence. Approximately 1,200 kilometres, in other words a half of the half or a quarter of the whole, requires defence against flooding and roughly one quarter requires some defence against erosion. That is the scale of the issue we are dealing with. The NRA has a responsibility to carry out surveys and in the last 18 months or so has been doing a survey of sea defences, leaving out the coast protection defences for the moment. When they looked at their own defences, which are about 800 kilometres, they found that 86 per cent of those were in a satisfactory condition or better. Roughly 14 per cent needed some attention and that 14 per cent is included in the capital programmes which they are required to prepare both for corporate planning purposes and for use in terms of planning defences. As far as the remainder of the defences are concerned, those are the responsibility of some other authority like a local authority or a private owner. The results of that work are really coming to the point of analysis. They have not got to the point where they have done an analysis. I have not seen them myself. Early indications were that the sort of pattern would be similar. It is a very small percentage which needs that attention. However—and this lies behind the funding provision that my Ministry have made available—there is a threat of global warming. We cannot stand still, so we have to look at whether these defences are going to be adequate to withstand whatever global warming might bring. Hence there is an urgency to make sure that those repair works, refurbishment works, are taken forward in an appropriate timescale and not left until there is a real risk.

30. Your memorandum in paragraph 32 suggests that planning policy guidance on coastal planning will be available by the end of the year. Perhaps you could let me know to what extent this will simply interpret the existing procedures and systems and to what extent it will put forward new mechanisms and new policies?

(Mr McQuail) We still do aim to get our draft of the planning policy guidance note out by the end of the year. A word of background about planning policy guidance notes. A new series of so-called PPGs started in 1988 which replaced an old and long-standing series of circulars. Some of these, like much legislation, are consolidating matters that bring together and modestly update guidance on particular aspects of planning. There have been 17 so far, starting with one on general policy and principles and No. 17 is sport and recreation. We expect that by the end of the series there might be somewhere in the mid 20s all together. They will themselves need updating from time to time. The updatings may take place because of some new event or because of a new policy initiative or a new legislation. There is another one in the pipeline that is mentioned in the memorandum which is about sources of renewable energy, which will have a reference to coastal aspects. Reverting to the promised PPG about the coast, this is still in draft and it has not yet been offered to Ministers. It will, however, be an important document in highlighting the significance of the coast in the planning system and thus raise its profile. In doing so it will certainly elaborate on some of the points about the planning system set out in the memorandum. The basic text of it is likely to draw on existing policies and mechanisms and to interpret them. It would not in itself be likely to be a major restatement of policy. It would no doubt elaborate in some respects on what is said in the memorandum; it would elaborate, if I could put it like this, on some of the thinking that in preparing ourselves for talking to the Committee we spotted as issues that need guidance to the public and planning authorities. It might help if I mentioned a few of the topics that look pretty certain to be covered: they would include something about the distinguishing characteristics of the coast in relation to the planning system; collaboration and consultation between neighbouring planning, and that would include district, authorities; interaction with other key agencies, and that for example would include the points that we touched on earlier about groupings around estuaries. It would certainly include special interests in the conservation of the coast, draw attention to areas of the coast requiring special attention, including those subject to coastal flooding. It looks to us as though existing development plans, for example, are rather light on the question of areas subject to flooding. It would also certainly include the balance to be struck between environmental and economic considerations in developments affecting the coast. We will let the Committee have a draft as soon as it is ready to issue and I would only say that experience suggests, particularly with a live and current issue, that any document issued for consultation attracts its own set of proposals which would necessarily be taken into account and would very likely require some modification to what is in the draft and possibly to policy.

Mr Jones

31. That seems a good set of questions but will there be a set of answers in the same document?

(Mr McQuail) There will be some answers. There will be answers. As the course of the conversation and indeed the content of the memorandum suggest there will not be answers that meet all the critics of

Mr Jones *Contd]*

the present position as reflected in the memorandum but they will be sensitive to those criticisms.

32. There will be a way of reconciling conflicting policies being pursued by neighbouring authorities?

(Mr McQuail) There will be encouragement to processes that would help to bring those to resolution. If I could put it like this, the planning system in one sense has the objective of protecting the environment and in another a vital function of the statutory planning system is to provide a means of resolution, providing public forums in which conflicting policies are brought under the spotlight and in which there is a tribunal that weighs one against the other and reaches a decision. It will be among other things an aid to clarity of thinking, to promoting clear discussion about the nature of issues and about the nature of conflicts, for example between economic and environmental considerations as they affect the coast.

33. That is precisely the point I am aiming at. I can well see, to take the north Wales situation, that because there is high unemployment in Clwyd they may regard job creation as being the most important object of their planning policy whereas in Cheshire, on the other side of the River Dee, they might regard leisure as the main thing that they want to see enhanced. How does one reconcile those conflicts? Is that what it is going to be about?

(Mr McQuail) It will be one of the things it will be about. A characteristic constraint on the way in which decisions are made about planning applications is the set of environmental policies that are in place for the time being. They include—not confined to the coast—Green Belt policies, policies related to Sites of Special Scientific Interest, special protection areas for birds. These are all existing policies. Those that are relevant to the coast would be drawn together and it would not necessarily be new policy but in establishing a hierarchy between these policies it would give pretty strong guidance that some of these policies as they related to the coast, policies for example to protect heritage coasts, were to be given high priority. How the balance is expressed will come down to fine judgements, partly political partly legal, about how these are put in the document at the end. It will certainly be possible to draw a clear story from the PPG; quite where the balance will be struck at each point in the frontier and how firmly I think remains to be seen. Since much of it does draw on existing policies and many of those protective policies are very strong indeed and create a very strong presumption against development it will be, to say the least, very unlikely that any such presumption would be overturned in the course of preparing this PPG.

Mr Steen

34. There are some deficiencies, believe it or not, in the planning system. One of them of course is that developments below low water mark are beyond the jurisdiction of the local planning authority and are not subject to the same planning controls as development on land. Will you consider extending the local planning authorities' jurisdiction out to sea? If so, how far? If not, how then will you seek to link the planning and management of the landward and seaward sides of the coastal zone? The second question is particularly relevant to my own constituency of South Devon and that is that although we have a lot of sensitive areas—in fact there are more areas of outstanding natural beauty than in most other constituencies in Britain—nonetheless there is nothing in the local plan that can prevent development on the grounds that the area may get flooded or that the cliffs may fall down or that there is some reason to do with coastal preservation that homes or developments should not be built. I am wondering whether that is going to be addressed as well.

(Mr McQuail) To take the last one first, I do not think there is any reason why a structure plan or local plan should not have policies of that kind. Those would seem to be perfectly legal and proper policies to have in structure plans. I would expect that there are some that have them.

35. The point is that if a lot of the coast is in a state of change or erosion or flooding that is not a reason in itself, as I understand it, for refusing planning consent. The second point flowing from that, is that houses that are given planning permission perhaps should not be if there is going to be a danger to those houses and the structures are going to be affected. The planning authority should have regard for those matters before planning permission is granted.

(Mr McQuail) The memorandum does acknowledge that some of these physical constraints have not been properly reflected in many plans in the past but I believe that they have in some. No. 14 of the PPGs, Development on Unstable Land, is relevant to the point. Going back to your main question about deficiencies in planning below the low water mark, we do not have proposals for extending the jurisdiction of local planning authorities or for introducing a system like the planning and development and control system on the seaward side of the low water mark. Just a few points to help to explain that, though they are briefly set out in the memorandum. On the landward side there are a number of different systems with many agencies involved contributing to environmental quality. Development plans refer to a wide variety of matters and they help bodies other than local authorities in carrying out their responsibilities in the coastal zone. Really the planning and development control system is concerned with the physical development of land and with changes of use which are development for the purposes of the Acts. It deals with specific parcels of land which have owners and occupiers who can be enforced against if they do not carry out the requirements. That position is obviously very different on the seaward side. The Crown Estate Commissioners own the seabed but the water itself does not have an owner and there are very ancient rights of navigation and fisheries which can only be interfered with at present by Act of Parliament. Control, particularly of the use of the sea as distinct from the use of the land, is not straightforward and a simple translation of the land regime to the sea does not seem to be a practical solution in itself. That is not to say there are not adaptations of it that might

Mr Steen *Contd]*

be practical but it would not be something that simply extending the planning system would do the trick of solving. There are some other points I could make but it is not a straightforward proposition.

Mr Jones

36. What is your solution to the problem then?
(Mr McQuail) It is to rest for the most part on adaptations of the present system.

37. In what form?
(Mr McQuail) We are, for example, in the case of the requirements of the forthcoming EC habitats directive considering what adaptations might be needed to the existing law, since SSSIs cannot extend beyond the boundaries of planning authorities, what legislation would be required to enable us to comply with the directive when it comes into force as we expect very soon. I think that we would on the whole rely on *ad hoc* arrangements of that kind.
Mr Jones: It does not sound very convincing to me.

Mr Steen

38. Could I ask you about the bylaw powers of local authorities and when they were last updated? Are there plans to review them? I have had a lot of bad experiences about local bylaw plans to do with estuaries because the Home Office tends to sit on them for years. I wonder whether you have any ideas about what is going to be done to deal with them and update them and speed them up?
(Mr Summerton) Obviously Mr Steen has the advantage of me since he has some specific cases in mind which are obviously of great concern to him. I certainly am not in a position to answer for the Home Office's administration of particular cases. I am not quite sure what is behind the question in terms of updating bylaw powers. So far as I am aware—and subject to correction if I am wrong—I do not believe there has been any general review of the bylaw-making system or the powers of bylaws in the recent past. As I will explain in a moment, there have been some specific updatings of bylaw powers. Of course there is a very wide range of such powers which can be exercised by local authorities and by other bodies such as harbour authorities, bodies of that kind, the Ministry of Defence, and the National Rivers Authority. One is speaking about a very wide range of possible powers.

39. May I just explain? What I am talking about is that local authorities at the moment cannot legally use existing powers to control coastal activities on conservation grounds or navigational grounds like jet skiing and water skiing.

(Mr Summerton) Naturally that is one area. But there are two areas in relation to coastal policy at the moment—perhaps you will be able to indicate others. As you say, jet skiing and water skiing of one kind or another is one area; the other perhaps is the impact of boating activities more generally on matters of marine conservation. Those are two very clear areas. In the case of the first, the position is that local

authorities do have powers under the Public Health Acts to apply bylaws to pleasure craft operations within 1,000 metres of the shore. As I understand it, in the Home Office there was a review of these specific powers last year. The Home Office have advised that authorities may make such bylaws to prevent such craft operating near bathers and to regulate their speed, but there is a problem about the nature of the powers. Obviously a bylaw cannot go beyond the power that has been given by Parliament in respect of making that bylaw in the first place or it would be unlawful to attempt to do so. What the power cannot do, as things stand at the moment, is to prohibit the entry of craft into defined zones. One might be looking for some kind of zoning system for such craft which would make it an absolute prohibition to enter those zones either in the interests of safety of bathers or conceivably in the interests of the amenity more generally, the noise of such craft, the impact on other people enjoying the water and on the countryside.

Mr Jones

40. It is not just the people, it is nesting birds and things like that.
(Mr Summerton) That is why I drew a distinction between the bylaws to which I was talking about now and marine conservation. If I may I will come on to those in a moment. I am simply saying there is no zoning power in respect of the impact on people. In fact I understand that there are certain local authorities that take local powers to do this. I believe the Essex authorities are one of them but in general there are no such powers. On the marine conservation side, I understand that both English Nature and local authorities have powers under the National Parks and Access to the Countryside Act which gives them power to make bylaws for the protection of the nature conservation interest of and the good conduct and safety of the persons within national nature reserves—that is the power of English Nature—and local nature reserves—that is the power of local authorities. I believe that model bylaws have been produced as a framework for such bylaws. Beyond that English Nature has bylaw making powers under the 1981 Wildlife and Countryside Act to protect marine nature reserves. These allow for the control or prohibition of entry to persons or vessels to the reserve, the protection of plants and animals and prevention of damage to the seabed or any object within the reserve, also for the prevention of the deposition of rubbish and finally for the licensing of entry to and activities within reserves. That is in fact quite a wide power. The question is in the end the willingness of the authority concerned to exercise that power. In respect of both items there are powers. In respect of the first case there is not this absolute prohibition of entry into the zone under the existing power. I understand that to be the position.

Mr Howell

41. What plans do you have to review the role of the Crown Estate Commissioners?
(Mr McQuail) The Crown Estate Commissioners in England and Wales have a key role in the control

Mr Howell *Contd]*

of aggregate extraction. Certainly that is the issue in England and Wales which gives rise to most concern. The position of course is different in Scotland where fish farming has been a recent subject of investigation and inquiry. Aggregate extraction is very important in the construction industry and maintenance of coast defences. It provides a big proportion—as the memorandum says, 16 per cent—of the supply of sand and gravel for the construction industry in England and Wales as a whole. As the memorandum explains, it is the Government's policy to encourage the use of marine aggregates, provided it can be done without the risk of coast erosion or unacceptable damage to sea fisheries and the marine environment. The way in which this activity of the Crown Estate is regulated or controlled is a non-statutory procedure called the Government View procedure. There is only a small number of new cases year by year; five or six in each of the last few years have come forward as new cases in addition to the existing number of licensees. The Crown Estate's consent is required, as owner of the land, for any extraction of aggregates below the low water mark but they regard themselves as bound not to issue a licence if there is a negative Government View. They accept in every case all the conditions that are included in a favourable Government View by incorporating them in the licence. In effect it is the Government and not the Crown Estate Commissioners that even under the present regime take the decision after considering it, among other things, from an environmental standpoint. It is not altogether fair, as often alleged, to say that the Crown Estate is in a position of wholly conflicting interests. I can say that the Government are conscious of criticism on this score, as the Commission itself is, and we are willing to give further thought to the role of the Crown Estate in relation particularly to the extraction of aggregates even though the Government View procedure was reviewed and revised as recently as 1989. I should just add, going back to some of the earlier discussion, that some of the alternatives that have been proposed, such as the extension of the planning authorities' jurisdiction below low water mark, are unlikely to prove straightforward. Just to add one point on Mr Jones's dissatisfaction with our last exchange, it is to stress the very small number of development type of applications that take place in the coastal zone. The five or six major aggregate extraction ones are not of course the whole story though they are actually, in England and Wales, the most important from an environmental point of view. But if one compares that with the figure of half a million planning applications that the development control system in England handles annually, one can see we are in different orders of magnitude, different types of case.

Mr Jones

42. At the moment.

(Mr McQuail) At the moment.

43. One has to bear in mind that the problems with fish farms occurred after people had looked at that situation several times and said it is only on a small scale, it is not a problem. Of course that particular

stable door had to be locked after the horse had bolted. I certainly do not want to see that happen again in this particular case. I think therefore you should be considering a structure which is capable of dealing with a considerable increase in fish farms and wind farms, not to say aggregate extraction, as it becomes more important to look at alternative energies and also to look at alternative aggregate sources other than the traditional ones on the mainland.

(Mr McQuail) What I have said is that the Government are willing to look at the role of the Crown Estates in relation to the extraction of aggregates.

44. Please do not dismiss it as *de minimis*, is what I am asking.

(Mr McQuail) I have stressed the importance of this subject. I was making related points and I entirely accept that it is not a *de minimis* subject in itself.

Mr Steen

45. Would you also look at the role of the Crown Estate Commissioners in relation to the rents they are charging and the effect that will have on the conservation of the environment? In the River Yam they are talking about a 1,000 per cent increase in rent which they will charge to the actual local harbour authority which will have a devastating effect on the environment because the harbour authority will have to increase by tenfold the number of boats that are going to have moorings there. I wonder whether you have considered that aspect?

(Mr McQuail) That is a different subject and I would need to refer that elsewhere.

Chairman

46. Is this not typical of the *ad hoc* way the whole of this area relating to the coast and the seawater, high water low water mark, has evolved and why there is such a multiplicity of interests which have been established for historic and probably anachronistic reasons today and we have just let them jog along because somebody had a vested interest that nobody really had the courage to examine and seek to take away. Why are the Crown Commissioners the owners of the land below the low water mark? What is the rationale for that today with everything that flows from land ownership? The Crown Commissioners are merely looking at this land in terms of an owner of land and what can be exploited or derived from that land in terms of rent. That is their sole interest. Is that a valid reason for leaving the control with them simply because once upon a time King Canute wanted to regulate the waves? The reason for Crown ownership goes back to the kings and queens of England and for no other reason.

(Mr McQuail) I am afraid I cannot give a comprehensive answer to that question.

Chairman: We will have to think about it and probably come to some view on it with all the other interests that there are.

Mr Steen

47. When you said that you would have to refer this question of the Crown Commissioners charging a 1,000 per cent rent increase and the effect that would have on the environment of an estuary elsewhere, to whom would you have to refer it?

(Mr McQuail) To the Commission itself in the first instance. Then I would see where my inquiries led me. Going back to the Chairman's question, I do reiterate the point I made about the extraction of minerals, that effective control of the extraction of minerals round the coast is with the Government through the Government View procedure.

Mr Howell

48. To what extent are you confident in the ability or inclination of the port and harbour authorities to take into account the environmental effects of their activities?

(Mr Summerton) I am obviously conscious that in certain one or two cases in the past, and one in particular which has exercised the House of Commons for a good deal of time, there have been concerns about the environmental impact of proposals for port development. I am not sure that it is obvious that there is anything special in this respect. With respect, it seems to me that the question seems to imply that there is something special about port and harbour authorities in terms of their environmental responsibility. I am not clear why they should take less account of environmental concerns than any other operational body in the public or indeed the private sector. There is, I believe, increasing consciousness as attitudes change in society generally. There is increasing consciousness of the importance of environmental concerns and I believe that those authorities, like the rest of British industry, will become more sensitive about these concerns in the future than they have been in the past, if it is necessary for them to be so. Certainly they are open to public pressure on these matters in the same way as other similar operational bodies. As far as development proposals are concerned, they are subject to whatever system of development approval is relevant to the particular proposals. Frequently because such developments entail interference with freedom of navigation they are approved by Private Bill as things stand at the moment. Certainly, as I indicated in relation to the case that I mentioned, Parliament was very concerned about the environmental impact of that particular development. I have in mind the Felixstowe Dock and Harbour case. I believe that for the future that consideration by Parliament will be assisted by the requirement to undertake a full environmental assessment before submitting the Bill to Parliament. That is a change in Standing Orders for this coming session. It is also the case, and will be increasingly the case, that harbour works will be authorised by harbour revision and harbour empowerment orders. Those orders are already subject to environmental assessment procedures and it is open to environmental interests and others to object to those orders, and those objections and the related

environmental assessment will be the subject of public inquiry, assuming the application was significant enough. I conclude by saying that of course port and harbour developments will continue to be necessary from time to time in the future. They will not be welcome to those who oppose any further development on the coastline and in estuaries, and the authorities concerned may no doubt be accused by others of being careless about the environment in bringing forward the proposal in the first place. I would say that from the Government's point of view the aim must be to ensure that the procedures are such that the environmental considerations are given proper weight in any decisions that are made. I certainly would want to assert that as an official of the Department of the Environment.

Mr Jones

49. Do you not think that is just a little bit hopeful? That these bodies are likely to take into account the environmental impact of the various activities in their area and listen to public opinion. I am sure that with the best will in the world there will be port authorities just like there are other authorities that are not too bothered about that and then can ride roughshod if they so wish over environmental considerations. I am not one who is opposed to any development; I do not want you to get the wrong impression. I just think that at least within a local planning authority system there is a chance for people to make their views known to elected representatives and for them to determine an application which does not apply at the moment. If I may draw a comparison, British Rail—which also has an exemption from a lot of planning—has just chopped down virtually every tree along the railway embankment in my constituency on the grounds that they are causing leaves on the line, as though somehow or other we had not had another autumn in the last 100 years. I find that they cannot be held to account for that. The same applies to the port and harbour authorities and I think that is what we are suggesting is unacceptable.

(Mr Summerton) If I may seek elucidation, is your reference to their operational activities as distinct from their development proposals? It was not the only leg of my argument, in relation to the latter, that they were subject to public pressure like everybody else.

50. Perhaps I could knock your legs away one at a time.

(Mr Summerton) What I was asserting was that there were other procedures which were being strengthened which would concentrate their minds wonderfully if they were neglecting these matters. As far as the operational side is concerned, clearly that is a more difficult question. Whether the operational activities of bodies of this kind should be subject to outside control raises a very large question indeed if they are to do the job that they have been asked to do.

Mr Steen: I do not think that is right. I support Mr Jones entirely. I have five estuaries in my constituency, of which three have ports and in one of them they have to consider—that is not the Crown

Mr Jones *Contd]*

Estate one but the other one—how are they going to increase their income so they can deal with the so-called environmental effects. For example, they are concerned about the amount of sewage and rubbish which is thrown into the water; they want more and more staff to clear it up and everything and more loos. So what are they doing? They are having to increase the number of boats so that instead of seeing water you can just see boats and marinas everywhere. That is of course exempt from any planning control and that affects the whole estuary, the environment and everything else. That is what I think you should be looking at because it does dramatically affect the environment. The whole landscape is being changed.

Chairman

51. Most of these procedures now for dealing with coastal development, whether it be port authorities or otherwise, are by way of Private Bill. Private legislation does not seem to lend itself to the kind of environmental impact study that we are becoming used to in dealing with the planning law in this country. How does the EC directive impinge upon the Private Bill?

(Mr Summerton) On that last point, the decisions of national parliaments are exempt under the terms of the directive but Parliament has agreed—this is not for the Government, Parliament has agreed—to a change in its Standing Orders for next year which will require that development proposals submitted to Private Bill procedure will require an environmental assessment which meets the standards of the directive. That is what I understand the position to be.

(Mr McQuail) Yes; that was one of the main recommendations adopted from the overview by the Joint Committee on Private Bill Procedure.

52. Do you still feel the Private Bill procedure, which is pretty cumbersome and the ability of people to raise objections are much more difficult through way of formal petition than attending a public planning inquiry? Do you feel a Private Bill is the right way to deal with matters of this kind?

(Mr McQuail) That was one of the main recommendations; the other one was that there should be a new procedure. There is to be forthcoming legislation under which railway and light rapid transit and harbour projects would be taken out of the Private Bill procedure and approved under an order making procedure. That would include provision for an environmental assessment and public inquiries as with other systems. Other categories of development may be transferred to that in due course. The legislation is not through yet. On those two major advances in the development of environmental assessment in relation to a Private Bill or comparable activities, Parliament and the Government between them do seem to have made big procedural advances. I would only say that there does seem likely always to be scope for the use of Private Bill procedures for some really unusual forms of development or forms of activity but that really is a matter for Parliament to decide for itself.

53. Is that related to the ownership by the Crown Commissioners of the land? That excludes the planning authorities from dealing with the matter essentially because the Crown Commissioners are outside the planning law and therefore you have to approach by way of Private Bill do you not?

(Mr McQuail) No, I do not think that is so. I was thinking of the really unusual forms of development such as the new island off Poole Harbour. So far as development affecting the seabed and the Crown Estates is concerned, some of those are done by Private Bill but not entirely, because of the matter of ownership, because of the nature of the development. To come back to the major class of cases around the coast of England and Wales, the Government View procedure offers, although we are reviewing it or going to look at it, certainly an alternative to the Private Bill procedure that was found more satisfactory in the review of 1989.

Mr Summerson

54. May we move from planning to pollution and talk about coastal pollution? This is particularly topical because of course there was recently a spillage of oil—I think 1,000 gallons of oil—into the Thames estuary with some pretty disastrous consequences. Who takes responsibility in the event of an oil or chemical spill in an estuary? When such a thing happens who actually takes charge of clearing it up? What happens if the spill results from a hose linking a vessel with a shore-based installation? Who is then responsible for prosecution? Who decides who should be prosecuted?

(Mr Summerton) As to the general principles, I believe that these are clear enough. The position is that the Department of Transport in the form of its marine pollution control unit takes the lead in all incidents of oil or chemical pollution from ships; the NRA takes the lead in the case of a spill from an industrial discharge or other land-based source; and the local authorities have the responsibility for cleaning up the consequences when they impact on the coast, that is to say when oil for example arrives on the coastline then it is the local authority's responsibility for clearing up. As you implied from the case you mentioned, in practice of course incidents do not necessarily divide neatly in that particular way and operationally it is not necessarily the lead body, as I just described it, according to principle, which is best placed to take action. In practice obviously the lead responsibility depends on a number of factors, principally the source, scale and location of the spillage. The position is that the Marine Pollution Control Unit has the major role with responsibility for making national contingency plans, for assistance to local authorities and for managing stockpiles of equipment for release to local authorities. As I implied, as with other types of emergency, pollution problems occur in many shapes and sizes and guises when they actually emerge and the fact is that they do require action by a number of bodies. That is simply the nature of the beast with many emergencies. The important thing is to have plans which define the roles of the different bodies and specify means of co-ordination and adaptation

Mr Summerson *Contd]*

of those plans to the specific circumstances of the events. That describes the general position. As I understand it for the clearup in the case that you specifically mention, the arrangement was that the terminal operators themselves, Mobil, have taken responsibility actually for dealing with the incident which lay within the boundaries of their facility. That perhaps illustrates the way in which it is important to have plans but for the bodies concerned to co-ordinate their actions so that they adapt to the particular incident and deal with it in the best possible way.

55. Are you satisfied that the plans you have told us about do indeed provide for co-ordination of activities?

(Mr Summerton) It seems to me that those arrangements, particularly with the Marine Pollution Control Unit, which is a specialised unit very much prominent in relation to that kind of spillage, are likely to secure the results that are required. That does not mean to say, of course, that there will not be an incident tomorrow which disproves it. The only thing you can do is to plan, have contingency arrangements and adapt them in the event and review each particular event to ensure that you have learned the lessons of that event.

56. Do exercises take place to test the validity of these co-ordination measures?

(Mr Summerton) I am sure that the answer to that is yes, but if you wish to explore that further than we can certainly provide further information on that specifically, about the kind of contingency planning and exercising that takes place. I must say that you are at the limit of my knowledge of the subject.

57. In view of the importance of this subject I should be grateful for those further details. May I move on then please to talk about possible rises in sea levels? We have of course already touched on this earlier in our proceedings. Nevertheless, may I ask how you are advising local authorities to plan for probable sea level rises as a result of climatic change? What is the Government's timetable of action on response strategies?

(Dr Park) Could I broaden the question very slightly by reference to the earlier discussion and indicate that NRA as well as local authorities and other drainage authorities will be covered by whatever advice we might develop? It goes beyond the local authorities if we are thinking about coastal defence. As long ago as two years we developed a strategy for dealing with sea level rise in particular which has been submitted to scrutiny by various organisations. I will not go through the detail of that strategy but really it involves identifying the state of the defences currently and refurbishing them as a matter of urgency, developing the understanding through R&D of the processes, following the international developments in terms of understanding of global warming and its impact and being ready to readjust in the light of what those predictions are suggesting. Two things that we recognised very quickly were the need in designing defences to design them in the knowledge that there might be a change in sea level or in wave activity or

something of that nature. So there was an encouragement to design defences in a flexible way so they could be adjusted when the predictions were clearer or when the outcome of the predictions was clearer. The other point that has come out since that point of time is that the Intergovernmental Panel on Climate Change, IPCC, has carried out an awful lot of work and has reported in the autumn of last year. Within that report they have indicated a series of predictions for global—I emphasise the word "global"—sea level rise. They are not in a position, because of the state of development of computer modelling actually to make predictions for a regional sea level rise. Of course it is all based on the best assumptions that are around at that point of time. The understanding will develop over a period of time. We have taken those predictions and are currently discussing with National Rivers Authority what allowances ought to be made in defence for sea level rise. At the moment an allowance of up to 30 centimetres per century—it is very important to get the timescale, because otherwise it could be quite frightening—is made in the current design and the thought is that that allowance should be increased slightly according to the area of the country where defences are to be installed. It will be a matter of weeks before we resolve that guidance. As soon as it is resolved that will be issued to all authorities, not just to NRA but to local authorities, as guidance for them in designing defences. Just one last point because I did not deal with the timescale point. The timescales that are being suggested by the IPCC group are that by 2030 there will be a change in sea level on a global scale of something like 20 centimetres and by 2100 it could be something of the order of 65 centimetres. That indicates the timescale over which we have to be ready to respond.

58. If I may say so, those figures sound suspiciously precise to me in the context of what we are talking about. How anyone can work out that there is going to be a rise in sea levels of 30 centimetres over the next hundred years is really beyond me. May I ask whether you liaise with such organisations as the British Antarctic Survey which is probably in the best position to give early warning of possible sea level rises?

(Dr Park) All the work that has been done by IPCC draws on the best expertise available in all countries not just in the United Kingdom. That will have been a contribution to the process of deciding what is the appropriate allowance to make or what is the most reliable prediction to make. The particular working group which made these predictions was of course led by Dr John Houghton—I think he is now Sir John Houghton—who is the Director General of the Meteorological Office. He was drawing in all the expertise that was available. We, on our own part, in interpreting those predictions for the design of coastal defences, are drawing on the work of institutes within the Natural Environment Research Council who are expert in this area. These institutes are interpreting these predictions so as to inform the practical design standards. Could I make one last point about the current standard of 30 centimetres? Extrapolation of historic movement of sea level

Mr Summerson *Contd]*

suggests a 15 centimetre allowance. Being good British people we have doubled that to make sure that we are coping with what will be coming in the future.

Chairman

59. Could I relate it another way because the historic extrapolation may not be correct in view of the anxieties that are expressed at the rate by which global warming has suddenly been perceived to be increasing? There is a posit between 1 per cent and 5 per cent centigrade increase in the warming of the globe suface. Your 30 centimetres relates to what in that kind of scale? Is it the equivalent of a one degree centigrade or two or three or has it no relationship to that at all?

(Dr Park) Can I draw a distinction? The 30 centimetres is what has been the allowance that we worked to and that was clearly based on an extrapolation of what had happened over the last century and we then doubled it.

60. If you look at the predictions of between 1 per cent and 5 per cent degrees centigrade in the increase of heat of the earth's atmosphere, have you worked out or tried to come to any view as to the effect on the coastline of the United Kingdom of increases in the temperature between that particular range? Has that exercise been done?

(Dr Park) Yes, but the IPCC has done the exercise at a global level, not at regional level. They actually made some predictions about change in temperature. They then tried to relate that to the likely movement of sea level and they have set out three scenarios that could come about, one of which has very low implications, one of which has very high and they took the middle one and they called that their best prediction. We have based our work on their best prediction. I must hasten to say that it is a global prediction not a regional prediction because the knowledge base does not allow regional predictions at this stage. What we are planning to add into this is to take what we regard as the IPCC's best predictions on a global scale and we are planning to make allowance for the fact that the country is tilting. There are certain parts of the country which will need a slightly higher allowance in terms of defence design and certain parts of the country which will need a lower. That refinement is going to be introduced into the agreement that we are about to strike with the National Rivers Authority.

Mr Steen

61. Which is higher and which is lower? Where is the tilt?

(Dr Park) The tilt is that London is at greater risk than Blackpool.

62. What about the south coast?

(Dr Park) The south coast would be about the same as London, but the south west coast would be sinking slightly less.

63. You said that you had taken the best prediction and you are now doubling it. Suppose you took the worst scenario, what would that be?

(Dr Park) Could I just clarify? The doubling related to the extrapolation of the historic situation. The new situation is that we take the best predictions from IPCC. We do not apply any factors to that save to allow for the fact that England is tilting.

64. But you say you took the middle of the possible scenarios.

(Dr Park) Yes.

65. Suppose you took the worst of the possible scenarios what would the figures have been then?

(Dr Park) I do not carry them with me, I am sorry.

(Mr McQuail) We could perhaps provide a little more.

(Dr Park) It is all published information so there is no difficulty.

Mr Summerson

66. May I then move on to nature conservation and ask why there has been a delay in designating Special Protection Areas and Ramsar sites?

(Mr McQuail) What the Government have given priority to has been the designation of Sites of Special Scientific Interest. There is now a pretty complete protection of the Sites of Special Scientific Interest through that designation. The programme of designating Special Protection Areas and Ramsar sites has deliberately been taken to follow on from that because in substance the protection that is accorded to the site by designation as an SSSI gives virtually the whole of the protection that is available through designation as an SPA or a Ramsar site. All of the sites in those latter categories are included within the 5,000-odd SSSIs that are designated in England. We are, however, making substantial progress with designating both SPAs and Ramsar sites and have a forward programme that will ensure that all of the ones that are identified as being appropriate to those designations are considered for designation over the next few years. We have so far designated 22 coastal Ramsar sites and 27 coastal SPAs which is in both cases rather more than half of the total of those two categories that have been designated so far.

Mr Jon.

67. Do any of those extend beyond the low water mark?

(Mr McQuail) The SSSIs do not.

68. I know that. That is why I was asking about SPAs and Ramsars.

(Mr McQuail) Part of the review of the legislative framework is to consider how we are going to amend the legislation—and the Government accept that they will have to—in order to be able to extend protection under the SPA and Ramsar regime beyond that in SSSIs.

Chairman

69. It seems an age ago since this Committee looked at the Wildlife and Countryside Act and we made some criticisms of the fact that marine nature reserves were not being given any consideration, it seemed, at all. I do not know what progress has been made in that particular field since we raised that criticism.

(Mr McQuail) I am not quite sure of the date of the criticism.

Chairman *Contd]*

70. 1984.

(Mr McQuail) What was certainly required in order to proceed with a designation of marine nature reserves, which we acknowledge is quite a difficult process, was detailed survey data that will come from the completion by the JNCC of the marine nature conservation review which was begun in 1987 but will not be scheduled for completion for another 10 years.

71. So in the seven years since this Committee had observations to make precious little has been done.

(Mr McQuail) Two sites have been designated. Seven sites were identified by the former Nature Conservancy Council. Beyond the two that have already been designated, three more are undergoing the extensive consultation process required to obtain agreement and the remainder are either operating on a voluntary basis or are part of wider nature conservation proposals.

72. The problem as we perceived it at the time was trying to reconcile fishing interests with a nature reserve concept. I do not know whether MAFF have made much progress in resolving that particular dilemma in the interval of time. Can you help me?

(Mr McQuail) It is not the right part of MAFF represented here today.

73. So the right hand does not know what the left hand is doing. Is that what we are being told?

(Mr McQuail) No. I go back to my earlier answer that we are now making better progress and that we could not have made any progress without the survey which the JNCC——

74. So two instead of none is a 200 per cent progress.

(Mr McQuail) It is an infinite progress but that is logic chopping and I would not wish to do that.

Mr Summerson

75. Perhaps I had better move on to my final question on information and research. What is being done to establish a co-ordinated national information base and research programme on coastal issues and how is the Government encouraging and supporting such initiatives?

(Mr Wilson) As has been evident from our conversation here, we are addressing a wide range of policies and activities which impact on the coastal zone. These include town and country planning, pollution control, wildlife conservation and uses ranging from industry, transport and defence to tourism and amenity. All these are subject to research from time to time. With so many activities it is of course essential that research is commissioned and managed to focus on the needs of specialised users

and policymakers. Nevertheless it is also vital that those commissioning and carrying out research are aware of other relevant projects. The way in which this is achieved is broadly first of all by developing information bases which are of general relevance to a wide variety of users. These frequently aim at pulling together information in easily accessible form and providing pointers to more detailed information sources. Recent initiatives here include the United Kingdom digital marine atlas and the North Sea coastal directory. The second way for achieving cross-sectoral and inter-disciplinary awareness is through specific projects which take an integrated and multi-disciplinary approach in pulling together information from a wide variety of sources. A current very good example of this is the two-stage project which we have on planning policy for the coast, which is the first stage, and on earth science information which will be used in support of coastal planning and management, which is the second stage. The first stage of this project is one to which my colleague Mr McQuail has already referred and will be reporting shortly. We will be able to make this available to the Committee. The second stage of the project will be looking in much greater detail at earth science information on what is happening to the coastal zone. This would cover stability of the coastal structure and the effects of erosion on the coast. The third way is by formal administrative arrangements of co-ordination at the research level; for instance the research councils review their research activities, both among the research institutes and the universities. A single Government sponsored research programme on coastal issues could be rather bureaucratic and unwieldy. Given the range of sectors to be covered it is by no means clear that overall co-ordination would produce the best results, indeed the research councils although funded by central Government are not controlled by it and the same goes for the academic institutions. Such bodies therefore have a degree of freedom in pursuing new ideas in research which after all may be inimical to Government policy and that is not necessarily a bad thing. The Government's approach has been that the mechanisms which I just described and the use of co-ordinating machinery clearly in progressing work in specific policy areas is a far more effective way of aiming for and achieving what would otherwise require a rather monolithic co-ordinating machinery.

Chairman: May I thank you, gentlemen, for your attendance this evening and for the help you have given us. We will have a lot to think about. I am not sure that we are much the wiser at this point about the way in which our coastal protection is being organised and run at the moment but hopefully by the end of the inquiry we will come to some firm conclusions. Thank you very much for your help.

WEDNESDAY 20 NOVEMBER 1991

Members present:

Mr John Battle   Mr Anthony Steen
Mr Robert B. Jones   Mr Hugo Summerson
Mr Ralph Howell

In the absence of the Chairman, Mr Robert B. Jones was called to the Chair.

Memorandum by Royal Town Planning Institute

COASTAL ZONE PROTECTION AND PLANNING

TERMS OF REFERENCE

1. The House of Commons Environment Committee has decided to conduct an inquiry into coastal zone protection and planning to complement its previous report on pollution of rivers and estuaries and the pollution of beaches.

2. The present inquiry will consider the dynamics of coastal change, the risks to coastal settlements and coastal ecosystems, the mechanics of coastal protection systems, planning policy in coastal areas and the co-ordination of the many authorities responsible for protection and planning in the coastal zone.

3. The Royal Town Planning Institute considers the inquiry to be timely and anticipates that the eventual findings will overcome what was seen as a deficiency in the Government's environmental strategy set out in its White Paper " This Common Inheritance " (Cm 1200) to which further reference will be made in paragraph 17 below.

4. The Institute therefore welcomes the opportunity to submit written evidence on this critical subject. It would be pleased to supplement this with oral evidence if the Committee so wished.

5. The evidence is structured as follows:—

A. *Background:*—

    (i) Comment on the proceedings of a symposium in Southport, 1989, " Planning and Management of Coastal Heritage " (published by Sefton Metropolitan Borough Council in 1990),

    (ii) A summary of Institute responses to recent planning documents,

    (iii) The European dimension.

    (iv) The Institute's current stance.

B. *The Institute's views on the matters being investigated*

C. *Conclusions*

6. The Institute is aware that the Government is contemplating the preparation of a Planning Policy Guidance (PPG) Note on the coast and that it is also conducting a survey relating to Planning Policy for the Coast and Earth Science Information in Support of Coastal Planning and Management. It presumes that the latter and this Committee's inquiry will be an input into the final Planning Policy Guidance Note.

A. BACKGROUND

I. **Southport Symposium: " Planning and Management of the Coastal Heritage "**

7. In 1989 the North-West Branch of the Institute co-hosted a Symposium of " Planning and Management of the Coastal Heritage " in Southport. The keynote speaker, Adrian Phillips (then Director-general of the Countryside Commission) suggested a Council of Maritime Local Authorities, similar to a Council for National Parks could be a useful advisory body, an aspect of which will be dealt with later. He also suggested that " we should be aiming at promoting a more holistic view of the coastline as a national resource ", in respect of which he listed five elements:—

(1) Consolidating the heritage coasts achievements.

(2) Extending the heritage coasts achievements to other parts of the coast.

(3) Looking at the marine and terrestrial aspects of the coast together.

(4) Placing coastal protection firmly in the broader context of land use planning and environmental protection, and

(5) Emphasising the importance of the coast as a whole as a resource of national, and indeed in international importance.

8. The Institute supports these five elements as an agenda for improved protection and planning of the coastline. However, it would make two comments on this:—

(1) Perhaps a sixth element should be improved co-ordination of the responsible authorities, which is referred to in the Committee's terms of reference and indeed to which Mr Phillips referred in his paper; and

(2) in referring to the importance of the coast as a resource (fifth element) it is necessary to recognise the valid (ie sustainable) demands on it by maritime communities, maritime based industry and tourism including recreation. This requires both effective planning and effective management.

9. What is quite evident is that the coastal planning and management deserves a much more co-ordinated approach through the medium of regional plans, of development plans (including subject plans) and of comprehensive management plans subsuming the current heritage coast plans.

10. Mr Phillips concluded that "there have been setbacks. . ., but on balance. . . achievements have exceeded disappointments. This provides us with a strong base from which to move forward: no longer concentrating exclusively on particular parts of the coastline, but speaking with confidence, of the precious, finite, national resource represented by our entire coastal heritage."

11. As the Symposium report concluded, " Britain has perhaps more reason than any other nation to take a lead in developing a better integrated approach to planning and managing its precious coastal resources."

12. The Symposium resulted in an Agenda for Action which is reproduced here as Appendix I. It is an agenda supported by the Institute.

II. Summary of Institute Responses to Recent Planning Documents on the Coast

1. *"Heritage Coast Policies and Priorities" (August 1990)*

13. The Institute responded to the Countryside Commission on its draft policy statement, in November, 1990. The Institute's general comments in the response are attached as Appendix II. As will be seen the Institute endorses the role of the heritage coasts and recognises their undoubted success as areas of managed coastline. Its main concern is to see that those benefits are extended to other lengths of coastline as referred to in the next paragraph.

14. The Institute which was generally supportive of the proposals in the document put forward six points which it considered the Commission should pursue. It should:—

(1) Take a much firmer line on protection and enhancement of the coastline as a whole, and press the Government to do likewise.

(2) Press for the statutory definition of heritage coasts *per se* or as AONBs (where appropriate).

(3) Ensure that regional advice includes a strategy and policies for the coast.

(4) Ensure that the extent of heritage coasts is firmly defined.

(5) Press for better funding particularly in those heritage coasts outside National Parks and AONBs.

(6) Provide for monitoring of progress in heritage coasts.

15. The Institute placed much emphasis on point (a) suggesting that the final policy statement should be far more positive about " the wider coastline ". The Institute was therefore pleased that the final document published in January, 1991, expanded the section dealing with the wider coastline and firmly endorses the Commission's three proposals as set out on page 9 of the policy statement.

16. The approach of the Commission contrasted strongly with the Government's White Paper on the Environment published in September 1990, and on which the Institute submitted its observations to the Government in July 1991.

2. *"This Common Inheritance": Cm 1200: (September 1990)*

17. The Institute, when submitting to the Government its observations on the White Paper, commented on paragraphs 7.60 and 7.61, which dealt perfunctorily with the coast, as follows:—

" The Institute was profoundly disappointed with the scanty attention given to a vital environmental resource, the coastline. It was subsequently pleased to see that the final document on heritage coast management, published by the Countryside Commission, recognised the need not only to manage the attractive length of coastline but to upgrade the less attractive length outside the heritage coast designation. It was further pleased to see that the Government has commissioned research on the planning policy for the coast and earth science information in support of coastal planning and management. The nation cannot afford just to look after those lengths at present unspoiled. It is presumed that the study will consider the need to protect not only coastal land but also coastal waters and estuaries."

3. *"Planning Policy for the Coast and Earth Science Information in Support of Coastal Planning and Management"*

18. Following the limited references to the coast in the White Paper, the Institute welcomed the intention of the Department of the Environment (DOE) to carry out the above research project, the terms of reference for which were contained in a draft circulated in March 1991. A copy of the response is attached as Appendix III, particularly because it sets out some information sources which may provide helpful evidence to the Committee.

19. The response it will be noted placed emphasis on the need to define the "coastal zone". It also was considered necessary to refer to the potential impact from renewable and non-conventional sources of energy.

20. Dealing with Task 1: Review of coastal planning and management policy and responsibilities, emphasis was placed on the need to upgrade lengths of the coastline in accordance with the BATNEEC approach ("best available techniques not entailing excessive cost") instead of concentrating on the best lengths of unspoiled coastline. The Institute felt that in reviewing existing policies there was a need to examine their implications for existing communities, residents and businesses in the coastal areas, particularly those subject to coastal preservation policies—there needs to be a balance between conservation and development.

21. Turning to Task 2: Earth science information in support in coastal planning, it was suggested that the number of disciplines and the number of aspects studied needed to be extended, referring particularly to archaeology, biology and ecology. The importance of environmental audit particularly in the context of monitoring was also referred to.

4. *"Turning the Tide: A Future for Estuaries"*

22. In December 1990, the Institute responded to an invitation by the Royal Society for the Protection of Birds (RSPB) to comment on the above document. In that response the Institute, in supporting the Society's general objective for protecting wildlife habitats, made the following points:—

(a) Support for the view that "the Government must produce and implement a national coastal strategy to guide planning and wise use of the coasts".

(b) Welcome for the Society's support for the production of statutory coastal plans to implement any national coastal strategy. However, it was considered that this could be achieved within the context of the present development plans system (but see paragraph 23 below).

(c) Regional planning should enable cross boundary issues "relevant to regions, counties and districts" to be addressed.

(d) Need for balance in considering the need for conservation and development proposals, special reference being made to tidal barrages. The Institute nevertheless supported the tenet of sustainable uses and development.

(e) Support for better co-ordination of the multifarious organisations dealing with coastal matters, particularly at Government level.

(f) Support for the Society's views on the role of the Crown Estate Commissioners.

(g) Agreement about the extension of planning controls into coastal waters.

(h) Support for wider use of environmental education.

5. *"A Future for the Coast"*

23. This document produced by the Marine Conservation Society in 1990 sets out proposals for a United Kingdom coastal zone management plan. The institute supported the concern about the lack of a national policy or strategy for the coastal zone and expressed support for the extension of planning powers below low water marks. However, it felt unable to support proposals for the establishment of yet another body, a coastal zone management unit, to manage the coast, bearing in mind that parts are within National Parks with their own administration, and that other lengths are within AONBs and/or Heritage Coasts which are managed quite successfully by local authorities.

24. The Institute again felt that positive coastal policies included in regional guidance (or preferably regional plans) together with similar policies in structure and local plans as appropriate was the best way forward.

III. The European Dimension

25. The Institute is aware of moves by the European Community to have a greater involvement in coastal zone management, beyond that currently relating to pollution of coastal waters. A European Coastal Charter has been developed by the Conference of Peripheral Maritime Regions. Action programmes have been initiated for countries on both the Mediterranean and the Atlantic seaboards (MEDSPA and NORSPA).

26. The Ministerial Conferences on the North Sea and the recently published reports of the Irish Sea Study Group have drawn attention to the need for linkage between marine and coastal resource management and co-operation between nations to achieve this. Further there is an important European conference on coastal conservation to take place in November 1991 in the Netherlands at which the United Kingdom will be represented.

27. The Institute considers that, despite the issues raised in this submission, the United Kingdom has a considerable amount of experience to offer on coastal matters. It could, and should, take the lead in Europe in the development and implementation of comprehensive strategies, planning policies and management techniques for the coastal zone.

IV. The Institute's Current Stance

1. *Rural Planning in the 1990s*

28. In July 1991, the Institute published a draft Rural Planning Policy Framework. This has not yet been fully adopted by the Institute and the intention is that in due course the policy framework will be elaborated. Part three of the document dealing with the diversity of landscape includes a section on coastal areas as follows:—

> 3.15. Coastal areas represent a vital resource. There is a need for mechanisms to enable the formulation of plans and policies which deal with the immediate marine environment as well as the coast and its hinterland. Control is vested in a number of agencies which can lead to conflict and a lack of co-ordination. Further there is a need not only to protect the unspoiled coastline, but to pursue policies for upgrading areas of coastline that have been developed or despoiled. In particular there have been demands from local authorities and others to bring marine fish farming under express planning control.

> 3.16. The Institute considers that policies for coastal areas should address:—

> * the need for a national strategy and policies relating to the whole coastline and a co-ordinated legislative framework;

> * the need for these policies to relate to coastal waters and estuaries as well as the coast and its hinterland;

> * the need both to protect the unspoiled coastline, and to consider the opportunities for the restoration of those lengths which have been environmentally damaged;

> * protection of coastal areas from inappropriate development especially caravan and other residential uses; and removal where feasible of intrusive developments; discontinuance of disposal of waste in coastal areas;

> * extension of heritage coast protection and support for the Enterprise Neptune initiatives;

> * public access; achieving levels consistent with protection of the coastal environment;

> * planning control over marine fish farming;

> * protection of marine and estuarine wildlife habitats;

> * protection of marine and maritime archaeology;

> * the need to regulate port development including marinas;

> * the need to regulate energy development, nuclear, thermal, tidal and wind;

> * the need to provide and regulate the nature of coastal defences;

> * the sources of pollution and methods of controlling it;

> * the desirability or otherwise of extracting sand and gravel from marine and estuarine sources.

29. The views expressed above will be elaborated in the section which sets out the Institute's evidence relating to the issues identified by the Environment Committee.

2. *The Regional Planning Process*

30. In May 1991, the Institute published for consultation, the above document, which dealt, in that part dealing with the content of regional plans with the environment. In referring to DOE Circular 12/72 (Welsh Office Circular 36/72) (The Planning of the Undeveloped Coast), the document commented " it is equally appropriate now, that the coastline as a whole, a scarce resource to which insufficient attention is being paid, should have particular consideration in the regional planning process. In particular, there should also be an awareness of the consequence of rising sea levels ". (On regional planning Circular 12/72 suggests: " The planning of the undeveloped coast needs to be considered in conjunction with the planning of the surrounding area and within the context of a regional strategy ".)

B. The Institute's Views on the Matters Being Investigated by the Committee

31. The preceding paragraphs set out background information to the evidence, now following, which responds directly to the matters specifically indicated by the Environment Committee as the subject of its Inquiry.

32. Before responding in detail the Institute would strongly request the Committee to define the " coastal zone " as those land and sea areas which are related in visual, social and economic terms to the littoral including river estuaries. The preceding paragraphs emphasise the perceived importance of this approach, irrespective of the outcome of the Inquiry.

33. For the sake of clarity the term " management " in this submission is used in a broad sense, to include the activities of all the agencies operating within the coastal zone.

34. Coast " protection " is also used in its broad sense although it is recognised that the term has a very specific meaning under the Coast Protection statutes.

1. **Dynamics of Coastal Change**

35. The Institute has considered this under two headings:—

(i) Development in the Coastal Zone

(ii) Physical Change to the Coastline

(i) *Development in the Coastal Zone*

36. Firstly the Institute wishes to make the point that the coastal zone is a fixed and vital resource with less flexibility than the wider countryside. Thus whenever development is contemplated the question must be asked whether it is necessary to be carried out in the coastal zone. Further the question must be asked as to whether it falls within the ambit of sustainable development (Sustainability: " Catering for the needs of the present without compromising the ability of future generations to meet their needs "—"Our Common Future—The World Commission on Environmental and Development "—The Bruntland Report 1987).

37. It recognises, in the context of the last paragraph, that there are developments which have to be located in the coastal zone and which are important to a maritime nation:—

(i) *Military defences*—in the current international climate, however, this appears to be a contracting need.

(ii) *Ports*—these have tended to be rationalised in recent years, and with the abolition of dock labour scheme ports, it is considered that there is scope to develop existing ports, if required, rather than developing new ones.

(iii) *Transport*—within this category are ferries, and at the macro-level the Channel Tunnel. The consequences of the latter for the former in the South East have to be firmly established. Obviously the economies of the channel ferry ports, and probable changes thereto, will be a matter that will have to be addressed positively in planning terms. (There may also be consequences for ports elsewhere along the eastern and southern coasts.)

(iv) *Maritime industry*—clearly those industries relating to fishing, off-shore oil and gas, shipping and boating need to be located in the coastal zone. An increasing threat particularly in sheltered waters is fish farming, the need for the control of which is increasingly being recognised.

38. In addition to the above there is development for which the coastal zone is favoured:—

(i) *Energy;* Both conventional energy and renewable energy generation present a potential threat to the coastline in visual and pollution terms. In particular it has to be recognised that renewable energy also presents environmental problems. Coastal sites are favoured for windmills, giving rise to both visual and noise pollution. Bearing in mind the comparatively low output, it has to be asked whether the environmental cost is justified. Tidal barrages similarly have their drawbacks. Clearly they too present an intrusion and there are fears about their ecological consequences. The crucial element in all these developments is the environmental impact they have on the coastal zone, both as individual proposals and cumulatively. Having said that the view of the Institute is that the development of methods of renewable energy generation are preferable to those associated with conventional methods (fossil and nuclear fuel). More preferable still is to reduce demand by investing more resources into energy conservation (paragraph 43(d) below).

(ii) *Extraction of minerals* in the coastal zone falls into two categories, dredged aggregates, and coastal mines/quarries.

Dredged aggregates have the advantage of reducing pressure for their production from land based quarries, with resulting environmental benefits. However, the effect of dredging on the coastline itself and its protection from erosion needs careful consideration. Dredging in estuaries also has its problems in terms of its effect on their regime with possible consequences for protection from flooding. Again to permit this to continue needs very careful consideration. Coastal quarries are now either being developed or contemplated because of the ease with which the output can be "exported" by sea—the price is visual intrusion, perhaps on a considerable scale. It is suggested that it reduces the need for distribution by road, but since such minerals have to be distributed by road at their destination port, and since most minerals are distributed within a short distance of their origin, this alleged benefit it dubious.

39. *Tourism/Recreation*

The other major industry which, because of the British tradition, is firmly established on the coast is tourism/recreation:—

(i) *Tourism* to a large extent has stabilised in recent years and indeed in some resorts this has led to pressures to change their character by conversion of hotels to other uses. This contraction is partly due to successful efforts to extend the holiday season. Figures produced at the Sefton Symposium suggests that the United Kingdom has one of the longest if not *the* longest holiday seasons of the countries in the European Community. This has also reduced demand for sites for non-traditional types of accommodation, except perhaps for touring caravans.

(ii) *Recreation* gives rise generally to management problems although provision for moorings is also a planning problem, and, unlike tourism, the demand for provision is on the increase. Apart from those merely enjoying passive recreation—sight-seeing and walking—are those who enjoy more active pursuits associated with boating and water sports—dinghy sailing to speed boats in the first category and windsurfing to water and jetskiing in the second category. These new forms of sea-based recreation create conflicts of use on inshore waters which will increasingly demand some form of control both to reduce or avoid conflict and to protect certain coastal areas from inappropriate use.

(ii) *Physical Change to the Coastal Zone*

40. At the present time there are areas where the coastline is being eroded and others where it is accreting. It is, however, the former which gives rise to major problems which are more difficult to combat. These are two-fold. Firstly, there are the threats to property, and secondly the methods used to counter erosion—sea walls etc, often ugly in their execution. Clearly there is a need for a more sensitive approach to coast protection.

41. In the longer term there is the anticipated threat from global warming. This will clearly have to be monitored so that there is not a premature reaction. On the other hand there are no grounds for complacency. Certainly the opportunity, in the meantime, should be taken to ensure that the design of preventative work is environmentally acceptable.

42. Physical change to the coastal zone is also brought about by visitor pressure. This may have consequences in terms of the effect on flora and fauna or it may result in the character of an area being changed. This may of course be due to an ill advised planning permission but more usually it is just the result of increased popularity. Usually management measures are used to counteract such threats.

43. *In dealing with the dynamics of coastal change the Institute recommends* as follows:—

(a) It is important to determine whether development is acceptable in the coastal zone:—

(i) Heritage coasts should continue to be protected, and development in these zones should only be in the national interest.

(ii) Beyond heritage coasts all the remaining undeveloped or open coast should be considered as a precious resource worthy of protection (and where necessary improvement). It is essential that the estuarial resources are properly valued and protected in this context.

(iii) For the developed parts of the coast (seaside towns and villages, ports, coastal cities and their industrial areas) there needs to be equal attention to the proper evaluation of such areas, their potential for renewal, improvement and capacity to accept new development.

(iv) A holistic approach to the coastal zone means that every area should be valued as an integral part of a national resource.

(b) There should be a recognition that, in terms of maritime military defences, port activity and need for non-traditional holiday accommodation, existing facilities are generally sufficient to meet current and future needs. If a need for a certain type of the latter is identified, eg touring caravan sites, it should be met inland away from sensitive coastal locations.

(c) There is a need to bring under planning control marine fish farming.

(d) Before pursuing proposals to provide for increased energy generation in the coastal zone, whether from conventional or renewable sources, more effort should be made to increase energy conservation, and also the feasibility of the redevelopment of existing but redundant generation sites investigated.

(e) Before marine aggregates are extracted by dredging, it should be absolutely certain that such activity, particularly in estuaries and close inshore, will not deprive coasts of essential replenishment material.

(f) New quarries in coastal locations should not be permitted. In this case and in the case of (e) above, the use of mineral waste, eg china clay waste, and recycling of waste materials should be fully considered as an alternative source.

(g) Water based recreation activities of a kind which requires either development (needing to be subject to planning control), or which gives rise to potential conflict, should be controlled by co-ordinated management measures.

(h) Coastline subject to erosion should be protected by methods which are sympathetic to the coastal zone. If these are not feasible, the need to counter natural forces should be carefully considered before incongruous defensive works are constructed.

(i) Sea level changes need to be carefully monitored to ensure that coastal protection works are not prematurely undertaken or that such work is unnecessarily delayed.

(j) The importance of the Environmental Impact Assessment procedures should be recognised in the context of sub-paragraphs (b) to (i).

2. **Risks to Coastal Settlements and Ecosystems**

44. The short term risks to coastal settlements and ecosystems primarily arise from ill considered development and tourist pressures. These risks can be contained by effective planning policies which are dealt with elsewhere. This not only occurs on land. Marine ecosystems are vulnerable to changes arising from marine dredging and from pollution deriving from land-based development (see also paragraphs 58–61 below). Associated with ill-considered development are modern pollution problems, especially in coastal waters. These primarily arise from conventional generation of energy, untreated sewage, discharges from ships and agricultural slurry. An increasing problem is marine litter which disfigures the coastline particularly between low and high water marks. Adequate resources and policing are the solutions although litter originating from ships is difficult to control.

45. Pollution brings particular risks to those settlements dependent on coastal tourism, especially if bathing beaches are affected.

46. Ecosystems are also threatened by visitor pressures—sand dune areas and popular cliff top areas are particularly vulnerable. Management techniques can, however, counteract and restore these areas, given the resources to carry these out.

47. Physical change to the coastal zone also gives rise to the risks both to coastal settlements and ecosystems. This has already been dealt with in paragraphs 40–42 under the heading of the Dynamics of Coastal Change.

48. *Action required to reduce risks to coastal settlements and ecosystems* includes:—

(a) Stricter development control based on effective regional and development plan policies.

(b) Continuing development of effective management techniques especially to control visitor pressures.

(c) More effective pollution control. The proposed Environmental Protection Agency (EPA) is a positive step to achieve this.

(d) Availability of more resources to overcome sources of pollution. In particular these are needed to meet the EC Directive relating to bathing beaches.

(e) Increased protection for SSSIs and increased resources to develop marine nature reserves.

(f) Monitoring of changes in sea-level (see paragraph 43(i) above).

3. **Mechanics of Coastal Protection Systems**

49. The Institute has considered this aspect under a number of headings:—

(a) Physical protection of the coastline, ie coast protection works.

(b) Protection of the coastal zone's environment from visual intrusion.

(c) Protection of the coastal zone's environment from pollution.

(d) Protection of habitats in the coastal zone.

50. The Committee will appreciate that there is some overlap in relation to the five main issues which are being identified for the Inquiry and it is asked to consider under this heading the contents of the preceding paragraphs.

(a) *Physical Protection of the Coastline*

51. Coast protection and flood defence policy are the responsibility of Ministry of Agriculture, Fisheries and Food (MAFF). Much of the work on the main waterways and sea defences is the responsibility of the National Rivers Authority (NRA). Internal drainage boards carry out the works in areas of special drainage need. Local authorities look after minor waterways and undertake coast protection works.

52. These works, which are grant aided by MAFF, must satisfy economic and environmental criteria and its engineers provide technical guidance and advice " during the development of flood defence policies. They advise on economic, financial and environmental aspects that have to be taken into account ". (The order in which these aspects are listed by MAFF in its publication " Balance in the Countryside " may or may not have significance.)

(b) *Protection of the Coastal Zone's Environment from Visual Intrusion*

53. Protection of the coastal zone above low water mark from inappropriate development will in the main be the responsibility of planning authorities through the regional guidance, development plan and development control systems. There are exceptions including the Government's own development and electricity generating proposals, which are dealt with by the Department of Energy under the Electricity Act 1908 and those with powers under private Acts, eg port/harbour authorities. Further, the effectiveness of the development control system is reduced by deemed planning permissions granted under the Town and Country Planning General Development Order 1988 (SI 1988 No. 1813), examples being 28 day tented camping; small touring caravan sites; development by drainage authorities, the NRA, dock, pier, harbour, water transport, canal or inland navigation undertakings and development under local or private Acts or Orders (see Schedule 2 of the General Development Order).

54. Under this heading will come the need to regulate those recreational pursuits which are intrusive whether by virtue of visitor numbers or noise. Much of this activity occurs in estuaries, and it will be necessary to identify one body as responsible for regulating those activities through byelaws or other methods, which will however have to be actively policed.

55. Associated with protection afforded by planning legislation is that which is secured by local authority management whether in National Parks (for which there are management plans), Areas of Outstanding Natural Beauty or Heritage Coasts. This is achieved by persuasion rather than by statute.

56. The most effective protection is afforded by sympathetic land ownership, particularly by the National Trust or by local authorities with well balanced objectives. Examples of such authorities are Lincolnshire County Council which owns several miles of dune coast, and Torbay Borough Council. There will be many more. The continuing success of " Enterprise Neptune " is the best guarantee, with sympathetic management, of protection which is available.

57. None of the protection is available below low water mark and examples of intrusive development or " uses " include marine fish farming, erection of oil rigs and " parking " of redundant shipping particularly in estuaries. These uses fall at present within the responsibilities of the Crown Estate Commissioners.

(c) *Protection of the Coastal Zone's Environment from Pollution*

58. The control of water-borne pollution is at present the responsibility of the National Rivers Authority and the water companies. MAFF also has a responsibility. For instance it regulates disposal of waste at sea under the Food Act and Environment Protection Act 1985 and " licences are not granted if the waste would cause harm to the marine environment ". Progress is being made on this aspect; the Institute is pleased to note that dumping of fly ash and liquid industrial waste will be banned by early 1993 and dumping of sewage sludge will be banned by the end of 1998.

59. The dumping of any form of plastic waste from ships and boats anywhere at sea has been banned since January 1989. However, many coastal areas are still heavily polluted by plastic (and other detritus) originating from ships and boats.

60. The Government's proposals for an integrated Environmental Protection Agency (EPA) will hopefully reduce still further marine pollution.

61. It is reassuring that more resources are becoming available for the reduction of pollution of bathing waters, although considerable work is needed to ensure that once declared pollution free they remain so.

(d) *Protection of Habitats in the Coastal Zone*

62. This will be achieved primarily by measures to reduce visual intrusion and pollution. Protection will largely be afforded by the designation of Sites of Special Scientific Interest, but these need to be effectively " policed " and protected from development having deemed permission under the General Development Order and which has the potential for adverse effects on the coast. Resources and greater powers are required to secure the establishment of marine nature reserves generally below low water mark.

63. *Suggested action to deal with the above problems identified under the heading* **Mechanisms of Coastal Protection Systems** *is largely dealt with in the sections which follow.* The Institute does, however, consider that more needs to be done to discourage the deposit of marine waste. Effective policing is clearly difficult, but a modest tax imposed on shipping within territorial waters could be used to finance litter removal. Such a tax might also act as a deterrent in the future.

4. Planning Policy in Coastal Areas

64. There are suggestions that new authorities responsible for new mechanisms, ie coastal zone plans, should be introduced to control development on the coast. The Institute would question the necessity for this and indeed is opposed to the creation of a super body for the purpose.

65. The way forward for **Planning Policy in Coastal Areas,** it suggests, should embrace the following principles:—

(a) Planning control should relate to the whole of the coastal zone, use the littoral and an area of land and sea on either side: (see paragraph 32 above).

(b) Planning control in the whole of the coastal zone should be exercised by local authorities whether at county, regional or district level.

(c) Associated with planning control should be effective coastal management, again the responsibility of local authorities.

(d) Acknowledging that local authorities do not possess all the expertise required: they will rely as the appropriate planning authorities (as they do at present inland of the low water mark) for much of the advice required from authorities/bodies also involved in coastal matters, albeit that some rationalisation is required (see Issue 5 dealing wih co-ordination).

(e) Planning control should be exercised within the ambit of regional guidance (or preferably plans), structure plans and local plans, all of which will contain coastal policies and proposals as appropriate to each level of the plan making process.

(f) bearing in mind the need for consistent plans particularly in maritime waters where there are no boundaries (or barriers) to activity, the Institute sees merit in subject plans (similar to those produced by some county councils for the landscape generally) to identify consistent policies. The Institute regrets that provision for these has not been made in the new legislation relating to plan making. Nevertheless structure and local plans are an effective way of formulating coastal policies, particularly now that local plans are to be district-wide.

(g) The Institute considers that national policy is an important element in the planning process. In particular it looks forward to a positive and effective Planning Policy Guidance Note which will not only identify present methods of planning in the coastline but will indicate positive ways forward.

(h) At the more detailed level the Institute has suggested ways of evaluating the coast, including present heritage coasts, the remaining open coasts (especially the estuarine areas) and the developed coasts to avoid inappropriate coastal development. Generally speaking, only essential development should be permitted in open coastal areas, ie outside existing urban areas.

(i) Also on a detailed note the coastal zone should be afforded the protection from proposals granted deemed planning permissions by the General Development order. These permissions should be reviewed with a view to withdrawing those which have the potential to make an adverse impact on the coastline: (see paragraph 62 above).

(j) The Institute supports the management arrangements in heritage coasts, where local authorities, with support from the Countryside Commission, together with landowners and inhabitants have achieved considerable environmental improvement.

(k) The Institute in its initial response on the Government's consultation paper "Local Government Review: The Structure of Local Government in England", expressed support for unitary authorities and for a system of strategic planning at regional level. In relation to the latter the Institute considers that much more positive planning would be achieved by regional plans rather than regional guidance.

(l) From the above, it will be seen that the Institute identifies two strands to planning arrangements to the coastal zone:—

(i) TOWN AND COUNTRY PLANNING

DOE	National Guidance
Either a regional authority or consortium as at present	Regional Plan
County or unitary authorities	Structure plan/coastal subject plan
District or unitary authorities	Local plan

(ii) MANAGEMENT

The above arrangements need to be both supported and informed by appropriate arrangements for managing the coastal zone. These arrangements must recognise the need to co-ordinate the activity of a wide range of agencies. This is especially problematical in some estuaries or intensively used parts of the coast. Management may need to be regulated through informal Management plans which may well span across the jurisdiction of several authorities and their preparation could also be overseen by the regional authority or consortium noted above.

5. Co-ordination of the many Authorities responsible for Protection and Planning in the Coastal Zone

66. As suggested throughout this evidence there should be rationalisation of the decision making process within the coastal zone, apart from extending the area to which town and country planning legislation applies. Table 1 summarises the present arrangements for planning, management and provision of advice in the coastal zone.

67. The rationalisation referred to in the previous paragraph would include:—

(a) The elimination of the division of planning responsibilities above and below high water mark.

(b) The reduction in the number of bodies concerned with various aspects of coastal protection and planning.

(c) The reduction in planning powers of some authorities/bodies at present having those powers.

68. The effect of paragraph 67(a) would be to vest plan making and development control primarily in local planning authorities, resulting in a reduction of the powers of other bodies, particularly the Crown Estate Commissioners, to "grant permission" for development below low water mark.

69. The Institute's views with regard to paragraph 67(b) are as follows:—

(a) Central Government—all decision making on development proposals in the coastal zone should be vested in the Department of the Environment.

(b) Local planning authorities—would have the responsibilities referred to in paragraph 68 above.

(c) Pollution control should be vested in one authority. The Government's proposals for the Environmental Protection Agency is therefore supported.

(d) Coast protection and flood defence procedures should be reviewed with a view to rationalising the present arrangements which are anachronistic.

70. The Institute's view with regard to paragraph 67(c) is as follows:—

(a) All decisions on development proposals by Government in the coastal zone should be the responsibility of the Department of the Environment, which should be given overall responsibility for coastal planning and management matters as already suggested at paragraph 69(a) above.

(b) Powers of the Crown Estate Commissioners to permit development below low water mark should be removed.

71. The Institute recognises that there must be a continuing role for agencies currently involved in the various aspects of coastal zone management. There may well be difficulties in co-ordinating some of these activities however, and the role of the Departments/Authorities set out in paragraph 65(l) will be crucial in assisting and, where needed, accelerating this process.

72. Further it considers that there should be more positive " cross-fertilisation " of input by professionals engaged in coastal zone protection and planning.

C. CONCLUSIONS

73. The Institute hopes that the inquiry will result in the improved planning and management of the coastal zone as a whole and that the Government will grasp an opportunity which was missed when " This Common Inheritance " was published. Subsequent developments suggest that the Government will now be rectifying this.

TABLE I

Present Administrative Arrangements for the Coastal Zone

Town and Country Planning		*Management*
Plan Making	*Development Control*	
Department of the Environment; Regional Authorities (Scotland); County Councils; National Park Authorities; Metropolitan Boroughs; District Councils.	(a) *Above and Below Low Watermarks* Department of the Environment (and the Planning Inspectorate); Department of Energy (for electricity generating proposals); Department of Transport; Other Government Departments and Public Bodies in respect of their development (including Ministry of Defence). (b) *Above Low Water Mark* Regional Authorities (Scotland); County Councils; National Park Authorities; Metropolitan Boroughs; District Councils. (c) *Below Low Water Mark* Crown Estate Commission.	(a) *Government Bodies* Government Departments; Countryside Commission; Nature Conservation Bodies; English Heritage; Forestry Commission; Crown Estate Commission; National Rivers Authority. (b) *Local Government* County Councils; National Park Authorities; Metropolitan Boroughs; District Councils; Parish Councils. (c) *Other* National Trust; Port/Harbour Authorities (not included above); Trinity House; Water Companies; Other Landowners.
Pollution	*Coast Protection/Flood Control Etc*	*Advisory Bodies/Enablers*
(Prior to formation of Environmental Protection Agency) Her Majesty's Inspectorate of Pollution; National Rivers Authority; Health and Safety Executive; Ministry of Agriculture, Fisheries and Food; County Councils; Metropolitan Boroughs; District Councils.	Ministry of Agriculture, Fisheries and Food; National Rivers Authority; Inland Drainage Boards; County Councils; Metropolitan Boroughs; District Councils; Water Companies.	(Other than those already listed) British Tourist Authority; National and regional Tourist Boards; Sports Council; Rural Development Commission; Hydrographical Establishment.

Appendix I

SEFTON SYMPOSIUM

Using the 1970 report by the Countryside Commission as a model, survey the current issues for coastal planning and management and recommend action.

Lead Agency: Department of the Environment.

Co-ordinating Agencies: National Rivers Authority, NCC, Countryside Commissions, Coastal Local Authorities. Key staff from these agencies to form working groups.

Immediate targets: Commitment by government to coastal planning in Environment White Paper, 1990 and establishment of regionally-based working groups to undertake national coastal zone survey and report within defined timescale.

In parallel with (1) above issue planning guidance for structure and other development plans (there is no guidance at all available at present).

Lead Agency: Department of the Environment.

Co-ordinating Agencies: Coastal Local Authorities in developing strategic guidance as a strategic issue.

Immediate Target: DOE commitment to coastal planning in current statements on Structure Plans and Regional Planning Guidance.

Link existing data-bases and ensure access to information; identify gaps in knowledge and promote integrated studies. Improve both central and local government co-ordination.

Lead Agency: Department of the Environment.

Co-ordinating Agencies: (1) Regional/national groupings comprising NRA, NCC, Countryside Commissions and coastal Local Authorities. (2) Local Authority Organisations (ACC & ADC). (3) Professional Organisations.

Immediate Targets: (1) Rapid establishment of further groupings such as Irish Sea Study Group to cover whole coast, with specific planning input. (2) Establishment of coastal adviser groups within Local Authority organisations. (3) Joint RTPI/RICS/ICE statement on Coastal Zone Management.

Use "key-worker", multi-agency models to help cross-boundary planning and management; improve co-ordination between coastal authorities, professional, academic, voluntary and business organisations.

Lead Agencies: Coastal Local Authorities.

Co-ordinating Agencies: NRA and other government agencies, voluntary sector "umbrella" organisations such as RSNC, professional institutes eg RTPI, ICE, RICS.

Immediate Target: Secure extension/establishment of coastal planning and management schemes at local level in all areas.

Stop the coast being "taken for granted" and reinforce the drive towards better integrated planning and management.

Lead Agencies: Professional and voluntary sector organisations.

Immediate Target: Getting the coast on to the political agenda.

These objectives are interdependent. Action on one will assist another. All can be carried forward in parallel and every individual and agency has a role to play—no one should wait for someone else to act first. Action by Government is necessary but not sufficient. Local initiative can set examples for others to follow.

Appendix II

HERITAGE COAST POLICIES AND PRIORITIES

(Extract from the response to the Countryside Commission on its draft policy statement)

November 1990

GENERAL COMMENTS

1. The Institute welcomes the fact that the Countryside Commission has, in looking at heritage coasts, the wider coastline in mind. It is pleased to be able to offer its comments on the draft statement.

2. It believes the starting point is to express its great disappointment that the Government did not take the opportunity to consider in its White Paper "This Common Inheritance" (Cm 1200), one of the nation's most limited resources, the coast, in a much more positive way. It is really quite inexplicable that the document dismisses the coast in two anodyne paragraphs, numbered 7.60 and 7.61. To introduce them with a reference to Britain's coast including "some of the most outstanding landscapes . . ." and then dismissing them in two paragraphs is quite extraordinary.

3. The Institute is therefore looking to the Commission to fill the gap and to produce a policy statement for the coastline as a whole. Whilst the Institute acknowledges the need to protect and manage the finest lengths of unspoiled coastline, the opportunities to protect and indeed manage the further lengths of unspoiled coastline should not be missed.

The Institute is conscious that resources are limited, but it sees no reason to rest on its laurels now that the original " programme of definition is virtually complete ". The opportunity should be taken to identify further heritage coasts with a view to improving their quality. It is noted that the Commission does not rule out further definitions in the longer term.

4. On the other hand the Institute is bound to comment that over emphasis on heritage coasts can give rise to the impression that other lengths of coastline can absorb displaced pressures. Heritage coasts remain a partial measure for the tackling of the full range of coastal issues.

5. In any event the Institute considers that the original concept of coastal protection or conservation areas as set out in the Ministry of Housing and Local Government Circular 7/66 was very worthwhile. Policies for such areas are contained in some structure plans and have proved very effective. It is worth the Commission reiterating the importance of such development control policies and the need to include them in development plans. They do not detract from or undermine the purpose of heritage coasts.

6. As was pointed out in Circular 12/72, " The Planning of the Undeveloped Coast ", " as pressures on limited resources of land go on increasing, safeguarding the undeveloped coast by suitable planning and management policies will need vigour and imagination and must be a high priority " (paragraph 3). It goes on to say that the " planning of the undeveloped coast needs to be considered in conjunction with the planning of the surrounding area and within the context of a regional strategy " (paragraph 4).

7. The Institute considers that the reference to regional strategy to be apposite at the present time. Clearly the opportunity is once again there to deal with the coastline and the surrounding area at that level of the planning process. The Commission is recommended in its final statement to offer some positive advice on the inclusion of a coastal strategy and policies in regional advice which is now emerging.

8. The draft deals with the wider issues in one, penultimate paragraph. However, the Institute found that the contents fell short of the initiatives which, for instance, the Commission is taking in respect of the countryside generally. It feels that the Commission should give consideration to the planning of the coastline as an entity. Indeed even where it has been developed the opportunities must be there to improve its appearance and management. Ideally the document should have been part of a wider remit paralleling the work which went into the Commission's earlier studies " The Planning of the Coastline " and " The Coastal Heritage " which in turn lead to Circular 12/72.

9. The Institute considers it appropriate to refer to the keynote address on the coastal heritage by the Director General of the Commission at the symposium on " Planning and Management of the Coastal Heritage " held in Southport in October 1989, hosted by Sefton Metropolitan Borough Council and the North West Branch of the Institute. In dealing with the way forward he asked " against the record of past achievements and the current trends, what ought to be the strategy for the future? ".

10. He went on to suggest that we should be aiming at promoting a more holistic view of the coastline as a national resource and suggested that there are five elements of this:—

(a) " Consolidating the heritage coast achievements.

(b) Extending the heritage coast achievements to other parts of the coast.

(c) Looking at the marine and terrestrial aspects of the coast together.

(d) Placing coastal protection firmly in the broader context of land use planning and environmental protection.

(e) Emphasising the importance of the coast as a whole as a resource of national and indeed international importance ".

11. The Institute wholly endorses the above view and suggests therefore that the statement should deal with heritage coasts in the context of the wider coastline and in particular the section dealing with the latter and inserted (as well as developed) at the beginning of the statement.

12. Turning now to the need or otherwise to make heritage coasts a statutory designation, it would appear that the Commission sees less reason for statutory recognition because most of those in the current programme are in either National Parks or Areas of Outstanding Natural Beauty (AONB). The Institute feels that there is an opportunity to bring further heritage coasts under the AONB umbrella. The Department of Environment in Circular 12/72 did say that " where a heritage coast does not already have a statutory designation, the Countryside Commission will consider whether it should be designated as an area of outstanding natural beauty (paragraph 12). The White Paper " This Common Inheritance " at paragraph 7.53 states that " . . . the Government is willing to support further (AONB) designations if suitable areas are identified ".

13. Whilst Circular 12/72 may have fallen into disuse in recent years it is surely once more of great relevance. The Commission will no doubt consider the advice in those two documents, and in the light of its recent Policy Statement on Areas of Outstanding Natural Beauty, which does however rely on heritage coast definition to look after coastal areas whilst making a number of proposals for further AONB designations primarily inland.

14. Whatever mechanism is used the Institute considers that heritage coasts should carry a national designation, and other parts of the undeveloped coast a regional (or local) designation, with appropriate policies aimed at its conservation, and, where resources allow, its management. Reliance should not be placed on development plans to protect the coastline of national importance.

15. Finally, the Institute emphasises the need to consider not only the littoral itself but also its inland extent. Also critical are the offshore areas, not only where they are contiguous with a heritage coast but with the coastline as a whole, for reasons set out under the heading "Control of Development". As the Commission in its penultimate paragraph "the Wider Coastline" points out "heritage coasts policies now need to be regarded as an important component of wider policies of the coastline and the seas around it". What the Commission clearly is promoting, and the Institute supports it, is an integrated approach to the planning of the coastline and the policies relating thereto. It ought to be spelt out more positively.

Appendix III

Copy of a Letter to John Zetter, Department of the Environment

PLANNING POLICY FOR THE COAST AND EARTH SCIENCE INFORMATION IN SUPPORT OF COASTAL PLANNING AND MANAGEMENT

Thank you for your letter of 15 March inviting the Institute's comments on the draft specification of the above research project.

Last year the Institute responded to the Countryside Commission on its draft policy statement on "Heritage Coasts: Policies and Priorities". A copy of that response is appended for your information. On 14 December last, I wrote to Des Coles in PDC1A drawing the Department's particular attention to the statement in paragraph 41 of the memorandum commending the Countryside Commission's own view that "the time is right for a new look at the coast and its problems—for a national review leading to strengthen the policies for coastal conservation".

The Institute has sought to encourage the Commission in promoting this review both through its own policy statement and its dealing with Government. Indeed, the Institute was greatly disappointed that the White Paper "This Common Inheritance" did not take the opportunity to consider one of the nation's most limited resources—the coast—in a much more positive way. It therefore saw the Commission's policy statement as a contribution towards filling that gap and hopes that it might lead towards policy guidance for the coastline as a whole.

It therefore welcomes the Department's intention to commission research with the aims of securing a basis for:

(a) Planning Policy Guidance on coastal zone planning; and

(b) advice on the use of earth science information in coastal zone decision making.

The brief seems to be a very promising start to a most interesting research project. The Institute looks forward to the results with some anticipation and would be willing to assist the Department and the contractors in their investigations and deliberations.

The following comments are offered on the draft specification:

Introduction

Paragraph 1.1 needs to recognise there need be a *definition of* what is meant by the *"coastal zone"*. For example, is this intended to cover estuaries and tidal rivers. In the Institute's view these should be included.

Background

On 28 March statements were made on the Government's policy on renewable energy and it was indicated that draft planning policy guidance is being prepared. It would therefore be appropriate in paragraph 2.2 to make specific reference to increasing pressures from *renewable and non-conventional sources of energy* such as barrages and wind farms and their environmental balance sheet.

Task 1: Review of Coastal Planning and Management Policy and Responsibilities

Special emphasis should be placed on proposals to *upgrade lengths of the coastline in accordance with the BATNEC approach,* ie the project should not just concentrate on the best lengths of unspoiled coastline as implied in paragraph 4.3(a).

The reference to the Town and Country Planning Act in paragraph 4.3(a) should be updated to include the amendments to be introduced by the Planning and Compensation Bill and the Wildlife and Countryside Act 1981 and the National Parks and Access to the Countryside Act 1949.

The scope of the study should review existing policies, as set out in paragraph 4.3(d), by examining their *implications in respect of the needs of existing communities, residents and businesses in the coastal areas,* particularly those which are subject to coastal preservation policies. The particular concern here is that a correct *balance needs to be achieved between conservation and development* so that such rural communities are sustained in a viable and lively form.

It is noted from paragraph 3.1 that the work will be based primarily on existing sources of information. As requested, the Institute's Librarian is putting together a list of reference material which could be made readily available to the contractors. However, it suggests that, perhaps, as a general approach it might be more useful for the contractors first to talk to key individuals in a variety of agencies. A list of such contacts might be a more helpful introduction to the current state of coastal policy than a bibliography, and visits to such individuals would give much quicker access to current thinking.

The Institute would be willing to offer some suggestions of whom might be included in such a list. In this sense paragraph 4.2 of the brief ought perhaps to make specific reference to interviews and discussions with recognised experts and practitioners as a valid source of information.

A primary source of reference ought to be the recently published report of the Irish Sea Study Group of Liverpool University, "The Irish Sea Study—An Environmental Overview", especially the volume on Planning, Development and Management, since this tackled the policy and legislative background set out in the brief in some depth and with specific reference to the coastal zone.

There is also some work being done on a management plan for the Mersey Estuary, co-ordinated by Ron Dawson at Liverpool City Council's Environmental Services Department; a series of topic papers summarising the great variety of interests concentrated on the Mersey/Liverpool Bay is being edited at present. The Mersey Barrage Company's studies would be relevant as well.

Clarification is sought as to what extent the study will examine the coastal policy and funding issues emerging from the European Community. MEDSPA is already operating and it is understood that NORSPA is likely to emerge in a year or so and will be of direct relevance to Britain.

Task 2: Earth Science Information in Support of Coastal Planning

The limitation on the list of disciplines set out in paragraph 5.3 is queried. It is appreciated that there has to be some boundary otherwise the study would become unmanageable, but the Institute considers that there are sound practical and intellectual reasons for a more holistic view. Perhaps, a "core/periphery" approach might at least allow some of the more productive lines of current research to be explored.

It would suggest for instance that archaeology and its specialist sub-disciplines such as palynology is very important for coastal research in providing hard evidence of earlier coastal changes, climatic change, and the impact of human influences on natural processes. This is possibly more relevant to soft coasts, but generally speaking it is the softer coasts where such changes are critical. Understanding the long term trends in coastal development is crucial to reasonable decisions about future action.

Biology and ecology are also important in understanding coastal processes. For example, on the Sefton coast, the ecology and geomorphology are inextricably linked, both at present and historically.

The importance of the ecological dimension is even more significant when dealing with estuarine systems which should be an important facet of the brief. This aspect was heavily stressed by two scientists from Acadia University, Nova Scotia who visited England last year and who had been engaged in extensive work on the Fundy Bay tidal power schemes. They described how information about the coastal environment "cascades" down from the complex ecological levels to the simpler levels of hydrology and sedimentology, providing ultimately more reliable models to assist design and decision-making.

The coastal environment is subject to both short and long term change and needs to be constantly monitored. This aspect is not mentioned in the brief, but ease of monitoring could well influence conclusions reached under paragraph 5.2, sections b, c, d, and f, and could lead to the assessment of relatively cheap data-capture techniques such as remote sensing using aerial photography or satellite imagery, as well as the possibility of obtaining large quantities of low order, but long term, data using a network of volunteers in a structured way.

In this context the *environmental audit work* being undertaken by some local authorities, for example; Cheshire and Lancashire, may well be of interest. Co-ordinating coastal databanks (both scientific and policy-related) is another very important issue not specifically mentioned. The contractor should at least be asked to consider the matter and perhaps the potential role that could be played in this respect by the Department itself or one or other of the public agencies such as English Nature or the National Rivers Authority.

Task 3: Dissemination of results

A reasonable realistic timescale has been set for the study in section 9, but the Institute wonders whether there ought to be an intermediate reporting stage to a wider audience than the steering committee noted in paragraph 8.2. This could, perhaps, take the form of seminars (see paragraph 7.3), but, say, halfway through the study.

I hope you find these comments helpful but if any points require clarification, please do not hesitate to contact me.

David Rose
Director, Public Affairs
October 1991

_____

Examination of witnesses

Mr Chris Shepley, Senior Vice-President and Chairman of the Development Planning Panel, Mr Philip Watts, Planning Policy Officer and Mr Michael Welbank, Chairman, from Royal Town Planning Institute's Management Board, were examined.

Mr Jones

76. Mr Shepley, we welcome you and the other members of the Royal Town Planning Institute, and offer the apologies of our Chairman, Sir Hugh Rossi, who is chairing the international equivalent of this Committee in Geneva at the moment. So my colleagues have asked me to take the chair. Perhaps I could ask you to introduce your colleagues and, for the record, yourself as well.

(Mr Shepley) Thank you, Chairman. Thank you for your invitation to take part in this discussion. I am Christopher Shepley. I am a past president of the Royal Town Planning Institute and I am the City Planning Officer of Plymouth City Council, as Mr Steen may know. My colleagues are Mr Michael Welbank, who is a Director of Shankland Cox, which is one of the largest firms of planning consultants and he will be the President of the Royal Town Planning Institute next year. My other colleague is Mr Philip Watts. He is a member of the staff of the Royal Town Planning Institute. Previously he was the head of the planning division within the county engineering and planning department of Devon County Council.

77. Thank you very much indeed. You have sent us a very helpful memorandum and perhaps I could tackle you on that straight away, because you say that you are opposed to the creation of yet another body—a coastal management unit. But obviously also you are in favour of a co-ordinated approach to these issues. I wonder how you can reconcile those? How do you see the simplification of the present procedures without going for some kind of super body?

(Mr Watts) Thank you, Chairman. We feel that the procedures can be streamlined and we take the view that we do not need another body, certainly in planning terms. We have several points to make on this. The first is, as you know, that as a general point planning control should be extended below low water mark and including, where applicable, estuaries. Then to ensure that planning decisions are taken by only one planning authority, by that we mean the local planning authority; by ensuring that at national level decisions are taken by only one department of Government—the Department of the Environment, alternatively, the Inspectorate which is part of that department; by providing for decisions to be taken by

one authority, the local authority in estuarine areas; and in regard to ports and harbours by placing details within the purview of the local planning authority, once the principle has been established by parliamentary procedures, by making one body responsible for flood protection in estuaries and drainage, possibly the National Rivers Authority; by making one body responsible for coast protection, possibly the local authority, and finally by making the point that the procedures are streamlined—which is really what we are looking at in terms of the first question which you posed to us—by having well developed and positive policies in regional and development plans or management plans.

78. I am slightly confused about what you have been saying about estuaries, because you referred to the local authority. Are we talking about a county authority or a district authority in that context?

(Mr Watts) We would suggest that as things are at the moment, the powers that rest with the county planning authority would rest with them in terms of estuaries and those powers that rest with the district planning authorities in estuaries would rest with them. What comes out of any reorganisation remains to be seen.

79. What would happen where the estuary is the boundary between two county authorities?

(Mr Watts) We appreciate the difficulties of that and that is why, in our evidence, we are suggesting that the use of subject plans for an area is a useful mechanism for dealing with that kind of situation. We do appreciate the difficulties there, not only with two county boundaries, but possibly having one or more district authorities. That is where we think that the development plan system is important in terms of achieving that holistic approach for estuaries.

80. As far as these subject plans are concerned, what if an authority on one side of an estuary wishes to pursue an entirely different approach to the subject from the authority on the other side? The example I always give is the Dee estuary, because the planning approach and the policies of Cheshire and the Clwyd County Council are totally different.

(Mr Watts) The Department of the Environment and central Government have the ability to call plans in. This is another benefit of the plan system. If two

Mr Jones *Contd]*

authorities are so opposed then, if I can use a colloquialism, it is up to central Government through the Department to knock their heads together and to try to achieve a co-ordinated approach in those circumstances.

81. So you are relying on the Department of the Environment and the Welsh Office to be the referee?
(Mr Watts) Yes, sir.

82. As far as coastal protection and flood defence is concerned, again you referred to the local authority. I assume from what you said that you want to see a lead authority in that regard in each area?
(Mr Watts) Yes.

83. Again what happens where there are bordering local authorities in an area which is threatened by flooding?
(Mr Watts) Again it depends which local authority is responsible. It may be a service provided in the present instance by the county authority. I would suggest that wider than that—I know this is not the present situation—the Institute takes the view that there should be regional plans and a regional approach to these matters. That is not the case at the moment, but you cannot have perfection in these matters. But where one has a series of district authorities, then probably the county authority might be the appropriate one. But we only put these views forward tentatively. What we are looking for is the streamlining of the procedures.

84. Yes, but for us to recommend any streamlining we have to have ideas put forward. If all our witnesses said that, yes, it would be advisable and indeed desirable to simplify this but did not put forward any positive suggestions, it would be very difficult for this Committee to come to any conclusions, would it not?
(Mr Watts) Our suggestions are positive, sir. We are saying that the local authority should do this.

85. Positive but tentative. Can I ask about how all this would fit in with the NRA? Obviously as far as the present arrangements are concerned, the NRA act as an agency in terms of sea defence.
(Mr Watts) We see the National Rivers Authority as being responsible for the estuaries and for drainage matters. We think there can be a clear distinction there between what happens in estuaries and what happens on the coast.

86. So you would rationalise their powers, as far as other coastal areas are concerned, between them and the county authorities?
(Mr Watts) Yes.

87. Now perhaps you could also explain to me what the necessity is for taking planning powers over port and harbour authorities?
(Mr Watts) There are frequently problems with ports and harbours. With existing ports there is the proposal suddenly to introduce marinas and marinas have with them housing proposals. There is always a question mark over the extent to which a planning authority has control over matters which not only

have consequences in local environmental terms, but also have consequences in traffic terms which very often planning and highway authorities have no control over. That is an instance where it is felt that it should be clear that the planning authority has power to control that aspect and also in terms of moorings, which is another instance.
(Mr Shepley) Perhaps it is worth saying at this point that one of the key issues which we raise—I know this may be covered in later questions—is the need to extend planning power beyond the low water mark. We believe that quite strongly and we see no major obstacles to doing that. There will be some procedural difficulties, I have no doubt. That would include not just port and harbour authorities but all forms of development, moorings, oil rigs, gas installations, energy installations, barrages or whatever it may be. So ports and harbour authorities will be part of that and many local authorities on the coast have problems with development below the low water mark at the present time, as you probably know.
Mr Jones: Yes, I am sure we shall return to that later.

Mr Howell

88. I wonder whether you are conversant with the situation in North Norfolk where we have sections of the coastline where the land behind is below sea level and where different arrangements occur. Where the land is above sea level it is funded, I understand, partly by the local authority and partly by the Ministry of Agriculture, Fisheries and Food. A different arrangement occurs where land immediately behind the coastline is below sea level and the National Rivers Authority come into the equation. Did I understand Mr Watts to indicate that in a situation where there is no estuary he thought that all that kind of territory should come under one ministry?
(Mr Watts) What I was suggesting was that drainage and estuarine matters should be vested in one authority, the National Rivers Authority is what we are suggesting to you. From what you say that seems to support the view, because there seem to be several elements in the North Norfolk case, depending on the level of the land.

89. What is fussing me is where the division between the cliff areas and the low lying areas occurs?
(Mr Watts) Is it not a fact that coast defence is just about the littoral? We have a similar situation in Lincolnshire where the county council owns all the sand dunes and they were bought originally for flood protection purposes, but that is a very narrow area which is clearly vested in them and which is quite clear on the ground. It seems to us that inland, which is what you seem to be talking about behind the initial defences, it is vested in several authorities. We are saying that the National Rivers Authority should do that.

90. Can I explain myself in a little more detail? Most of the village of Happisburgh has high land behind it. There is a division between what is happening there and what is happening in the village

Mr Howell *Contd]*

of Eccles where massive sea defences are being put in place. But the cliffs are just eroding and the sea is coming in in the Happisburgh area, because of this division of responsibility. What I have been calling for for the last 20 years at least is that such a division should not occur. Have you any views on how this situation could be improved?

(Mr Watts) Our suggestion is that it should be reduced to the two authorities. You have mentioned three at least and we know that MAFF are involved and the National Rivers Authority, local authorities are involved. I believe the water companies have a role in this and also drainage authorities, those five. We are suggesting two. Is that not going towards what you are trying to achieve?

91. I want it all to finish up in one ministry. But you are still suggesting it should be in two. Is that right?

(Mr Watts) Yes.

92. I cannot quite understand the logic——

(Mr Shepley) I thought you said the environment? I think our view is that the environment department should be responsible for these things, not the Ministry of Agriculture, Fisheries and Food.

Mr Jones

93. Partly through a large number of local authorities and partly through the National Rivers Authority?

Several: Yes.

Mr Steen

94. What is going through my mind is, is there any useful purpose for a national coastal strategy? I am very fortunate to have both Mr Shepley here who, as you know, directs the planning in Plymouth, surrounded as he is by water, and Mr Watts who used to be in Devon and knows some of the coastline which I represent, which is about 88 miles. I just ask myself, what advantage is there to a national coastal strategy? What has gone wrong up until now—looking at my coastline, or the south west coastline—which would have been better if we had had a national coastal strategy? As I recollect it, the whole of those areas are already under coastal preservation and all the rest of it. Why do we have to go into more detail and designate more things? I just do not quite see what the advantages will be. Let us suppose we are walking along that coastline now and we see it as it is. What will happen if we have a national coastal strategy, except lots of people with yellow bands round their arms?

(Mr Shepley) Shall I start? That is a broad question and there are quite a number of answers to it. It partly depends what you mean by the word "strategy". I do not think any of us envisages somebody going away and producing some enormous document that looks at each inch of the coastline and says what shall happen to it. We are talking partly about the strategy for the administration of the coast, the definition of what the coast and the coastal zone are. Of course two of us know the South Hams coast very well and none of us wishes to see it change very much. But there are other

parts of the coast—as I am sure you will have anticipated—which are not quite so fortunate. We have heard about one where there are administrative problems. There are other parts of the coast where there are environmental problems or problems of erosion, or whatever it may be, where change is needed. We see the need to rationalise the administration. We have talked about that a little. We see the need to look at areas which are not defined as heritage coasts where sometimes, because they are not so defined, pressures tend to be concentrated. Most of the South Hams coast is protected in one way or another, but in some parts of the country there are bits that are not and everybody immediately focuses in on those. We need to look at the relationship between the areas which are protected and the areas which are not.

95. Can you give us some idea on mileage? Do you know what areas are not protected?

(Mr Shepley) I could not give you a figure, no. I do not know whether Mr Watts could.

(Mr Watts) We can provide those figures.

96. That would be very helpful. It would also be useful to know, because they are not protected, what has happened there or what is happening.

(Mr Shepley) There are a number of points in the evidence. There is the question of the General Development Order powers and things of that sort which all come under this whole issue. There is to be a planning policy guidance note—we shall probably talk about that later—which will go some way towards what we are talking about.

97. Following on from that—I am slightly wicked in this question—as you know a lot of my coastline is covered with caravan sites right on the heritage coast which were put there between the 1940s and the 1950s. If we had a national coastal strategy would you have a view about that in your coastal strategy and would you give any powers to the local authority to remove them?

(Mr Shepley) I would. I think all of us would.

(Mr Welbank) Yes, I would.

98. And having a coastal strategy you feel would help?

(Mr Shepley) I believe so.

Mr Jones

99. And that would mean presumably that compensation would have to be paid?

(Mr Shepley) It is a matter for debate, the extent to which that would extend.

(Mr Welbank) These are commercial set-ups and inevitably would require compensation. That would express the conviction which we put into the national heritage coast.

100. That is in the view of the local authorities for whom you work, that they would be happy to find the money for that sort of scheme?

(Mr Welbank) That is a very leading question.

(Mr Shepley) I think it is a view they would be happy about, but not necessarily the view that they would be able to——

Mr Steen

101. It seems to me that there is no point in having a national coastal strategy and heritage coastline if it is covered with lots of caravans and the enormous amount of development that was allowed 20 or 30 years ago.

(Mr Shepley) I absolutely agree. There are two points. The first point is to make sure that it does not get any worse, which on the whole happens certainly in the protected areas of coastline, although it may not in others. The second question is to retrench more slowly than you and I would wish back to the situation which we all wish to see.

(Mr Welbank) But is that not the point about having a national strategy, because there are areas where we may want to put national resources? That would guide us in matters like the removal of eyesores from areas which we have designated as protected areas of heritage coast.

102. So you could argue that the national coastal strategy would be accompanied by funds from the DOE?

(Mr Welbank) I think that would be our thrust, yes.

(Mr Watts) Can I add to that the point that those powers are already there in terms of discontinuance orders? They have been used in Devon. We tried, unsuccessfully, Mr Steen, to get rid of Stoke beach many years ago, but we were frustrated by central Government on that.

Mr Steen: That was before my time, probably.

(Mr Watts) But the point is that also there are funds and contributions available within national parks and AONBs, however it needs central encouragement concerning some of the worst eyesores and it means that funds have also to come from central sources via the Countryside Commission. The strategy would help with that.

103. Can I follow that up in the question of drawing boundaries? You talked in paragraph 32 of your memorandum of drawing the boundaries. Why is your definition needed and what is happening anywhere which means that you have to change that or redefine the coastal zone? I am trying to think of the sections below the low water line and what is actually happening there and what could be happening. Why do we need to bother with a redefinition?

(Mr Watts) Where there are clearly defined AONBs, national parks and heritage coasts, it is not necessarily a problem. But where you do not have those definitions one has to define where those policies should apply. We could suggest a definition of what is a coastal zone, if you so wish, in a little detail.

Mr Jones: I think that would be helpful.

(Mr Watts) What I am relying on is what is used in Devon, and Mr Steen will be familiar with this. The coastal protection area in Devon is defined as the area immediately visible from cliff tops, beaches, sea or estuary, or viewed from access roads, paths etc leading to and from the coast. So that is the definition in the rural parts of the coastline. In urban areas we would suggest that it is those parts of urban areas directly associated with maritime activities. We would give as the important elements there ports, harbours and those areas associated with tourism and obviously the adjacent land areas. Then, so far as the sea is concerned, we see the importance of extending planning powers over the maritime areas, we would suggest that a distance of three miles is appropriate.

104. In urban areas which are holiday resorts, such as Brighton, how do you reconcile your definition of maritime involvement with what is a holiday resort with blocks of flats, hotels and things like that?

(Mr Watts) I think you have to take a broad brush approach. One cannot say that the whole of Brighton seafront is given over to maritime activities. There are obviously flats and houses interspersed among the hotels, but in a place such as Brighton or Torquay—as has happened in Torbay—they have to define the important hotel areas which need to be protected. One would look to those to define what might be regarded as the coastal zone.

105. But are you not reinforcing the case that it is impossible to deal with this by national definition? It has to be something which is tailor made to the individual areas. Mr Howell earlier gave a very good illustration of an area which might not be considered part of the coastline but which was clearly affected by coastal policy in that it was a low lying area inland. Would it not be more appropriate to leave this to the local plan so that they could define areas according to what local people know and understand by the areas related to the sea and its activities?

(Mr Watts) You are absolutely right, sir. You asked me to give a definition which I attempted to do. I illustrated it with a local situation, but the important thing is to make sure that in a particular situation the zone is tailored to the area and that the policies and strategies flowing from it are similarly tailored to the requirements of that area.

106. When you say we asked you for a definition, that is perfectly true, but you urged this Committee to define it, so it is hardly surprising that we should pose the question to you.

(Mr Watts) You asked me to be specific about that. You are quite right. We are absolutely of the view that before one can start with coastal planning and strategies we need to define what is meant by the coastal zone.

Mr Howell

107. I want to revert back to a previous interchange. Mr Watts indicated that he wants to boil the five or six bodies that are involved down to two. I am not clear what those two are.

(Mr Watts) There is the National Rivers Authority in relation to estuaries and drainage matters and inland generally. When I say inland I mean that land which is immediately to the rear of the coastal areas. But the littoral and the problems of coast erosion the Institute is suggesting should be a local authority responsibility.

Mr Howell Contd]

108. But both these bodies come under the Department of the Environment, do they not?
(Mr Watts) Yes.

109. Are you saying that the Ministry of Agriculture, Fisheries and Food should be out of this area altogether?
(Mr Watts) I would not quite put it like that, but we are suggesting that in rationalising this, perhaps the Department of the Environment is the appropriate department to deal with it.

110. Unless we get down to the nitty gritty of this, I think we are all waffling around the point, are we not?
(Mr Shepley) The Ministry of Agriculture, Fisheries and Food may need to be consulted on certain issues by the Department of the Environment, but we are suggesting that that department should be the controlling department.

Mr Steen

111. There is one problem, though. MAFF give all sorts of grants to farms and fields on the heritage coastline. You are saying that they can go on doing that, but under the lead of the DOE?
(Mr Shepley) Yes.
(Mr Watts) Yes, within Government the Department of the Environment should be the central body. From that there will be two direct bodies, NRA and local planning authorities, but there can be the involvement of other agencies, but under the direction or the guidance or lead of the Department of the Environment.

Mr Jones

112. So these grants should be led by environmental requirements and not by farm income requirements?
(Mr Welbank) Yes, in this instance.
Mr Jones: I can see that going down well with the farmers!
(Mr Watts) Can we distinguish between what is agriculture and what is coast protection. I do not think there is any suggestion that the DOE should suddenly become responsible for the agricultural aspect of it.

Mr Steen

113. In the better local authorities there are local plans and there will be local plans throughout the country in due course. Each local planning authority will look at its coast. Are you saying that a national coastal strategy should give guidelines to every local planning authority and they will ultimately make a decision through their local planning committees and council?
(Mr Shepley) Yes, I think it would give two things to them. I think the first thing it would give to them is encouragement to look at the coast perhaps in a more positive way than many local authorities have done in the past. Experience varies, but our experience has shown that when a Committee such as this produces a report of the kind you intend to produce or when the Department of the Environment produces a

guidance note of the kind it intends to produce, that gives a new focus to local authorities to examine issues that perhaps they had given priority to among all the things they have to deal with.

Mr Jones

114. The national coastal strategy would be guidance to planning authorities, but the planning authority will make the final decision?
(Mr Shepley) Yes, in co-ordination, no doubt, with its neighbours in many cases.
(Mr Welbank) The vehicle for this could be the regional guidance which is issued by the Department of the Environment to planning authorities, strategic guidance in coming to their local plan preparation or structure plan preparation.

115. Do you see any distinction then between a national coastal strategy and what should be said in the planning policy guidance note on coasts?
(Mr Shepley) We see a distinction between a coastal strategy and what would be said if a policy guidance note were produced now because a policy guidance note can only be couched in terms of the current legislation and the currrent practice. In a number of respects we hope that that will change. It needs to go beyond what a current PPG might say.

116. Let me rephrase that. Given that the policy framework evolves in terms of the structure of the planning system, would you see this strategy as being pursued through the instrument of planning policy guidance notes?
(Mr Shepley) I think the national strategy, yes.
Mr Jones: I will return to that later.

Mr Summerson

117. In paragraph 9 you say: "What is quite evident is that the coastal planning and management deserves a much more co-ordinated approach through the medium of regional plans, of development plans (including subject plans) and of comprehensive management plans subsuming the current heritage coast plans". Who is going to co-ordinate this hierarchy of plans?
(Mr Shepley) I think that we first talked about the planning policy guidance note. What I was about to say was that I think the planning policy guidance note, as it does in other areas, would define the role of the various levels of plan: the regional guidance, the structure plan and the local plan, which are the three levels. I see the regional guidance, as in other areas, as having a relatively limited role. It would be particularly important, in cases such as the Dee estuary which has been mentioned or other estuaries which involve more than one county, but it would be a very general level of guidance. The detailed guidance would appear in local plans co-ordinated perhaps, as we have suggested, through subject plans where there is more than one district authority involved. I do not think there is great difficulty in distinguishing between those different levels of plans. It is done in relation to housing development or other forms of developments and I think that is fairly easy

Mr Summerson *Contd]*

to work out. Management plans are a separate area. When we talk about planning control and planning powers, we obviously are not talking about planning control over navigation or over the use of water, sailing, jet skiing or waterskiing, within particular areas of water. Management plans are required to deal with that kind of thing and they are not normally statutory plans and they are normally more flexible. They evolve as changes in habits and recreational practices and so on change. They are a separate kind of operation.

118. Taking it a little further, you have, to some extent, defined management plans. But how would management plans for heritage coasts or coasts that are privately owned be integrated within this arrangement?

(Mr Shepley) A management plan would be rather different normally in the case of a heritage coast. I was thinking of management plans for water. Management plans for the coast itself would resemble the sort of management plans that national park or areas of outstanding natural beauty authorities produce, dealing with access arrangements and things of that kind. Again there would be an overlap with the local plan. I would think probably in those cases—I am thinking aloud—the majority of those issues would be part of a local plan and the managment would be carried out by the local authority quite straightforwardly. I do not see any great problem there. It is part of normal practice in many parts of the country, that there are management plans and local plans which overlap.

119. You do not think all these different plans will be tripping over each other?

(Mr Shepley) No, I do not think so. I think it is a normal part of practice, particularly on land. It is more difficult when you get on to the water.

Mr Steen

120. You said that you thought that global low water mark should be three miles out. That is an awful long way. Why go that far? Why not go just half a mile to cover marinas and so on?

(Mr Welbank) I am not sure that half a mile would cover some of the marina plans which are envisaged. I think the three mile line has been taken because it has been a traditional line and therefore it is well known to people. Therefore that is the first boundary line which is known beyond the coast. Obviously it is within easy view and access.

121. I am thinking of dumping. You would include dumping of soil from the land into the sea as being for the planning authority?

(Mr Welbank) Yes.

122. So that would be a new feature?

(Mr Welbank) Yes, the seabed in that area you would see as submerged land and subject to planning permission, but not the use of the water over it which is part of the management process.

Mr Jones

123. Can I return to the question I was asking you about the planning policy guidance note? We had established that if it was issued under the present framework it would be unsatisfactory and that you would like to see a simplification down to two lead authorities. Can I ask what else you would like to see in that planning policy guidance note?

(Mr Shepley) I do not know that I have a great deal to add to what has already been said. I think it needs to define the roles of the different levels of plan. I think Mr Summerson's question is an important one and can be answered quite well, but it needs to be clarified so that everybody understands it. It needs to explain the importance of seeing the coast as a whole and seeing it in the kind of terms we described as a zone and it needs to talk about the kinds of environmental improvements that Mr Steen raised and to encourage local authorities to pursue these issues with vigour. That is what I see happening. That could then be carried forward through the other mechanisms that we have discussed. You may have other things to add.

(Mr Watts) Yes, one of the most important aspects which the Institute looks at when PPGs are issued is the extent to which the PPG addresses the importance and value of the development plan system in achieving proper strategies and policies. I have to say that some PPGs are thin on this. Others develop it very well. For example, the sport and recreation PPG develops the role of development plans and their importance exceedingly well, so we would look to the PPG to recognise that and to remind planning authorities of the importance of the coastal zone and to develop strategies and policies relevant to them. We would wish to see addressed, as Mr Shepley has said, not only the protection of the already attractive coastline, but the upgrading of coastline, and again, on a point which Mr Steen raised, advice on the removal of eyesores. These are the kind of issues that we would expect to see addressed in the PPG. As you will know we have expressed the hope that issues which will have arisen through the work of your Committee may also find their way into the PPG as important matters which arise during your consideration of the matter.

124. Earlier you referred to the General Development Order and problems with the GDO. I meant to pick you up at the time, but perhaps I can give you a chance now, Mr Shepley, to put some flesh on those bones and perhaps you can tell us how that would fit in with the PPG because presumably an amendment to the GDO is what you are suggesting in some regard or another?

(Mr Shepley) Yes, paragraph 53 of the evidence we have submitted lists what we see as being the problems. I am not quite sure what the procedure would be. I imagine that an order would be needed to change the General Development Order. It could not simply be done through a planning policy guidance note, but we list in that paragraph: " 28 day tented camping; small touring caravan sites; development by drainage authorities, the NRA, dock, pier, harbour, water transport, canal or inland navigation undertakings and development under local or private Acts or Orders ". Mr Watts listed that and I ought to give him some credit for the research he did on it. All

Mr Jones *Contd]*

those are things which can have a significant impact, which I am sure you realise.

Mr Summerson

125. I should like to ask you if you think that coastal management plans should be made statutory, now that subject plans have been abolished?

(Mr Shepley) I would say no. I draw this distinction again between planning and management. I do not think management plans need to be statutory. I think they need to be the subject of a great deal of consultation and debate locally, but I do not think they need to be statutory. I think that would be over-elaborate. The question of subject plans is a slightly separate one and more to do with planning than management, if I can pursue the point. The problem with the loss of subject plans is that there was an opportunity through that mechanism to produce what would have been a statutory plan for, say, an estuary where three, four or more local authorities would have been involved. That opportunity no longer exists. What I expect local authorities will do is to produce non-statutory plans, where they feel there is a necessity to do it, and then to incorporate their own sections of those statutory plans within their district wide local plans. That is a workable system, but it has the problem that the estuary plan itself is non-statutory and therefore is not subject necessarily to the same examination as a whole that it might have been as a subject plan.

126. Does it also mean that the subject plans get out of the cycle of the statutory plans, because obviously they are considered at separate times? By the time they can be incorporated they may be more out of date than the rest of the plan and therefore that may undermine both.

(Mr Shepley) Yes, it does mean that. Of course, it means that, whether it is called a subject plan or whether it is a non-statutory plan. There will always be that kind of timing problem. The advantage of having it as a subject plan—by definition as a subject plan it will only cover certain issues—is that the public can relate to it. There can be an inquiry related to those issues related to that estuary all in one inquiry, so that any person or body, such as the Royal Society for the Protection of Birds, who may have an interest in the whole of the estuary can talk about that at that inquiry. That can be accepted and implemented by the authorities concerned. If it is being dealt with as part of two or three separate local plans, that debate may have to be repeated two or three times and there may be different inspectors and different weights of evidence on different occasions, and possibly different answers, although one hopes that would not happen, but it is a danger.

Mr Battle

127. I apologise to you, Chairman, and to our witnesses for being late for the Committee, but I am serving on another Committee simultaneously. I hope that if my questions have been answered, you will forgive me. I am not just repeating it for the sake of it. The Department of the Environment have made clear that they reject the idea of extending local authority jurisdiction beyond that low water mark. What kind of difficulties do you envisage that creates and what other options are there for integrating the landward and seaward planning systems?

(Mr Welbank) I fail to see what the problem is. They may have stated it is a problem, but I am sorry, I fail to see what it is. If the only other option that we can envisage is the creation of another authority, we have already rejected that because we want to simplify and streamline the system and not make it more complex. Although they may have said it is a problem, I think it is their problem. I do not think it is a problem, but they are making it a problem because they are refusing to address it seriously. It is land. It is submerged land. We have given it a defined extent of three miles from the shore in our views. That is submerged land over which the planning authority should have the same controls as it has over other land.

128. You do not envisage local authorities dealing with fisheries and navigation, for example, do you?

(Mr Welbank) No, not fisheries and navigation. We have covered this slightly, but we see the use of the water—which covers boats on the surface and non-fixed things—as not being subject to planning control. That should be applied to fixed elements.

Mr Jones

129. Such as floating pontoons?

(Mr Welbank) I would think that if they are intended to be a permanent structure, yes. They are fixed in location, whereas a boat, unless it is anchored is moving, yes, but a floating pontoon we would regard as a fixed structure.

130. May I change the focus slightly? In your submission of evidence in section 4 *Planning Policy in Coastal Areas* in paragraph 65(h) you state: " Generally speaking, only essential development should be permitted in open coastal areas, ie outside existing urban areas ". What would you see as being the definition of " essential "?

(Mr Watts) We see that as development which is required for the benefit of the public at large, where the need has been established, where there is a national need and where the need cannot reasonably be met elsewhere. We make this definition particularly in relation to development which requires a marine location, so we are talking about ports and harbours, marine recreation and matters relating to the exploitation of offshore oil and gas. Having said that, if a need has been established, one then tries to find the least sensitive locations to meet that need.

(Mr Welbank) Our view is that in open coastal zones there really should be a presumption against development, unless it can be demonstrated under these headings as being essential. This should be a characteristic of these zones: a presumption against development.

Mr Howell

131. This question may have been asked while I was away, but who gives the authority for aggregates to be taken from less than three miles out to sea?

(Mr Welbank) At the moment the freeholder of the seabed is the Crown Estate. They deal with the licences for extraction on a commercial basis with those extracting it and they have an advisory body who advises them if there are any adverse environmental impacts of so doing. But it is not subject to planning control. It is merely a commercial contractual arrangement between the freeholder and the extractor.

132. So what is going on, again off Happisburgh, is not under the control of the Government at all?

(Mr Welbank) It does not come under the planning system. It is not under the control of the planning system and we suggest that within this area it should be. We see no problem of doing that, as distinct from the views of the DOE, I gather.

(Mr Watts) The Institute sees the question of minerals as one that needs to be co-ordinated. It ought to be part of the minerals local plan and, therefore per se, the planning system extended and it has consequences not only for extracting material in terms of its effect on coastal protection, but it also has consequences in terms of landing the materials and their distribution. These are points which ought to be looked at comprehensively, as they are on the land.

133. So when I am assured that there is no problem being created off Happisburgh by the Ministry of Agriculture, they really do not know? At least neither they nor the Department of the Environment have any control over what is happening and that means, since there is no other ministry involved, the Government have no control over what is happening.

(Mr Welbank) Except the Crown Estate Commissioners are a responsible body and therefore they are giving licences in relation to the criteria which they believe will not produce any adverse environmental impact.

(Mr Shepley) I do not think you would accuse them of not knowing, but I think we would say that we do not know whether they know, because, unlike the planning system, this is not an open public debate. One of the great advantages of the planning system is that everything is done openly and everybody can see what evidence there is and can ask questions and debate it. That process may go on in terms of dealing with the licences and the investigations may take place, but people like yourself do not have the opportunity to become involved in that debate.

Mr Jones

134. That you say is the weakness of the system, because when we had the DOE officials here and we were cross-questioning them on this point, they were saying in effect that the Crown Estates Commissioners simulate the planning system. They would not go ahead with a project where they were advised that it would not meet with planning requirements.

(Mr Welbank) Except when it comes to seabed activities they do not consult the planning authority. How can they, because the planning authority has no jurisdiction on that area? What they said, sir, relates to the Crown Estates Commissioners on land because in that case they would consult the planning authority and it is their policy not to go ahead with a development which was not approved and agreed by a planning authority, although in a formal sense they do not have to get planning permission. But the Crown Estates Commissioners cannot do that on the seabed because the planning authority does not have any jurisdiction.

135. They do not consult even though there are clear implications in terms of traffic and on land back up activities?

(Mr Welbank) If it was on land activities——

136. No, I am talking about spin off effects.

(Mr Shepley) Mr Jones is right. It is not just the Crown Estates Commissioners. There are developments going on offshore with other organisations, perhaps a port or harbour authority, as you have already mentioned. I have had experience of those. The Duchy of Cornwall owns a great deal of the seabed around my part of the world and there are other similar kinds of situations elsewhere. This is not simply an attempt to criticise or to become involved with the Crown Estates Commissioners. They are the biggest of these organisations, but there are many others. The two issues—the first is this question of openness and everybody knowing what is going on and having their opportunities to beome involved and the second one, which you rightly raise, which is the question of implications on shore for developments which are taking place offshore. This certainly can generate the need for car parking, traffic or whatever it may be.

(Mr Watts) May I just add one point? The planning system is subject to the requirements of public participation and consultation. That is not pursued in terms of what goes on in the off-shore maritime areas.

Mr Battle

137. What in your view are the current environmental impact assessment procedures? Would they be adequate, do you think, for assessing and helping to identify the impacts of various developments on the coast or not?

(Mr Welbank) That is a very pertinent point. One of the aspects of defining a coastal zone and having policies for it is that within those areas it might be appropriate to think about having an environmental impact assessment of the policies applied to a particular coast as a whole before we come to projects. But essentially within those zones any projects of significance should, in our view, be subject to such procedures rather than the categories which are listed under the normal environmental assessment procedure.

138. Are they adequate to cope with the types of project?

(Mr Welbank) Yes, the expertise is available in the country and it is up to those who are undertaking the work to make sure that the expertise is inserted into it.

Mr Summerson

139. Can we continue to talk a little about the Crown Estates Commissioners? If their powers to permit development below the low water mark are removed and, going a little farther, if marine fish farming is brought under planning control (which will directly affect the Crown Estates Commissioners) and if port and marina developments are regulated, would local planning authorities have the resources in the true sense of that word—staff, expertise and finances—to cope with their added responsibilities?

(Mr Shepley) It would be dishonest to say that most local authorities have the staff resources to take on extra responsibilities on any scale. They may need extra staff resources to do it and it is no good ducking that issue. I do not think those would be large for two reasons: first, the number of applications and cases that we are talking about is not large. Secondly, in my experience, many of those cases—even though currently they are not subject to planning control—involve the local authority in a great deal of work anyway because people start to complain and grumble and demand that action is taken and the local authority has to explain why it cannot take it. A good deal of involvement and work is already involved. It would be wrong to suggest to the Committee that this is something that can just be taken on by most local authorities without any problem.

140. In many ways I am heartened by your reply. I thought you would take the opportunity, as President of the RTPI, to say that of course there would be a need for lots more staff, all of whom should be chartered town planners.

(Mr Shepley) They should certainly all be chartered town planners!

141. As a chartered surveyor, I cannot agree with you. This is not really a fair question, but it would be very interesting to ask whether you have any idea at all what financial effect this would have on the Crown Estates Commissioners?

(Mr Welbank) I think you are asking a question which cannot be answered. It does not mean necessarily that the activities of the Crown Estate in this area would stop. It would merely mean that such activities, if they took place, would have to be agreed by the planning authorities. It does not mean that they would stop, they would just be controlled or limited.

142. I understand that, but as Mr Shepley has already said, many of their activities do involve large numbers of the public making complaints. Planning authorities, like any democratically elected body, are very sensitive when it comes to public complaints. I suggest to you that if these powers were removed from the Crown Estates Commissioners, it would have quite a drastic effect on their operations.

(Mr Welbank) I think it would have an effect on their operations, yes, but on the way they did things, not necessarily on the basic fact of fish farming or mineral extraction. It would just mean that they would have to improve their operations.

(Mr Watts) Can we add to that: as on land the viability of an organisation is not necessarily a planning factor, but obviously it is a factor as far as the Crown Estates Commissioners are concerned. Then presumably, if they were subject to the same rigours as on land, they would have rights of appeal, so if planning authorities were dealing with them unreasonably they would have the right of redress. I think this point has to be made.

(Mr Shepley) I was going to make a similar point and also to add that the presumption in favour of development, unless there were material considerations and so on, would also apply. Something like 80 per cent of planning applications are approved. They may well be subject to conditions which would not otherwise be imposed about access and so on, which may affect the profitability of particular schemes. I do not think this means a halt to development offshore. It simply means public involvement in it and some control over it.

143. You sound very sanguine about it. I am not quite so sure that I share your views on that. However, before we start getting involved in a debate about planning in general, may I ask you if the planning profession would be able to respond to the additional responsibilities involved in coastal zone management regardless of extensions out to sea?

(Mr Shepley) Yes, do you mean in terms of the resources or the expertise or both?

144. Both.

(Mr Shepley) The answer on resources is the same as the answer that I gave before. It would be dishonest to suggest that that could be taken on board without some additional resourcing. In that case in many authorities it would involve work which is not currently being done. I do not think that we are talking huge numbers of people, but extra resources would be required. On the question of expertise—and this applies to all the things that we have been talking about—we see it as being very similar to many of the other areas of work that planners are involved in which may require some form of specialist knowledge. It may not always require that. As you know, planners are involved in everything from nuclear power stations to gypsy sites, if those are extremes. I do not know that they are. They seek advice from the relevant expert bodies and organisations and they involve consultants and so on where necessary. So I think the same thing would apply. I do not see a great problem in terms of expertise at all.

145. Are you satisfied that the Private Bill procedure, with its forthcoming amendments, is capable of addressing the environmental implications of coastal developments?

(Mr Watts) We see the new arrangements for the Private Bill procedure to be advantageous. We are pleased that it will, in the great majority of cases, allow greater public consultation, which has always been a concern with the Private Bill procedure. Generally we welcome them. What we say as an Institute, and we have said this in our evidence, is that once the principle has been established in terms of ports and harbours many of the details surrounding

Mr Summerson *Contd]*

them ought then to be subject to the normal rigours of planning control. They should not have *carte blanche* to do all things within harbour and port areas. For example, we mentioned marinas and recreational harbours which are not directly related to the prime reason for establishing the port or harbour in the first instance. We have said that. The other thing we would mention which we are not quite clear about is that in dealing with rail the Government discussion paper talked about environmental assessment. When it came to harbours it was silent. We feel that it should have addressed environmental assessment and we hope that the final proposals of the Government will address that matter. Those are the two things really. We understand and we agree with the move towards—if I may say this in this company— a more open debate of these matters, a greater involvement of local planning authority in nitty gritty, in related matters, and more emphasis on environmental assessment.

Mr Jones

146. Can I tidy up a point about environmental assessment? Can someone tell me whether coastal protection works require environmental assessments to be done?

(Mr Shepley) There are two categories, I believe. There are categories where environmental assessment is compulsory. I do not believe that coastal protection falls within that category. There is then a second category where environmental assessments may be required by the local authority and they do fall into that category.

(Mr Watts) I suspect from memory that it probably depends very much on the extent of them and the size of them, but that is off the top of my head. Do you wish us to come back to you on that?

147. No, I think that is probably a sufficient answer. We can check that out ourselves if necessary. Can I also ask you about the economic consequences of coastal protection, because some witnesses have suggested to us that it might be more cost efficient to move the line of defence back and abandon certain areas, in effect, to the sea. Would our planning system cope with re-grouping, as it were? Is it

prepared for that kind of eventuality? Or is this just a theoretical point put forward at the moment?

(Mr Shepley) It depends on what terms you mean. I think the planning system could cope in terms of the land allocations and whatever may be involved in that. There may be financial implications which might be difficult to cope with in terms of compensation to people. But in terms of the planning of that and the provision for that within local plans or whatever the mechanism might be, I do not see any problem. One of the points that we made in our evidence is that there may be cases where the traditional heavy engineering approach to coastal defence has fairly unpleasant and unsatisfactory environmental consequences and that there may be cases, certainly not universally, because there are obviously areas where human settlement needs to be protected, but there may be cases where a softer approach to coastal protection and a more natural approach to the evolution of the coast would be appropriate. The engineering profession is very skilful and enjoys challenge and is able to meet most challenges, but sometimes only by means of works which render the coast almost impossible to use or enjoy. We would favour a move in that direction without suggesting that that should be the universal solution.

(Mr Welbank) Returning to your question on potential abandonment of areas of land, that seems to be a sort of national issue which would come into our national strategy and then come out through strategic guidance to regions and then be implemented. But a decision of that magnitude seems to me to be part of a national strategy, not to be taken locally. If it is not accepted that it should be abandoned, then that should be a national strategic decision.

148. It would depend where the costs fall, because it is right that the cost benefit analysis should be carried out by the body that would have to foot the bill.

(Mr Watts) Which would be unlikely to be the local authority.

Mr Jones: I thank you for coming along and for so fully answering our very wide ranging questions today.

WEDNESDAY 15 JANUARY 1992

Members present:

Sir Hugh Rossi, in the Chair.

Mr John Battle Mr Robert B Jones
Mr Barry Field Mr Hugo Summerson
Mr Ralph Howell

Memorandum by the National Rivers Authority

COASTAL ZONE PROTECTION AND PLANNING

CONTENTS

Section

1. INTRODUCTION
2. DYNAMICS OF COASTAL CHANGE
3. RISKS TO COASTAL SETTLEMENTS AND ECO SYSTEMS
4. MECHANICS OF COASTAL PROTECTION SYSTEMS
5. PLANNING POLICIES IN COASTAL AREAS
6. CO-ORDINATION OF AUTHORITIES RESPONSIBLE FOR PROTECTION AND PLANNING IN THE COASTAL ZONE
7. ENVIRONMENTAL ASSESSMENTS

1. INTRODUCTION

1.1 The National Rivers Authority (NRA) was established in July 1989 under the Water Act 1989 as a Non-Departmental Public Body sponsored by the Department of the Environment with strong links to the Ministry of Agriculture, Fisheries and Food and the Welsh Office. The 10 regions of the Authority are shown on the map at Appendix 1.

1.2 The NRA has a central role in the management of inshore coastal water quality. Under the Water Act 1989, the NRA has statutory responsibilities relating to the environmental quality of controlled waters which are both general and specific. Controlled waters in this context include territorial waters within three nautical miles of their baseline, and coastal waters between the baseline of territorial waters and the landward limit of high tide (this includes estuaries). The NRA has overall supervisory responsibility for Flood Defence, and is responsible for a large number of sea defences and some tidal barriers including the Thames together with associated barriers and that at Hull. The NRA is not responsible for coast protection which falls within the remit of maritime authorities. The NRA's navigation duties in coastal waters around England and Wales are extremely limited, its principal responsibility as a Harbour Authority being at Rye Harbour, Sussex.

1.3 The Water Act 1989 imposes upon the NRA, a duty to " **further** the **conservation** and enhancement of natural beauty and the conservation of flora, fauna and geological or physiographical features of special interest; to have regard to the desirability of protecting and conserving buildings, etc; to take into account any effect which the proposals would have on the beauty or amenity of any rural or urban area or on any such flora, fauna, features, buildings, sites or objects ", (Section 8(1)). The NRA is required generally to promote conservation and enhancement of the natural beauty and amenity of inland and coastal waters and of land associated with such waters.

1.4 As part of its responsibilities to **maintain, improve** and **develop fisheries,** the NRA has substantial duties and powers in respect of salmon and migratory trout to a distance of six nautical miles around the coasts of England and Wales. In those Regions where it is appropriate for the NRA to manage sea fisheries, the duties and powers are contained in the 1966 Sea Fisheries Regulation Act and the 1967 Sea Fisheries (Conservation) Act.

1.5 *Funding*

1.5.1 The Authority is funded through various means. For Flood Defence, monies are mainly raised through levies on local authorities, through a general drainage charge and through contributions by Internal Drainage Boards. The Minister of Agriculture, Fisheries and Food contributes towards expenditure incurred by the Authority for the improvement of existing works or the construction of new works. The NRA charges for water abstractions and discharges and receives grant in aid from the DOE. It also receives income from licences for freshwater angling and from navigation.

Footnote: References to the Water Act 1989 and Sections thereof will be superseded by the Water Resources Act 1991 when it becomes statute on 1 December 1991.

1.6 *Flood Defence Legislation*

1.6.1 The Water Authorities' functions relating to Flood Defence were transferred to the NRA under the Water Act 1989 (Section 136(2)). This imposed upon the NRA the requirement for all functions carried out under the Land Drainage Act 1976 (1976 Act) relating to flood defence, to be carried out through Regional Flood Defence Committees (Section 136(3)).

1.6.2 The NRA has the power under the 1976 Act to maintain existing defences and to improve and construct new defences against sea water and tidal water (Section 17(1),(2)). Under Section 34 of the 1976 Act, the NRA, by virtue of Schedule 15 1989 Act, can make byelaws.

1.6.3 The Water Act 1989 requires the NRA to undertake surveys of the areas in relation to which it carries out flood defence functions (1989 Act Section 136(1)).

1.6.4 The NRA has the power to provide and operate a flood warning system (1976 Act Section 32(1)).

1.7 *Water Quality Legislation*

1.7.1 The NRA, under part III, Chapter 1, the Water Act 1989, is specifically responsible for water quality in controlled waters. These responsibilities include the determination and issuing of consents for discharges of waste into controlled waters, (except for those discharges subject to the provisions of the Environment Protection Act 1990), monitoring of the extent of pollution in such waters, control of diffuse pollution and achievement of water quality objectives and classification requirements as determined by the Secretary of State.

1.7.2 Part I of the Environmental Protection Act (1990) (EPA) provides for the introduction of Integrated Pollution Control for " prescribed " processes and the regulation of discharges from these processes to air, land and water, including discharges to controlled waters, by Her Majesty's Inspection of Pollution (HMIP). The EPA provides for the NRA specifying such consent conditions for discharges to controlled waters as appear appropriate for the purpose of achieving any water quality objectives which apply in those receiving waters, and a memorandum of understanding between HMIP and the NRA describes the operational interface between the two bodies.

1.7.3 A wide variety of EC Directives which impact upon aspects of coastal water quality directly, or indirectly, also apply. The NRA is responsible for achieving some or all of the requirements of these Directives.

Important examples include:

Bathing Waters Directive (76/160/EEC),

Shellfish Water Directive (79/923/EEC),

Various titanium dioxide industry directives (78/176/EEC, 82/883/EEC, 89/428/EEC),

Urban Waste Water Directive (91/217/EEC),

Dangerous Substances Directive (76/464/EEC) and associated daughter Directives.

1.7.4 Pollution from ships is of concern to the NRA because of the potential for damage to the environment. The NRA maintains a close involvement despite such discharges being both governed by legislation and monitored by other authorities, not least because of the resource implication of mitigating any adverse effects.

1.7.5 Because of the complexity of this legislation, informal working arrangements have been developed to ensure co-operation betwen the appropriate bodies, and in certain instances formal Memoranda of Understanding have been or are being agreed.

2. DYNAMICS OF COASTAL CHANGE

2.1 The coastline contains much evidence of man's activities and some of which have had great influence on the stability of beaches by interrupting the littoral drift generated by long-shore currents. The supply of beach-building material from the littoral drift has often been restricted by groynes, protruding harbour arms and piers, and structures which divert coastal currents. The NRA is limiting adverse effects of wave action and long-shore currents by building off-shore islands or shore-connected islands.

2.2 The Anglian Region of the NRA has recently completed a major study of the coastline between the Humber, and the Thames. The report outlines the extensive range of data gathered during the study. It explains various analyses of the relational data bases, and adds to the existing knowledge and understanding of coastal processes.

2.3 This survey confirmed the beach erosion, and drew attention to the fact that beaches were steepening, because the sea-ward limits of beaches are being eroded whilst the landward limits are often constrained by the proximity of a hard defence structure. This constraint prevents the beach from migrating landward in order to maintain the natural beach slope. Appendix 2 gives the location of the eroding foreshores, which shows some 29 per cent of the NRA's defences are fronted by these problem beaches.

2.4 Sea Defences have a limited life, and refurbishment programmes are of paramount importance. It is likely that general dilapidation will be aggravated by the effects of climate change if they bring about a rise in sea level.

2.5 When designing new sea defences the NRA is making allowances, where benefit/cost justified, for sea level rise in accordance with the rates of change forecast by the Intergovernmental Panel on Climate Change in their report, which was endorsed by the government in 1990. This approach is supported by MAFF.

2.6 Beach erosion and the related phenomenon of decay of salt marshes puts at greater risk many sea defences by allowing increased wave action to reach the primary structure. The NRA is undertaking beach nourishment schemes and regeneration of saltings.

2.7 The NRA seeks readily obtainable sources of sea-dredged aggregates, or removes material from beaches assessed as having surplus material, to build-up beach levels. In this way, the NRA avoids utilising scarce land-based sources and minimises environmental damage.

2.8 The NRA's policy of establishing high-level beaches provide environmental, amenity and technical benefits.

3. RISKS TO COASTAL SETTLEMENTS AND ECOSYSTEMS

3.1 *Flood Defences*

The coastline of England and Wales is approximately 4500 km long. A recent survey of sea defences, comprising some 1214 km of the coastline, has been undertaken by the NRA. The purpose of the survey was to establish, from a visual examination, the condition of both the defences and their elements, and to establish certain physical and tidal data. The landward limit of the survey was determined by those limits defined for coast protection defences in Schedule 4 of the Coast Protection Act 1949, with one exception. The survey included the defences around the Wash as the nature function and importance justified inclusion.

3.2 Because of the varied ownerships the survey was undertaken in three phases, each phase relating to different bodies or groups of organisations perceived as having a responsibility for maintenance or improvement of the defences either as a result of declared ownership or by custom and practice.

—Phase 1 relates to those defences maintained by the NRA. (Appendix 3.)

—Phase 2 relates to those defences maintained by local maritime authorities.

—Phase 3 relates to those defences owned by private individuals, corporate bodies (ie British Rail, Crown Commissioners) or others.

3.3 Appendix 4 lists the distribution of the defences between the three responsibilities. The NRA maintains some 800 km, Local Authorities maintain around 240 km and private individuals, corporate bodies and others some 170 km.

3.4 The findings of Phase 1 of the survey were published in July 1991. Similiar information from Phases 2 and 3 has not yet been collated, but should be available by the end of 1991.

3.5 No accurate evaluation of the number of people nor extent of lands protected by any of the defences has been made. However, in a paper presented at a recent conference,[1] an assessment was made that some 8000 km² of land lay below the 5.0 metre contour related to the Ordnance Datum ($+5.0$ m AOD(N)), the zero being established from the sea level at Newlyn. The map in Appendix 5 shows these high risk areas, together with some limited Tidal Data.

3.6 The NRA is in the process of evaluating and categorising land protected by their sea defences. Each defence will be classified according to the use made of the land protected by that defence. (Appendix 6.) Some 66 per cent of the Authority's defences protect those lands which are more urban or higher-quality agricultural areas.

3.7 A key finding from Phase 1 is that only 14 per cent of the NRA's defence elements are in need of moderate or major works to bring them back to the " as built condition ". Defences in need of these works are included in current capital programmes.

[1] (The Greenhouse Effect. An Assessment of Lands at Risk. I. R. Whittle. Loughborough Conference of River Engineers 1989.)

3.8 The NRA spends £26 million annually on improvement works on sea defences. Further monies are spent on sea defence works by way of maintenance of embankments, revetments, groynes and beach replenishment etc.

3.9 The disposal of solid waste at sea, including sewage and dredge spoil dumping, licensed by MAFF under the Food and Environment Protection Act (1985) (FEPA), can result in the release and/or remobilisation of pollutants into the water column.

3.10 *Fisheries*

One of the NRA's primary operational duties is to protect fish stocks. There are over 1,000 licensed commercial salmon and sea-trout instruments in rivers, estuaries and off the coast and, except for privileged engines, these are subject to limitation orders. The type and use of gear and seasons are all regulated. In some areas **illegal fishing** is a major threat to fish stocks and coastal fisheries protection involves the NRA's own offshore fisheries protection vessels.

4. MECHANICS OF COASTAL PROTECTION SYSTEMS

4.1 Section 105 provides paragraphs for the Secretary of State as set for controlled waters.

4.2 The specification and achievement of Statutory Water Quality Objectives is fundamental to NRA's water quality management strategy. Section 104 of the Water Act identifies three criteria for classifying controlled waters:—

(a) General requirements relating to the use of water.

(b) Specific requirements concerning the presence or concentrations of certain substances in the water and

(c) specific requirements relating to other characteristics of the water.

4.3 The NRA is preparing a proposal for submission to the Secretary of State which suggests that statutory quality objectives should consist of three elements:—

(i) Beneficial uses of the receiving waters.

(ii) Relevant EC Directives.

(iii) Target class based on an absolute classification of quality.

4.4 The NRA's conservation duties pervade all of its activities. Sea defences under the direct responsibility of the NRA may provide habitat of high nature conservation value. During necessary maintenance of these defences, the conservation value is protected and, wherever possible, enhanced by implementing appropriate environmentally-sensitive engineering methods. Moreover, in the light of potential sea-level rise associated with possible climate change, there are significant opportunities to enhance the environmental value of certain coastal areas. The protection and conservation of features of physiographic, landscape and archaeological interest is also of paramount importance.

4.5 Continuing improvements in coastal and estuarine water quality, through pollution control activities of the NRA, will help to improve fisheries and in particular, the distribution and abundance of migratory species. In addition, the impact of estuarine barrages is becoming an increasingly important issue with respect to migratory fish species.

4.6 Rye Harbour represents the only significant coastal location over which the NRA has controlling rights, and the NRA's responsibility toward recreation in coastal areas is therefore largely confined to promotion in a general sense. However, water quality, particularly with respect to bathing beaches, has a significant impact on the promotion of coastal recreation activities which involve contact with, or immersion in, water.

4.7 Discharge from sewage, industrial and other outfalls unless adequately controlled by the consenting process can pose problems to, for example, the use of beaches for bathing. Many such discharges were first granted consents during the mid 1980s and the determination of appropriate standards is underway. A significant investment programme by the Water Companies to improve sewage outfalls is also in progress.

5. PLANNING POLICIES IN COASTAL AREAS

5.1 Defences are designed for a specific point in time and the value of the " damage avoided " by any defence is commensurate to the sum of the capital investment, plus discounted maintenance costs. There is general perception that the NRA is able to respond rapidly to provide new defences. This is incorrect in that the current need for consultation meetings and committees results in a long " lead " time. Sufficient time is needed to obtain sufficient levies and adjust works programmes.

5.2 The Authority is concerned that the existing planning guidelines are based on a general presumption in favour of development. It considers that planning authorities should be more accountable for, and remedy the consequences of, any rejection of NRA recommendations. If the NRA advice is ignored, it is suggested that the Planning Authority should make a public statement on why the advice was ignored.

5.3 The jurisdiction of the Local Planning Authorities within the coastal zone only extends to low water. The NRA is a statutory consultee for specific types of development within this jurisdiction under article 18 of the General Development Order 1988. There is no Planning Authority jurisdiction below low water and the NRA is unable to influence and/or comment on proposed developments in this area at the planning stage.

5.4 The NRA has no powers of direction in relation to the planning affecting water quality or other NRA interests. In most instances Planning Authorities take on board NRA advice when arriving at a decision on Planning Approval, but in some instances, the advice has been disregarded to the possible detriment of the receiving waters.

5.5 The NRA is routinely consulted during the production of Structure Plans and Local Plans by the Planning Authorities. Coastal zones to low water mark are included within these plans as appropriate and the NRA comment accordingly on water quality issues, particularly in relation to the provision of adequate effluent disposal facilities. It is clear, however, that specific concerns relating to the coastal zone are often diluted or lost within the overall context of these plans.

5.6 A number of planning proposals in the coastal zone fall within the requirements of the EC Directive on the Assessment of the Effects of certain Public and Private Projects on the Environment (85/337/EEC). The NRA is a statutory consultee for Annex 1 projects, for which an environmental statement must be provided, but is not a statutory consultee for Annex 2 projects, for which an environmental statement may be required. A welcome development is that an increasing number of development proposals are now accompanied by environmental statements, produced either on the initiative of the promoters of the project or at the suggestion of the Planning Authorities, irrespective of whether or not they fall within the requirements of Directive 85/337/EEC.

6. Co-ordination of Authorities Responsible for Protection and Planning in the Coastal Zone

6.1 The Authority is working in conjunction with the DOE and MAFF to provide new guidelines to Planning Authorities which are more robust and will require those authorities to have a high regard to NRA objections, or recommendations against development, in flood risk areas.

6.2 The Authority works closely with maritime authorities and has been active in establishing many of the coastal committees, comprising representatives from Local Authorities, central government and other bodies, set up to consider and review common coastal problems. Elsewhere, the NRA has joined with established groups, maintaining close collaboration with MAFF and sea fisheries committees on coastal fisheries issues.

6.3 Maritime authorities' are able to undertake coast protection works to prevent erosion of high land (Coast Protection Act 1949). The NRA is a statutory consultee. This is important as the effects of defences may adversely affect the NRA's interests in adjoining or nearby sea defences or beaches.

6.4 The NRA recognises the national and international Nature conservation importance of many coastal areas such as salt marshes, mud flats and estuaries by collaborating and liaising with organisations including English Nature, the Countryside Council for Wales and the Royal Society for the Protection of Birds. The NRA also has a nature reserve on land that it owns in Rye Harbour. The NRA's objective of improving water quality through pollution control will have a significant rehabilitation effect, improving the nature conservation status of coastal flora and fauna in many areas degraded in the past by poor water quality. Protecting features of physiographic, landscape, historical and archaeological interest is of paramount importance. The NRA maintains close liaison with organisations representing these interests such as the Natural Heritage Coast Forum.

6.5 Many other bodies and organisations have some interest in the coastal zone and a report by the Coastal Engineering Research Advisory Committee (CERAC) of the Institution of Civil Engineers identified such interests. That report was submitted in evidence to the review by the Select Committee for Marine Science and Technology chaired by Sir John Mason (1989).

6.6 Crown Estate Commissioners grant leases and licences for activities such as fish farming and aggregate removal which may have significant water quality implications.

6.7 The specification for material for use on beaches means that the Authority is competing with demands of the aggregate supply industry. There is a limit on the number of sea-bed sites licensed by the Crown Estate Commissioners for gravel extraction. The Authority is undertaking research to find alternative ranges and sources of materials.

6.8 The NRA is responsible for maintaining and improving water quality in controlled waters, including coastal waters, in order to achieve Statutory Quality Objectives for these waters. However, many activities which may influence water quality are beyond the control of the NRA. Some activities are authorised by other agencies with no requirement in some instances even for consultation with the NRA before granting authorisation. Some rationalisation or additions to existing legislation may be necessary to enable the NRA to guarantee compliance with Statutory Quality Objectives in such situations. It may prove necessary for the NRA to seek powers of direction in relation to Planning approval in order to be able to ensure that statutory quality objectives in controlled waters including coastal waters can be achieved and where appropriate compliance can be maintained.

6.9 The absence of any independent statutory Planning Authority below the low water mark is a matter of grave concern to the NRA. This could be resolved by extending Local Planning Authority jurisdiction beyond low water to embrace at least the area encompassed by controlled waters. Special arrangements would be necessary to secure consultation between adjacent Planning Authorities on offshore development proposals in recognition of the holistic nature of the marine environment.

6.10 Management of coastal water quality is essentially a matter of reconciling the different uses of those waters and their implications for receiving water quality. Some of these uses are inevitably in conflict, as in the case of recreational activity and effluent disposal. The NRA supports the need for the zoning of uses to facilitate water quality management.

7. ENVIRONMENTAL STATEMENTS

7.1 The NRA welcomes the increased role that Environmental Statements can and should play in the Planning Process. However the NRA is concerned on a number of counts:—

(i) The quality of many Environmental Statements and the associated Environmental Assessments is extremely poor, often missing or misrepresenting the key environmental issues.

(ii) In general, Planning Authorities do not have the technical expertise to determine the validity of the conclusions in an environmental statement by a rigorous examination of the supporting evidence in the environment assessments.

(iii) Environmental Assessments and Statements to some extent cut across the statutory responsibilities of the NRA and other regulatory agencies by indicating whether or not the impact on the environment is acceptable. No formal mechanism exists to indicate to the Planning Authority whether or not the regulatory agencies concur with the conclusions of all Environmental Statements; this must devalue their role in the planning process.

7.2 In order that Environmental Assessments and Statements play their proper role in the Planning Process in the coastal zone and elsewhere, the NRA would recommend:—

(i) There should be an independent audit of the quality of Environmental Assessments prior to submission of an Environmental Statement to the Planning Authorities in order that the Planning Authorities can make objective decisions on the basis of the statement. The audit body should include representatives of statutory regulatory agencies.

(ii) There should be a formal requirement to consult with all relevant statutory agencies prior to the production of an Assessment and the Environmental Statement should indicate the extent of that consultation, the issues raised and the extent to which they were addressed.

(iii) The Planning Authorities should be required to formally consult all relevant statutory agencies on any Environmental Statement and the associated Environmental Assessment to establish their views on the validity and acceptability of the conclusions. It is essential for statutory agencies to have access to the Environmental Assessment which contains the detailed technical arguments on which the broad conclusions of the Environmental Statement are based.

(iv) Environmental Statements should always be formally considered when arriving at a decision on a Planning Approval.

Appendix 1

National Rivers Authority

HEAD OFFICE (LONDON)
30-34 Albert Embankment
London SE1 7TL
Tel: (071) 820 0101
Fax: (071) 820 1603

HEAD OFFICE (BRISTOL)
Rivers House
Waterside Drive
Aztec West
Almondsbury
Bristol BS12 4UD
Tel: (0454) 624400
Fax: (0454) 624409

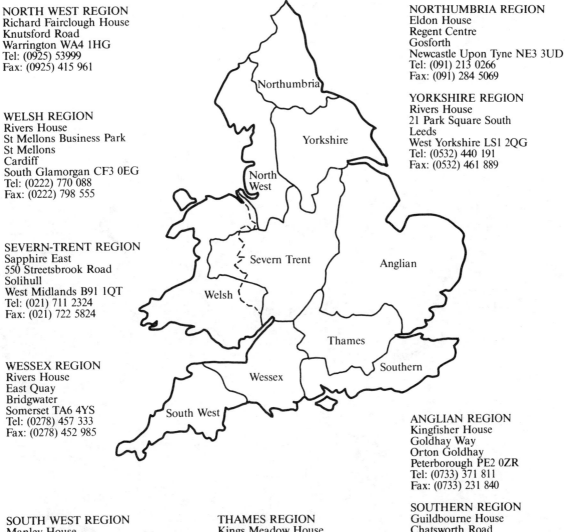

NORTH WEST REGION
Richard Fairclough House
Knutsford Road
Warrington WA4 1HG
Tel: (0925) 53999
Fax: (0925) 415 961

WELSH REGION
Rivers House
St Mellons Business Park
St Mellons
Cardiff
South Glamorgan CF3 0EG
Tel: (0222) 770 088
Fax: (0222) 798 555

SEVERN-TRENT REGION
Sapphire East
550 Streetsbrook Road
Solihull
West Midlands B91 1QT
Tel: (021) 711 2324
Fax: (021) 722 5824

WESSEX REGION
Rivers House
East Quay
Bridgwater
Somerset TA6 4YS
Tel: (0278) 457 333
Fax: (0278) 452 985

NORTHUMBRIA REGION
Eldon House
Regent Centre
Gosforth
Newcastle Upon Tyne NE3 3UD
Tel: (091) 213 0266
Fax: (091) 284 5069

YORKSHIRE REGION
Rivers House
21 Park Square South
Leeds
West Yorkshire LS1 2QG
Tel: (0532) 440 191
Fax: (0532) 461 889

ANGLIAN REGION
Kingfisher House
Goldhay Way
Orton Goldhay
Peterborough PE2 0ZR
Tel: (0733) 371 811
Fax: (0733) 231 840

SOUTH WEST REGION
Manley House
Kestrel Way
Exeter EX2 7LQ
Tel: (0392) 444 000
Fax: (0392) 444 238

THAMES REGION
Kings Meadow House
Kings Meadow Road
Reading RG1 8DQ
Tel: (0734) 535 000
Fax: (0734) 500 388

SOUTHERN REGION
Guildbourne House
Chatsworth Road
Worthing
West Sussex BN11 1LD
Tel: (0903) 820 692
Fax: (0903) 821 832

Appendix 2

NRA National Sea Defence Survey 1990

National Extent of Defence Lengths
adjacent to an Eroding Foreshore

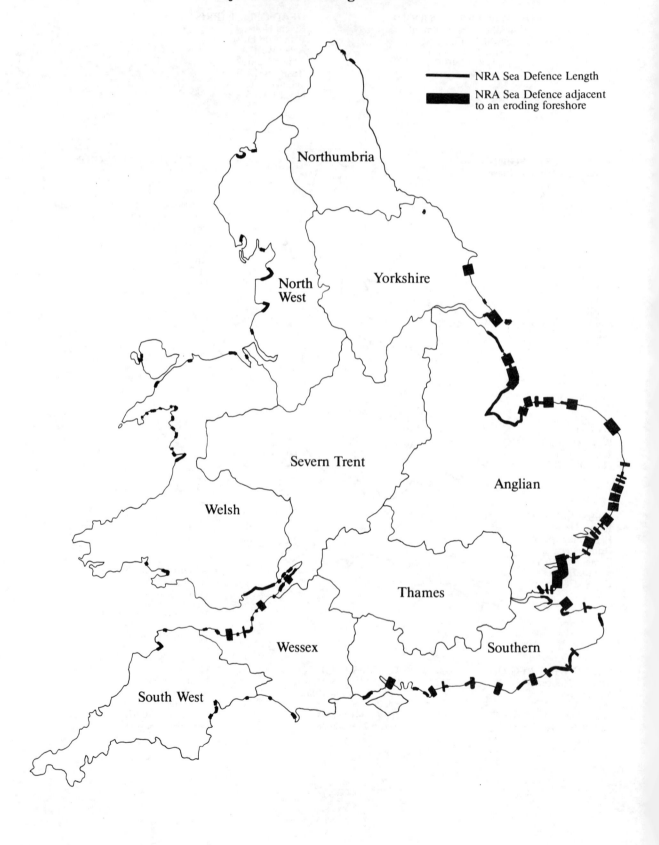

Appendix 3

NRA National Sea Defence Survey 1990
National Extent of Defence Lengths

NRA Sea Defence Length

Appendix 4

Regions	Sea Defence Lengths in Phases 1, 2 & 3		
	Phase 1	Phase 2	Phase 3
Anglian	363·13	12·42	50·94
Northumbria	7·65	1·06	1·33
North West	68·64	51·25	12·85
Severn Trent	30·63	0·00	6·35
Southern	143·80	40·48	10·93
South West	23·06	33·10	18.86
Welsh	111·98	73·32	32·65
Wessex	43·40	30·06	24·10
Yorkshire	12·55	0·31	9·11
Total	**804·84**	**242·00**	**167·12**

15 *July* 1991

Appendix 5

High Risk Areas

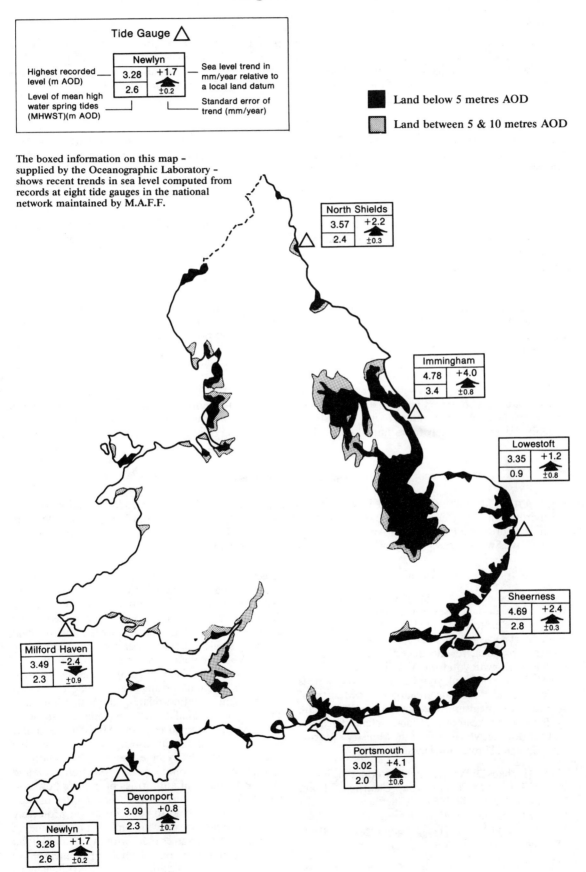

Tide Gauge △

Newlyn	
3.28	+1.7
2.6	±0.2

Highest recorded level (m AOD)

Level of mean high water spring tides (MHWST)(m AOD)

Sea level trend in mm/year relative to a local land datum

Standard error of trend (mm/year)

■ Land below 5 metres AOD

▨ Land between 5 & 10 metres AOD

The boxed information on this map – supplied by the Oceanographic Laboratory – shows recent trends in sea level computed from records at eight tide gauges in the national network maintained by M.A.F.F.

North Shields	
3.57	+2.2
2.4	±0.3

Immingham	
4.78	+4.0
3.4	±0.8

Lowestoft	
3.35	+1.2
0.9	±0.8

Sheerness	
4.69	+2.4
2.8	±0.3

Milford Haven	
3.49	−2.4
2.3	±0.9

Portsmouth	
3.02	+4.1
2.0	±0.6

Devonport	
3.09	+0.8
2.3	±0.7

Newlyn	
3.28	+1.7
2.6	±0.2

Appendix 6

Table 3.3 Typical Nature of Land Use by Band [Robertson Gould Report]

BAND A

A reach containing the urban elements of housing and non-residential property distributed over a significant proportion of its length, or densely populated or developed areas over some of its length. Any agricultural influence is likely to be over-ridden by the urban interests. Amenity use such as parks and sports fields may be prominent in view of the floodplain's proximity to areas of population density.

BAND B

Band B category reaches will contain either housing or non-residential property distributed over a concentrated in part on its length but not of the same density as band A. Agricultural use could be more intensive in the less populated areas of band B reaches.

BAND C

Isolated rural communities at risk from flooding, with both residential and commercial interests, will be found in band C reaches but in limited numbers. Consequently, farming interests will be more apparent than band A and B reaches.

BAND D

Isolated properties at risk from flooding, both residential and commercial, will be found in band D reaches but in limited numbers. Agricultural use will probably be the main customer interest with arable farming being a feature. Where band D reaches are found in undeveloped pockets of largely urban use, amenity interest may be prominent.

BAND E

There are likely to be very few properties and roads at risk from flooding in these reaches. Agricultural use will be the main customer interest with extensive grassland the most common land use in the floodplain. Amenity interests are likely to be limited to public footpaths along or across the river.

Examination of Witnesses

MR MICHAEL MADDEN, Grade 3 on Environment Policy, MR PETER BOYLING, Grade 5 Head of Marine Environmental Protection Division, examined and DR JIM PARK, Grade 5 Head of Flood Defence Division, Ministry of Agriculture, Fisheries and Food, further examined; DR JAN PENTREATH, Chief Scientist and DR CLIVE SWINNERTON, Technical Director, National Rivers Authority, examined.

Chairman

149. Good morning, Gentlemen, may I welcome you to this evidence session of our inquiry into coastal zone protection and planning. I wonder if we could start as usual on these occasions by inviting you, Mr Madden, to introduce the team that you have brought with you from MAFF.

(*Mr Madden*) Thank you, Chairman. My name is Michael Madden and I am in charge of the Ministry of Agriculture's Environment Policy Group. With me are Dr Park, who is head of the Flood Defence Division, and Mr Boyling, who is head of the Ministry's Marine Environmental Protection Division.

150. Perhaps Dr Pentreath will introduce the NRA.

(*Dr Pentreath*) Thank you, Chairman. I am Dr Jan Pentreath, the NRA's Chief Scientist, and I have with me Dr Clive Swinnerton the Technical Director who has special responsibilities for matters relating to flood defence.

151. Thank you very much indeed. My first question is addressed to both teams as you have

separate interests in the matter. A number of witnesses have suggested to us that because the distinction between coastal protection and sea defence is blurred they should be redefined under another title, such as coastal works, and brought under a single funding supervisory and executing body. What are your respective reactions to that?

(*Mr Madden*) The Government's general view is that the present arrangements reflect two-tier involvement in matters relating to coastal protection. There is the national level, the regional and the local level. At the national level the Minister has policy responsibility and then lower down the operating authorities actually execute the work. The maritime district councils promote coast protection work and the NRA and local authorities both have power to undertake sea defence works. As a matter of fact, local authorities tend to concentrate on protecting built up areas. In drawing up its programme of works, the NRA agrees them with the regional flood defence committees on which local councils have majority representations. The arrangements generally reflect the Government's view that

Chairman Contd]

decisions on schemes should have a very strong local involvement because they affect local people, but these decisions should be taken within an overall framework of priorities set nationally.

152. Does the NRA have any view of this?

(Dr Swinnerton) We very much support the view expressed by Mr Madden. We do not see a need for the works to be executed by the same organisation. There needs to be a good understanding of how they interrelate, but the sea defence work done by the NRA can sensibly fit together with the coast protection work done by other organisations, as long as that national overview is retained.

Mr Howell

153. I have mentioned before the situation in my constituency where in the areas where flooding occurs massive defences are being built. In the other areas where the cliffs are the sea is coming in. Lack of co-ordination could not be more obvious. Can I have your comments on the situation at Happisburgh and Eccles?

(Mr Madden) I am not personally familiar with the precise details of the case you mention, Mr Howell, but perhaps I can speak in general terms. We have encouraged the formation of groupings of local authorities concerned with coastal defence matters. Very shortly about 98 per cent of the coastline in England and Wales will be covered by those arrangements. The advantages of those arrangements is that they enable the local authorities, whose boundaries may include natural hydrological phenomena, to operate together and deal with the kind of problems that you outline. For example, and it is certainly known historically, defences can be constructed which may defend one local authority's area and impair those of another. The idea of these groupings is to bring the areas together and take advantage of the common aim to achieve the best possible form of defence.

154. But at this moment, as Dr Park well knows, we have the situation where defences have been abandoned in one village and massive defences are being built up in the next village. There is no co-ordination whatsoever. It is just chaotic.

(Mr Madden) In deciding whether to go ahead with any form of defence, whoever is responsible in those circumstances has to have regard to the cost-effectiveness and environmental implications of those defences. There is no obligation on any of those bodies to carry out defences. There is only a power that they may do so.

155. But the lack of action by the district council is putting at risk the huge defences which are being built up by the other authority. Surely this cannot be right or sensible?

(Dr Park) Perhaps I may add to what has been said. The operating authorities are following a set of national priorities for which areas are most appropriate to defend. We give priority to protection of people and property. The urban communities will have priority for protection within that set of priorities. The various operating authorities—in this case I think you are talking about a particular district council and probably the National Rivers Authority—are required to co-ordinate their arrangements for defence. Certainly the National Rivers Authority, although Dr Swinnerton is better able to speak for them than I am, would not be relying on the defences of Happisburg and Winterton if they were aware that there was any chance of those defences being outflanked by action or inaction on the part of another authority. There are arrangements in place to make sure that those authorities are consulting about their plans, or lack of plans, to make sure that the defence of the coastline is regarded as an entity rather than as individual and piecemeal contributions to defence, but there is no requirement on individual authorities to defend. If they do they have to satisfy national criteria. Those criteria, as Mr Madden mentioned, include economic worthwhileness. It is possible—I do not know the details of the case you are referring to—that there is a question about the worthwhileness of a particular defence, but that is a matter we can look into.

156. Can I ask for a report on this particular situation?

Chairman

157. Would it be possible to let Mr Howell have a memorandum on this particular matter?[1]

(Dr Park) Can you please give me the district council boundary that you are referring to?

Mr Howell

158. It is all in the North Norfolk district, but Happisburgh comes under the district council primarily and, of course, the flooded area at Eccles and beyond is handled by the NRA.

(Dr Park) So it is Happisburgh district council inaction that you are concerned about?

159. It is the inaction of the North Norfolk district council in the village of Happisburgh.

Chairman

160. When you are preparing the memorandum, would you like to deal with it for the purpose of the view of the Committee as a whole; as an illustration, as it were, of the criticisms we have been receiving as to the split of functions? This seems to be an example of where the existing arrangements have been causing problems. The Committee has not yet formed a view, but if we decide to accept the evidence that we have received, as I mentioned earlier, from several witnesses that there should be a single funding, supervisory and executing body, that would really require the merger of the functions that MAFF has for funding and supervising with the functions of the NRA and the district councils for executing. If we are minded to accept that evidence and recommend that one single body be responsible for all these functions, what we shall then have to consider is whether the body that takes on this responsibility should be either

---

[1] Not printed.

Chairman *Contd]*

MAFF or the NRA or even DOE, or whether we recommend the creation of a new body altogether, a new coastal zone unit, for example. If we come to the conclusion that one body is correct, which of those options would appeal to you most?

(Mr Madden) The Ministry of Agriculture is already responsible for both flood and coastal defence matters. To vest the responsibility in a single body is unnecessary because it is already there. It would be more a matter of who would be carrying out the works on the ground. As I have said already, strictly coastal works—that is works of coast protection—are matters which the maritime district councils carry out. Flood defence works can be carried out either by the NRA or the local authorities. So there is a good deal of overlap already. The only distinction is between coast protection and flood and sea defence; the latter is concerned with flooding. If you were minded to recommend the amalgamation of those two types of function, the interest of the maritime district councils in coast protection is so strong that they would have to be represented. That is quite clear. The future lives—as Mr Howell implied in the question he posed earlier—of many people could be very much influenced by the nature of the work carried out on that coastline. It is not clear to us that there is a case for putting all these responsibilities together. I cannot speak in any detail on the case cited by Mr Howell, but we have been making steady progress in the direction I indicated earlier in bringing local authorities together in a formal collaboration between themselves. The effects of that will take time to work through and we shall see an increasing improvement in the way these problems are tackled. There is this division between administrative boundaries and natural geological features and the groupings are intended to take care of that. There is a need for a great deal more research to be done on the precise effects of such things as littoral drift, for example. We are funding a great deal of this at the moment, but the results obviously take time to come through. It is not obvious to us that making one authority responsible for all these things with the multifarious range of views that have to be taken into account is an answer to the problem.

161. So the main problem, as you see it, to the proposition I was putting is the relationship between the single body and the maritime local authorities?

(Mr Madden) That is one of them and also the problem of dealing with the geology of the area as well as the administrative boundaries.

162. You can see no way of overcoming those problems?

(Mr Madden) I am sure there is always a way round these problems, but the present system in general works extremely well. It has been reviewed in recent years and it is not clear to us that changing it, perhaps in the direction you have indicated, would necessarily secure any marked improvement.

163. Has the NRA any other view on this?

(Dr Swinnerton) No, I would share that basic view that there is no obvious argument for a single body. It is a matter of making sure that the parties involved

in those activities are brought together sensibly and are aware of each other's activities. There is no reason at all why the NRA should not be aware of activities by maritime district councils and vice versa.

164. The questions of sea defence and coastal protection are covered by two separate pieces of legislation. One of them is quite old. The Welsh Affairs Committee has recommended that there be consolidation of the two Acts in question. I understand that the Government have responded that this is desirable. Can you tell me what steps have been taken to fulfil the recommendation of the Welsh Affairs Committee? How far have we got in consolidating and updating the Coastal Protection Act 1949 and the land drainage Acts generally?

(Mr Madden) Both those Acts have been amended to take account of various changes, for example, in local government and as a result of the establishment of the NRA and the introduction of conservation duties on the NRA and internal drainage boards. The legislation relating to water, which includes land drainage, has been consolidated following an exercise initiated by the Law Commission. That appeared last year in the form of a number of consolidated Acts.

Mr Field

165. There has been some criticism by maritime authorities that they could receive adequate capital funding from your Ministry, but that they would not get any contribution towards maintenance. There was a strong feeling that a small amount of maintenance would, in the long term, save considerable capital expenditure. Was that point addressed in the review which took place?

(Mr Madden) Not in the consolidation itself. That was a plain, straightforward consolidation of various pieces of legislation into one. But perhaps I may deal with that. It has been a long-held principle that once a capital asset has been created it is the responsibility of the area which enjoys the benefit of that investment to maintain it in a form which continues to provide that benefit. But the distinction between capital and maintenance is not quite as clear as it sometimes appears and, in a sense, has become less clear as a result of the greater use of " soft " defences. For example, we will provide financial assistance where major repair involving a massive investment may be needed. In schemes, principally of soft defences, where beach rechargement is needed, that can also qualify for grant. So the dividing line is not clear, but the principle is as I set out at the very beginning of my explanation.

166. It has been my experience that with very large structures it has often been necessary to allow them to fall down completely in order for the local authority to be able to afford to rebuild them, rather than to maintain them as an interim measure.

(Mr Madden) I would need to look at particular cases to be able to comment on that, but as a general proposition I would find it difficult to understand why a local authority should resort to that approach because it would be losing the benefits for which the capital asset was created.

THE ENVIRONMENT COMMITTEE 71

15 January 1992] MR MICHAEL MADDEN, MR PETER BOYLING, DR JIM PARK *[Continued*
DR JAN PENTREATH AND DR CLIVE SWINNERTON

Chairman

167. Continuing with the topic of the maritime local authorities, are there any proposals for placing environmental duties on them?

(Mr Madden) There are environmental obligations on both coastal maritime authorities as there are on the NRA and on internal drainage boards. They come about in various ways. In the case of coastal protection works they are subject to planning permission and as such are caught by the planning requirements including environmental assessment. But there are also statutory obligations which apply generally to bodies concerned with flood defence and, as a matter of general policy, the Ministry would require any major work to be subject to environmental assessment and would need evidence that it had received planning permission and that the Countryside Commission and English Nature had been consulted.

168. So really you are saying that you see no need for any further environmental duties being placed on maritime local authorities? Is that what you are saying?

(Mr Madden) The present arrangements work very well indeed. I am not aware of any major shortcomings in them.

Chairman: I think Mr Summerson wants to press you on the planning matters.

Mr Summerson

169. Yes, I would indeed like to pursue these planning matters. The NRA welcomes the role of environmental assessment in the planning process, but we have been informed by the Royal Town Planning Institute that coast protection works are not covered by the Town and Country Planning (Assessment of Environmental Effects) Regulations, while flood defence works are. Do you consider that this is acceptable?

(Mr Madden) This is not my information, Chairman. My information is as I expressed earlier. I should be interested to see the basis on which that evidence to the Committee was provided.

Chairman

170. We had a memorandum from the RTPI. We can certainly send it to you and we would like your written comments on it. There is a clear conflict of evidence now before this Committee.

(Mr Madden) Yes, I would remind the Committee, of course, that planning matters are not strictly for the Ministry of Agriculture.

Mr Summerson

171. Yes, I understand that, but the Ministry, like everyone else, is subject to them.

(Dr Park) May I add something on flood defence? There is a special set of regulations that relate specifically to land drainage, as defined in the Act, which, in the terms of our discussion today, encompasses flood defence. All flood defence schemes are required to go through an environmental assessment procedure. If they are new flood defence schemes they are caught by the planning system and therefore, to that extent, there is agreement between our understanding of the system and the memorandum you referred to. It is our understanding that projects which go through the planning system require environmental assessment under the appropriate regulations, which are a different set of regulations from the flood defence regulations.

172. But according to the RTPI that does not apply to coastal protection works?

(Dr Park) We will look at their memorandum.

Chairman

173. May I put the question another way? Within your knowledge it is necessary to apply to a local planning authority for permission to carry out coastal defence protection works?

(Dr Park) It is important that we get the terms right. "Coast protection works", which are promoted by the coast protection authority, are required to gain planning permission before they can proceed. All coast protection works also have to receive the Ministry's permission before they proceed.

174. Which Ministry?

(Dr Park) The Ministry of Agriculture in England and it will be the Welsh Office in Wales.

175. Although the final arbiter on appeal on all planning matters is the Secretary of State for the Environment?

(Dr Park) Yes, but this is approval for the scheme as a whole. One of the conditions in considering that approval will be that the scheme has achieved planning permission. If it has not then our Minister will not consider it for approval or for grant.

176. I see. So it would necessarily have to go through the whole gamut of a planning application to the local planning authority. If the applicant is dissatisfied he has a right of appeal to the Secretary of State who, if necessary, will hold a local planning inquiry. Then after all that is through and the inspector makes recommendations which are endorsed by the Secretary of State for the Environment Department, it then has to come to your Minister who finally decides whether or not to allow the thing to go ahead. Is that what you are saying the system is?

(Dr Park) Yes, and there is a range of other considerations that our Minister would be required to take account of, which would not have necessarily be considered within the planning application. There are recent examples on the south coast where this process has had to be followed through.

Mr Jones

177. I think there is a slight difference between the two questions, in that the first question that Mr Summerson addressed to you was concerned specifically with the Town and Country Planning

Mr Jones *Contd]*

(Assessment of Environmental Effects) Regulations, which is, if you like, the environmental impact assessment, as opposed to the planning system. I want to establish that when you respond to the Committee on what is contained in the RTPI's memorandum to us you will deal specifically with those regulations as well as with the more general planning process.

(Mr Madden) Yes. My response was based on the assumption that those regulations applied in the cases cited.

Mr Summerson

178. May I move on to a slightly more general question on development control? The NRA's memorandum states that "the existing planning guidelines are based on a general presumption in favour of development". Do you see this as a problem?

(Dr Swinnerton) We need to make sure that any development in a flood plain does not increase the risk of flooding. Therefore either works which would increase the risk of flooding should not be given permission, or compensatory arrangements must be made for that.

179. Are you satisfied with the existing arrangements for making your fears known to the planning authorities?

(Dr Swinnerton) We have the opportunity of making our views known, but that is advice and that advice can be balanced against other interests and can be ignored and developments can go ahead.

180. Are you happy about that?

(Dr Swinnerton) We would like, in certain cases, a stronger line to be taken on making sure that compensatory arrangements are put in hand before that permission is given. If such compensatory arrangements are not put in hand, then we do not think that development should be allowed, either in flood plains or in areas which will increase the problem of flooding in flood plains.

Mr Jones

181. Would you like to be able to issue a directive comparable, for example, to highways directives that highways authorities can issue?

(Dr Swinnerton) We would not wish to be able to issue a veto. I think it is right that we have to put our views in and due regard is given to those views.

Mr Summerson

182. But how do you deal with the situation if you put in your views to the effect that such and such a development should not be permitted and it nevertheless went ahead and landed you with all sorts of problems?

(Dr Swinnerton) That is why I am emphasising the fact that if a development is to go ahead against our advice, usually changes in engineering works or compensatory arrangements can be made to offset the effect of that and therefore greater weight should

be put to our remarks and advice on that. Perhaps if, for some reason on a balance of interests, our advice is ignored, a statement should be made why they felt it was necessary for that development to go ahead, so at least people could see why it was being done and could also recognise the fact that we had advised against it.

Mr Jones

183. That would create an unusual precedent if any local authority had to state why any development should go ahead. Obviously they are obliged to state the reasons why things should not go ahead if they reject them, but not the other way round on any occasion. This would be a wholly new concept in planning law, would it not?

(Dr Swinnerton) There is a justification for us being able to know the balance of interests and if there was an increased risk in the flood plain, why that was considered justified.

184. In which case you have reversed the general presumption in favour of development and said that there should be a presumption against development and that local authorities should have to justify any departure from it.

(Dr Swinnerton) I would return to the point I made that development should only be permitted where it will not have a detrimental effect on a flood plain area.

Mr Field

185. Also it defends one of the principles of common law, that you never explain a decision when you refuse it. This is one of the reasons the Speaker never explains his decisions because they could always be open to challenge.

(Dr Swinnerton) Certainly I feel that if a development is to take place and we have advised that it will have an adverse impact in a flood risk area, compensatory work should be required from that developer so that it is not the NRA and, in due course, the general public who are landed with the bill for protecting that area.

(Mr Madden) You will recall, Mr Chairman, that when the evidence was taken from my colleagues in the Department of the Environment they advised the Committee that these guidelines were under review and it was hoped to reissue them at a later date.

Mr Howell

186. What advice would the NRA give on dealing with proposals for development in coastal areas which are not directly at risk at present, but which might become so as a result of sea level rises?

(Dr Swinnerton) As you know we have carried out a survey of all sea defences of England and Wales, not only those owned and maintained by the NRA but others as well. It is an ongoing requirement to monitor the effectiveness and state of repair of those defences. Areas which are not currently at risk which may come under risk in future due to sea level rise will be monitored and we have agreed with MAFF a joint

Mr Howell *Contd]*

policy on dealing with sea level rise. The essence of that is to accept the IPCC prediction for the sea level rise figures and to build those in, making an allowance for tectonic effects, to the design of schemes where that is cost justified. That does not necessarily mean that we will be building sea walls and defences to a higher level at this stage, but it would be a flexible approach so the designs of the schemes would be such that if the monitoring of climate change and sea level rise confirms that sea level rise will take place we can extend those schemes without having to reconstruct the whole defence. That is the approach we are following.

(Mr Madden) The Ministry revised its guidance fairly recently on allowances for sea level rises. I think the Committee was given some evidence on that on 4 November. I should add, with perhaps a touch of pride, that that guidance was issued ahead of other departments. We were anticipating the need for these changes, we are glad to say, ahead of some of our colleagues.

Chairman

187.   Dr Swinnerton, with regard to the survey that is being carried out, at the back of the very helpful memorandum which you have submitted to the Committee in the appendices there are some maps. The thin lines suggest that they cover the NRA sea defence length. I am entitled to assume by inference that everything that is not marked has nothing to do with the NRA. The responsibility is with some other authority.

(Dr Swinnerton) That map was produced after the first phase of our sea defence survey which was the defences owned and maintained by the NRA.

188.   The parts marked with thick lines are those where you adjoin an eroded area?

(Dr Swinnerton) There will be other sea defences operated by the local authorities and other bodies which will be the results of phases two and three, which will be released shortly. The coastline which is not covered by the thin line is a mix of sea defence, which is not covered in phase one and coast protection sites for which we are not responsible.

189.   Is there a map available that shows which different bodies are responsible for parts of the coastline so that the whole of the coastline is then covered and so that we could see at a glance what is NRA, what is a particular local authority and so forth?

(Dr Swinnerton) There is not a map available at the moment. We have a database now from phases two and three of the sea defence survey which provides that information.

190.   Could that be made available to the Committee?

(Dr Swinnerton) We have not produced it in the form of a map, because there are a very large number of very small lengths of privately owned defences, so it would not show up on a map of that scale. We have the facility of producing it on maps of larger scale, but it would not be particularly meaningful at that scale you have for the NRA-owned defences.

191.   It would be very helpful to the Committee for us to have a bird's eye view of where the responsibility lies at different parts of the coastline. From that we can get a view as to how much of a hotch potch there is or otherwise of the situation in terms of responsibility for sea defences.

(Dr Swinnerton) We will have all that information at the end of phases two and three.

192.   When will that be?

(Dr Swinnerton) We hope to release results of that some time during February. We can certainly look at the map information to see what it would look like. I repeat my own view that it will look a rather odd map, but we can see what it would look like and let the Committee have it.

193.   If you can produce something for us, even a larger map, by the end of February that would be very helpful. Thank you very much.

(Dr Swinnerton) Yes.

Mr Field

194.   Is it true to say that some defences are disputed concerning whether or not they are NRA responsibility?

(Dr Swinnerton) I am not aware of any. We have gone through all the sea defences of England and Wales and surveyed all of those. We have come to agreement with the other parties about who is responsible for doing that survey.

195.   And sluice gates?

(Dr Swinnerton) As I say, I am not aware of any existing disagreements, but if there are any you are aware of I should be grateful if you pass them to me and we will look into that.

Mr Howell

196.   How can the executive powers of the NRA's Regional Flood Defence Committees be justified when the central exchequer pays the greater share of sea defence and land drainage schemes?

(Mr Madden) The Regional Flood Defence Committees are a modern adaptation of a much older body that was reformulated when the NRA was established. The legislation which governs the constitution of the Regional Flood Defence Committee provides that local authorities will appoint a majority of members, the NRA appoints two members per committee and the Ministry appoints the chairman and a number of other members who include environmental representation, in the course of which we would consult the Nature Conservancy and so on.

197.   The Ministry of Agriculture?

(Mr Madden) The Ministry of Agriculture appoints the chairman of each committee, plus a number of other members. It varies according to the committee, but it always includes at least one person concerned with environmental issues.

Mr Jones

198.   Can I follow that up, because although that is very interesting, is it not the case that local people are paying only a very small part of the cost of

Mr Jones *Contd]*

such defences but through their local council representatives that they have a majority position on these committees, a departure from the usual " he who pays the piper calls the tune "? As there are national environmental issues at stake, the fact that there is a token environmental person on each committee does not, in any way, counter the argument that the wrong balance is drawn on these committees.

(Mr Madden) There are several questions there. First of all it is not always true that the local people are paying the lesser part of an investment made by the committee. The rates of grant vary from, speaking from memory, 35 per cent upwards, so in some cases they could actually be meeting much the greater part of the cost. Secondly, we would not accept that the appointment of someone concerned with environmental interests was either token or the end of the story. To start with it does not follow that other members of the committee would not have an interest, even though they may not be there specifically for that purpose. As I tried to explain earlier, our system builds into it a good deal of requirement to ensure that any project is environmentally sensitive.

Mr Howell

199. I should like to clarify that question. It seems to me that people living in constituencies such as mine are not only paying the average that everybody pays, but a bit extra. The idea that Mr Jones' constituents are having a bad deal is quite wrong. It is we who are paying more than the average.

(Mr Madden) I would not like to get involved in a dispute between Members of the Committee, but as a general rule we try to arrange that the level of grant available from the Ministry reflects the degree of need which that area needs, so, for example, the Lincolnshire area would attract the highest rate of grant in the country. We have a formula, which I confess I cannot produce off the top of my head, which broadly takes account of the needs of the area as exemplified by the amount of investment that is necessary to protect it.

200. Why does Lincolnshire get more than Norfolk?

(Mr Madden) Their investment programme for the near future is a higher one and our formula produces the result that they will, on that basis, qualify for the highest level of grant. Norfolk qualifies for a very high level of grant.

Chairman

201. Would that be related to the map in appendix 5 of the NRA submission, which indicates the high risk areas? They are defined in two categories; land below five metres and land between five and ten metres. Lincolnshire on that seems to show a very large extent of land at risk and therefore attracts the greatest amount of support or help. Would that be the reason?

(Mr Madden) That is one of the reasons. Perhaps I could add that as a result of the latest public expenditure survey settlement, we are proposing—in fact the Minister has already announced—an increase in certain grant rates. I will not take up the time of the Committee by explaining them now, but if it would be helpful I am very happy to put in a paper which compares the existing rates with the proposed rates.

Mr Field

202. Concern has been expressed by some local authorities that these committees have the right to precept the local authority with no recourse by the local authority as to whether their budget allows for it in the current year. Have you had representations on those lines? What is your view?

(Mr Madden) They certainly have the legal right to precept. The Regional Flood Defence Committees go to great lengths in their deliberations to try to match the flood defence needs against ability to pay, taking account of grant available. We know of cases where there has been difficulty or reluctance on the part of the local authorities concerned to raise their share of the money. I hasten to add that this has happened in a relatively smaller number of authorities. I suppose it is understandable, given all the pressures on local authority finance. There is also further assistance going to local authorities through the general Exchequer support provisions which take account of their commitments on flood defence as they do on other matters.

203. Can I hear the NRA's view on this, as they are doing the asking?

(Dr Swinnerton) The general arrangement works well. When we look at the end product, which is the important thing, where we are spending the money and whether it is going in the right place and addressing the high risk areas and whether there is any indication of a bias coming out of an agricultural nature, we find that is not the case.

204. How many local authorities is this true of in your experience, given Mr Madden's evidence?

(Dr Swinnerton) It is very small number.

205. One? Five?

(Dr Swinnerton) I would not want to go into a precise number, but it is a handful and is that sort of number. I would say it is half a dozen or less, that order of things. Generally the arrangements work well. Generally we are able to get the required funds and when you look at the distribution of those funds, the kinds of schemes which are being done, it is a majority on urban defences rather than agricultural.

206. And how many maritime districts is this half a dozen out of?

(Mr Madden) It would not be maritime districts. It would be flood defence districts.

(Dr Swinnerton) I will be corrected if I am wrong, but I suppose on the flood defence district numbers, it is round 20.

207. So it is half a dozen out of 20?

(Dr Swinnerton) But it is indication and it is not as if they are not voting through reasonable funds. When we are going for 12 per cent, they might go for something like 10 per cent.

Mr Field *Contd]*

208. With respect, Dr Swinnerton, you are telling the Committee that a third of the flood defence committees are in dispute on the precept for flood defence work.

(Dr Swinnerton) We are currently going through the rounds at the moment. Yes, as Mr Madden has said, there are indications that some of the flood defence committees are becoming more reluctant to vote through the precepts which we have put forward to achieve the programmes which we would like to see going forward. If we do not get those precepts, normally to date it has been possible to adjust the programme and still have a sensible programme. Particularly on the sea defences, the survey of our sea defences indicates that the level of maintenance and level of construction of new defences is keeping pace well with the problem.[1]

Mr Jones

209. What did you mean when you said there was a dispute about whether it be 10 per cent or 12 per cent? 10 per cent or 12 per cent of what?

(Dr Swinnerton) Increase in relation to the current rate of levy. There is a tendency, quite understandably, to expect increases to be round about inflation unless there is a programme of work which justifies going above that.

(Mr Madden) Of course what Dr Swinnerton was reflecting was the state of affairs last year before we announced the increased rates of grant. We are not quite sure how beneficial those increases will be. We are obviously hoping that the increased contributions in many cases will ease the problems that local authorities may have.

Mr Howell

210. It is not my view, but it has been suggested that the Regional Flood Defence Committees have too much of an agricultural bias. Would you like to comment on that?

(Mr Madden) In a sense I anticipated your question. The structure of the committees, which is set out in statute, would allow us to dispute that to some extent. Basically the appointments are two from the NRA, the chairman plus certain others from the Ministry and the majority from local authorities. As in many cases, probably the majority of cases, the work of the Regional Flood Defence Committees is concerned with urban areas it is a reasonable assumption that the local authorities concerned are more involved with urban areas than with rural ones. But one has to accept that agriculture covers 80 per cent of the land surface and so one would expect a considerable agricultural interest.

Chairman

211. I think the problem arises with the difference between the agricultural interest and conservationist or wild life interests. The wild life interest is where the

criticism is coming from. They are sometimes at variance with farmers.

(Mr Madden) I understand that. We have our procedures, as I have explained at some length now, to try to take care of those concerns. As I think the Committee was told when my colleague gave evidence in November, the Minister is very sensitive to the concerns of many organisations and has instituted a series of regular reviews with bodies concerned with conservation. If I may remind myself by looking at his statement, his aim in doing so—this is particularly concerned with coastal matters—is that he wanted to see whether the concerns of flora, fauna, landscape, wild life and bird populations are being properly taken into account. He has held the first of those meetings just before Christmas and proposes to continue them on a regular basis.

Mr Jones

212. Your Ministry appoints the chairmen of these bodies. Can you give us some idea of the percentage of these chairmen who are farmers? I do not necessarily expect you to answer that off the cuff.

(Mr Madden) Speaking from memory, I believe they all have farming interests, but whether farming is their principal occupation I do not know.

213. All the chairmen have farming interests?

(Mr Madden) I think so. I had better check that.

214. It is scarcely surprising, is it not, that it is suggested that there is a bias in the system? You cannot possibly expect me to believe that the only people who have expertise and public service to contribute in these areas are farmers. Come on!

(Mr Madden) I think it is perfectly reasonable to dispute it. The reason why the present body of chairmen are in that position is because of their long association with and experience of land drainage matters. As I say, the work of these committees predates their names as Regional Flood Defence Committees. They go back much further than that and the Minister is obviously very influenced when he is considering suitable people for appointment with whether or not they have the relevant background knowledge and experience.

Mr Summerson

215. Is there any evidence to show that these chairmen are not doing their jobs properly?

(Mr Madden) Quite the reverse. The committee system works extremely well and most of the flood defence work these days is concerned with urban areas anyway and not rural areas.

Chairman: Can we move on before we have the Committee at odds with itself?

Mr Field

216. What is the justification for the continued existence of the internal drainage boards? We are aware of the almost universal view that there has been a considerable loss of wetland habitat.

(Mr Madden) In the years preceding the formation of the National Rivers Authority, Ministers gave a good deal of thought to whether or

---

[1] *Note by witness:* 7 out of 24 Flood Defence Committees in England and Wales approved levies less than those proposed but the total shortfall was £0·5 million out of a total budget of £250 million.

Mr Field *Contd]*

not IDBs should be continued. It was decided that they performed a very useful function and they were not out of date, mainly because they secure the involvement of the local interests who need to be concerned. IDBs are responsible for drainage infrastructure in areas of special drainage need outside the main river system. They are subject, concerning conservation matters, to Section 12 of the Land Drainage Act. Many of them in one form or another are of great antiquity. In fact the early drainage systems in the Fenlands were constructed on the basis of local committees. Many of the systems we have today go back many years. It is the view of Ministers that they work very well indeed and, as I said, a review was carried out in relatively recent times. The Minister has not been persuaded that they should make any changes. IDBs have been put under the same sort of obligations as other organisations, such as the NRA, with regard to conservation matters.

217. You are aware that considerable research is being done currently on river basin management and the general collection of water and the aquifers. What is the objection that this is not under the umbrella of the NRA?

(Mr Madden) In a sense the IDBs are broadly under the umbrella of the NRA. The NRA has functions with regard to IDBs, but many of them operate on a huge scale. Some are tiny, but many operate on a huge scale. The Middle Level IDB in East Anglia is a very good example. It covers an enormous area.

218. There is no logical reason why problems of scale should prevent or provide a transfer to the NRA. What are you trying to tell us when you say that some of them are very large? What is the point of saying that?

(Mr Madden) Only that they are self-contained areas operating on a big scale which would have to be taken over by someone. The NRA I suppose could take it over, but since the system works extremely well at the moment, there has been no reason to do it. In one area in Essex there is not an IDB, so there are examples of areas where IDBs do not exist, but where they do they have operated extremely well.

Mr Jones

219. When you say that it all works extremely well, of course that is a point you have made about every aspect that we have questioned you on so far, which is very impressive, but some of us may be slightly more sceptical than others. The IDBs are not without their critics and undoubtedly there are environmental critics who feel that these IDBs have been insensitive on occasion to environmental considerations. I shall not ask you whether all the chairmen of the IDBs are farmers——

(Mr Madden) I do not know the answer to that.

220. I can speculate on the answer to that. Indeed I shall probably put down a parliamentary question later on. But I should like to know what your response is to the criticism by the environmentalists, such as the RSPB, that these IDBs have not necessarily a poor record, but on occasions are insensitive on these issues?

(Mr Madden) I think it would be difficult for anybody operating on any sort of scale—as I have said already, some of them operate on a very wide scale—to avoid criticism completely. But the criticism is, from my knowledge, confined to certain areas, particularly of acute interest. I have in mind especially the Somerset Levels. It is indisputable that there are problems there, but I think I am right in saying that there are over 200 IDBs, the vast majority of which do not seem to come in for any sort of criticism. I believe that bears out my remark that by and large they work very well. There are occasions when they come in for criticism. We have issued guidance which we hope will help to reduce or even remove that criticism in the future.

221. I do not think it is an argument that just because in the vast majority of cases things work without a conflict that one does not have to address the problems where conflict occurs. After all the vast majority of people do not murder their neighbours, but we have laws to govern the occasions when they do.

(Mr Madden) But there are laws governing the conduct of IDBs in relation to conservation.

222. But clearly there is a lot of criticism as far as the Somerset Levels are concerned and a very poor relationship between conservationists and the IDB for that area.

(Mr Madden) I accept that there have been a lot of difficulties in that area, yes.

Mr Field

223. I do not know whether you recall the scheme, Mr Madden, to drain the Amberley Wild Brooks behind Arundel Castle. When that went into a very big public inquiry it was said that that was the only wetland habitat left of its type in Northern Europe. Nobody ever disputed that. Surely the point is that once this particular unique ecology is destroyed it is not re-established. It is unique in its own right.

(Mr Madden) I cannot comment on that case, I am afraid.

224. But the generality of destroying a wetland habitat is that each one is unique unto itself?

(Mr Madden) But there are other examples—of which I saw one for myself the other day—where wetlands are preserved as a result of the collaboration between the IDBs and local bodies concerned. The one I am thinking about is just south of Kings Lynn where there is an agreement between a local wild life trust, I believe, the conservation body concerned and the local IDB to maintain the water levels in the interests of the birds. That is not an isolated example.

225. Indeed the substantial part of the Amberley Wild Brooks is now in the ownership of the RSPB. Would the National Rivers Authority be prepared to take over the inland drainage boards?

(Dr Swinnerton) I should like to begin by saying that undoubtedly the work done by the IDBs has to

Mr Field *Contd]*

be done by somebody. It is fundamentally important work to the water system. Yes, if it was decided to do away with IDBs, I would say that the NRA is the logical organisation to take on that work. But in some areas it is significant and the resource implications of that would have to be assessed. But I have to comment on what has been said. Generally there is a good working relationship and the work of the IDBs on the smaller water courses dovetails in well with the work on the main water courses done by the NRA. Yes, there are examples and you have quoted some where there have been problems on conservation issues, but even on those you have quoted such as the Somerset Levels much work is being done to bring those parties together to a greater degree. That is gradually proving successful.

(Dr Park) I should like to come back on the conservation point. These guidelines, of which we can make a copy available to the Committee, were revised towards the end of the year. It was promised in the Government memorandum. The guidelines give an account of all the responsibilities of the drainage authorities, including the IDBs, so it sets out very clearly what their duties are within the current legislation. It goes on to identify and suggest ways that they would take account of those conservation interests and in effect interprets their duties. It draws attention to the very important environmental protection system that operates in the UK through designation of sites as being very special in the way that Mr Field was suggesting. Those sites are given special emphasis within this guidance note. We have a system which places duties on the authorities, guidance which helps them interpret it and gives particular emphasis to those very special sites which the environment community has designated as being particularly sensitive and therefore require a particular type of protection. These guidelines go to all the authorities for them to follow. The evidence we have with the previous set of guidelines is that in general they are well used by authorities. I know Mr Jones is going to ask what about the exceptional ones, the odd ones that do not follow them. There are general requirements placed on all the bodies. There is an expectation that they take full account of their duties and we place particular emphasis on a close working relationship at a local level between the authorities and the environmental interests. We do not expect to resolve the local issues centrally. We expect those to be resolved locally and a good dialogue between a promoting authority and the environmental interests is essential in order to make sure that these interests are properly protected.

226. With respect, you have not addressed the point I have made, which is that if you accept the generality of the statement that each individual wetland is a separate and contained ecology in its own right which is unique to that area. It is not replicated elsewhere in the country, so once you have destroyed it, it has gone. It is not sufficient to say that you have issued the guidelines and therefore they are being observed in 99·9 per cent of places. It may be the 0·1 per cent that has destroyed the wetland for ever which will not re-establish itself.

(Dr Park) The point I was hoping to bring out without getting into too deep a discussion was that the environmental and ecological significance of the wetland would be recognised by some form of designation. That level of designation then draws from the authorities a particular response if they follow these guidance notes. That is the point I was making. Mr Jones made the point about murder. We have legislation. Most people follow that legislation, but the odd ones do not. We do our best through guidance and through the terms of legislation to make sure that it is clear to the authorities what their responsibilities are. We believe the majority are following them.

227. Perhaps we can move on to the protection of agricultural land. Is the protection of agricultural land through coastal defence works generally cost-effective?

(Mr Madden) All our schemes have to be cost effective, but one must take into account cost beneficiality. May I elucidate this? There is a difference—I am not seeking to be pedantic here—between cost effectiveness and cost beneficiality. Our schemes not only have to be cost effective and also technically effective, but they also have to be cost beneficial. In taking into account the sort of benefits that arise, we try to evaluate environmental benefits. This is not an easy area. It is often easy to identify something that might be regarded as an environmental benefit. It is not always easy to put a value on it, but we now always require them to be identified. In at least one case that comes to my mind—that is the defences at Aldeburgh—the benefits which were identified, in broad terms, to conservation matters, carried the day and made up the beneficiality side of the equation, if I may put it that way.

228. There has been a suggestion that generally the protection of agricultural land is not economic.

(Mr Madden) When we look at something as a benefit it has to be regarded as a benefit in the national interest, not simply to an individual. Obviously the fluctuating fortunes of agriculture will be reflected in the value that you put on agricultural land. It is significant, as I have said perhaps rather boringly a number of times, that most schemes these days are very much more involved with urban protection than with agricultural protection.

229. Can you tell us your attitude towards the concept of paying farmers to manage land that is reclaimed by the sea as areas for conservation and wildlife to replace lost habitats, or of paying compensation to property owners in areas where managed retreat is desirable?

(Mr Madden) Compensation is a difficult question because, as has come out in what we have said already, there is no obligation on maritime district councils, on the NRA or local authorities to provide defences. They may, but there is not a " must ", so the question of compensation does not readily arise. Paying land owners to accept a different line of defence is not an easy one. When we approach a problem requiring a defence we look for the most satisfactory technical solution. In many cases the

Mr Field *Contd]*

solution may be on a different line from the one that exists already, if indeed one does exist. Or it may be a soft solution which would produce different results from a hard solution. The concept of setting up a line on a managed basis is one which we are currently doing quite a lot of research into because it raises questions about what is the most suitable line. At the moment we tend to approach it on the basis of the best technical solution.

230. But you do not make payments at the moment?
(Mr Madden) We do not make payments at the moment. We would not have any statutory powers to do so, except possibly in the case of environmentally sensitive areas where they impinge on the coastline.

231. What is the objection to not doing so when the Nature Conservancy Council enter into management agreements with farmers for particular areas, and you are paying farmers set-aside not to go into production any longer because of surpluses? Where is the objection to allowing them to receive an income to manage a habitat where the sea is allowed to ingress?
(Mr Madden) This is an area where we are doing more work on at the moment. That is development which may take place in the future, but at the moment we do not feel that we have sufficient technical information about how to approach it.

232. Can you tell us who pays farmers for management agreements with what is now English Nature in our terms?
(Mr Madden) They would be paid by English Nature.

233. Who is funding English Nature for that work?
(Mr Madden) That would be the Department of the Environment. In the case of environmentally sensitive areas, it would be the Ministry of Agriculture of course.

Mr Battle

234. I want to ask about coastal management plans, because in their memorandum to the Committee the Institute of Civil Engineers made the point that MAFF has drawn the attention of the coastal protection authorities to the need for coastal management plans but has not provided any specific guidelines for those plans or for their preparation. Who do you think should provide the guidelines and who should resource the preparation of what may be quite detailed work?
(Mr Madden) I think I mentioned earlier that we are encouraging the formation of groupings of authorities concerned with coastal defences. About 98 per cent of the country is now covered by those. We are very strongly encouraging them to draw up management plans which cover their stretches of coastline. The resource implications are quite small and are met by the local authorities concerned from their own resources, but in drawing up a plan, for example to deal with a particular problem, the work required to draw up a scheme which might lead to

defences is very likely eligible for grant from the Ministry of Agriculture. I cannot think of a specific case, but if perhaps half a dozen authorities have a joint problem and they appoint engineers to solve that problem the work of those engineers can qualify for grant aid as part of the general scheme of assisting in the production of the defences.

235. Would it be a 100 per cent grant to cover the work?
(Mr Madden) No, as I have said earlier, the level of grant varies from one area to another so it would depend on the authority.

236. And the rest of the work would then have to be funded from the local authorities?
(Mr Madden) Yes.

237. It would fall back on the local authorities?
(Mr Madden) Yes.

238. So although, as Dr Park said before the Committee on a previous occasion, you are encouraging the setting up of the coastal groups, the resourcing of them could be at local level not from your department or from the Department of the Environment?
(Mr Madden) The resourcing of the defences will be met partly by the Ministry and partly by the local authorities.

239. Which coastal activities authorised by other agencies have no requirement for consultation with the NRA and what impact do they have on achieving statutory water quality objectives?
(Dr Pentreath) Perhaps I should take that question. The principal agencies which have no formal requirement to consult with the NRA, with respect to matters that would affect water quality, are, first of all, my colleagues on my right, MAFF, through the Food and Environmental Protection Act. This is obviously for historical reasons, but they previously did not have any statutory duty to consult local water authorities and therefore not with the NRA. The second example would be the Crown Estate Commissioners who have no statutory responsibility to consult the NRA and the third most important group would be the port authorities, because there are many local and general enactments and so on. At the moment where water quality standards have been put under pressure as a result of these activities, for example, as a result of dredging disposal activities carried out by MAFF, then we have set up informal arrangements to do something about it. What concerns us most is looking to the future, rather than looking to the past, because, as you are fully aware, the NRA has a consultation document out on statutory water quality objectives as a scheme which would cover the marine environment, or at least the first three miles of it. Therefore if the NRA or the environment agency was to be in a position where they have to deliver the scheme, it would clearly be extremely difficult for such an organisation to guarantee to deliver compliance with statutory water quality objectives at some point in the future, if it found that activities were being carried on which would put compliance in jeopardy but for which there was no formal means of consultation.

Mr Field

240. Can you tell me whether the NRA has a view on the effect of the licensing of aggregate dredging by the Crown Estate?

(Dr Pentreath) Have a view with regard to what? That is another activity which is carried on.

241. Do they consult you?

(Dr Pentreath) No, they do not. There is no statutory requirement for them to do so.

242. You can be busy putting up flood defences and the Crown Estate, in conjunction with MAFF, could be busy licensing aggregate dredging which could be undermining your work?

(Dr Pentreath) My answer was in relation to water quality. I will ask my colleague to answer with regard to flood defence.

(Dr Swinnerton) In theory that could be the case, but we would hope that our contacts and liaison would pick up that kind of activity.

243. Is it true to say that the licensing of sea dredging of aggregate is only looked at in terms of the effect on fishing and not in terms of the littoral drift?

(Mr Madden) I will ask Mr Boyling to answer that as he is the specialist on these matters.

(Mr Boyling) The procedure for approving proposals for aggregate extraction includes an arrangement whereby the Crown Estate seek the technical advice of Hydraulics Research as to any possible coastal erosion effects of an application to extract aggregate. Only if Hydraulics Research advise that there would not be an effect on coastal——

244. Is that organisation part of your Ministry?

(Mr Boyling) No, it is a private company. It used to be part of the public sector, but now is a private consultant.

245. Do you get to see that report?

(Mr Boyling) The Ministry does not see that report. What we do know is that we do not get an application unless Hydraulics Research are content that there would not be a coastal erosion problem.

246. So you are busy funding works to defend the United Kingdom's coastline and the Crown Estate are busy licensing aggregate sea dredging which can or may undermine it, but you never get to see the report and the quality of it knowing whether it does not undermine it?

(Mr Boyling) Until now we have been content to rest on the expertise of Hydraulics Research.

247. Despite the fact that you do not see what that expertise is?

(Mr Madden) This is a very respectable organisation which, until recently, was a public sector research organisation and is still largely dependent on public funds and has a world wide reputation. Whether or not we see the report is secondary to the standing and the quality of the kind of advice that we can reasonably expect from them.

Chairman

248. Hydraulics Research was not within your Department, was it? It was the Department of the Environment.

(Mr Madden) Yes.

249. Is that probably why historically the reports have never come your way?

(Mr Madden) It probably is the reason, but I am afraid I do not know from personal experience.

Mr Field

250. But it is not outwith your knowledge that their information on the effect on coastal defences of aggregate dredging is often disputed by other recognised hydraulic researchers?

(Mr Madden) I suppose in any technical matter there is room for more than one point of view. But this particular establishment has a world wide reputation and they are our principal advisers on this. Obviously in any circumstances you frequently find that there are opposing points of view.

Mr Jones

251. I hate arguing by analogy, but there are so many different situations one can think of where the expertise may be absolutely terrific, but a mistake occurs. The proper thing is to encourage as much debate as possible about the conclusions that have been reached so that if there is a mistake or if there is another point of view this can be taken into account before decisions are reached. Just because the advisers to the Department of Transport on traffic flow, for example, are extremely able and well respected does not mean that their figures and conclusions are not open to scrutiny and to challenge at public inquiries. I am surprised at the Ministry's attitude to the reports, which is basically a very *laissez faire* attitude. Would it not be better to look at the reports and preferably, from time to time, get some alternative advice?

(Mr Madden) I should like to ask Mr Boyling to comment on this. But can I make one comment? As I said earlier any major works on the coastline require planning permission and the planning process itself will give an opportunity for various points of view, including different types of expert opinion, to be expressed, so there is a safeguard there.

Mr Field

252. But there are some coastal defence authorities who, because of a peculiarity in the legislation, have a right to licence aggregate dredging themselves without recourse to the Crown Estate or your Department.

(Mr Madden) I am afraid I am not competent to deal with that.

Chairman: What I understand from the answers you have been giving and your total reliance on Hydraulics Research stems from the fact that you do not have within your Department the expertise to be able to gainsay Hydraulics Research, if they produced a report that you wished to question. One

Mr Field *Contd]*

can understand that, but it is not necessarily always black and white in their reports whereby you can say: " If they say black it must be black and if they say white it must be white. We are not in a position to challenge it ". But if they produce a grey area where there is a balance of probabilities on which a judgment has to be made, perhaps that is something which ought to be brought to your attention for your consideration and further thought.

253. Can we have a report on those coastal authorities who still retain the right to licence aggregate dredging themselves without recourse to the Crown Estate or to your Department? Can we also know how often the Hydraulics Research information is disputed by accepted centres of learning in these matters, to the knowledge of your Department?

(Mr Madden) We will see what we can come up with.

Chairman

254. We shall be having the Crown Estates before us and no doubt we can pursue some of these questions with them when they give their evidence.

(Mr Madden) Mr Boyling has something to add to what has been said.

(Mr Boyling) I do not think we would be arguing that we should not in the future see these reports by Hydraulics Research. The point I was making was that in the past we have not seen them and that we were content to rely on Hydraulics Research. There are possible developments in the future which will make it increasingly appropriate for the Ministry to see these sorts of reports, in particular an increasing number of applications to dredge closer and closer inshore and with thus perhaps an increased risk of an effect on coastal erosion. I do not think it will cause a difficulty for the Ministry to become more closely involved in this activity.

Mr Field

255. Dr Swinnerton said that the National Rivers Authority were looking at a three mile strip. In that context, would it not be right and proper in view of the very considerable sums of coastal defence money that is being spent and the work being done by the NRA that the NRA should have the opportunity of commenting on aggregate dredging?

(Mr Boyling) This is something that might well be taken up in the review which the DOE has already announced it will be making of the Government procedure for aggregate extraction.

Mr Jones

256. I should like to take up a couple of points from the evidence that the Norfolk and Suffolk Broads Authority have submitted to us. They have told us that the current financing arrangements for coastal defences are not conducive to programmed investment on a staged and sequential basis. I was wondering what the NRA thought about that, particularly with its borrowing for long-term programmes.

(Dr Swinnerton) It comes back to the point we were making earlier that in most cases we are able to obtain adequate funds to take forward an appropriate programme. There is a major review of what defences are required for that area so that all the options are being considered and a plan is being taken forward. But that would fall into the area where, while we would be able to achieve the programme of works, the time scale would be delayed by the indication that it is becoming more difficult to get the necessary funding for that.

257. The second thing they said to us was that they think the levels of central Government grant should be 80 per cent for coast protection and 90 per cent for sea defences. I believe it is 60 per cent at the moment, is it not? What is your reaction to that evidence from them?

(Dr Swinnerton) No doubt Mr Madden will want to comment on that, but the grant rates are being reviewed and they are being increased where considered necessary. Quite clearly from our point of view when we are trying to achieve very significant programmes of work, where it is justified and grant rates can be increased, they are very welcome.

(Mr Madden) In a sense I anticipated the question in some of my earlier remarks when I said that we have announced increased grant rates and we will send details to the Committee. Our principal aim is to try to provide funds in areas of greatest need. The new ranges will go up as high as 85 per cent in certain districts of the NRA and up to 75 per cent on coast protection. I am afraid I do not have at my fingertips what they will amount to for individual authorities, but we can let you have that.

258. So there is a general recognition that individual local authorities have a problem with the funding of some of these works?

(Mr Madden) A problem not only with the funding but also an increasing need to do extra work, which is one of the influencing factors on this.

259. It amounts to a problem with the funding?

(Mr Madden) It can easily amount to a problem with the funding, yes.

260. The other question I wanted to ask you— I do not know whether you will be able to answer it off the top of your head or whether it would be better to send us a memorandum—is can we have the latest estimates of needed capital and maintenance expenditure on coastal works over the next 10 years?

(Mr Madden) Yes, we would not really have it in that form. The National Rivers Authority seek advice from their regions for five years ahead. They can speak for themselves, but they produce summaries of this in their corporate plan. We work on the basis of the Public Expenditure Survey. Although we encourage further looking ahead, our views on what is needed are reflected in the announcements of the outcome of the Public Expenditure Survey, which, as you know, is for three years ahead and which I have promised the Committee already.

Mr Jones Contd]

261. And the estimates for Government expenditure on this, contributions for grant, as in the Public Expenditure White Paper, do they include the revised grant rates?

(*Mr Madden*) Yes.

262. So those figures are wholly up to date?

(*Mr Madden*) Those you have in the memorandum are out of date and we will send you a note which will cover that.

263. Basically anything that you can give us on that in the form of an update would be helpful.

(*Mr Madden*) Yes.

Chairman

264. There is just one final question I should like to ask the National Rivers Authority, if I may revert to the matters relating to flood risk and planning and development. Have you contemplated negotiating agreements with local planning authorities for exclusion zones where no development should take place because of flood risk? Could these be included in the statutory development plan where appropriate?

(*Dr Swinnerton*) We have not done so. We would certainly have areas which we consider to be at high risk level and therefore we would be extremely doubtful whether any proposal could be put forward which would be acceptable to us from a flood risk point of view. But, there is always the possibility that somebody can come forward with a technical proposal which would make it possible and acceptable to us. We would certainly make it clear to local planning authorities that in the critical flood plains we would certainly be extremely doubtful whether planning and development would be acceptable, but we would always be willing to consider proposals to see what alternative compensatory arrangements could be proposed.

Chairman: I think that concludes all the questions we have to ask you this morning. I am afraid we have left you with a number of questions still outstanding that you have said will answer by way of written memoranda which we look forward to receiving. I thank you for your attendance this morning and compliment you, Mr Madden, on your robust defence of the status quo. I wonder why we are holding this inquiry if everything is working so well. However, we shall listen to other witnesses and come to a conclusion at the end of the day. Nevertheless thank you for your help this morning.

WEDNESDAY 22 JANUARY 1992

Members present:

Sir Hugh Rossi, in the Chair

Mr John Battle   Mr Robert B. Jones
Mr John Cummings   Mr Tom Pendry
Mr Barry Field   Mr Anthony Steen

COASTAL ZONE PROTECTION AND PLANNING

Memorandum by the Marine Conservation Society

The Marine Conservation Society (MCS) is a non-governmental environmental organisation working towards the protection of the seas for both wildlife and future generations.

The Society has had a long-standing interest in coastal matters through its work on conservation designations at the coast, marine pollution and, more recently, coastal zone management. We welcome this review as a timely and necessary opportunity to examine coastal planning and management. Many problems with the current arrangements have been brought to our attention during our recent research into the subject which is summarised in the report " A Future for the Coast? Proposals for a Coastal Zone Management Plan " (Gubbay, 1990).

Our evidence concerns the whole approach to planning and management of the coastal zone and *all* the major uses to which this is put (Figure 1). This is an important distinction because it takes the perspective of an overall integrated use of the coastal zone rather than the perspective of specific users. Our evidence therefore encompasses the subjects which the Committee has indicated it wishes to address but in the context of a broader framework.

This paper presents our evidence in four parts:

1. The coastal zone and coastal zone management

2. Concerns about coastal planning and management in the United Kingdom

3. Concerns specific to nature conservation as a part of coastal zone management

4. Areas for progress.

The evidence is presented from a United Kingdom perspective as it is our view that the problems it raises apply equally to all parts of the United Kingdom and that they cannot be resolved effectively in a single country context.

1. THE COASTAL ZONE AND COASTAL ZONE MANAGEMENT

1.1. *The coastal zone*

There is no universally accepted definition of the "coastal zone". In the context of planning and management, however, it is widely accepted that it includes elements of coastal land, the intertidal zone *and the adjacent sea*. This type of holistic view is considered necessary because although it covers two very different environments they are closely linked at the coast. Activities taking place in either part can affect the adjacent area making it inappropriate to deal with the planning and management of these environments in isolation of each other. **All references to the coastal zone in our evidence support these views and therefore take the coastal zone to be a band along the coast which includes both land and sea areas.**

The value of the coastal zone, and the demand for the space and resources it provides, have put a great deal of pressure on this environment. This alone justifies giving special attention to coastal matters. The case is even stronger when one considers that the coastal zone with its landward and seaward components, is a distinctly different environment. It therefore requires an approach taking elements from both land and sea planning disciplines, reinforcing the view that there is a need to give the coastal zone special consideration.

Current policies in the United Kingdom do not match these circumstances. The coastal zone is not singled out for attention nor is it dealt with in a holistic way.

1.2. *Coastal Zone Management*

The idea of integrated planning and management of activities and uses of coastal areas is often described as "Coastal Zone Management" (CZM). The approach has been given increasing attention since the 1960s and it is currently being implemented or considered by a number of maritime nations. The precise approach differs from nation to nation but a common factor is that it has been implemented following problems in managing the demands being made of coastal resources. A review of approaches and techniques for coastal and sea use management used by various countries has been carried out by the MCS and is enclosed with this submission (see Gubbay, 1989).

The basic ideas behind CZM are outlined in Figure 2. This makes it clear that nature conservation objectives are only part of the remit of CZM but basic environmental principles underlie the whole process. The main aims are to balance demands for coastal zone resources, promote their sustainable use and, as far as is possible, resolve conflicts of use by integrating planning and management. The basic tenet of a CZM programme is invariably to promote environmentally sensitive use of the coastal zone and to introduce an element of strategic planning for coasts. This type of approach is relevant to coastal protection works but also to all other uses of the coastal zone. **The MCS would like to see the ideas behind CZM providing the framework for all coastal zone management practices and planning in the United Kingdom.**

2. CONCERNS ABOUT COASTAL PLANNING AND MANAGEMENT IN THE UNITED KINGDOM

Research carried out by the MCS has shown that many organisations involved with coastal planning and management have concerns about the current arrangements. It is important to emphasise that the issues are common to all the user groups.

The MCS carried out a questionnaire based survey in 1990 to identify concerns about the current arrangements for the planning and management of activities in the coastal zone. The target groups were individuals working in nature conservation, fisheries, recreation, navigation, coastal engineering, pollution control and mineral and energy extraction. The results showed seven underlying concerns which were common to these groups.

* The lack of a national policy or strategy for use of the coastal zone
* The lack of a planning regime for areas of sea
* The large number of government agencies involved in marine matters
* The poor level of integration between land- and sea-based users in the planning and management of coastal activities
* Increasingly congested water space and crowding on the coast
* The increasing level of pollution in inshore waters
* The role of the Crown Estate Commissioners.

The findings of this research are summarised in a report for the World Wide Fund for Nature from the Marine Conservation Society "A Future for the Coast—Proposals for a United Kingdom Coastal Zone Management Plan". Further detailed examples of 22 problems which illustrate these points are given in a report to the Department of the Environment prepared by Wildlife Link.

This work shows a lack of integrated planning for the coastal zone and an *ad hoc* approach to dealing with problems.

3. CONCERNS SPECIFIC TO MARINE NATURE CONSERVATION

A large part of the work of the MCS is on marine nature conservation. The Society's view, however, is that we cannot succeed with this work by concentrating on areas of nature conservation importance to the exclusion of all other matters. Equally we believe that nature conservation cannot succeed if it is not set into the context of general environmental principles which are applied widely. The priority area for this type of approach must be the coastal zone because this is where many of the problems which face the marine environment originate. It is also one of the most productive parts of the sea and the area which is under most immediate threat from human activity.

The MCS view is that the current arrangements for the planning and management of activities at the coast do not provide a background against which nature conservation can be successful. The case of Marine Protected Areas is used below to illustrate this point.

Marine Protected Areas

In the United Kingdom there is only one type of statutory designation which provides for the conservation of marine areas, the Marine Nature Reserve. Progress with their establishment has been extremely slow (only two sites in 10 years) due to difficulties with the legislation, the procedures and the approach. Even having established these two sites their effectiveness in achieving marine nature conservation is questionable. Part of the problem lies with the limited controls adopted for these areas but other important considerations are that they are not set into any wider management context for the surrounding sea areas and that they are not perceived to have any benefits beyond nature conservation.

A system of CZM, based on the ideas outlined above will have considerable benefits for MPAs in the United Kingdom.

Provide a context

The success of MPAs in achieving nature conservation objectives is very dependent on sensitive use of all marine areas and maintaining the quality of the marine environment as a whole. A MPA cannot be isolated from surrounding waters and although buffer zones can increase the chances of success, ultimately they must be set into the context of a wider management regime. The Lundy MNR, sits outside any general management system for the surrounding sea areas for example, where controls are on an activity by activity basis rather than a co-ordinated approach. Equally the sensitive coastline on either side of the Minch is covered by many conservation designations but there is no management regime for the waters of the Minch where the threat of oil pollution incidents and where fishing practices have the potential to undermine the conservation value of the region. CZM would provide a framework for the management of inshore waters and coastal lands.

Links with conservation on land

The current system does not allow for statutory conservation designations to cross the land/sea divide. Marine Nature Reserves extend to the high water mark whereas land measures generally stop at local authority boundaries (low water). The legality of SSSIs which do go beyond this, for example in the Wash, is questionable and indeed raises doubts about whether the United Kingdom can fulfil its obligations under the Ramsar Convention because SSSIs are the principle mechanism of implementing this Convention. This split at the coastline means that planning and management is not prepared for the whole area. Such an approach is unrealistic because it lead to situations where seabirds are protected on cliffs to ensure successful nesting but this is not linked to the protection of their food source in the surrounding waters. Similarly seal colonies are protected at their breeding and haulout sites but not the food they depend on in adjacent waters. Problems can even arise with landscape conservation. Heritage Coasts provide for the quiet enjoyment of unspoilt coastline yet the wardens have no remit or powers to control noisy jet skis just offshore.

Providing an overview

A United Kingdom, and indeed European, perspective is very important in establishing a satisfactory network of MPAs. The situation over the loss of coastal wetlands which has been documented by the former NCC and the RSPB shows that the lack of broad perspective has allowed the slow deterioration and loss of wetland habitats to the point where the integrity of the whole resource is threatened.

Public participation. This is a key part of CZM because it involves people in the process of decision making about their coastal zone. One of the reasons MPAs have failed is the approach to consultation and the fact that there is no involvement prior to a proposal. This encourages opposition. The regular forum of CZM makes it an ongoing process so there is not a major upheaval every time a MPA is proposed. The Cardigan Bay Forum and the Dorset Marine Forum are two examples of the sorts of groups that will come together with the focus of having to develop CZM plans.

Working together. The development of CZM plans will require local authorities to work together. This is especially important for congested or threatened areas such as estuaries because this needs to be in place before a conflict situation arises. The years of difficulty over designating Strangford Lough is an example where the requirement to produce a plan for the lough would have avoided much delay. Instead the quality of the lough has deteriorated, whilst the arguments continue. In its report on " Environmental Issues in Northern Ireland " (1990) the House of Commons Environment Select Committee recognised the need for urgent action to safeguard the unique sea-bed ecology of the lough (Recommendation 6 of the report).

A unified approach to MPAs. CZM brings together user groups and tries to find common management solutions. This would certainly help to resolve the *ad hoc* approach to establishing MPAs. There are many different types of MPAs, eg fisheries, navigation, archaeological importance, nature conservation. A unified system could be developed with the context of CZM. This has been proposed by conservation organisations.

Focus on the coastal zone. This would undoubtedly benefit MPAs by bringing attention to the inshore waters where pressure is great and environmental problems particularly apparent. Progress with MPAs for nature conservation has been painfully slow and will improve if consideration is being given to management options for these areas.

4. AREAS FOR PROGRESS

A number of key areas where progress needs to be made are highlighted in the MCS/WWF report " A Future for the Coast ". We would like to draw special attention to six of the recommendations.

4.1. **Strategic Guidance for planning in the coastal zone.** The view of the MCS is that the pressures, problems and special nature of the coastal zone make it worthy of special attention. A broad United Kingdom-wide perspective is needed to provide a common base of aims and direction. The preparation of a Planning Policy Guidance Note on coastal matters by the Department of the Environment cannot fulfil this role as it is simply aimed at interpreting existing procedures. These make no provision for dealing with the coastal zone as a whole, and are only aimed at Planning Policy in England.

4.2. **The establishment of a CZM " unit ".** The formation of a unit would provide an invaluable focus for developing, co-ordinating and supporting CZM work in the United Kingdom. This would draw together the many agencies and authorities currently involved in aspects of planning and management at the coast and would act as a link into European and wider international CZM matters. The MCS has prepared a discussion document on this subject, included here as Appendix 2.

4.3. **The preparation of Regional CZM plans.** A regional perspective is important for coastal zone planning. This is particularly true when dealing with matters such as coast protection works and water quality management as the processes need to be managed on the scale at which they operate. These can be large areas—coastal cells in the case of coast protection and watersheds in the case of water quality. Regional planning is already carried out to a certain extent, requiring liaison between planning authorities but will need to be more widespread for a coherent system of CZM as current administrative boundaries do not necessarily reflect the geography of the coast.

4.4 **Focus on intensively used coastal areas and those under threat.** The pressures on parts of the coastline of the United Kingdom are such that there is a need to give priority to planning the use of these areas. Estuaries are the focus of a great deal of this pressure but other parts of the United Kingdom such as the Channel also face this type of problem. Good liaison and a duty to prepare coastal zone plans is needed for these areas as are powers to implement them. In north-west Scotland the Highland Regional Council have prepared " Framework Plans" for some of the sealochs and on the south coast Hampshire County Council have produced a coastal strategy. The ability of these bodies to put the plans into practice is however questionable, as they do not have the powers or the remit, yet they have clearly identified a need for this process.

4.5 **Linking land and sea planning.** Attention must be given to ways in which holistic planning of the coastal zone can take place. This current situation encourages the coastal zone to be treated as a boundary area on the edge of the jurisdiction of coastal authorities. An extension of the jurisdiction of planning authorities is one way of linking the landward and seaward components.

4.6 **Statutory support for CZM.** The need for improvements in the system of coastal planning and management has been stated on many occasions (see Table 1). Some progress has been made but this is limited. The MCS believes that the preparation of CZM plans requires statutory backing.

CONCLUDING COMMENTS

A more integrated approach to planning and management of activities in the coastal zone can only be beneficial. There are clear benefits for nature conservation but also for other interest groups because it would provide a framework in which activities could be planned and managed much more effectively.

The conflicts for space and resources of the coastal zone in the United Kingdom will not decrease. There is a need to deal with the further difficulties which will result from this as soon as possible.

Many of these concerns have been expressed before; what is needed now is definite commitment to change.

Annex I

KEY FEATURES OF COASTAL ZONE MANAGEMENT

A DEFINITION

" A dynamic process in which a co-ordinated strategy is developed and implemented for the allocation of environmental, socio-cultural and institutional resources to achieve the conservation and sustainable multiple use of the coastal zone."

(CAMPNET, 1989)

Coastal Zone Management **aims** *to:*

* promote **sustainable use**
* **balance** demand for coastal zone resources
* **resolve conflicts** of use
* promote **environmentally sensitive use** of the coastal zone
* promote **strategic planning** for coasts.

Coastal Zone Management **recognises:**

* the "coastal zone" as a unit for planning purposes
* that planning and management of coastal land and waters cannot be dealt with separately
* the coastal zone is an area which requires special attention for planning and management.

Coastal Zone Management **requires:**

* a **national** perspective
* a **long-term** view
* an **integral approach** to planning and management
* **communication, collaboration** and **co-ordination** between planners, managers and users
* **public involvement**
* a **flexible** approach
* a specific **agency** to deal with coastal zone matters.

Annex II

KEY DATES FOR COASTAL PLANNING

1936	Council for the Protection of Rural England (CPRE) report highlights pressures on the coast.
1938	Coastal Preservation Committee set up (CPRE, National Trust, Open Spaces & Footpaths Preservation Society).
1938–42	Reports from the committee noting urgent need to conserve remaining areas of beauty and scientific interest on the coast.
1943	J A Steers appointed advisor to Government and started coastal surveys of England and Wales.
1946–53	Steers survey of the coast of Scotland.
1947	Town & Country Planning Act introduced giving development control to local authorities but not for inshore waters.
1949	National Parks Act and Access to the Countryside Act. No specific mention of coast. Suggestions for a national authority rejected in favour of individual National Park Authorities.
1962	Annual report of the CPRE highlighted lack of Government action and ineffective coastal planning. Suggestion for a new Government advisory committee rejected.
1963	DOE Circular 56/63 "Coastline Preservation and Development" called for coastal local authorities to survey their coastlines.
1965	Letter from planning ministers to local authorities expressing concern about continuing spread of coastal development.

1966	DOE Circular 7/66 asks for clear statements from each local planning authority on their policy for coastal areas.
1966	National Trust alarmed about loss of unspoilt coast launch Enterprise Neptune to buy coastal land.
1966–67	Nine regional conferences set up by National Parks Commission at the request of the Ministry of Housing and Local Government to assess policy and form basis for planning on the coast.
1968	National Parks Commission became the Countryside Commission and published reports of the coastal conferences.
1970	Publication of " The Planning of the Coastline " by the Countryside Commission which put forward the idea of Heritage Coasts and Maritime Industrial Development Areas. The latter idea was not taken up.
1970	Publication of " The Coastal Heritage " by the Countryside Commission.
1972	DOE Circular 12/72 Planning of the undeveloped coast.
1973	Three pilot stretches of Heritage Coast set up with Heritage Coast officers.
1974	Scottish Development Department publishes Coastal Planning Guidelines for the oil and gas industry identifying Preferred Development Zones and Preferred Conservation Zones.
1977	Countryside Commission for Scotland publishes first report on the resources of the Scottish coast.
1980	Relaunch of Enterprise Neptune.
1981	European Coastal Charter recognises the value and problems of the coast and calls for a European policy to protect this environment and develop the coastal economy.
1988	World Wide Fund for Nature and Marine Conservation Society initiate a project on coastal zone management.
1988	Heritage Coast Forum set up to promote the concept and act as a focus for Heritage Coast interests.
1989	The Organisation for Economic Co-operation & Development (OECD) instigates studies on coastal management. This includes a United Kingdom case study.
1989	Sefton Conference of the Royal Town Planning Institute calls for better integrated coastal planning through its " Agenda for Action ".
1990	43 out of the 44 proposed Heritage Coasts defined by the Countryside Commission.
1990	Publication of the Marine Conservation Society/World Wide Fund for Nature proposals for a United Kingdom Coastal Zone Management Plan.

COASTAL ZONE PROTECTION AND PLANNING

Memorandum by the National Trust

Thank you for your recent letter requesting written evidence from the National Trust for the Environment Committee's proposed inquiry into coastal zone protection and planning. The Trust is grateful for the opportunity to submit its views which are as follows.

INTRODUCTION

1. The Trust very much welcomes the decision of the Environment Committee to conduct this inquiry. We believe the issue of coastal protection and planning to be one of the most important items on the environmental agenda today and one requiring urgent action to address the problems involved. In this context we obviously welcome the current study of coastal matters being conducted by the Government and its intention to publish Planning Policy Guidance on the issue in the near future.

2. The Committee's inquiry comes in the wake of a large number of similar initiatives over the last two years, itself an expression of the concern which is felt over this issue. This concern and the desire for positive change have gathered momentum in 1991 and it is evident from the various reports, statements and calls for action which have resulted that a consensus is developing between the parties concerned as to a way forward.

3. We provide an overview of this process below and make recommendations accordingly. However, we hope that, in considering the evidence before it and in its eventual report, the Committee will bear in mind the pattern of responses of which the Trust's forms part and of the strength of opinion which lies behind them.

BACKGROUND TO THE TRUST'S RESPONSE

Ownership

4. The National Trust for Places of Historic Interest or Natural Beauty was founded in 1895 as an independent charity to hold and manage in perpetuity for the benefit of the nation countryside and historic buildings in England, Wales and Northern Ireland. Under the National Trust Acts, Parliament granted the Trust the right, which it shares only with the National Trust for Scotland, to hold land and buildings inalienably. This means that it can acquire land for permanent preservation with a legal presumption against development or acquisition by other bodies. The Trust is the largest conservation organisation in Europe, supported by over 2·1 million members.

5. The Trust has been involved from its inception in the acquisition and management of the coastline of the United Kingdom and in 1965 launched its Enterprise Neptune Appeal with the aim of acquiring 900 miles of unspoilt coastline and adjoining land. This appeal, which last year celebrated its silver jubilee, has helped augment the Trust's ownership of coastline to 473 miles and of coastal land to 91,000 acres. The Trust protects a further 52 miles and 23,500 acres by covenant. The Trust leases a further 61 miles of foreshore from the Crown Estate Commissioners. The value of this estate in terms of its natural beauty is highlighted by the fact that fully 356 miles of the total coastline protected fall within Heritage Coasts as defined by the Countryside Commission—some 38 per cent of the total Heritage Coast so far defined.

Management

6. Through its acquisition and management of coastline and associated hinterland, the Trust strives to prevent change which it judges to be detrimental to the qualities of those landscapes and wildlife habitats. At the same time, it seeks to understand and allow the process of natural change to continue—a process nowhere more rapid than on the coast. It therefore aims to strike a balance between the legitimate use of coastal land by tenant farmers and those who seek access for recreation on the one hand and the Trust's own efforts to enhance landscape and wildlife qualities on the other.

7. To this end, the Trust has developed overall management guidelines for its coastal land reinforced by a series of management plans for each property. Central to their philosophy is the need to recognise the coastal zone as an integral unit consisting nonetheless of a series of different habitats and landscape formations extending from below the low-water mark to the boundaries of the coastal hinterland, and to manage it accordingly.

8. In this capacity the Trust has undertaken extensive biological and archaeological surveys of its properties which give it a better understanding of its coastal resource; it has a growing body of advisory staff to help put management policies into practice; and it employs in addition 80 coastal wardens who help manage the landscape and facilitate and control recreational use of the coast, mainly on and around its many amenity beaches. A prime, practical example of this philosophy in action can be found at Strangford Lough in County Down where the Trust is a major landowner and where it has a leading role in the Department of the Environment's recent proposals for integrated management of the whole area. Other important examples can be found on the Norfolk marshes, the Suffolk coast at Dunwich, the Dartmouth estuary and the Ceredigion coast in Wales.

Development pressures

9. In order to help protect its coastal properties and as part of its legal responsibilities, the Trust will also resist development on or adjacent to that land which it regards as unsympathetic to the qualities for which the property was acquired and is managed. Examples might include:

—road schemes (such as the A20 road improvements near Dover in which we were nonetheless unsuccessful) which involve disturbance from noise, severance of habitats and pollution;

—offshore fish farming which affects the view and may pollute the marine environment (or, as is the case with scallop-dredging in Strangford Lough, which may damage the sea-bed);

—proposals for wind farms or radar installations on exposed cliff tops;

—sewage treatment works and related marine outlets which harm the landscape or affect amenity considerations;

—and tourist-related development such as applications to construct marinas in estuaries important for birdlife.

10. In the face of such threats, the Trust will nevertheless act in a responsible fashion, considering each case on its merits, including the alternatives (or lack of them) and recognises legitimate national interests where they exist. We will also seek to work closely with the various public and private concerns with an interest in the coastal zone in order to help mitigate such threats. Some recent examples of this attitude are as follows:

—the National Rivers Authority (NRA) and the Ministry of Agriculture (MAFF) in the case of proposals for coastal defence works. At Porlock Beach in Somerset we have commissioned an assessment of such a proposal;

—British Petroleum (BP)'s proposals for oil exploitation off the Dorset coast. The Trust sought and received assurances that the company would observe strict environmental conditions in the development of its plans for an offshore oil rig installation in Poole Bay which will require landfall for pipelines on Trust property;

—the Countryside Commission (CC) and Countryside Council for Wales (CCW) whom we assist in gaining a better appreciation of the purpose of Heritage Coasts through the Heritage Coast Forum;

—local authorities, with which our regional staff work closely in many parts of the country to help develop visitor facilities at coastal properties, maintain access, ensure adequate signposting, provide car parking, etc.

National policy and legislation

11. Development of these public contacts at a policy and legislative level helps to complete the Trust's work in protecting and enhancing the coastal land in its care. In recent years the Trust has made a number of submissions and responses to initiatives involving the coast by Government and public agencies. These include our recommendations to the Department of the Environment (DOE) during the production of its White Paper, "Our Common Inheritance" in April 1990 and a similar contribution to the Countryside Commission's report, "Heritage Coasts: policies and priorities", published in January this year.

12. The Trust, together with the RSPB, also sought changes to the Ports Bill during its passage through Parliament earlier this year. The amendments we sponsored, though unsuccessful, would have placed environmental conservation duties upon trust ports and their successors in case of their eventual privatisation and sought safeguards on the transfer of land by them in line with the Water Act 1989. The Trust's concerns were for the possible impact of commercial development by newly privatised ports upon surrounding land and estuaries, in particular in relation to Poole Harbour in Dorset. Here the Trust owns such properties as Studland Bay, Shell Bay, Brownsea Island and Middlebere Heath which together attract millions of visitors a year for their beaches and unparalleled landscape but also for their great wildlife and historic value. These internationally important assets may still be threatened by unregulated port development and associated harbour works.

13. Two other initiatives in which the Trust was involved this year are important in providing the context of our submission. In April we joined nine other conservation bodies, with the backing of the CC and the NCC (as it was then), to submit a coastal zone initiative to the Secretary of State for the Environment. This statement, which we enclose, calls for an integrated national policy framework for the coastal zone. In the process it asked for an assessment of existing threats to the coast and consequently for a review of the extent to which planning controls, designations, local authorities, public and private agencies and indeed the legal framework are currently adequate to address the problems in question. This was followed by a list, submitted by Wildlife Link, of 22 case studies illustrating these threats and hence the need for a United Kingdom coastal zone management policy.

14. In the same month the Trust, the CC, the DOE, and BP with the involvement of the European Commission, organised a workshop on coastal zone management in the European context at which a number of other organisations from the United Kingdom, EC states with a coastline and Sweden were represented. Consideration was given to viable and sustainable approaches to coastal zone management and the need for concerted action to secure this objective. The delegates called on the EC to play a key role in developing strategies for the coastal zone and to stimulate Member States to develop comprehensive strategies of their own.

THE TRUST'S GENERAL CONCERNS

15. The above has, we hope, provided an overview of the Trust's work and philosophy in relation to the coastal zone and a number of specific illustrations of the threats to the land in our ownership. We have in consequence a number of general concerns:

—there is, in the Trust's opinion, a lack of a co-ordinated national policy for the United Kingdom coastal zone;

—there is, by implication, a plethora of agencies with responsibility for various aspects of the coastal zone. In many cases these responsibilities appear to be overlapping, unclear or not publicly accountable;

—in addition, those controls which they exercise over development of the coastal zone are fragmentary, inconsistent or lacking in perspective;

—there is, finally, a conspicuous failure on the part of public policy to treat the coastal zone as an integrated unit.

16. *As its first recommendation the Trust therefore urges the Government to accept in principle the need to develop a co-ordinated policy for the coastal zone which treats the latter in an strategic and integrated fashion.* Central to the aims of this policy should be the conservation of the remaining unspoilt coastline of the United

Kingdom and a presumption that development proposals for other parts of the coast should be sustainable and should respect the precautionary principle. We develop these ideas with respect to a number of specific issues below.

COASTAL RESPONSIBILITIES

17. It is estimated that as many as 240 Government Departments and public agencies exercise some form of responsibility for the coastal zone of the United Kingdom. This includes the jurisdiction exercised by the Crown Estate Commission by virtue of its ownership of the seabed and the local control over development which is shared by planning, port and harbour authorities. The forms of control exercised by these authorities, their implementation and geographical coverage are almost as varied.

18. A number of ingredients can be added to this recipe for confusion. The first is the perceived lack of consultation between many of the parties involved in decisions affecting the coast and marine environment. There is, in particular, no formal mechanism by which public agencies can regularly inform each other of their actions affecting the coast. The second is the failure by many of these parties to develop an adequate policy for the exercise of their responsibilities as they impinge upon the coastal zone or even to recognise the need for such an approach. Those local authorities which have published coastal strategies as part of their development plans, such as Devon County Council and, more recently, Hampshire County Council, are to be commended but are in a minority.

19. The Trust is also concerned about the degree to which agencies concerned with the development of the coast are publicly accountable and subject to environmental or planning controls. We do not believe that such agencies should be subject to less stringent scrutiny of the developments they promote just because they take place under the sea or far from shore. This is particularly worrying in the case of the Crown Estate Commissioners and private ports, where approval for development or commercial exploitation of their property brings direct financial benefits but may have a damaging impact on the coast or seabed. In a number of areas, however, such as dredging activities, development in association with harbour duties, the exacavation of marine aggregates and licences for fish farming, planning approval is not necessary and the process of consultation is voluntary and informal.

20. *The Trust therefore recommends that the Government conducts a thorough review of the briefs held by the various Departments, public agencies and local authorities with an involvement in the coast.* Such a review should include the degree to which they currently communicate with each other with a view to improving the mechanisms employed. Government Departments and public agencies already have a general duty to protect the environment in the formulation and implementation of their policies. *We believe this should be tightened through legislation in order to make specific reference to the need to protect and enhance the coastal resource. Such a duty should, we strongly urge, be extended to all port and harbour authorities operating around the United Kingdom coast and to the Crown Estate Commission.*

21. Within the review described above, *the Trust would also welcome consideration being given to the need for a lead body or Department to take overall responsibility for matters affecting the coastal zone.* The NRA has been promoted as a candidate for this role by a number of organisations and individuals, not least the House of Commons' Welsh Affairs Select Committee in 1990. We have no firm views on this matter but are very supportive of the work so far undertaken by the NRA for whose establishment the Government is to be congratulated. We believe in addition that the regulatory principle thus far adopted by the NRA has great potential for wider application. *The Trust would consequently view with concern any attempt to undermine this work or potential and we will be studying the Government's proposals for a new pollution agency very carefully in this respect.*

COASTAL PLANNING

22. From the background to our work provided above it will be obvious that the Trust is concerned about those threats which are beyond our ownership and control but which can have a great impact upon its coastal properties. For this reason we would like to see the Government adopt a range of new planning controls, both strategic and detailed.

Strategic concerns

23. *In the Trust's opinion, there should be a presumption against development on or which may affect unspoilt coastline. Proposals for already developed coastline should be sustainable, of a scale and design sympathetic to their surroundings and should respect the precautionary principle.*

24. *The Government, we believe, should also provide a strategic overview for the planning of the coastal zone,* not just through national planning policy guidance, as it has already promised, but also through inclusion of specific policies for the coast within regional planning guidance and the requirement that local authorities with coastlines should publish coastal strategies within development plans.

Offshore development and the Crown Estate

25. There are a number of activities or developments which take place away from the shoreline but over which planning authorities cannot at present exercise control with any confidence. They include:

—**pollution** from sewage and sludge disposal, oil spillages and the detritus associated with shipping, and run-off from chemicals, fertilisers, industrial activities and other domestic effluent. All of these affect the cleanliness of beaches and the quality of bathing waters and may impact upon the wildlife of coasts and estuaries;

—**fishing activities** such as fish farming, the exploitation of the seabed for shellfish and crustaceans, and the use of mono-filament nets which can be fatal to sea mammals and birds which dive for fish;

—**offshore development** such as oil rigs, pipelines, dredging and extraction of marine aggregates. These activities have various, not fully quantified effects on the seabed and shore. They may include damage to landscapes because of the effect which extraction may have on cliff erosion and deposition of material along softer shores and destruction of sunken archaeological remains where navigation channels are widened;

—**recreational pressures** such as noise and physical disturbances from jetskis and powerboats.

26. We deal with the first category in more detail in relation to water quality and bathing beaches below. As a consequence of the remainder, however, *we believe that local authorities should have the power to exercise planning controls below the low-water mark and out to the three-mile limit.* In addition to extending the range of planning controls *per se*, such a move would allow certain proposals near the coast or in estuaries (such as housing development associated with a marina) to be considered as a whole, something which developers as much as local authorities might welcome.

27. For the most part, however, offshore and seabed activities are controlled by the Crown Estate Commission, an anomalous position which we have described above. The Trust would support and would be willing to co-operate with more detailed assessments of the impacts of offshore mineral extraction in the long term. As a more immediate measure, however, *we believe that planning controls should be extended to encompass the land and activities of the Crown Estate. Complementing such a move should be a general presumption against licensing mineral extraction within three miles of the shore.* Any such application should also show due regard for the implications of the proposed operations upon coastal erosion (resulting from interference with the movement of sediments which supply the coastline) and marine archaeology and ecology. The extension of such controls would of course encompass fish farming, dredging for shellfish and the establishment of mussel beds.

Environmental Assessment

28. *The Trust fully supports the concept and application of Environmental Assessment (EA) within the planning process and believes that it has a role to play in helping to protect the coastal zone.* We see it, however, as having a wider and potentially more effective role than solely as a means of predicting the effects of a particular development upon the coastal environment.

29. In general, *we believe that EAs should be conducted of prescribed processes and developments at the earliest stage possible and in a strategic fashion.* Road proposals and the network of which they form part, for example, should in our view be considered in their entirety and not in a piecemeal fashion as is present practice. In the Trust's experience, this usually means that short stretches of road, the improvement of which would impact upon sensitive landscapes or habitats, are left till last in a road scheme by which point financial logic will dictate that they should go ahead.

30. EAs could also usefully be conducted of local authority development plans or the corporate strategies of commercial concerns as the EC is presently proposing. This would allow development proposals, general policies and practices to be considered as a whole and in relationship to each other and permit early scrutiny of potentially damaging proposals so that alternatives may be drawn up. *Above all, we believe EAs should be acted upon.*

31. In this way individual policies and proposals may be better examined for the impact they may have on the coastal zone in particular, allowing the problems arising from different media to be addressed appropriately. This approach could, for example, uncover previously unsuspected sources of pollution affecting coastal waters, such as those arising from industrial activities further inland, or improve understanding of the effects of offshore developments on the coastline. This would allow the necessary remedial action to be taken or help authorities in promoting activities in certain areas only. For port and harbour authorities, it would also highlight the damaging effects of some of their operations or force them to think again about proposed development.

Renewable energy proposals

32. A number of recent proposals for the siting of development associated with renewable energy sources have highlighted the need for greater consideration to be given to such issues and for guidance to be provided for local authorities as a consequence. In this context, the Trust welcomes the Government's intention to issue Planning Policy Guidance on this matter in the near future.

33. We are concerned, however, that such guidance will consist of no more than a description of the various forms of development which are being proposed and of the mechanisms which exist to control them. A number of these proposals, such as those for barrages to harness tidal energy; on- and offshore wave energy schemes, and wind farms are by their nature associated with the coastal zone and have potentially damaging implications for it.

34. The Trust has no desire to restrict the development of such proposals **in their appropriate setting** and is itself investigating the suitability of a number of its properties for similar purposes. As a consequence, however, *we believe that the Government should review the demand for and requirements of alternative energy and in particular their potential impact upon the landscape and ecology of the coastal zone.* The results, we suggest, should be also incorporated in the proposed coastal PPG.

Recreational pressures

35. There are a number of problems associated with recreational demands upon the coast and close to shore which the Trust feels the planning system could more adequately control. *We feel that local authorities can do this by accepting two principles.* The first is the presumption recommended above, *that new development should not normally be permitted on unspoilt coastline or be of a nature and scale that clashes with the local landscape or damages wildlife habitats.* In the Trust's experience, a number of recent proposals for harbour and marina developments, especially within estuaries, have fallen foul of this principle and should be rejected or amended accordingly.

36. The second principle which local authorities could usefully adopt is *that all areas have a limited capacity for bearing the recreational demands put upon them.* Demand in excess of this point may threaten the local environment, endanger personal safety, or simply curtail the enjoyment which visitors expect. Proposals for development which threaten to help exceed this capacity, say in terms of encouraging more boats to a harbour by providing extra mooring facilities, should therefore be recognised as such and restricted. With positive planning, however, it may be possible in some cases to identify and help locate such proposals more appropriately.

37. Certain recreational activities which are noisy and intrusive, such as motorised watersports, may also be more appropriately dealt with in this fashion. It will be necessary, however, for relevant local authorities to make careful assessments of the capacity of the areas of coastline under their control to absorb the demands put on them in order to carry out such policies. In some areas it may be more appropriate to strike the right balance of uses and to restrict numbers of pleasure craft accordingly. *We would additionally support the judicial use of byelaws by local authorities* to impose speed restrictions, engine sizes and so on.

38. *We suggest, finally, that local authorities should give greater consideration to the provision and location of car parks near coastal amenities.* Where provided, these should be sited sensitively, if possible away from the shoreline itself. In sensitive or over-visited areas, alternatives—such as the encouragement of public transport—should always be considered to increasing the capacity of car parks or building new ones which will merely encourage more car-borne visitors.

DESIGNATIONS

39. *Greater consideration will also have to be given to the nature and objectives of designated areas for the coastal zone.* These are necessary to give protection to and even enhance marine and coastal areas which are still relatively unspoilt or which would be extremely sensitive to some of the development and polluting activities described above.

40. Sites of Special Scientific Interest (SSSIs) are of limited value in protecting the coastal zone because their remit does not extend beyond the low-water mark. Heritage Coasts have similar limitations. We are naturally concerned about the slow pace of designation of Marine Nature Reserves and have expressed our disappointment that the Environment White Paper promised nothing more imaginative to speed this process or to put in its place.

41. *The Trust believes that much could be done to improve this situation were the Government to find means of speeding up the designation of MNRs and also to give a higher priority to the designation of candidate Ramsar Sites and Special Protection Areas (SPAs)* of which many remain outstanding. Ultimately, however, consideration may need to be given to new forms of designation which encompass the whole coastal zone. One such proposal, for **Marine Protected Areas**, has been worked up in considerable detail by the Marine Conservation Society among others and has, in the Trust's opinion, considerable potential. *We believe the Government should give serious thought to the practicalities of adopting this proposal.*

CLIMATE CHANGE AND RISING SEA LEVELS

42. A number of local authorities, public agencies and conservation bodies like the Trust are now regularly having to plan ahead and make difficult judgements in the face of potential encroachment by the sea upon land or coastal defences within their jurisdiction. They have to decide, both in practice and within their policy statements, when it may be appropriate to abandon land to the elements and when it may be prudent to attempt to retain it. These decisions sometimes have serious implications for houses or settlements in the areas concerned, often require economic assessments of the options and always necessitate intelligence regarding the physical processes involved.

43. The latter are obviously associated with the normal, age-old actions of the sea and storms in causing losses and gains to our coastline. They include, for example, the erosion of cliffland; changes in the shape and size of popular beaches; and flooding of low-lying agricultural land. The phenomenon of rising sea levels and the increasing unpredictability of the weather, however, has introduced a new and potentially serious dimension to the problem which requires effective, concerted action. It will be difficult for the Trust, in common with other bodies, to take the necessary action without considerable guidance and assistance from Government. Reform in certain, crucial areas is also needed.

44. The Trust acknowledges that the Government has commissioned a number of research studies into this problem, most notably by East Anglia University. Following the conclusions of this work, *we believe the Government should act urgently to set overall guidelines for the implementation of a policy of "managed retreat" by the various agencies involved on the coast.* These general, economic as well as physical, criteria would help lend authority to the practical decisions being carried out at a local level.

45. We feel such guidelines would also provide a better yardstick against which to measure the need for coastal defences and their appropriate size and quality. *In some areas it may be that large-scale, capital-intensive sea walls are no longer appropriate and that consideration should be given to smaller or more temporary defences, or to the use of natural sea defences* such as beach replenishment, restoration of sand dunes, etc. In the Trust's opinion, more thought should be given to these considerations, both at a policy level and according to local circumstances. *Where natural or temporary defences are constructed which nonetheless require regular maintenance, we urge the Government to make grant-aid available on a revenue basis as well as or instead of in the form of a capital payment.*

46. We would like to refer briefly here to the question of co-ordination of responsibility for the protection of the coastline. In the main, coastal defences are the responsibility of the Ministry of Agriculture (MAFF), operating through local authorities, and the NRA. But there are, as a result of a jumble of legislation, both public and private, a number of other bodies who may exercise responsibility for sea defences according to the area in question and rights of ownership. Repairs to sea defences along part of the North Wales coast following the floods of 1990 were carried out by British Rail as an illustration of this problem.

47. The Trust acknowledges that this situation has improved in recent years with the creation of the NRA and the establishment of coastal groups of authorities on a regional basis. However, in keeping with our earlier recommendation for consideration to be given to the need for a single body or Government Department to co-ordinate coastal zone activity, *the Trust also believes the time has come to place responsibility on a national basis for coastal construction of defences and protection from the sea under the auspices of one organisation.*

BATHING BEACHES AND WATER QUALITY

48. The Trust very much welcomes the NRA's intention to regulate to ensure the achievement and maintenance of EC quality standards for coastal waters appropriate to their use. We support the continued devotion of effort and resources to maintaining and improving bathing water quality and as a consequence to the Government's programme of upgrading sewage disposal facilities and its commitment to the ending of the dumping of sludge at sea.

49. These commitments, however, present other problems in terms of the siting of new sewage treatment works and what to do with the sludge which can no longer be dumped. *We believe careful consideration will need to be given by water and planning authorities alike to the location and design of new facilities,* many of which will be near the coast, if they are not to harm the local landscape, archaeological remains or amenity. With population increases, some of these facilities may need eventually to be expanded. Provision should also be made for this factor within the design of the treatment works so that development upon another site is not subsequently required.

50. The Trust is disappointed that nearly all water authorities have opted to incinerate surplus sewage sludge rather than investigating other, more environmentally-friendly alternatives. Emissions of smoke and ash from incineration will simply result in another form of environmental pollution. *We urge the Government to encourage water authorities, with grant assistance if necessary, to look again at the alternatives, including horticultural and agricultural use and power generation.*

MANAGEMENT OF THE COAST AND NGOS

51. The physical management of the coastline to maintain and enhance the landscape and the diversity of wildlife, to repair the damage of countless feet and control inappropriate activities is a complicated and expensive business as the above has implied. It nonetheless reaps great rewards in terms of the preservation of our natural, coastal heritage and also in financial terms such as tourism and local employment opportunities.

52. The Trust and other non-governmental conservation bodies play a vital role in this process. This has been recognised by the Countryside Commission in its previously mentioned report on Heritage Coasts. Indeed the CC has committed itself to assisting the Trust in financial terms in the acquisition and management of the coast. The Trust very much welcomes and values this commitment which will undoubtedly assist us in the work already described.

53. The management burden borne by bodies like the Trust, however, grows heavier each year as leisure activity increases and ever higher standards are demanded of its management. Added to this are new responsibilities acquired as knowledge of the damage we do to the environment improves and remedial action is required. Some of these responsibilities have already been alluded to, such as coastal defence work and monitoring and wardening activities. Others arise directly from legislation such as the litter control duties of the recent Environmental Protection Act. *We hope very much that the Government will continue to pay close attention to these increasing demands upon the resources of NGOs in the allocation of funds to grant-making bodies such as the CC, CCW and English Nature.*

A EUROPEAN COASTAL PROTECTION POLICY

54. We would like finally to turn the Committee's attention to the urgent need for an effective protection policy for the coastline of continental Europe. Though the threats which we have described to the United Kingdom coastline should not be under-estimated, they are of an altogether different dimension on the continent and in particular along the Mediterranean coast where a planning system worthy of the name is missing. Since the Second World War especially, pressures of population, tourism and associated development have combined, it has been estimated by the Council of Europe, to leave barely half of the entire Mediterranean coast unspoilt. In the case of Spain, Italy and France barely a third remains unspoilt.

55. Various national initiatives exist to combat this threat, such as the French " Conservatoire du Littoral " and the recent tightening of planning controls by the Spanish Government. The EC has also introduced a number of schemes to help combat these threats, including the proposed NORSPA scheme for the North Atlantic coastal zone which the Trust welcomes. *What is still urgently required, however, is concerted action by EC Member States to develop and implement a European-wide coastal protection policy,* building upon many of the lessons learnt in the United Kingdom over the last 30 years or so and some of the principles espoused above. *We therefore feel that the United Kingdom Government has a vital role to play in the development of such a policy* and note, in this context, the support which the CC has also recently given to such an initiative in its report, " Caring for the countryside ".

Thank you once again for the opportunity to submit our views to the Committee. We would be willing, with sufficient notice, to give oral evidence to the Committee if this were so required or to elucidate on any of the points we have raised above.

14 *October* 1992

Examination of witnesses

DR BOB EARLL, Head of Conservation, the Marine Conservation Society; MR JULIAN PRIDEAUX, Chief Agent, MR JO BURGON, Adviser on Coast and the Countryside and DR JOHN HARVEY, Chief Adviser on Nature Conservation, The National Trust, examined.

Chairman

265. Good morning, Ladies and Gentlemen, may I open this inquiry session into coastal zone protection and planning? We welcome here representatives of the Marine Conservation Society and the National Trust. I thank you for the written evidence you have submitted which is very helpful and upon which we will be addressing you some questions. For the purposes of the record, would you be kind enough to introduce yourself. First of all Mr Prideaux, would you like to introduce the representatives from the National Trust?

(Mr Prideaux) Sir Hugh, thank you very much. On behalf of the National Trust may I thank you very much for inviting us to expand on some of the things

we have said in our memorandum. I have with me Dr John Harvey, who is our Chief Adviser on Nature Conservation and Mr Jo Burgon, who is our Adviser on Coast and the Countryside. I am the Trust's Chief Agent.

266. And Dr Earll?

(Dr Earll) Yes, good morning, I am the Head of Conservation for the Marine Conservation Society.

267. Thank you very much. I should like to start the questioning by addressing the first question to you, Dr Earll. Your society has argued the need for integrated coastal zone management. You suggest the formation of a coastal zone unit as a way of co-ordinating coastal zone management. Can you

Chairman *Contd]*

give us some idea of what is in your mind about how this unit would work? Would it be part of the Department of the Environment, for example, or free standing or part of some other organisation? What sort of powers do you think this unit should have and how would it relate in its work to the responsibilities of the National Rivers Authority and also to the Ministry of Agriculture, Fisheries and Food on such issues as flood defence, for example?

(Dr Earll) Thank you, Mr Chairman. The Marine Conservation Society is a national non-governmental environmental organisation working towards the protection of the seas for wildlife and future generations. We are committed to the rational and wise use of the marine environment and our work has developed over the last four years in relation to the issue of coastal zone management in this country. This has led us to explore the possibility of the creation of a unit at a national level to provide a national perspective in relation to this topic. We have in our evidence to you submitted some ideas on the role of such a unit and where that unit might be based and how that unit might be organised. We feel that it would be up to government to decide how best to organise it, but our strong feeling is that it should be based within the Department of the Environment or strongly related to it. There are benefits in relation to a number of arrangements in terms of accessibility and accountability and this would be in a sense for the Government to judge, but we feel that the Department of the Environment should play a very strong role in relation to the unit itself. We feel that that unit could achieve a number of things which are lacking at present. It would certainly provide a strong focus for issues within the coastal zone and we see it as facilitating effective coastal zone management. We see it providing a national perspective. We see it providing long-term continuity, which is something that is very greatly lacking in terms of the *ad hoc* responses to a wide range of new and existing issues. We see it providing an integrating role in relation to the management of the marine environment. We see it also undertaking a number of regulatory and monitoring functions and, in certain sti tuations, providing funding for particular works. In relation to the powers that the unit might have, it would impinge significantly on two organisations in particular: MAFF in relation to coastal works and also the relationship with the NRA. We feel that it would be more consistent if a DOE-based organisation took over the functions of MAFF in relation to coastal protection. We also see some of the existing powers of the Crown Estate Commission as being undertaken by this new unit.

268. The National Trust, Mr Prideaux, also seems to be in favour of a lead body or department. How does your idea fit in with what we have just heard from Dr Earll?

(Dr Harvey) We accept the idea of some kind of national co-ordinating, policy making, overseeing unit. We have no strong views about what system that should be set within. We see certain desiderata. We think it needs a certain amount of support from Government, it needs the weight of the Government behind it, but perhaps it ought not to be tied too closely to Government. It may be an evolutionary thing, that one starts perhaps with the MSC option 2, which I believe is a unit operating under the auspices of the Department of the Environment, and see how it goes from there into the future.

269. We have been receiving a great deal of evidence from outside Government that the present set up is unsatisfactory. It is too fragmented, overlapping responsibilities with a whole variety of ownerships and bodies with responsibilities and there seems to be a very strong feeling that there ought to be one body responsible for the whole of the coastline. The authorities we have seen so far seem to be quite content with the status quo and they think that most of the overlapping anomalies can be cured by way of negotiated agreements between them so that they themselves provide the demarcation of their own areas and their own interface. So we shall have to reach a conclusion as to which is preferable. Where the Committee is reluctant to proceed is to go towards creating another bureaucratic tier, and imposing that somewhere within the existing structure, because that might only complicate matters. I wonder whether you have addressed your minds to that particular possible objection?

(Dr Harvey) We would support the objection. I think there will be a problem there and I do not think we see any unit as having any great function on the ground. Many of the functions will continue to be fulfilled by the local authorities, by whatever regional systems are in existence, but we would argue strongly that there is need for some overview to ensure that agreement between adjoining regions and adjoining units operates in practice, and also that over the coast as a whole there is some kind of view in that inevitably one is dealing with a system where things can pass over one boundary into another.

270. Do you wish to comment on that at all, Dr Earll?

(Dr Earll) Yes, I would endorse what Dr Harvey has said. We do not see the unit as being another tier. We see it much more as being an integrating function within what happens at present, with a much more identifiable basis. This is one of the problems that people have outside in terms of their perception. They find it very difficult to understand who does what within a Government context, even though Government and the civil servants seem to be able to cobble together arrangements between them. This came out very strongly within the research that we did for the report that we have circulated to the Committee. There is a great deal of uncertainty about who is doing what out there. It quite clearly needs a home and a focus within Government to pick up on certain issues, given the increasing pressure on the coast and given things like climate change and that sort of thing.

Mr Battle

271. You have stated that there is need for an integrated function, an overview and co-ordination which is fine, as it were, as lines on paper when the original plan is drawn up, but how would that work in practice? If we were to focus on the Marine Conservation Society's stress on the need for regional

Mr Battle *Contd]*

coastal zone management plans, what areas and topics do you think those plans should cover? Who should prepare them? Who should carry them out? What would be the role of the coastal unit in that process?

(Dr Earll) The coastal unit we feel would give guidance in relation to that planning process. The planning process would be carried out by the planning authorities that exist at present, so we see no new operation on the ground at all. What those plans would be and how they would be managed rather depend on the area under consideration, but they would be regional plans of a large scale based on a variety of factors. We feel that the environmental factor is one of the main points that has not been taken into account at present. One simply has to include economic factors, administrative factors and a variety of cultural factors as well. So the arrangements on the ground would depend to some extent on the regions and the counties that were covered by the plans. The planning process itself would both be a physical planning process, but also it would cover major uses of the environment not necessarily covered by physical planning. Those uses might include recreational pressure and the uses might change in flavour in different areas and this is where guidance for the unit perhaps from national government would be extremely helpful. We have discovered in the work we have undertaken a very large number of problems in the coastal zone. However you count them and you can work from larger or smaller groups. To give an example of the central problem areas: in our *Future for the Coast* document we identified 111 concerns from the different sectoral groups. We have presented evidence to the Department of the Environment on 22 specific issues. We perceive nine major sectoral uses of the coastal zone. All these groups have different problems and there are differences of use. We see those uses being covered by guidance notes from this unit and some of those being included in a local authority physical plan.

Mr Jones

272. Can I press you on that because plainly the coast does not present a totally consistent set of problems? They are different in different areas.

(Dr Earll) Absolutely.

273. We have been dealing with flood defence and that does not exist as a problem in some parts of the country. That means that any guidance from this coastal unit either has to be on a topic by topic basis or its has to be fairly broad brush, because it would not have universal applicability. So how do you see these, what you call, regional coastal zone management plans reconciling conflicting priorities on the part of neighbouring authorities? I will perhaps come back to that in the estuarial context later on, but I should like to know how you see this emerging when you may have people saying different things in neighbouring local authorities?

(Dr Earll) It is a very important point. You probably realise that conservation organisations quite often are dragged into what is effectively an

adversarial process. We are particularly enthusiastic about the idea of coastal zone planning because it sets a framework for constructive debate about how the area of the coastal zone, both land and sea, without this rather odd edge in the centre, is to be used. We feel that is the constructive way forward for many people, but it was not simply a view expressed to us in relation to our work that came from just the conservation side. Many groups felt that a format and framework for planning, as happens on land in the present time, would be a far more effective way of resolving conflicts both as they arise, but also anticipating and resolving long standing problems. When it comes down to it, the regional plans will have to be based on a combination of factors. Some judgment will clearly have to be made, but that is rather preferable to a number of smaller units where there are minor problems which are just passed along from one to another, certainly in relation to coastal protection. A problem is solved in one unit, but that creates a problem for another unit and so on and so forth. We would see the regional plans taking a rather broader view in relation to the overall view of that area.

274. I can understand that. That is almost self-evident. I think that was rather a woolly answer, if I may say so, to my question about who would make the decision where there were irreconcilable priorities. Do you envisage the Department of the Environment refereeing, for example, or had you some other process in mind?

(Dr Earll) No, we would see the unit looking at these plans once they were constructed by the people on the ground in the regional context. How one merges local authority and other interests is a matter for debate.

275. I appreciate that it is a matter for debate. That is why we are all here. What I want to know is your views?

(Dr Earll) Our views are that the local authorities on the ground would draw up these plans—bearing in mind the local involvement interests and the public interest involvement on the ground—and they would then come back to the unit for review.

276. So the unit would have veto power?

(Dr Earll) I suspect that the plans would be far more complex than that anyway, and that there will be a number of issues to be resolved and discussed in that context.

277. It is not about discussion in that sense. It is about how to get a firm framework. If these plans are to have any effect at all they must have very firm decisions made about particular topics. It can well be, as we have already experienced, that one part of a community takes a totally different view from another. Someone has to decide which prevails. I am just putting you a perfectly straightforward question: who makes that decision?

(Dr Earll) At the end of the day if the situation cannot be resolved, then it comes to the Secretary of State, but via the unit, and we would hope that the guidance from the unit would help to clarify that and we see it, on a geographic level, helping to do that.

Chairman

278. We are getting into a situation similar to our planning procedures, in having local inquiries, local discussions and then, if anybody is unhappy with the local decision, it goes to the Secretary of State on appeal who has his inspectorate to advise him and make recommendations. It is something analogous to that that you appear to have in mind?
(Dr Earll) Yes.

Mr Cummings

279. Last week representatives from the NRA and MAFF stressed the role of the coastal groups in dealing with coastal defence. We were informed that there are some 17 coastal groups in Great Britain, consisting mostly of engineers, focusing their attention upon coast protection work. In your opinion are these groups the appropriate agencies for dealing with integrated coastal zone management?
(Dr Earll) The short answer will be no because, as we outlined in our report, there are at least nine major sectoral interests in the coastal zone. These groups will only consider the interests of one of those sectors and so in a sense they are rather short of expertise. We have had quite a lot of experience, and it is certainly clear from our review of work done overseas, that effective fora for debate of these issues is essential at a local level. We have had some experience of dealing with voluntary fora on the ground in Cardigan Bay and in Dorset with the Marine Environmental Information Forum and have found that a wide cross-section of coastal users have come together to discuss issues at the coast. This is why we are quite keen on this non-adversarial approach. It helps to resolve a lot of issues before they ever become problems.

280. Do you concur with that view, Mr Prideaux?
(Mr Prideaux) I think, generally speaking, yes.

281. For the same reasons?
(Mr Prideaux) Yes, in broad outline certainly.

Mr Steen

282. I want to talk about "controlled retreats". You know the policy of moving coastal defences further inland so that it will expose previously protected land. The question I want to ask is how do you think that should work? Who will decide what particular coast is to be protected and managed? How would you suggest the balancing of the loss of habitats and archaeological sites against the presumed benefits of such an approach? The RSPB in their evidence has suggested a controlled retreat response. I am just wondering what your feeling is about that and which land should be allowed to be reclaimed by the sea? Can you give us some views as to where you think that might be?
(Dr Harvey) I think we touch here again on another problem which has been identified—who makes the decisions at the end of the day? I would support Dr Earll in saying that it may be possible to develop criteria. For example, the old Nature Conservancy Council developed criteria for the identification of Sites of Special Scientific Interest which could be applied across the country and which would allow this kind of problem to be addressed. One would have to decide what desiderata there are in making these decisions and the kinds of factors which one might take into account. For example what would be the fate of the area if there were no protection? If one is talking about protection, then settlement patterns, value of agricultural land, a whole range of factors need to be taken into account. It would be possible I suppose—although cost benefit analyses do have certain weaknesses—to include these kinds of things in a cost benefit analysis. The concern from the nature conservation side is always that it is difficult to cost in what we feel is an appropriate fashion the nature conservation/wildlife aspects of that. It could be possible to develop a mechanism which would then allow a fairly objective decision to be made as to which areas should be allowed to retreat and which should not be allowed to retreat.

283. Can you give me some idea of which areas are candidates?
(Dr Harvey) In national geographical terms East Anglia is an obvious candidate. One is dealing with low-lying coasts which already depend on major sea defence works and which would be at risk if faced with sea level rise. The problems on the west coasts will be much less where you are dealing with hard rocks and high cliffs. One can define it in geographical terms of that kind and then, automatically one goes into salt marshes, sand dunes, reclaimed agricultural land on silt, as the kinds of habitats which would be affected.

284. Are we talking about hundreds of miles at risk?
(Dr Harvey) I have not done the calculation, but approaching hundreds of miles I suppose, yes. But to go on with the question about loss of archaeological features and balances which come in there, the point I would stress is that we are dealing with a dynamic system. It always has been a dynamic system. Areas have always been lost. In the case of the National Trust we own Dunwich in Suffolk and the town of Dunwich is now supposedly three miles out to sea and that kind of erosion is still going on. In the past it has been balanced by gains and we would hope that it would be balanced by gains in the future. A controlled retreat is better than a catastrophic failure. It allows one to manage the system in some way. If one is dealing with archaeological features, for example, one could build in some kind of protection for them so that one gets archaeological features under water, as one does in the Isles of Scilly, or as one does with Roman or Greek towns in the Mediterranean. That would be a much better situation than simply seeing it all disappear.

Mr Field

285. You own Whitecliffe headland in Whitecliffe Bay on the Isle of Wight. It is one of the National Trust properties and one of the adjoining owners is the operator of a very large caravan and holiday camp site. At his own expense he wished to erect sea defences to stop erosion at the bottom of the cliff. The Nature Conservancy Council objected to that and

Mr Field *Contd]*

prevented him from building those defences because the sea was exposing the best fossils of their type in Northern Europe. If one takes it to its absolutely ludicrous extreme one could say that one should allow the whole of the Isle of Wight to be washed away in order to expose the best fossils in Northern Europe. He received no compensation for that objection and he was prevented from protecting his property. How would you reconcile that sort of dilemma?

(Dr Harvey) I would not like to be held responsible for the activities of the Nature Conservancy Council. My concern would be that any work done on any small patch of land has an impact on other land immediately adjacent obviously, but also further down the coast. From the Trust's point of view I would be very concerned about what those works did to our particular section of cliff. One needs therefore an overall view of this coastal management zone. To say that this section of coast is eroding, it is something in the long term which we may be able to do very little about. Perhaps the caravan site is in the wrong place. We need to address that problem rather than the protection problem.

Mr Steen

286. Can I ask you about this controlled retreat? Do you imagine yet another layer of bureaucracy arising now where you would have all these inquiries all over Britain and the Secretary of State would ultimately make the final decision as to which bit of land would be abandoned and which bit of land would not; then the compensation elements would come in and then the archaeological and the wildlife people. It could be a most enormous bureaucracy.

(Dr Harvey) I would hope not. I would hope that it would all fit into what we have described as this unit, which we have talked about already.

287. Then there would be appeals. It would be a huge thing, would it not?

(Dr Harvey) The Chairman has already suggested that we perhaps need on the coast the same kind of planning structure as we do inland, which allows for this kind of appeal.

288. Then you would have to deal with the local plans, the structure plans?

(Dr Earll) When we talk about the responsibilities of the local authorities, we mean those responsibilities extending out seawards. Yes, that process would work. This process is happening anyway and what we have is a series of very unsatisfactory solutions to it at present—and there are numerous examples of this thing—where one council with a very small perspective of the coast makes certain decisions, then another council adjacent to it makes another set of decisions with a separate set of contractors and possibly a third council may be involved. This happened on the North Wales coast in Towyn. It is not an easy west/east split because there are these problems all over. In the example of Towyn, I believe there were three or four councils, all with adjacent bits of coast, all of whom were investing in coastal protection, all

of whom had separate contractors within, in effect, what was one dynamic system. It is not as if there are no problems in the system at present. That is clearly not the case. What we are saying is that by having regional plans one would be able to take a proactive view of all of this and people would know quite clearly where they were or were not to put certain things, or it would be a good idea to put things like caravan parks, agriculture or a whole variety of different things on the landward side, and also on the marine side as well.

289. I want to ask you about the timescale of this whole business of controlled retreat. You can imagine the defences suddenly stopping and the councils no longer putting money in that line of defence and they go back five miles in East Anglia. What is your time scale on this? When do you see the decisions have to be made and are we wasting a lot of money at the moment by going on putting sea defences in areas where we should have abandoned?

(Dr Earll) I am no particular expert on sea defence, I have to admit that, but it seems to me that the situation at present is unsatisfactory. There are large investments going on and the decisions are having to be made on a routine basis and this will become an increasing problem if we are to believe the impact of climate change, not only in relation to sea level rise, but perhaps the greater incidence of large storms. This will be an increasing problem. If we are not continuously to be on the back foot in relation to this, reacting all the time, creating more work in that context then surely we have to take a proactive view where many of these problems are resolved constructively before rather than after the event.

Chairman

290. The more I listen to your evidence, the more it seems to me that fundamentally what you are criticising is the way in which our existing planning system fails to address the particular problems of coastal defence and coastal management. You talk about the fragmentation of the responsibilities of different local authorities on a particular stretch of coastline. I wonder whether you are, apart from this inquiry, taking any steps to submit evidence to the Government which, as you know, is in the course of looking at the structure and responsibilities of local government. It is considering the setting up of unitary authorities, for example, which may or may not help address the problem that is concerning you, but to which no doubt some input should be made in the Government's consideration of these matters. Have you put in a paper to the Government suggesting the shape of unitary authorities and what powers they should have and how they should co-ordinate in dealing with these particular matters?

(Dr Earll) I have to admit that I am not expert in that particular field. All I would say is that we perceive the problem as being one of all of the users of the marine environment of which coastal protection is just one of nine—even if you draw the broadest classification—major uses of the coastal zone. Clearly this is a massive interface, and we have expertise within the organisation in relation to the

Chairman *Contd]*

local authorities and the planning sector. But again each of the fisheries, navigation and a variety of other sectors all have the same common problems. So although we started with a perspective that was very much more oriented towards nature conservation, we suddenly realised there was a common core of concern which related to both physical planning as well as use management.

291. May I leave this thought with you that whatever may be recommended regarding the setting up of a national coastal unit—I use that as a neutral term—all the evidence you have given us this morning seems to accept that the local authorities will still have a very important public relations/managerial role to play within the structure and therefore, as they will be the people at the coal face, you might address your considerations as to what recommendations you should be making elsewhere as to how the local authorities should be structured with their responsibilities to meet this particular problem. So rather than receiving your evidence, could you receive my suggestion and take it away with you?
(Mr Prideaux) We will indeed.
(Dr Earll) Yes.

Mr Jones

292. Can I ask Mr Prideaux about the suggestion that the National Trust has put forward that there should be a presumption against development in coastal areas? We have quite a complicated planning edifice at the moment and wherever there are heritage coasts and AONBs and SSSIs and so on, in effect there is a presumption against development. I wondered why you thought there should be a change in the basis of planning law, because that is what it amounts to?
(Mr Burgon) The general planning system, as far as I understand it, is not towards the presumption against development. The burden of proof is not on the person wanting to develop. AONBs and heritage coasts which fall within AONBs do not prevent development occurring within those areas. Heritage coasts were not originally set up as a planning tool. The vast majority of them fall within AONBs and that is the designation within the planning framework. Our view is that with the present designations—although particular care and attention may be paid within those defined areas—development will still occur within them which threatens the values for which those designations were created in the first place.

293. But when they draw up their plans, whether these are structure plans, topic plans or whatever, the vast majority of local authorities contain within them policies on development within AONBs etc. While these do not specifically turn round the presumption what they say is that new building will be allowed only for agricultural purposes or various specific exemptions which means that for the average applicant there is, in effect, a presumption against development. I fully understand that you may be concerned, as indeed I am, about the agricultural

planning system, but as far as most other people are concerned, is it not already the case that in effect you have your way?
(Mr Burgon) Of course there are other authorities within those areas that fall without the planning system, like harbour authorities which are not beholden to the planning authority.

294. I think we shall cover that in a moment.
(Mr Burgon) But I am just making the point that there are a whole lot of other planning mechanisms—and it comes back to the points we made earlier about the need for a better integration of all planning within the coastal zone.

295. Have you any other examples besides the harbour authorities?
(Mr Burgon) The MOD, the Crown Estate Commission, who are exempt from the planning system. The local authority planning system will cover only certain physical planning, land-based planning systems and of course their boundaries do not extend below the low water mark.

296. Reversing the presumption would not necessarily affect the Ministry of Defence, the Crown or agriculture anyway?
(Mr Burgon) No, because they fall without the system.

297. But that is not the problem. It is the fact that they are outside the system?
(Mr Burgon) Exactly.

Mr Field

298. Is the concept of marine consultation areas a satisfactory mechanism for protecting marine resources? And what is your reaction to the progress made with designating these areas so far? Is it a fact that the Kimmeridge Ledge is still the only marine conservation area in the whole of the south of the United Kingdom? Are you happy with the progress that English Nature are making in designating wetlands and salt marshes under the Ramsar Convention in recommendations to the Government?
(Dr Earll) Although we have submitted evidence to you on coastal zone management in the round and the central perspective of coastal zone management, we also have a hat to wear from the nature conservation side. Our organisation has been involved for a number of years, in fact over 10 years now, with the protection of marine protected areas at the coast, both voluntary and statutory. In relation to statutory reserve areas, the Government have, through its agency the Nature Conservancy Council, managed to come up with two marine nature reserves in 10 years. The marine consultation area proposals are new and only tried in Scotland. The experience in Scotland of marine consultation areas was rather unsatisfactory and recommendations from Scottish members of NCC suggested that they should have far more teeth. To answer one question, we are clearly very concerned about the lack of progress in relation to effective marine habitat conservation in our coastal waters. We do not feel that progress is at all

Mr Field *Contd]*

adequate. It would be true to say that Kimmeridge is one of the voluntary conservation sites which has attracted considerable attention and has given us a clear insight into the benefits of multiple use involvement in terms of getting an effective working arrangement together to confer some protection of an important site. There are a number of voluntary sites on the south coast in Sussex, also Wembury in Devon and in Cornwall there are a number of sites in the Helford River and in the Fal. We have been instrumental in promoting voluntary conservation and whether the local authority or the county trust takes a lead there are benefits, as we said, in terms of involvement that can come from voluntary designations. This is one of the weaknesses we feel with the MCA approach. It is neither one thing nor another, as we understand it; we are waiting to hear what the Government have to say on it. It is neither voluntary in the sense that it includes public involvement which is one of the major strengths of the voluntary approach to marine conservation areas, nor does it have sufficient teeth, judging from the Scottish experience, to protect those sites which are of the very highest nature conservation importance. It was for that reason and for our overall dissatisfaction that we submitted evidence to you in relation to a unified approach to marine protected areas which was supported by a number of other conservation organisations and the colleagues who will be giving evidence later. We feel in general and in relation to the Ramsar Convention in particular that while terrestrial measures for protecting birds, for example while they are on land, are in place and one can argue about the effectiveness of those, as soon as one gets on to the water there is no effective way of protecting their interests. That is quite clearly ludicrous. This is why we have pressed very hard for a more unified approach to marine protected areas which is simplified and which looks after the interests of a number of other groups of people, including the fisheries' interests, recreational interests and also the archaeological interests who have an interest in designating areas of seabed for a variety of protective functions. We think a unified approach to marine protected areas is the way forward and ought to be explored, rather than just adding on in an *ad hoc* and bolt on fashion another designation which simply does not work. The MCA is neither fish nor fowl, in effect.

299. Mr Prideaux, I may have missed it in the National Trust evidence, but I do not believe there is any comment on the effect that people have on the conservation area. Perhaps I may give you an example, the price of a mooring overnight in the Newtown river is very much cheaper than anything available anywhere else on the Solent, with the consequence that this very lovely area gets a tremendous amount of use at the height of the summer season by yachtsmen—of which I am one— and they have a very serious effect on the ecology of the area. The National Trust does not seem to have drawn any conclusions as to its pricing policy and the effect this can have on maintaining conservation.

(Mr Prideaux) The short answer to your question, Mr Field, must be that we are at fault in not

addressing this because every piece of land, every piece of water has probably got a certain carrying capacity for a particular activity. Those need to be identified and those need to be incorporated in the management plan for that particular property. That needs to be extended wider to talk to neighbours so that within a particular coastal zone there is an agreed policy that is looking at the wider issues, not just the National Trust looking at the piece of coast that it owns. Even as a yachtsman we must look at your mooring.

Mr Pendry

300. Mr Burgon, may I ask you to develop what you were about to develop earlier when Mr Jones asked you about how the whole question of coastal zone management plans relate to the statutory development plan system generally? It seems to me, and the Chairman has made the point, that this seems to be a less than clear area from both your evidence.

(Mr Burgon) Yes, indeed. At present coast zone plans would not fall within the present statutory system, because the system does not allow it. There would have to be some changes to be made to bring coastal zone plans into it. That may occur through changes to local government and unitary development plans and all the rest of it. It is whether then coastal zone plans in themselves become unitary development plans as such, taking in the range of issues which we have identified and discussed with you. But in order to do that we also have to look at the way that local authority boundaries address the whole of the coastal zone. At present the boundaries will stop at the low water mark. There are suggestions that that area should be extended out to the three-mile limit. Then you do start looking at the full impacts and planning issues related to the total zone, both sea use and land use.

301. As you have touched on the three-mile limit, that brings us to my next question. There seem to be some differing views among those who have submitted evidence and those we have heard about, this whole link between seaward and landward elements of coastal development proposals and the need for local planning authorities to have some jurisdiction to these levels. In fact it is demonstrated by you both being here today. As I understand it, the National Trust recommends up to three miles and the Marine Conservation Society recommends up to 12 miles. I do not want you to have a punch up, but perhaps you can tell me why you think this extension is necessary and, secondly, why do you differ in the lengths that I have just described? How can the extension be achieved?

(Mr Burgon) I will let the Marine Conservation Society speak for themselves, but I think our view is that within the three-mile limit that is where you are getting the most intense use, where you are getting the problems of marine aggregate extraction and where you are getting most recreational pressures and where most of those activities are having to come ashore immediately adjacent to the sea. One of the areas that needs to be addressed is the question of marine aggregates and licences and how one should develop a mineral planning system that is the

Mr Pendry Contd]

equivalent of and similar to the mineral planning system we have on land. The responsibility there lies with the local authority. Responsibility out at sea lies with the Crown Estate Commission and MAFF. Perhaps we shall come on later to the role of the Crown Estate Commission, but one of the areas in terms of sea use planning for the local authority to address is mineral planning. The Hampshire County Council coastal strategy document identified this as an area of concern from their point of view and others are doing the same. When they have started to look beyond their own jurisdiction at present, local authorities are beginning to identify key issues that are affecting their terrestrial interests as well.

Mr Field

302. If you accept your three miles, you would not actually cover all the Solent. You would have a strip down the middle of it in places which would not be covered by either authority.
(Mr Burgon) Either by the Isle of Wight or Hampshire County Council.

303. There would be a no-man's-land.
(Mr Burgon) I think you then have to look at each coastal area. If you go out to the twelve-mile limit you then would——

304. If you actually threw the limit, whether it is 12 or three, around every offshore island, and across every estuary mouth, you would come to a more sensible solution. In many estuaries you would have a wide swathe down the middle where it would be nobody's responsibility.
(Mr Burgon) I think that is where you would have to look at the configuration of the coastline and perhaps you would have to adjust your limits with regard to that configuration.

305. Would you accept that at the public inquiry into the draft Isle of Wight County Council replacement plan the planning inspector ruled that there is no requirement for county structure plans to interlock when there is water in the way? Would you accept that from me?
(Mr Burgon) No.

306. You do not? But that was his ruling and it is also confirmed by the Department of the Environment and it strikes me as pretty ludicrous that all county structure plans in the United Kingdom have to interlock until there is water in the way, so there is no requirement on the part of Hampshire to have any cognisance as to how people from the Isle of Wight get on and off.

Mr Pendry

307. Perhaps I could get back to my more macro question. Perhaps Dr Earll could express his organisation's views.
(Dr Earll) It is a rather good example of what we are talking about. The key point here which we made on the planning Bill last year which was rejected, was that coastal zone management involves both land and sea in both physical planning and in the sense of considering the uses. That is the essential concept

behind it. Where one draws boundaries both inland and out to sea varies, and our researchers found that there is a variety of national models. The reason that we advocate a 12-mile limit is for three reasons. It is the limit of our jurisdictional extent in relation to territorial seas. There are a variety of what in effect are *ad hoc* boundaries. Again our report points to a number of these but every legal instrument has a variety of these boundaries in them which again is one of the things that continuity might help to resolve in the long term. And, lastly, there seems little point if one is approaching this topic and trying to resolve the way forward and setting a sub-zone at three miles and having many other things going on beyond that up to 12 miles which fall outwith this area. We feel quite strongly that this would be a decisive step towards what is actually required within our territorial seas. That is why we take this rather stronger view than the three-mile limit.

308. Can I just ask you about the Sefton Borough Council zoning? What are they suggesting?
(Dr Earll) I cannot comment.

309. It seems to me that the National Trust has given this special mention in the document you have sent to us, and I was wondering what they mean.
(Mr Burgon) About Sefton?

310. You talk about the plan of the Sefton Borough Council " which includes policy objectives for a coastal planning zone ".
(Dr Earll) Zoning is a mechanism that is widely practised throughout the world in terms of deciding where particular things should be put. Sefton, like Hampshire, have tried to advance this particular area of thinking and are to be commended because of that. They have thought certain of the applications through in the context of zoning.

311. I just really wanted to know what was so special about Sefton.
(Dr Earll) The special thing about them is that they are showing some initiative.

312. When we go there we shall see for ourselves.
(Mr Burgon) There is the point that there are a number of examples, such as the Northumberland coast. I have just seen a recent one that the Suffolk County Council has drawn up for the heritage coast there where they are looking out to sea. As I said earlier, Hampshire has done the same. They are looking at a wider range of issues than they may have done five or 10 years ago.

Mr Steen

313. The Royal Society for Nature Conservation has stated that the greatest threat to the coastal zone from developments requiring planning consent is from marinas. The RSPB have produced a report on the marina industry and they came to the conclusion that recreational demand would grow. That is quite a fair comment, but how would you suggest that the regional coastal zone plans address this demand? The other question at the same time is, is there any scientific evidence that suggests that marinas destroy wildlife?

Mr Steen Contd]

(Dr Earll) On the specific point about marinas being a major threat, I mentioned earlier, but it is worthwhile stressing that every user group you go to has different perceptions of what the major threats and concerns actually are. We have just produced a guide and marinas are just one of 60 different major impacts on the coast, all of which have environmental consequences to a greater or lesser degree. There are 60 major different impacts. I will leave it to colleagues from RSPB and WWF to respond directly to the marina question. That would be more appropriate. But it is important to realise, as I said earlier, that if one is to address the issues of these problems of development, which are increasing apace from many sectors of society, in a constructive and proactive way the framework provided by the central unit and the regional plans will help to do this and resolve these conflicts, we hope in a far less adversarial way than happens at present when one gets one-off ad hoc developments which cause an awful lot of work for all of us.

Mr Cummings

314. How would you tackle the awful problem we have off the Durham coast which has been despoiled now for well over 60 years because of coal tipping and slurry tipping? That is not just having an effect upon the sea shore, but also for several miles out to sea where a lot of marine life has been destroyed.

(Dr Earll) Our society has just completed a survey of the sea and coast and we can confirm very directly the evidence of gross seabed pollution a number of miles offshore from the Seaham coast. Our colleagues from the National Trust have a very ambitious plan to look at the renovation of the Seaham coast.

(Mr Prideaux) Our Enterprise Neptune campaign, which I am sure you are all familiar with, is still very active, particularly in the North East. It is interesting already how much the sea itself has done to clean up the coast there, simply by having a policy of non-intervention in this case. We shall continue with Enterprise Neptune to acquire these bits of the coast that may have been degraded in the past.

315. But with Enterprise Neptune, the area of land you have in my constituency is on the cliff tops. I am referring now to the despoliation of the actual beach, which has taken place over many years. Have you carried out an in-depth survey as to the problems which have been created for marine life and also for the general environment because of that despoliation?

(Mr Burgon) We have not.
(Dr Earll) But we have.

316. You are looking at the cliff top and you are looking at the cliff bottoms?

(Dr Earll) Yes.
(Mr Burgon) Yes.
(Mr Burgon) Is the problem that waste is still being poured into the sea from Easington colliery?

317. Yes, the waste is still being poured into the sea.

(Mr Burgon) And that will stop by 1995 or 1994.

318. I hope not because there are 2,000 jobs at risk if it does.

(Mr Burgon) Or they are looking at alternatives of how to dispose of the waste from it?

319. They are looking at alternatives, yes. I am fascinated to hear that you have just carried out a survey.

(Dr Earll) Yes, I have visited the site and we have had a survey running there last year and the effects on the sea bed are some of the worst, most obvious direct smothering contamination effects you will see anywhere in the country on an area of sea bed. It is not surprising that the fishermen complain bitterly. But this is precisely the problem. Pollution was one of the major concerns of the wide range of users. Precisely what we are trying to do is to have these regional plans so that situations like that do not arise by default, in effect, so you get major despoliation of what was I believe a very popular bathing beach at one time.

320. Indeed, it was.

(Dr Earll) It certainly is not now and it is certainly not a place that many of us would like to live for very long.

Mr Cummings: The Committee had a view of the beaches some three years ago.

Chairman: The Committee has looked at this problem in the past. I remember being amused by some evidence we received that successive junior ministers from the Department of the Environment fairly shortly after their appointment as part of their training were taken up there—and this has happened under successive administrations, not just recently— and each Minister has stood on top of this coal and said " Disgraceful. This has got to be put a stop to ", and that was the last anyone heard of this matter. There are complications in the coal industry, the cost of doing everything and that kind of thing which has been the main stumbling block. But in this day and age when there is now a much greater awareness of the value of the environment and a cost having to be put on that, corrective steps ought to be considered much more seriously than they have been in the past, even though this may require the coal industry to become more active in finding alternative means of disposing of what they are still producing.

Mr Cummings

321. One point I would make in relation to the Durham beaches is that the problem is further complicated with broken down sewerage outfalls at sea. Here we have a national agency again, the National Rivers Authority, along with the various water authorities and no one seems to know precisely what to do. They cannot renew the sewers until the beach settles down. The beach will not settle down because collieries are still polluting and so the situation just gets worse. In this new organisation we are talking about for coastal zone management, how do you see the new system linking in with existing authorities who have other responsibilities which operate over the beach and out to sea?

(Dr Earll) This is an absolutely classic example of how the land interests, quite a long way inland, have

Mr Cummings *Contd]*

an effect on the coastal zone. There is almost decision-making by default. As has been explained, if local authorities have no responsibility further out to sea, which they have not at present, then they are left powerless to influence what goes on further out to sea.

322. It is not so much local authorities, but we are talking about the National Rivers Authority now and the various water authorities?

(Dr Earll) We feel quite strongly that a combination of the National Rivers Authority and local authority mechanism on the ground would be the way the implementation and delivery would happen on the ground.

Mr Steen

323. I was talking about marinas and we were sidetracked, understandably. The reason I want to come back to marinas is because in my constituency I have a number of marinas sandwiched between National Trust land. I am slightly concerned that the National Trust has been silent, perhaps because they did not have a chance to answer, about their views on marinas and the National Trust land and whether they had a view about this, or are expressing a view, because the marinas are getting larger, they are having a greater impact on the environment, they are changing the face of the National Trust land—in fact, you cannot see it in some places because of the masts and so on. Is there something which you are considering about the marina industry, its growth and its effect?

(Mr Prideaux) I do not think we have particularly addressed the marina subject as such, but we are aware that the spin out from developments of this size on the coast is a sort of creeping development that can come from such things. That gives us some concern and again comes back to the need for there to be an overall plan for wider pieces of coast rather than just to say that that local authority thinks that something is appropriate. If I were not here before the Committee now, Sir, I would be at our committees where this week we have been considering two particular coastal acquisitions, both of which are heritage coast and one is an AONB. One of the areas has had planning consent given to it, but we want to acquire it in order to extinguish the consent. Both have planning type operations going on without planning consents and they are wonderful pieces of coast. Certainly there is a large area, not exactly a marina, but it could be a "marina" appearing on one of the bits of coast. We would not want to see that spread down the coast, if that perhaps more directly answers your question.

Chairman

324. Before we were distracted by marinas we were talking about the extension of planning powers out to sea. There was the discussion about the three-mile and the 12-mile limits, but during the course of that discussion there was a reference to the Crown Estate. We have received evidence of the ownership of the Crown Estate along the whole of the coastline

below the low water mark level. The National Trust I believe feels that the Crown Estate should be made subject to planning procedures and planning applications in respect of the land it owns right round the coastline below the low water mark level. Would you like to expand on that and give your reasons?

(Mr Prideaux) As I am sure the Committee was informed, the Crown Estate operate under the Crown Estate Act 1961, which, as I read it, is really financially driven. They are there to enhance and maintain the capital value of the Crown Estate. In our experience in certain places, that seems to be the fact that drives them. As I further understand it, they are not subject at the moment to what I may call the planning controls that are there for the ordinary person. I suppose three options are available. One could seek to alter the 1961 Act. One could seek to bring the Crown Estate under normal planning controls or perhaps it is a matter of tightening up the rules of the deemed planning consent procedure. I think we feel strongly that it would be enormously helpful if the Crown did operate in the same way as everybody else.

325. We were told—I think by the witnesses from the Department of the Environment—that although the Crown Estate is historically traditionally exempt from being subjected to the laws that affect the ordinary citizen of the land and therefore is outside the planning system, in fact it operates by having due regard to it. It will always take advice before embarking upon an operation and in fact behaves as though it were subject to the law which does not apply to it. Can you tell me from your experience whether that evidence is 100 per cent valid or whether you have come across instances where the result would have been different in so far as a private owner of land is concerned?

(Mr Prideaux) I am trying to rack my mind. I do not know whether my colleagues can give me a specific example.

(Dr Earll) I shall take one step back and then come to your question. In the research that we did leading up to the reports we produced, among a very wide range of users, one of the consistent concerns was the role of the Crown Estate Commission. They were perceived as driven by economics but as both poacher and gamekeeper. People felt quite happy about their role as landlords. There was no real problem with that, but where they found it difficult was to reconcile their role as the planning agency for the coastal zone. There is a distinct difference even within the Commission itself in relation to that role. Where that has been highlighted and brought very forcefully to the public attention has been in relation to the fish farming industry in Scotland where many, many people have complained about the role of the Crown Estate, the way they have operated in a high-handed fashion, where they even now have the responsibility under the environmental assessment directive to conduct environmental assessment for fish farms. That is seen by many people as totally unsatisfactory, totally outwith the original conception of the Crown Estate for them to have those powers and duties. Our feeling and the response of many, many other users of the marine

Chairman *Contd]*

environment—and it is an extremely widely held view—is that the role of the Crown Estate in the marine environment has to be looked at very hard and this dual role that they perform has to be reviewed very clearly.

Chairman: Thank you for that. That confirms some of the suspicions we had and you have used a trigger term, in so far as this Committee is concerned, when you talk about poacher/gamekeepers. That really sets us off in a major way.

Mr Jones: Can I just ask a supplementary on that though? If the Crown Estate are saying that they submit their plans to local authorities to comment on, what I would like is an example of where the local authority commented adversely on such a plan but the Crown Estates nevertheless went ahead. That really proves the point, because if obviously they take notice of the planning authority in every case then it does not make very much difference. I do not expect you to answer that off the top of your head, but I wonder if you can go away and think about any such examples and send them on to the Committee, particularly if you can do it in time for our session with the Crown Estate?

Mr Field

326. May I make the point to Mr Prideaux, I think I am right in saying that there is a tiny number of coastal local authorities who still retain the right to licence aggregate dredging themselves in their own right? You have not addressed that, I think, in your evidence. When you say that you want planning controls exercised over all their activities, I take it as read that you include aggregate dredging in there for obvious reasons. Are we also talking about capital dredging and maintenance dredging? Perhaps you could let us know about that. Can I also clear up a point? The National Trust raises the issue of unregulated port developments. Will you comment on that in relation to your point about the potential wider role of environmental assessments in your submission? As I understand it, the Local Authority Associations have drawn to our attention the fact that ports and harbours enjoy exemptions from planning controls under the Town and Country Planning General Development Order.

(Dr Harvey) We see no reason why they should enjoy that exemption and we think they should be subject to the same environmental impact assessment criteria as any other development. Indeed perhaps more so, because many of these developments are large and therefore they could conceivably have a larger impact.

327. Do you have any specific examples?

(Dr Harvey) The one which would affect the Trust most severely would be Poole, where the Trust owns Brownsea Island in Poole Harbour and owns the Studland Peninsula. Over recent years there has been considerable development of the ferry port there and increasing size of boats with the consequent problems of wash, erosion and so on.

328. Is there a by-law in Poole Harbour on the height of wash?

(Dr Harvey) I am afraid I do not know the detail of that.

329. I would suggest there was, but I am not positive. There is certainly a speed limit.

(Dr Harvey) Casual observation would suggest that the draw down and hence subsequent run back following the passage of the large ferries which are now in operation——

330. But that is suction not wash.

(Dr Harvey) Suction, yes, but that effect is significantly greater than it used to be and presumably it must have an impact.

331. Have you found that you have had to reinforce defences on Brownsea Island?

(Dr Harvey) We have indeed. There is severe erosion of one of the important protecting banks and the Trust has launched an appeal for £300,000 for the protection.

332. Is that on the Poole town side?

(Dr Harvey) The Poole town side of Brownsea Island.

Mr Steen

333. When you mentioned unregulated port development, I am just thinking of the ports I have in my area, Dartmouth and so on. They are subject to local planning requirements. When you said "unregulated port development", do you have an example of something that has happened which has not been subject to local planning control?

(Mr Burgon) Not off the top of my head.

Mr Field

334. Do you have a seat on the conservancy committee of Poole harbour?

(Mr Prideaux) Not that I know.

335. Is there a conservancy committee?

(Mr Prideaux) I do not know the precise answer.
Mr Field: I believe there is, but I am not positive.

Mr Steen

336. You raised the issue of unregulated port development, but if you have no examples why does it matter?

(Mr Prideaux) I sit here stuck.

Chairman

337. If you collect your thoughts later you might like to write to us with that information. That may help you. There is one final question I would like to ask you. I noted in your paper that you state very emphatically that you will resist the development of sewage treatment works and the related marine outlets which harm the landscape or affect amenity considerations. I can understand your concern and reservations, but the way in which you put it seems to be rather draconian. We have taken evidence, as you know, when we were looking at pollution of beaches which led this Committee firmly to the conclusion that there should not be the discharge of any raw sewage anywhere from our coastline. That has now become Government and it is certainly European

Chairman *Contd]*

policy. If that is right, one has to find a site that is related to the centres for production of the sewage, that is convenient so that it can be properly treated and, if it is not properly treated, we have all the problems of eutrophication possibly, we have the problems of viruses, pathogens being discharged out to sea, which were things on which this Committee felt very strongly. There seems to be a balance required there, certainly where there is a choice, not to put sewage work where it is going to look ugly or where it will cause offence, but if needs must what is your priority: to treat the sewage or to let it rush out raw into the sea? I should like the answer to that question.

(Mr Prideaux) I think it would be fair to say that our written evidence could have been better expressed on this point and that all the items that you have just outlined are fundamental. It would be wholly inconsistent for the Trust to object to some

building or pumping station being in a place where clearly it was the only place where it could do the job to ensure that, as you say, raw sewage did not go out to sea.

Chairman: I am glad to hear that flexibility from you. I am most grateful to you, Mr Prideaux, Gentlemen and Dr Earll for your attendance and your assistance this morning. Thank you very much indeed.

Mr Field

338. May I just make a point, Chairman? Mr Prideaux when I was talking about Newtown river, I was not talking about residential fixed moorings. I was talking about visitors who do not use the moorings. They use their own anchors.

(Mr Prideaux) That is what I have written down. Thank you for that clarification.

Memorandum by the Royal Society for the Protection of Birds

Contents

COASTAL ZONE PLANNING

1. INTRODUCTION

The Royal Society for the Protection of Birds is Europe's largest voluntary wildlife organisation with a membership of some 900,000. In encouraging the conservation and protection of wild birds and their habitats, the Society takes an active interest in a wide range of environmental and land-use issues and employs specialist staff to advise on such matters. In addition the Society owns or manages over 50 coastal nature reserves, an area in excess of 75,000 hectares. Our analysis of the problems of estuary conservation in the United Kingdom was published in 1990[1] and more recently the Society contributed to the production of a discussion paper on protection of the marine environment[2]. The RSPB regularly appears at public enquiries and in front of Parliamentary Committees to present the conservation case against damaging developments. In addition we contribute to the long-term monitoring of wildfowl and wader populations in the United Kingdom through the British Trust for Ornithology Birds of Estuaries Enquiry and Wildfowl and Wetlands Trust national wildfowl counts and our own research. The RSPB welcomes the invitation to present evidence on the Environment Committee's examination of coastal issues, an area in which the Society has developed considerable expertise.

Our evidence to the Committee will summarise the nature conservation interest of the United Kingdom's coastline and adjacent waters and the threats to them. An examination of the difficulties of protecting and managing the coastal zone given the existing control structures is presented. We look at the existing role of a number of managing agencies including local authorities, the Crown Estate Commissioners, the Ministry of Agriculture, Fisheries and Food and the NRA. Throughout the text we make recommendations for changes to the existing system which we believe will help to achieve a more integrated approach to the use and management of the coastal zone. Our recommendations are summarised below.

We conclude that the current system of coastal management is unable to deliver effectively the protection and sustainable use of one of the United Kingdom's few internationally significant wildlife resources.

2. SUMMARY

Criticisms of the existing planning and management system of the coastal zone centre on the presence of a large number of organisations with coastal jurisdiction and the lack of any co-ordinating authority. This has led to an inability to deliver and administer effective management or to resolve conflict.

Local authorities' powers are limited and central government has to date failed to provide national guidance on coastal zone issues. No central Government Department has actively taken on the role of overall co-ordinator although logic would suggest that this should be the Department of the Environment.

The development of an integrated approach requires action to be taken at local, regional and national levels. Currently, at a national level, as we have demonstrated, responsibility for coastal activity lies with several Government Ministries. There is a clear need for a discrete, well resourced Unit to be established within central Government to take the lead on coastal matters. Such a Unit should be charged with co-ordination of marine and coastal management and planning at a National strategic level. It should be able to integrate the activities of the various sectors of Government and develop comprehensive consultation and implementation mechanisms for the sustainable use of the coastal zone. Individual agencies and authorities could be given or retain management and implementation powers but they should also have strong environmental duties. Their activities would be integrated within coastal management strategies.

Day to day management of regional advice could be devolved to a National Coastal Authority involving the National Rivers Authority in England and Wales. The NRA already covers key areas like fisheries management, recreation, water quality, sea-defence and conservation. Coastal zone management at a regional level could be added to its remit without difficulty. The NRA would be well placed to oversee the implementation of national policy at regional and local levels. Its equivalent would be required in Northern Ireland and Scotland.

Clearly local authorities have a key role to play in the production and implementation of coastal management strategies and their relationship with a National Coastal Authority would have to be close. Some local authorities have already approached the subject of coastal zone management in depth.

The abilities of individual authorities to deal adequately with the problems of coastal management are limited in two major ways.

1. Local authorities lack the powers to fully achieve effective control of activities throughout the coastal zone. Zonation of recreational pursuits for example would be essential to the achievement of sensible coastal water management and yet is not possible under the current system. It is nonsensical that local authorities can control and develop onshore facilities and yet are unable to manage the immediately adjacent water space or even some intertidal activities such as shellfish gathering.

2. Such is the nature of coastal activities that they require a broader analysis than many County or District authorities are able to provide. Ecological units such as estuaries often form the borders between local authorities or even countries and the development of management strategies for individual authorities is limited in its value. The production of whole estuary management plans is not encouraged by the existing planning system and yet to manage what is in many cases wildlife value of international significance a whole estuary approach is essential.

The solution to these problems is straightforward. Local authorities should be furnished with the necessary powers to control the management of intertidal and adjacent water areas. In addition local authorities should be encouraged to work together to co-ordinate the production and implementation of regional coastal management strategies. The development of these should be in line with the national coastal strategy produced by a coastal unit of central Government and overseen by a National Coastal Authority.

The administration and financing of coastal defence and associated land drainage requires a considerable overhaul. Executive powers should be removed from the vested interests of agriculture and priorities chosen that reflect national conservation policy. The system must also be capable of more coherent response to the challenges and opportunities for site safeguard and creation brought about by rising sea-levels.

The legislation relating to wildlife in the United Kingdom is inconsistently applied. It demonstrably fails to protect important intertidal areas and cannot deal effectively with marine species or habitat conservation. It fails to meet the requirements of the EC Directive on the Conservation of Wild Birds (79/409). In addition the United Kingdom Government has failed to meet its obligations under international convention and law to designate and protect sites of international conservation value.

The RSPB seeks a coastal planning and decision making process which has clear policies, an overall national strategy and central guidance. Regional and local implementation must be in a manner which ensures the sustainable use of the coastal zone and protection of the natural environment. Due emphasis must be given to the international and national nature conservation needs in the national strategy and any local decision making process. Thorough Environmental Assessment should be mandatory for all significant developments which impact the coastal zone.

- Local authority powers should be extended to ensure adequate management of the sea surface, inside a defined coastal zone.

- The example of planning power extension in the coastal zones of the Northern Isles should be extended to give equivalent powers to all United Kingdom local authorities.

Permitted Development

- Permitted development rights under the General Development Order should not apply where there is a question of adverse effects on a notified SSSI, or a candidate or designated EC Special Protection Area or Ramsar site. Relevant legislation should be amended accordingly.

Article 4 Directions

- Revised Government guidance should be issued immediately expressing support for the application of " Article 4 Directions " on land designated as of SSSI status.

- Government should provide substantial assistance to local planning authorities on meeting any valid claim for depreciation on land value under the Article 4 provisions of the General Development Order and sections 108 and 109 of the Town and Country Planning Act 1990, in order to protect SSSIs.

Environmental Assessment

- EA for coastal developments should cover the whole coastal zone including both landward and seaward elements. The EA requirement criteria should be set down in one single Statutory Instrument. The responsibility for determining whether EA is necessary and analysing submissions should rest with a single local/planning authority, in consultation with other authorities, agencies or the National Coastal Authority as relevant.

- Size-only criteria for assessment thresholds should be abandoned in favour of environmental sensitivity criteria.

- The CEC's status as " competent authority " for EA should be removed.

Harbour Authorities

- Government should grasp the opportunity presented by the Transport Works Authorisation Procedures Bill to give all harbour authorities and conservancies a strong environmental duty.

- The Chichester Harbour Conservancy model in our view provides a helpful example; it should be recommended as such to all port and harbour authorities and conservancies.

- Fisheries legislation must be amended by giving fisheries Ministers duties to "further" conservation so as to allow protection to be given to marine wildlife at risk from fisheries operations. This would place fisheries on an equivalent footing to Agricultural matters.

- All estuaries meeting the internationally agreed criteria should be designated as Special Protection Areas under the EC Directive and as Ramsar sites under the Ramsar Convention as soon as possible.

- Government must as soon as possible publish a list of sea areas qualifying for designation under the EC Birds Directive and Ramsar Convention.

- Tanker routes should be strictly controlled to avoid sensitive coastal areas.

- Greater efforts must be made by Government to assess the full environmental impacts of marine sand and gravel extraction.

4. *The Coastal Environment*

The scenery and habitats associated with the coast are many and varied from the open sea to salt marsh, from Scottish sea lochs to cliffs and sand-dunes and from small estuarine creeks to the vast intertidal flats of the Wash or Morecambe Bay. The wildlife that these areas support is equally varied and of great significance. The United Kingdom's coastal waters are relied upon by huge numbers of birds throughout the year. The numbers of birds using the United Kingdom's coastline are monitored regularly by the BTO/RSPB/WWT Birds of Estuaries Enquiry.

Our seabird breeding colonies are of global importance with significant populations of the world populations of some species being present. Of the 261 internationally Important Bird Areas (IBAs) present in the United Kingdom 28 qualify because they hold over 1 per cent of the world's population of certain seabird species. Sixty-one IBA sites qualify because they hold over 1 per cent of the European (EC) populations of certain seabirds. (eg razorbill, gannet, Manx shearwater, and great skua).

The United Kingdom's estuaries are among the most biologically productive systems in the world. The intertidal muds and silts support many invertebrate species, often in large numbers. Wading birds in particular have evolved to exploit this rich resource. We are fortunate in the United Kingdom in having a wealth of estuarine habitats. These sheltered shores, bathed by a mix of nutrient rich waters from river catchments and saline water in shallow seas, have a range of habitat types from submerged mud-flats, intertidal mud and sand through to saltmarsh.

The main ornithological importance of the United Kingdom's estuaries is their vast populations of wintering and migrant birds. Birds come from breeding grounds in temperate, Arctic and sub-Arctic areas from northern Canada in the west across Greenland, Iceland, northern Scandinavia and Russia to central Siberia in the east. The United Kingdom is at the centre of this y-shaped migration route—the East Atlantic Flyway.

For migratory wildfowl and wading birds, the western European estuaries form vital links in a chain of sites from breeding to wintering grounds. Of the 155 recognised estuaries in the United Kingdom, 68 qualify for designation as Ramsar sites under the Ramsar Convention[5] to which the United Kingdom Government is a signatory, or as Special Protection Areas (SPAs) under the EC Directive on the Conservation of Wild Birds[4], which places strong legal obligations upon member states to protect these SPAs.

Estuaries also provide spawning and nursery grounds for fish, including many of commercial value and are key areas for waterborne and waterside recreation.

5. THE THREATS

Estuaries

In 1990 the RSPB published "Turning the Tide—a future for estuaries"[1] (attached). This document catalogues the conservation interest and problems facing the United Kingdom's estuaries, and many of the recommendations outlined in this evidence are also referred to in "Turning the Tide".

Estuaries have long been subject to interference from man. The Romans were active in claiming land for agriculture in the Thames estuary. As our technology developed, so the ability of man to alter the natural state of affairs in estuaries increased. During the last few hundred years very extensive areas of intertidal land have been lost to agriculture, ports and other developments. Today this trend shows every sign of accelerating. The number of threats, their widespread nature and the difficulty in predicting the effect of piecemeal development on birdlife, constitutes a complex problem.

The results of a recent survey of threats carried out by the RSPB in the United Kingdom gives a guide to the areas of greatest concern[6]. Of the 123 estuaries surveyed (80 per cent of the United Kingdom total), 80 were considered to be under some degree of threat, with 30 estuaries in imminent danger of permanent damage.

Land claim for barrages, bunding, infilling and other purposes is a major threat to estuarine birds which depend on sand, saltmarsh or mudflats as their feeding habitat. Significant loss of, or alteration to the estuarine habitat, means that the numbers and abundance of food species available to these birds would decrease[39,40]. The scale of the threat posed varies from scheme to scheme but there are many current proposals and each contributes to a decrease in the total amount of estuarine habitat. Large developments are increasing, but many smaller ones also take place and their total effect can be sizeable. Almost every estuary in Britain has some potential problem, either planned or under way.

A decrease in the feeding area or a reduction in the food availability in an estuary is likely to result in fewer waders being supported by that site. As the area decreases the birds could be forced to seek food in sub-optimal habitats, where there is less food, or in areas already fully occupied by others, thus increasing competition for resources. In these circumstances food may be depleted more quickly, resulting in increased competition and conflict for food between individuals, and a lower intake of food. Over-exploited food species may be reduced beyond their capacity to recover from predation pressure and disappear altogether. All these effects are most serious during severe winter weather when bird mortality is already higher[41,42].

The Nature Conservancy Council has also documented the extent of knowledge of the estuarine habitats of England, Scotland and Wales and the activities affecting them. The NCC's survey indicated that ·5 per cent of all estuarine habitats are being permanently lost *each year* to development or waste disposal[7].

Some developments, notably marinas, barrages, port expansion, land claim and industrial uses, are irreversible and reduce the extent of intertidal habitats available to birds. Others, such as bait-digging or cockle fishing, have more temporary impacts. If properly controlled, the damage they cause can be minimised, is relatively short term, and is reversible. Collectively they can deny large areas of suitable habitat to migratory birds. However such activities are largely outwith any controlling mechanism.

Habitat loss has been strongly implicated in the steep decline of at least one wader species, the dunlin. In this instance, the loss has been attributed to the spread of Spartina, an invasive saltmarsh grass which causes mud flats to dry out and lose their invertebrates. This has been exacerbated by built developments affecting the upper shore and consequent loss of mudflat. Similar losses are occurring in the south-east of Britain, where natural subsidence of the land, and the possibility of rising sea-levels, will also reduce the extent of intertidal land.

For some sites, the battles have largely been lost. The intertidal area of the Tees estuary has been reduced by around 90 per cent in the last 100 years, having been lost to land claim for port and industry related developments. In another north-eastern England site, the Tyne, all the intertidal flats have been lost to land claim. In Northern Ireland, 85 per cent of the intertidal area in Belfast Lough has been developed for industry and landfill.

One major problem revealed by our surveys is the piecemeal nature of many types of estuarine development. Although a proposal may affect only a small area, this contributes to habitat fragmentation and overall loss. When added to the numerous other similar incursions or impacts, the effect can be great. Furthermore as many local authorities often have only one part of an estuary within their area they are poorly equipped to look at the problems on a larger scale.

Marine

The United Kingdom coastline, as well as being of great scenic and recreational value, is of global significance for sea-birds as we have indicated above. One of man's most catastrophic impacts on this resource stems from our dependence on oil and its transport around the world. Oil pollution has the most visible and acute impact on birds, scenery and amenity. The number of pollution incidents in United Kingdom waters is increasing (116 pollution incidents in 1985, 395 in 1989[43]). Oil developments remain a serious threat to coastal ecosystems.

The Marine Pollution Control Unit in the Department of Transport is responsible for responding to oil incidents. Whilst their capacity to contain incidents has been upgraded and improved, their abilities do not extend to full control of a major incident. At present they have the capacity to deal with 14,000 tonnes of oil at sea within 48 hours and 5,000 tonnes of oil onshore within one week. This is considered to be sufficient for the "normal" spillage. An accident of the Exxon Valdez scale (38,143 tonnes) or the Amoco Cadiz (232,591 tonnes) would remain a catastrophe and the MPCU response would be extremely limited in its effectiveness. Even given the devastating impact of such a catastrophe on our coastline, efforts continue to be centred on a small scale clean up operation. No measures are taken to ensure that tanker routes avoid our most valued coastal areas.

Recommendation: Tanker routes should be strictly controlled to avoid sensitive coastal areas.

Marine aggregate extraction is also an area of considerable concern. The impacts of such activities are little understood and subject to a system of scrutiny that lacks scientific rigour. Management of the resource is carried out by the Crown Estate Commissioners[44] whose role in marine activity control is covered in section 9. Demand for marine aggregates is high and may increase as land based sources become scarce. Concern in the Netherlands has centred on the impacts of extraction on coastal sediment movement. For this reason the Dutch Government has barred any such activity from within 20km of their coast. There is limited information on this topic but grounds for concern exist and further work is required. Most criticism of the marine aggregate industry in the United Kingdom comes from fishermen who are concerned about impacts on fish migration routes and spawning grounds but little research is available to confirm or otherwise, impacts. As with coastal erosion further work is required.

Recommendation: Greater efforts must be made by Government to assess the full environmental impacts of marine sand and gravel extraction.

6. COASTAL JURISDICTION

As an island nation the coast and surrounding seas have always held a strong interest. The coastal zone is the focus of much recreational, fisheries and industrial activity. Substantial sectors of the United Kingdom economy rely on ports, harbours, off-shore resources or land claimed from the sea. The coastal zone is very much subject to the problems of the integration of multiple, and at times, competing uses.

The juxtaposition of high conservation value, and intensive multiple use for development or recreation has increasingly led to resource management problems. In recent months a variety of organisations have been active in promoting the need for an examination of the United Kingdom coastal zone planning system. For example the Countryside Commission have organised an international coastal zone planning conference as have the Royal Society for Arts. In addition to the RSPB, the Marine Conservation Society, the World Wide Fund for Nature, and the British Association of Nature Conservationists have produced critiques of the current coastal management system[9,10].

The Local Authority associations have also been active in policy development in the coastal zone. A consortium of local government associations (CPOS, ADC, AMA and ACC) has set up a coastal and estuary working party and the Welsh local authorities similarly have a coastal working group. At a local level Sefton, Wirrral and New Forest district councils and Kent, Devon and Dorset county councils are active in developing coastal management strategies although their scope is limited by their lack of powers. The most recent analysis of the problems of a coastal authority comes from Hampshire County Council who have produced their own coastal strategy[37]. This strategy graphically demonstrates the frustrations of local authorities grappling with the day-to-day problems of coastal management without the enabling powers they need.

The existing system of coastal management displays a number of inadequacies which together ensure failure in both planning of coastal use and in attempts to integrate terrestrial and marine management. The extent of these problems was detailed in a paper produced by Wildlife Link at the Secretary of State for the Environment's request. Detailed in this paper are 22 examples of the problems caused by multiple jurisdiction in the coastal zone (See appendix 2). The coastline is treated as a boundary of many different legislative and administrative systems and not as the focus of activity. The required holistic view cannot be taken whilst such a large number of diverse bodies lay claim to responsibility[38]. A table showing the range of bodies with coastal and marine jurisdiction is shown on page 21 of " Turning the Tide " (attached). Rationalisation seems long overdue.

The RSPB's view remains that the current system of coastal management is unable to deliver effectively the protection and sustainable use of one of our few internationally significant wildlife resources.

7. THE TOWN AND COUNTRY PLANNING SYSTEM

7.1 *Introduction*

Local authority jurisdiction ends at low water over most of our coastline. Structure and local plans covering coastal areas can play a vital part in guiding the use of the shoreline and avoiding conflict. However they can only go a limited way towards the required and logical holistic management process. Local plans can only deal with land-use issues. Many of the activities and processes impacting on our coastline lie outside the control of local planning authorities.

The use of both the land and sea along the coastal zone can have major impacts on sites of national or international conservation value. Town and Country Planning legislation and policy, together with the regulation of the sea bed and sea surface by statutory decision-making authorities, Central Government, public and private land-using sectors and other organisations is crucial in determining the fate of important areas. The RSPB seeks the establishment of regulatory systems in the coastal zone that are capable of dealing with development and other pressures. We seek to ensure that the cumulative impacts of development are fully considered in reaching decisions that affect, either directly or indirectly, important conservation resources, and that bird conservation (as required by, for example, the EC Directive on the Conservation of Wild Birds 79/409) is satisfactorily delivered as a result.

7.2 *Legislation*

One of the major problems with the Coastal Zone is the lack of an authority with overall jurisdiction over the combination of coastal land, the intertidal zone and inshore waters. Currently at least 33 different agencies have responsibilities and powers relating to the Coastal Zone, and generally powers are split between the land and sea. This division is artificial and fails to accommodate processes and developments that operate across this zone. Many landward developments and processes have seaward implications and vice versa, and some operate over both at the same time, for example marinas with integral housing developments, or port and harbour facilities.

Consultation procedures and practice between the bodies responsible for these areas of jurisdiction are currently inadequate. Even in instances where consultation does take place there can be conflicts between the landward and seaward " assessing authority ". The environmental consequence of this breakdown, is that the implications of proposed land developments upon the sea and vice versa are not fully assessed. Some applicants intentionally exploit this confusion, in the knowledge that certain impacts of a proposal fall outside the " assessing bodies' " jurisdiction.

In respect of physical planning, powers for the landward part of the coast fall within the provisions of the Town and Country Planning Acts[11] and the remit of relevant local planning authorities. The physical outer limit of a local authority in England and Wales is median low water mark[12] and by common practice this is also taken to be the outer limit of planning jurisdiction. In Scotland this correlates to " ordinary " spring low water mark, while in Northern Ireland the jurisdiction of the Department of the Environment (Northern Ireland) for planning along the coast remains relatively undefined. (Northern Ireland Town and Country Planning provisions are set down in The Planning (Northern Ireland) Order 1991.)

This variety of planning authority coastal boundaries, which in some cases is further varied by local/Private Acts of Parliament, is highly confusing. In the case of the Solway Firth, for example, the planning powers of local authorities extend further into the estuary on one side than on the other. Consequently, regulatory systems can become muddled and on occasions necessary consultations or assessments could potentially be omitted through misunderstandings allowing environmentally degrading or damaging developments to secure approval. The precise limit of where local authority control over development on " land " (defined in the legislation as " any corporeal hereditament ") ends and the sea begins is unclear with the result that in parts of the intertidal zone neither land nor sea agency would be entirely certain of its jurisdiction[13].

The seaward part of the Coastal Zone is complicated. No equivalent of Town and Country Planning powers apply to its surface although some control does apply to the seabed. The lack of surface control is exacerbated because the sea is a multiple-use surface where conflicts can regularly arise, whereas the terrestrial land surface tends to be limited to one or two uses at any one time—where conflicts are " mediated " by the operation of the planning system and the grant of planning permission.

It is the duty of the Crown Estates Commissioners (CEC)[14] to manage the seabed (*not* the sea surface) and also around 55 per cent of the foreshore up to median high water mark. This land is vested in the Crown. The CEC have a quasi-planning role by virtue of their ability to control and manage this large estate, through leasing and licensing operations on the seabed for operations such as marine mineral/aggregate extraction, marine fish farming, marinas, etc. However, this role is compromised by their requirement to have regard to the maximisation of profits as landlords of the seabed; the CEC is required " to maintain and enhance the estates value and the return obtained from it ". Thus, what is already a generally ineffectual " planning " control further fails to apportion the necessary importance to the environment. Such is the CECs remit towards profit maximisation that they cannot adequately balance conflicting demands for development in the coastal zone[24,25].

Sea surface use has considerable potential to affect the environment. For example, most waterbirds are wary and readily move away from human activity. At roosting times it is especially important that these birds remain undisturbed as disturbance affects their energy requirements and reserves with consequent demands on food intake. Waterside recreational pursuits such as sailing, windsurfing, jet-skiing, etc all have the potential for such disturbance if they are unmanaged. Yet open access to both foreshore and sea surface make water space and recreation management difficult and conservation conflicts almost impossible to resolve.

Recommendation: The artificial split of " planning " functions within the Coastal Zone into landward and seaward elements should be replaced by a holistic approach managed by local authorities to ensure more effective strategic planning and management of coastal issues.

Recommendation: To achieve this holistic approach Local Authority planning jurisdiction should be extended out to sea to cover marine developments. This should be achieved through an extension of the jurisdictional coastal envelope. The RSPB does not wish to be prescriptive in relation to this jurisdictional extension—there are merits in a number of alternatives—but consider that a three mile extension has benefits to offer.

7.3 *The Northern Isles*

The rapid development of the North Sea oil industry in the early 1970s required major developments in the coastal areas of Orkney and Shetland. In 1974 the Government recognised that it was desirable that such developments " be provided in an orderly, co-ordinated and effective manner under the administration of one authority "[23].

Accordingly, S 11 and S 12 of the Zetland and Orkney County Council Acts 1974 grant the Councils (now Islands Councils) the power to control, by licensing, works and dredging, respectively, in the Coastal Zone. This equates to *de facto* planning control as applications for licenses are considered, in public, by a committee of the Council, are available for comment by interested parties and refusals are subject to appeal to the Secretary of State.

RSPB experienced in Shetland and Orkney is that these systems of development control are, indeed, orderly, co-ordinated and effective. Although initially established for the oil industry, these procedures have since been used to plan the development of marine fish farming in Shetland with considerable success. Indeed, Shetland has seen an application for a fish farm refused, and upheld on appeal, on environmental grounds.

It is the RSPB's view that the procedures for development control in the Coastal Zone in Shetland and Orkney set a helpful precedent for the rest of the United Kingdom. The success of the system in the Northern Isles demonstrates that there should be no real obstacles to extending the planning role of local authorities to cover the whole Coastal Zone.

Recommendation: The example of planning power extension in the coastal zones of the Northern Isles should be extended to give equivalent powers to all United Kingdom planning authorities.

7.4 *Subject Plans*

Until the coming into force of the Planning and Compensation Act 1991, S 36(2) of the 1990 Act provided for " different local plans to be prepared for different purposes for the same area ". This enabled co-operative statutory development plans, more commonly known as " subject plans ", to be prepared by more than one local planning authority for *inter alia* estuaries and coasts. The Planning and Compensation Act in seeking to streamline the development plan system removed this important power and now there is no longer any means by which local planning authorities, both Districts and Counties, can group together to jointly prepare statutory, subject local plans for specific areas[15]. The RSPB views this as a retrograde step, as it removes such opportunities for sound planning in the coastal zone which have existed.

The resultant inability to address coastal problems uniformly results in isolation of policies, fragmentation, varying practice, inconsistency and duplication of efforts in local authorities' approach to coastal zone issues. Whilst in theory joint working on informal " bottom drawer "plans may be possible there is no provision to require policies in such plans to be incorporated into relevant District-wide local plans. Voluntary plans also have the drawback that without statutory requirement to prepare them different local authorities fail to find common ground or agree on a lead authority, with the result that the plans get bogged down in a " stalemate ".

" Bottom drawer " plans on their own carry little weight. The Draft Planning Policy Guidance (PPG12) on " Development Plans in the Planning System " (1991) states that (paragraph 3.22) in the case of " informal ' bottom drawer' plans and policies, which have not been subject to the statutory procedure of public consultation and formal adoption. . .the Secretary of State regards (these) approaches as unsatisfactory. . .". It goes on to say that they are *not*, " an acceptable alternative to the preparation and adoption of a statutory local plan after the normal process of public consultation. Informal documents carry less weight for development control purposes, and provide less satisfactory guidance to developers and reassurance to local people about the future pattern of development, than a properly prepared and adopted local plan ".

Recent collaborative exercises such as the Devon County Council Exe Estuary Management Plan and Kent County Council's North Kent Marshes Study are commendable but regrettably rare. Even on completion these management planning exercises can only partially fulfil their purpose. The lack of appropriate local government powers or co-ordination of relevant agencies required to implement policy works against the comprehensive holistic approach we believe is necessary[16].

Out of the 155 recognised estuaries in the United Kingdom[7] four cross country boundaries, 20 are covered by two or more counties or regions, 60 are covered by two or more local authority districts. Taking some of the more extreme examples, the Thames is surrounded by 20 district councils, the Humber by 14, the Severn by 12 and the Wash and Forth both by eight. It is not surprising that such a spread of local authority interest around estuaries of international conservation value leads to a lack of consistent approach to the treatment of the common resource, with all that implies.

Recommendation: Subject plans enabling local authorities jointly to prepare statutory development plans covering coastal, and especially estuarine, areas be re-introduced as a high priority.

Recommendation: It should become a statutory duty for all Coastal local authorities to prepare subject plans in co-operation with other authorities along the coastline/estuary. These could be known as " whole estuary " or " coastal " subject plans.

7.5 *Regional and National Guidance*

Considering the United Kingdom's large coastline it is surprising that current Town and Country Planning legislation has so little to say specifically about the treatment of the coast. Both the 1990 Act and the Planning and Compensation Act 1991 do not have any specific sections relating to the coast.

There is no national planning strategy or guidance for the coast in the United Kingdom. The last Department of the Environment Circular relating directly to coastal issues dates from 1972[17]. This concentrated predominantly upon the scenic quality of the coast, with no specific reference to nature conservation. The circular pre-dates many of the United Kingdom's international nature conservation obligations including European Community Bird Directive[4] and Ramsar Convention[5], as well as Sites of Special Scientific Interest/Areas of Special Scientific Interest (SSSI/ASSI)[18]; many of which relate to our most important coastal wildlife areas. The United Kingdom Government is party to the Ramsar Convention. Cmnd. 6465 details the Convention, although this requires updating as the Convention has been amended since Cmnd. 6465 was produced (1976). The circular is still extant but inadequate for today's complex environmental considerations and development pressures. In Scotland national planning guidelines for coastal development were issued by the Scottish Development Department in 1974[19]. This guidance is still in place although currently under review.

The coastline is an invaluable national resource, yet there are no national and regional policy guidelines for the sensible and sustainable use of this resource nor for the management of the conflicting demands of wildlife conservation, fisheries, development, business, leisure, etc. It is known that the Government is preparing Planning Policy Guidance on " coastal zone " issues, but early indications are that this will only highlight the existing inadequate provisions rather than prescribing new mechanisms. This will not address the problem of where two agencies conflict through the lack of a common plan. Such a plan would allow longer term strategic thought on the management of the coast by all relevant bodies and provide the regional context for " whole estuary " plans.

Recommendation: A Regional Coastal Zone Management Plan should be produced that brings together all relevant agencies. This plan will form the management and co-ordination tool for the coast preventing duality of action or conflicts between bodies or authorities. The Town and Country planning aspects of the plan should be included within countrywide Regional Planning Guidance and thus form the basis of " whole estuary "-type plans.

These Regional Coastal Zone Management Plans will be logical " cascades " down from a new National Coastal Strategy which is necessary to protect and manage this national asset. Scotland benefits from the national coastal strategy referred to above[19]. This allocated parts of all the Scottish coast as " preferred development zones " and " preferred conservation zones ". As a first attempt these guidelines were welcome but they have never been updated nor supplemented in spite of subsequent relevant legislation and the passage of time. There is a distinct and urgent need for a National Coastal Strategy to plan for the sustainable use of our coasts and one which has statutory " teeth ". In addition a body to oversee national and regional implementation of such a strategy is essential.

Recommendation: Government should produce a National Coastal Strategy for the sustainable use of our coasts. This would have provisions to prevent damaging activities/uses and habitat destruction where they affect sites that are recognised as of national or international importance.

Recommendation: A National Coastal Authority (reporting to DOE) should be established with responsibility for planning and co-ordinating matters affecting the sea surface as well as the seabed including strategic regional planning in the coastal environment.

7.6 *Permitted Development*

Some operations and developments are exempt from the need to obtain formal planning permission from the relevant local planning authority. They amount to " permitted development ", ie development permitted by virtue of the Town and Country Planning General Development Order 1988 (as amended). Some of these " permitted development " rights can have serious implications for coastal zone areas, especially where they affect internationally recognised sites of importance for birds[20]. The " permitted development " right for the deposit of dredgings by statutory undertakers is particularly open-ended having no qualifying conditions whatsoever—this can result in such operations damaging recognised internationally important sites which the United Kingdom Government has obligation to protect, for example Lappel Bank in the Medway Estuary in Kent (the " permitted development " rights in themselves do not obviate these obligations). In order to afford protection to these sites certain " permitted development " rights should be removed so that operations can be fully assessed by a planning application accompanied, where applicable, by an Environmental Assessment (EA). This will not necessarily prevent planning permission being granted but will ensure that any proposal receives due and proper consideration.

Recommendation: Permitted development rights granted under the General Development Order should apply where there is any question of adverse effects on a notified SSSI, or a candidate/designated EC Special Protection Area or Ramsar site. Relevant legislation should be amended accordingly[21].

The above recommendation provides the best means of dealing with this problem. An alternative is the use of Article 4 Directions, but local planning authorities are generally extremely reluctant to use these powers because of the compensation provisions that apply under these procedures.

As an absolute minimum the RSPB considers that the provisions under Article 4 of the GDO must be strengthened. SSSIs/ASSIs are recognised as being the most important sites in the United Kingdom for, *inter alia,* nature conservation. Circular 22/88, Appendix D indicates that " permitted development rights should be withdrawn only in *exceptional circumstances*: and that such action will rarely be justified unless there is a *real and specific threat*, ie there is reliable evidence to suggest that permitted development is likely to take place which could damage an interest of acknowledged importance and which should therefore be brought within full planning control in the public interest" (paragraph 1); SSSIs are such an "interest of acknowledged importance". They are capable of being irreparably damaged in a very short space of time, sometimes in a matter of hours. Whilst Appendix D does make reference to "Agriculture and the Countryside" (paragraph 4) there is absolutely no specific mention of nature conservation. This is clearly wholly unsatisfactory.

It is quite possible that damaging development will commence before the planning authority is aware of it, Article 4 Directions can only be made prior to works commencing on a site. Moreover, these permitted operations could also damage sites over which the United Kingdom Government has international nature conservation obligations, particularly the EC Birds Directive[4], and over which the United Kingdom Government could be challenged in the European Court.

Recommendation: As an absolute minimum the RSPB recommend that revised Government guidance should be issued expressing support for the application of " Article 4 Directions " on land nominated as of SSSI status or a candidate/designated EC SPA or Ramsar site. This guidance should indicate that such support will be given where any general threat is perceived or expected and will be accepted on any notified SSSI land or candidate/designated EC SPA or Ramsar site, no matter how large the area, if justified. The Government should indicate that where a planning application is subsequently submitted and refused by the planning authorities they will provide substantial assistance to local planning authorities towards meeting any claim for depreciation of land value under the Article 4 provisions, and S 108 and 109 of the Town and Country Planning Act 1990 in order to protect these recognised nationally important sites.

7.7 *Environmental Assessment*

In respect of EA the landward part of the Coastal Zone is covered by the Town and Country Planning legislation[22]. The CEC are responsible for leases and licences on the seaward side and are presented as the "competent authority" for EA on the seaward side for certain developments. They operate to an EA Statutory Instrument for fish farming for example, but the actual criteria upon which the CEC decide whether a marine fish farm proposal requires EA or not are formulated on the basis of a "guidance note" which uses size of development as its key criteria. In a complex marine environment size alone is not a satisfactory measure of effect, instead the need for EA must be based upon "sensitivity criteria" related to conservation interest, proximity to land based SSSIs, etc. Similar inadequate circumstances apply to marine aggregate extraction which can impact of coastal erosion or deposition. Given the range of expertise required there is considerable room for doubt as to the CEC's ability to act as "competent authority" in these cases.

The RSPB consider that the CEC should not have "competent authority" status in respect of Environmental Assessment. As landowners of the seabed and much of the seashore they are subject to an unsatisfactory "judge and jury" role in respect of such proposals as those illustrated above. Moreover, the requirement imposed upon CEC by the Treasury to have due regard to profit maximisation as landlords leads to a conflict of interests and makes their position untenable.

Recommendation: The CEC's status as " competent authority " for EA should be removed.

Recommendation: The " planning " responsibilities of the Crown Estate Commissioners should be passed to more competent elected authorities. The CEC should act solely as landlord of the seabed.

The jurisdictions and responsibilities of EA "competent authorities" are complex and diverse. There are a number of Statutory Instruments relating to Environmental Assessment which apply to different areas and organisations in the coastal zone. This confusing situation must be rectified by clear Government guidance as to who has "competent authority" status and under what specific provisions. Ideally EA "competent authority" status in the coastal zone should be invested with one single local/planning authority preferably under one Statutory Instrument. The Indicative Criteria and Thresholds for identification of projects requiring EA under the Environmental Assessment circulars (15/88; WO 23/88; SO 13/1988) will need to be expanded in order to accommodate such coastal zone matters.

Recommendation: EA for coastal developments should cover the whole coastal zone including both landward and seaward elements. The EA requirement criteria should be set down in one single Statutory Instrument. The responsibility for determining whether EA is necessary and analysing submissions should rest with a single local/planning authority, in consultation with other authorities, agencies or the National Coastal Authority as relevant.

Recommendation: Size only criteria for assessment thresholds should be abandoned in favour of environmental sensitivity criteria.

7.8. *The Town and Country Planning System and Private Bills in Coastal Areas*

Interface between the planning system and Parliamentary Private Bills occurs regularly in the coastal zone, particularly in relation to Ports and Harbour Facilities. The RSPB regularly petitions against private bills which seek to authorise developments upon sites of national or international importance for bird conservation. In 1987 the RSPB gave evidence to the House of Lords/House of Commons " Joint Committee on Private Bill Procedure ". We identified the problem that occurs with this procedure whereby a development that encompasses only a small element of development outside the jurisdiction of planning control, such as a port development, (and which therefore requires express parliamentary approval because of its interference with, for instance, navigation), tends to fall totally under the Private Bill procedure. This removes development that falls within the ambit of planning legislation from planning scrutiny. Indeed local planning authorities themselves often petition against some or all aspects of private bills in an effort to present their views.

We recommended to Parliament that Private Bills should only be required for those matters for which Parliament alone is privileged to grant authority—all other matters to be determined under the planning system. The Joint Committee broadly concurred with this view and recommended that " where the primary purpose may be authorised through other means, the Committee on the bill should insist that those means be pursued first, and that the approval of Parliament be limited only to the specific components which require its authority ", (paragraph 28, 1988). The Government did not concur with this view and consequently the problem of development being decided outside normal planning procedure still exists.

Recommendation: Our recommendation (section 7 above) that local authority jurisdiction be extended out to sea to cover development on the seabed would both overcome this problem and significantly reduce the Parliamentary workload that currently exists with regard to this form of development consent.

8. HARBOUR AUTHORITIES AND CONSERVANCIES

Around the United Kingdom coast there are many harbour authorities and harbour conservancies set up by various Acts of Parliament. Many cover estuaries of national and international value to nature conservation which are designated as SSSIs. The Harbour Authorities vary greatly in their powers but many are able to undertake operations without seeking planning assent. In doing so the protection conferred by conservation legislation is also overridden by way of the General Development Order (see above). They frequently have powers to control navigation, undertake dredging works, provide moorings and reclaim land from the sea but rarely have any explicit environmental duties.

The forthcoming Transport Works Authorisation Procedures Bill provides Government with a clear opportunity to give these coastal management agencies a strong environmental duty. Early discussions with the Department of Transport have however proved disappointing with no desire apparent to place environmental duties on these important authorities, or even require them to balance their functions with the requirements of the EC Directive on the Conservation of Wild Birds[4].

One harbour conservancy however, is a model for others with clear environmental duties. Chichester Harbour is the easternmost of three large natural harbours lying to the east of Southampton Water. It has suffered less in terms of organised recreational demand and encroachment by development than have the other two harbours (Langstone and Portsmouth). It is also less polluted. These conditions have resulted from 20 years of precautionary planning and management. Conservation orientated policies are applied by local planning authorities. Harbour functions are exercised by a conservancy board with a specific duty towards the conservation of amenity and nature. (See Appendix 1.)

Complications may arise from the Transport Works Authorisation Procedures Bill, which proposes that harbour powers can only be legislated for by order made by the Secretary of State for Transport—rather than by private bill. Consequently, the creation of " new " Chichester model authorities will be made immeasurably more complicated: an order and a bill would be required.

Harbour authorities are in a powerful position to influence coastal management yet remain largely outwith planning controls and have wide ranging powers which can seriously damage coastal conservation interests. There are a large number of such authorities but their role differs greatly from place to place. Harbour managers could play a key management role if their remit were to be extended to cover provision for recreation and conservation management.

Recommendation: The Chichester Harbour model in our view provides a helpful starting point; it should be recommended as an example to all port and harbour authorities and conservancies.

Recommendation: Government should grasp the opportunity presented by the Transport Works Authorisation Procedures Bill to give all harbour authorities and conservancies a strong environmental duty.

9. CROWN ESTATE COMMISSIONERS (SEE ALSO PLANNING SECTION AT 7.7)

The CEC owns approximately 55 per cent of the United Kingdom's coastline and almost all of the adjacent seabed. In Scotland (apart from Shetland and Orkney) the CEC has been given the role of quasi-planning authority in controlling the development of the marine fish-farming industry. This position has led to severe criticism especially from local authorities and conservation bodies[24,25]. In England and Wales the CEC generally complies with local authority decisions in coastal development planning.

The CEC also has a key role in the examination of environmental impact assessment with regard both to fish-farming and offshore sand and gravel extraction. As explained in section 7 of this evidence we consider the CEC should not be the " competent authority " to judge environmental assessments, nor should they have the quasi-planning role that Government has thrust upon them.

The CEC is beholden to the Treasury and has no broad environmental remit or duties. This situation should cease as soon as possible. The CEC estate should revert to being a landlord and given strong environmental duties so as to manage the nation's coastal estate in a sensitive manner in perpetuity. Such management should be in accord with national strategy and in consultation and agreement with the activities of our proposed National Coastal Authorities Regional Coastal Plans.

Recommendation: The role of the CEC requires critical examination, removal of planning and environmental responsibilities and charging with strong environmental duties.

10. SEA DEFENCE AND COASTAL PROTECTION WORKS

The provision of sea defences, their type and location can have major long term impacts on the coastal zone.

The RSPB has identified a number of deficiencies in the existing policy framework and arrangements for the delivery of sea defence objectives which require amendment or revision if nature conservation objectives are also to be achieved. These deficiencies centre around the decision making process, principally the institutional structures, the financing of works and future strategies in response to the prospect of sea level rise.

The defence of low lying land from flooding and the protection of soft coasts from erosion by the sea have been carried out for many centuries around the coast of the United Kingdom. The purpose behind such works has been varied—to protect life, housing, and infrastructure such as railways and roads; to protect existing agricultural crops; and to claim and drain intertidal land for development and agriculture. Since the last war great emphasis has been upon the protection, drainage and claim of coastal land for agriculture. The Government policy was to increase food production. Agricultural product prices have been maintained at a high level through the CAP price support mechanism in order to promote increased production. Specific grants have been, and are still, available to fund approved sea defence improvement schemes.

This policy has led to the " claiming " and destruction of intertidal land and coastal grazing marsh which supported internationally important wader and wildfowl populations. These losses of wildlife habitat have been most marked in the south and east of England where the coastal plain is most extensive[26] (up to 80 per cent loss since the 1930's). But extensive areas of the Ribble, Morecambe Bay and other estuaries have also been destroyed. As well as direct loss of intertidal land there has also been fragmentation of the grazing marsh habitat behind the sea wall, as land drainage infrastructure has been operated to minimise the occurrence of winter flooding and to lower summer water levels. These factors, which operate throughout the United Kingdom, have reduced the quality of the remaining habitat for breeding and wintering birds.

There have been changes in Government policy and priorities in recent years, brought about by alterations in public opinion and economic circumstances, large land claims for agriculture are irrelevant with cereals for example in large surplus. Legislation however, has not kept pace with these changes. These policy changes have led to a lesser priority being given to agricultural land improvement but this has not always been reflected in the decisions of the National Rivers Authority (NRA), Regional Flood Defence Committees (RFDCs) and Internal Drainage Boards (IDBs) who collectively control drainage practice. This is because in law they are given almost complete autonomy from central Government direction.

The improvement in defence standards which continues to this day, and is projected to continue around the coast of England and Wales in the NRA five year Corporate Plan, permits an increase in the intensity of agricultural management of low lying coastal land. This is, and will be, to the considerable detriment of nature conservation interests which depend upon extensive, traditional grazing systems, a high summer water table and shallow winter flooding.

Due to the potential for damage that sea defence and coast protection works can and have had upon nature conservation interests, all agencies which operate in the coastal zone should have statutory and very positive duties toward nature conservation. At present such positive duties " to further " the conservation of flora and fauna[27] only apply within England and Wales to the National Rivers Authority, Internal Drainage Boards, and privatised water companies. The duty also applies to the Secretary of State for the Environment and the Minister for Agriculture when using their powers in relation to the preceding bodies.

Recommendation: All agencies managing the coastal environment should have positive, statutory nature conservation duties applied to them. The duty to further conservation should be extended to apply throughout the United Kingdom, particularly to Scottish Regional Councils, River Purification Boards and the Secretary of State in Scotland and in Northern Ireland to the Department of the Environment and Department of Agriculture, especially the renamed Watercourse Management Division.

10.1 *The planning and decision making process*

In the existing legislative framework and decision making process over flood defence and land drainage works in England and Wales too great a discretion is left to local bodies, local interests and their lobby groups. Part of their restructuring must be to include the rigorous application to all decision making of the three tenets of Government policy, which at present apply only to the grant aid procedures. These tenets are that flood defence and land drainage works should be technically feasible, cost effective and environmentally sympathetic. This policy must also apply to the operation of the existing drainage infrastructure and when defining the need for maintenance works.

10.2 *Ministerial responsibility*

The position of MAFF as the lead Government Department upon sea and coastal defence matters results in the continued impetus to improve defence standards for agricultural land and the reluctance to acknowledge that policy and legislation must be changed if the United Kingdom's remaining wetlands are to be protected and managed for nature conservation interests. One of the consequences of MAFF appointing people to the posts of chairman and members of NRA Regional Flood Defence Committees (RFDCs)[28] is that the appointments have favoured those with the greatest interest in, and potential to benefit from, improvements to the standard of protection given to agricultural land. The RFDCs have not been seen to take the lead in promoting the management of water levels on coastal grazing marshes for nature conservation interests even in cases where the land has been identified as of international wildlife importance, and is notified as a SSSI.

The transfer of the role of lead Government Department from MAFF to the Department of the Environment will only reflect existing departmental functions, responsibilities and recent shifts in policy. In particular it should be noted that it is the DOE that has lead responsibility for town and country planning matters, local authority finances (the cost of coast protection, sea defence and land drainage works is met in part from local authority funds), environmental protection, nature conservation and as sponsoring department for the NRA. All these matters impact to a far greater degree on sea defence and coast protection policy than does food production. Even in the prioritisation system defined by MAFF for grant aid towards sea defences, food production comes at the bottom of the list below protecting urban properties and infrastructure. Thus MAFF grant-aided the Thames Barrier which is purely part of an urban flood prevention scheme. The RSPB concludes that responsibility for sea and coastal defences should pass to the DOE.

Recommendation: There should be restructuring of the decision making process and a reform of the land drainage and sea defence legislation.

Recommendation: The role of lead Government department upon sea defence, land drainage and coast protection should be transferred from Agricultural Departments to Environmental Departments.

10.3 *National Coastal Authority*

The RSPB has identified the need for a national body to co-ordinate strategic regional planning (See section 7). To be effective such a body would need to take on the existing functions of the following or equivalent bodies:

> *NRA*—Catchment management (pollution control, water resources, flood defence, land drainage, fisheries, recreation, conservation and navigation).
>
> *Maritime district authorities*—Coastal protection works.
>
> *HMIP*—Integrated pollution control and " Red List " chemical control.
>
> *DOE*—Regional town and country planning.
>
> *MAFF*—Licensing of dumping of wastes and dredgings in near shore waters and at sea.

This proposed National Coastal Authority would report to the DOE and be organised and managed regionally, following a national policy agreed with the DOE. The professional staff could be guided by regional advisory committees which contained a balance of local authority members, statutory bodies and representatives of voluntary interest groups.

This proposed National Coastal Authority is in effect an extension and broadening of the role of the NRA as the integrated catchment planning agency. The RSPB strongly supports the role of the NRA as an independent authority responsible for the regulation of the water environment and having strong environmental and nature conservation duties[29].

The RSPB would view with great alarm any proposal to create a closely focused agency responsible solely for sea defence, coastal protection and land drainage works, reporting to MAFF. The RSPB believe that there should be a merging of agencies to form a stronger, more effective independent agency to build upon the strengths of the NRA and its approach to integrated catchment management. This new agency proposed by the RSPB would be the basis upon which to build the National Coastal Authority.

In Scotland the different organisational structure will require a separate approach. A unified River Purification Board would act as the nucleus for a National Coastal Authority. Which would take on some of the functions of HMIP, SOEND, SOAFD and local authorities.

Recommendation: A National Coastal Authority should be established to co-ordinate strategic regional planning in the coastal environment. The NRA should be strongly considered for the role.

10.4 *Local decision making*

The continued existence of Flood Defence Committees as " executive " committees of the NRA, that is as the body having the final authority to determine which sea defence and land drainage schemes proceed, is usually justified on the grounds that those who pay for the works should have the final say. Under the present funding arrangements, nominally the local authorities fund the NRA's budget for sea defence and land drainage works. In practice, because of central Government grants to local authorities following the Standard Spending Assessment, capital grants from MAFF and the recently introduced " cost-pass-through " arrangements for additional expenditure determined by the NRA, the local authorities raise only a small proportion of the costs of works through local taxation. In effect the direct, local beneficiaries pay only a small proportion of the costs, the majority comes from the central exchequer raised from the nation as a whole. The RSPB has concluded that the basis of the argument for the RFDCs having " executive " status is false.

This special status is not accorded to those individuals and bodies who finance all other activities of the NRA. For example, freshwater anglers recreation interests or those who are charged for discharge consents. Their only route into the decision making process of the NRA is through *advisory committees*. That is, the decisions upon the direction and priority to be given to non drainage works and activities of the NRA, which amount to approximately half its annual budget, are taken by the professional staff of the NRA with the assistance of two advisory committees. By contrast, the decisions upon the direction of expenditure of the drainage budget of the NRA is determined by an appointed committee who have a direct interest in receiving the benefit of that expenditure.

Recommendation: the special status afforded to drainage interests through the existence of the Regional Flood Defence Committee is not justified and should be eliminated by making this committee advisory only.

10.5 *Internal Drainage Boards*

In some low lying areas around the coast of England and Wales, Internal Drainage Boards provide and operate drainage works, pumps and sluices behind the sea walls. Their primary purpose is to fulfil the wishes of the land drainage ratepayers within their district. From an examination of the operation of IDBs over a number of years, in particular their role as promoters of drainage works which have damaged and destroyed important nature conservation sites, the RSPB has concluded that IDBs have no justification for their existence which cannot be met by the NRA[30].

The abolition of IDBs would necessitate the transfer of all pumps and water level control structures to the NRA along with the staff who operate and maintain them. Responsibility for the principal internal watercourses would be transferred to the NRA in the form of main river and the minor internal watercourses would revert to riparian landowners. This process of transferring their responsibilities has already occurred in post-war years in several regions of England without detriment to flood protection standards afforded to housing and infrastructure.

The activity of the NRA within the former internal drainage district would be funded by more extensive use of general and special drainage charges. The RSPB cannot envisage how costs would rise under this system. Economies of scale in administration and a reduction in the intensity of management of some watercourses should bring savings to the owners and occupiers within the former districts.

Recommendation: Internal Drainage Boards no longer have any justification and should be abolished.

10.6 *Financial administration*

The existing rates of grant aid given by MAFF across the regions and districts of the NRA have been developed through a combination of historical precedence, perceived " need " for drainage and flood defence works and the ability of the local authority to contribute. The result is a grant rate which favours predominantly rural areas rather than being determined by an objective assessment of the national priorities. There should be a complete reassessment of the rates of grant paid across England and Wales with much better targeting of funds to works which satisfy defined national priorities. The grant supplement for sea defence works is an example which already exists. The process of prioritisation and justification of works should not be left to local interests who may have difficulty in recognising national priorities.

Recommendation: There should be much better targeting of grant aid to works which satisfy defined national priorities

There is a need for a test of cost effectiveness, through cost-benefit assessment (CBA) methodology, for all works undertaken by drainage authorities irrespective of the intention to seek grant aid. The RSPB is aware of RFDCs placing schemes into the forward programme with CBA ratios of less than one (and hence they fail this test). When this has been questioned, the RFDC has voted to proceed with the works without seeking grant aid and without regard to the effective use of public money. This results from a real imbalance of interests upon some RFDCs as local authorities have failed to nominate people to represent the financial interests of those who pay the drainage levy, rather they have often selected those with a direct knowledge of and benefit from land drainage, that is a landowner, farmer or contractor.

Recommendation: There should be a test of cost effectiveness, through cost-benefit assessment methodology, for all works undertaken by drainage authorities irrespective of the intention to seek grant aid

10.7 *Funds for conservation*

Unseasonal flooding of bird breeding sites and tidal inundation of freshwater based coastal ecosystems can be very damaging. The predicted rise in sea levels as a result of the greenhouse effect will make the latter problem of greater significance in the future. A provision is required for the funding of works to protect sites of high conservation value. Under the present CBA procedure such sites would be given little or no financial value and would not qualify for grant aid. In order to protect nationally and internationally important freshwater sites such as the Norfolk and Suffolk Broads from saltwater flooding, there is a need for " conservation grant aid " within the budget allocated for sea defence and coast protection works. The allocation of such "conservation grant aid" should be with the express approval of the statutory conservation agencies and based upon their determination of priorities, eg SSSIs

Recommendation: Specific grant aid should be available to protect, from saltwater flooding, freshwater sites of high nature conservation importance.

10.8 *Sea level rise*

A study carried out by the RSPB upon the potential impact of sea level rise on birds and their habitats concluded that up to 26 Red Data Bird species were threatened by rising sea levels[31]. In particular, wintering wildfowl and waders on estuaries are threatened by the loss of intertidal areas through an increase in erosion of soft shores and as the sea rises up against hard sea defences[32].

The proposal by the principal sea defence agency in England and Wales, the NRA, that they can, given the money (probably £100–£150 million per annum for 20 years) build up existing sea defences around all low lying parts of the East Anglian coast presents a significant threat to estuarine species. This NRA proposal fossilises the existing coastline and as the sea level rises the mean low water mark moves shorewards, reducing the area of intertidal habitat available. The principal conservation strategy to avoid such a loss of habitat is the " controlled retreat " option. This involves selecting stretches of coast where the existing sea defence line cannot be maintained and implementing new measures to protect life, infrastructure and housing at a point further inland.

In many cases the additional measures required will be minimal because settlements historically have ben located on the higher ground which was not prone to tidal inundation, the land which now lies to seaward of such settlements being in agricultural use only. The area seaward of the new defence line is not abandoned but can be managed for nature conservation interests. As such it is a " controlled retreat " response to rising sea levels which creates new areas of intertidal land and coastal grazing marsh. For such a strategy to be implemented, positive incentives are required for landowners to manage their land for nature conservation.

There are a number of examples in the United Kingdom of sites of high nature conservation interest today that at one point in time were reclaimed for agriculture. Due to deliberate or accidental breaches in sea defences these sites are now coastal marsh, lagoons or intertidal areas. These examples show that it is possible to recreate areas of high nature conservation interest, given sufficient time, from agricultural land and indicate that the controlled retreat option is one that is practicable and viable and should be explored[33].

Recommendation: A national strategic plan to respond to the predicted rise in sea levels is required which through a policy of controlled retreat retains habitat for the internationally important wader and wildfowl populations which occur in the United Kingdom.

11. NATURE CONSERVATION LEGISLATION

One key area in which the existing United Kingdom legislation demonstrably fails is in that of coastal and marine nature conservation. Under the Wildlife and Countryside Act, 1981[18] the country agencies succeeding the Nature Conservancy Council have a duty to identify and designate Sites of Special Scientific Interest (SSSIs). In Northern Ireland DOE (NI) administers the equivalent designation, Areas of Special Scientific Interest (ASSI). The importance of these designations is recognised in Government circulars[34] but problems with the existing SSSI/ASSI system limit its effectiveness in a number of ways. A detailed critique of the legislation as it affects estuaries is given in the accompanying RSPB policy document " Turning the Tide—a future for estuaries "[1].

Although many of the activities which could occur on estuaries notified as SSSIs/ASSIs are controlled by the Wildlife and Countryside Act or NI equivalent[18], some potentially damaging operations are not. In particular, those governed by common rights, planning controls and private Acts of Parliament specifically override the protection afforded by the SSSI/ASSI notification.

The geographical coverage of intertidal SSSIs in England and Wales extends only to median low water, thus excluding and failing to give any protection to those areas falling between this point and low water spring tide. This area is often large and can be important for birds. In Scotland, however, SSSIs extend to the extreme low spring tide mark. In Northern Ireland, no decision has yet been made as to the lower tidal limit of ASSIs. Such inconsistency is without sense and is being exploited by MAFF who refuse to consult English Nature over shellfish farming operations below median low tide in spite of possible adverse environmental impacts on adjacent intertidal areas

Problems also occur where the high tide refuges for birds are located on habitats which are not of any other scientific value (eg arable land adjacent to an estuary). Damage to or disturbance of these important areas can affect bird populations and should be covered by the designation and management process.

Below low water mark the situation is at its worst. In intertidal areas below median low water and in inshore and offshore water there is no protection for wildlife or co-ordinated management that could deliver such protection. Only two Marine Nature Reserves (Skomer and Lundy) have been declared under Section 37 of the Wildlife and Countryside Act 1981, after many years of negotiation between conflicting parties. Such delays serve only to demonstrate the inadequacy of marine conservation legislation. The Marine Nature Reserve legislation is widely regarded as a failure, relying as it does on complete agreement from all parties. A consortium of bodies including the Nature Conservancy Council, World Wide Fund for Nature, the Marine Conservation Society and the RSPB recently produced an alternative scheme for marine protected areas as a discussion document for Government[2]. A considered response is awaited.

For seabirds protection of breeding cliffs is of limited value. Over 75 per cent of their year is spent at sea and their food resources are all marine in nature. Seabirds remain vulnerable to pollution, disturbance and over-fishing. Important inshore feeding, moulting and gathering grounds should be protected by the EC Directive on the Conservation of Wild Birds 79/409. The United Kingdom Government is demonstrably failing these important populations in spite of obligations under international convention and law (Ramsar Convention and Birds Directive). The Wildlife and Countryside Act 1981 provides no powers to do this. Severe criticisms of marine and intertidal wildlife conservation legislation has been made not only by the RSPB[1] but also in reports from the Worldwide Fund for Nature and the Marine Conservation Society[35,16].

It is currently quite legal for synthetic gill nets (lethal to seabirds) to be set immediately below internationally important seabird breeding cliffs and accordingly reap an avian harvest[8]. The RSPB is currently examining the scale and extent of such " bird kills ". The United Kingdom Government has a duty under the EC Birds Directive to protect seabirds whilst they are at sea. At present the domestic legislation cannot fulfil this duty and coastal wildlife is demonstrably at risk.

The United Kingdom Government's responsibilities under the EC Birds Directive are detailed in DOE Circular 27/87 (Welsh Office 52/87). Article 4 of the Directive requires that, in the case of particularly sensitive species (detailed in Annex 1 of the Directive) special measures should be taken to conserve the habitat of these species of birds in order to ensure their survival and reproduction in their area of distribution. In particular, member states are required to clarify the most suitable areas for these species as Special Protection Areas.

Sixty-eight estuaries meet the criteria for designation as Ramsar sites or as SPAs under the EC Birds Directive. To date Government has only notified 19. The EC Directive was introduced in 1979 and member states have to have met its terms by April 1981. Ten years later the United Kingdom Government has still only managed to achieve 25 per cent of the necessary coverage. There is no indication that this position will change markedly. Unless major efforts are made we can look forward to further years of delay and failure to meet international commitment.

Areas of sea qualifying for such protection remain unidentified and unprotected by either domestic or national wildlife convention or law. The protection of marine and coastal birds remains woefully inadequate.

Recommendation: Designation of SSSI/ASSIs must be uniformly to spring low tide mark throughout the United Kingdom where conservation value merits it.

Recommendation: A workable system of marine wildlife protection through Marine Protected Areas or Marine Management Zones must be instigated as soon as possible.

Recommendation: Common law rights and traditional activities such as baitdigging, wildfowling, shellfish gathering and access can be shown, in certain circumstances to damage wildlife interests. Where this is the case, local or harbour authorities should control such activities through any existing powers (eg Local Nature Reserve bye-laws). Where no such powers exist, the appropriate authorities must be given powers similar to those contained in Section 28 of the Wildlife and Countryside Act 1981. Such controls should form part of whole estuary management plans and regional coastal strategies.

Recommendation: Fisheries legislation must be amended by giving fisheries Ministers duties to "further" conservation so as to allow protection to be given to marine wildlife at risk from fisheries operations. This would place fisheries on an equivalent footing to Agricultural matters.

Recommendation: All estuaries meeting the internationally agreed criteria should be designated as Special Protection Areas under the EC Directive or as Ramsar sites under the Ramsar Convention as soon as possible.

Recommendation: Government must as soon as possible publish a list of sea areas qualifying for designation under the EC Birds Directive or Ramsar Convention.

REFERENCES

(1)   Rothwell P I and Housden S D (1990). "Turning the Tide, a future for estuaries", RSPB. Sandy.

(2)   Warren L and Gubbay S (eds) (1991). Marine Protected Areas, a discussion document. MPA Working Group, c/o WWF(UK), Godalming.

(3)   Grimmett R F A and Jones T A (1989). Important Bird Areas in Europe. ICBP/IWRB Cambridge.

(4)   European Community Directive on the Conservation of Wild Birds 79/409 (1979).

(5)   International Convention on Wetlands of International Importance especially as wildfowl habitat. Ramsar, Iran 1971. Cmnd. 6465. HMSO London.

(6)   Rothwell P I (1989). Update on threats to United Kingdom estuaries, *Conservation Review*, 3:28–29. RSPB, Sandy.

(7)   Davidson N C *et al* (1991). Nature Conservation and Estuaries in Great Britain. NCC. Peterborough.

(8)   Robins M (1991). Synthetic Gill Nets and Seabirds. RSPB/WWF Sandy.

(9)   Gubbay S (1990). A future for the coast? Proposals for a UK Coastal Zone Management Plan. MCS. Ross-on-Wye.

(10) BANC ECOS (1991) Vol 12 No 2 Swimming against the tide.

(11) Town and Country Planning Act 1990.
      Town and Country Planning (Scotland) Act 1972.

(12) Local Government Act 1972, Section 72.

(13) Local planning authorities have powers over development. Section 55(1) of the 1990 Act (Section 19 TCP(S)A 1972 defines development as "the carrying out of building, engineering, mining or other operations in, on, over or under land, or the making of any material change in the use of any building or other land".

(14) Crown Estate Act 1961.

(15) The Hampshire County Council's Portsmouth Harbour Plan and River Hamble Local Plan and Lincolnshire County Council's "Reclamation of the Lincolnshire Coast" Subject Plan, 1982.

(16) In the Dee Estuary, adjacent local authorities have adopted policy stances diametrically opposed to each other. Local government on one side of this estuary promotes industrial development and infrastructure along the shoreline whilst on the opposite bank strong conservation policies are pursued. This different view of the future of a common resource prevents agreement on the management of a single ecological unit that is designated as a Ramsar site, Special Protection Area and Site of Special Scientific Interest.

(17) Department of the Environment Circular 12/72 (Welsh Office 36/72). The Planning of the Undeveloped Coast.

(18) Wildlife and Countryside Act 1981.
      Nature Conservation and Amenity Lands (Northern Ireland) Order 1985.

(19) North Sea Oil and Gas Coastal Guidelines, SDD, 1974.

(20) This is particularly, but not exclusively, the case with "development under local or private Act or Orders" (Schedule 2 Part 11, Class A); "Dock, pier, harbour, water transport, canal or inland navigation undertakings" (Schedule 2, part 17, Class B); "dredgings" (Schedule 2, Part 17, Class D), etc.

(21) The RSPB consider that in the case of "permitted development" rights granted under Schedule 2, Part 17 of the Town and Country Planning General Development Order 1988 (as amended), development should not be permitted if it would consist of or include works within a notified SSSI, or a candidate/designated EC Special Protection Area or Ramsar site. "Permitted development" rights should equally not apply to works outside of one of these sites which will have adverse effects on that site. Similar provisions should apply to all other "permitted development" operations. The General Development Order should therefore be amended by the inclusion under Article 3 (7) of a new condition (c) which shall stipulate that "Schedule 2 does not grant permission for . . . (c) development which involves or is likely to involve works or operations within a designated Site of Special Scientific Interest or within any area which would materially affect the scientific interest of that designated site". The equivalent provisions in Scotland and Northern Ireland should be so changed.

(22) Town and Country Planning (Assessment of Environmental Effects) Regulations 1988.

(23) The Zetland and Orkney County Council Acts 1974.

(24) The Crown Estate Commissioners, Marine Update. Winter 90/91. WWFN. Godalming.

(25) Fish Farming in the United Kingdom 1989. RSPB Evidence to the House of Commons Agriculture Committee.

([26]) In the United Kingdon as a whole the losses of wet grassland habitat since the 1930s have reached over 40 per cent by area. In the Greater Thames estuary (the north Kent, east London and south Essex coast) the area of coastal grazing marsh has fallen from 44,000ha in the 1930s to 16,000ha in the 1980s, a loss of nearly two-thirds. Conversion of grassland to arable production has been the major cause (69 per cent of losses) but in an area so close to London, losses to industrial and residential development have also been significant. Ref Ekins R 1990. Changes in the extent of grazing marsh in the Greater Thames Estuary. RSPB. Sandy.

([27]) Water Act 1989. Section 8.

([28]) Water Act 1989.

([29]) For this reason the RSPB views with grave concern the Government announcement (8 July 1991) to form a new Environment Agency whose functions could be limited solely to pollution regulation. This would require the existing water pollution regulation, monitoring and enforcement functions to be removed from the NRA. The RSPB regards this mentioned proposal as a retrograde step, further dividing responsibilities for the water environment amongst agencies and leading to the loss of the well co-ordinated integrated catchment approach being developed by the NRA since its creation in 1989.

([30]) The continued existence of IDBs appears to be justified on the grounds of satisfying particular local needs, the input of local knowledge, local funding of works and historical precedence. The argument that only they can satisfy some essential local drainage need is not borne out by the fact that large areas of low-lying coastal land in England and Wales continue to exist without the " benefit " of IDBs. There are no drainage boards in large parts of northern, south west and southern England and west Wales. In other regions a number of boards are managed by the NRA, providing the administrative, technical and direct labour input.

The concept of the value of local knowledge may have been useful in the 19th century but technological advances with hydrometric gauging stations, flood forecasting, aerial photography and satellite imagery have superseded such local input. The only value of IDBs which exists today is that they provide a mechanism for raising drainage levies directly from those who benefit, that is the owners and occupiers within the district. Given that the NRA has powers to raise general and special drainage charges through the provisions of the Land Drainage Act 1976 this argument is not a convincing one. The cost of maintaining the existing IDB structures, if passed to the NRA, would still fall upon the owners and occupiers within the former drainage district through the use of a special drainage charge.

([31]) The impact of sea-level rise on red data birds (1989). RSPB internal review.

([32]) On the Essex coast it has been estimated that a rise of 0·8 metre in sea level could lead to a loss of mudflat area of 20 per cent, with up to a 50 per cent loss for a rise greater than 1 metre (DOE 1988 *Possible Impacts of Climate Change on the Natural Environment in the United Kingdom*). From studies on many estuaries it has been shown that these estuaries cannot hold any more birds than they do at present—birds cannot simply " squeeze up " if mudflat feeding sites were reduced in area by rising sea levels.

([33]) These examples are the Blyth Estuary, Suffolk; Minsmere RSPB reserve, Suffolk; Titchwell RSPB reserve, Norfolk and Pagham Harbour, Sussex.

([34]) DOE Planning Circular 27/87 (Welsh Office 52/87, Scottish Office 1/1988, DOE(NI) to be published).

([35]) Gubbay S (1989). Using SSSIs to Conserve Seashores for their Marine Biological Interest. WWF(UK). Godalming.

([36]) Warren L (1989). Statutory Marine Nature Reserves in Great Britain, A Progress Report. WWF(UK). Godalming.

([37]) Hampshire County Council (1991). A Strategy for Hampshire's Coast.

([38]) Department of Trade (1977). Marine Activities—Guide to the Responsibilities of Government Departments and Agencies. HMSO London.

([39]) Evans P R, Dugan P J (1984). Coastal birds: numbers in relation to food resources. *In* Evans, P R, Goss-Custard, J D & Hale W G, (eds), Coastal waders and wildfowl in winter: 8–28. Cambridge: Cambridge University Press.

([40]) Goss-Custard J D (1985). Foraging behaviour of wading birds and the carrying capacity of estuaries. *In* Silby, R M & Smith, R M (eds) Behavioural Ecology: Ecological consequences of adaptive behaviour: 169–188. Oxford: Oxford University Press.

([41]) Evans P R & Pienkowski M W (1984). Population dynamics of shorebirds. *In* Evans P R, Goss-Custard J D, & Hale W G (eds), Coastal waders and wildfowl in winter: 82–123. Cambridge: Cambridge University Press.

([42]) Prater A J (1981). Estuary Birds of Britain and Ireland. Calton: Poyser.

([43]) National Audit Office (1991). Department of Transport: Oil and Chemical Pollution at Sea. HMSO London.

([44]) Licences for marine aggregate extraction are granted by the Crown Estate Commissioners. Prior to a licence being granted, the Department of the Environment undertakes consultations on behalf of the Commissioners. The consultees must include the successor bodies to the Nature Conservancy Council and the local authorities. How much attention can be paid to the impacts of dredging on marine wildlife, archaeology, fishing, hydrology and coastal erosion depends on the extent of knowledge available on present value and impacts. There remains considerable doubt that such matters are taken into account during the impact assessment process. In addition as with other criticisms of the CEC in regard of fish-farming[24,25] the position of landlord, benefactor and decision maker is inappropriate and should be changed (see section 9).

Appendix 1

PROVISION OF THE CHICHESTER HARBOUR ACT

Provisions which amplify general powers of the Conservancy, or provide safeguard for specific interests include:

- duty to have regard to the desirability of conserving the natural beauty and amenity of the countryside;
- duty to have regard to avoiding interference with fisheries;
- qualified ancillary powers to do or promote the provision, erection or maintenance of buildings and plants etc;
- permissive powers to hold exhibitions, regattas etc; provide accommodation and related facilities, places of refreshment, roads and parking places; disseminate publicity and provide information centres; and levy relevant charges;
- permissive powers to enter into agreements etc with other legal entities to undertake the Conservancy's ancillary functions;
- no exercise of functions on land without the consent of any such person who holds an interest in that land;
- no exercise of functions to the exclusion or restriction of any other enactment (including the Planning Acts re non-operational land) except under the provisions of such an enactment;
- permissive powers to promote or oppose legislation;
- permissive power to acquire and dispose of land or of rights or interests over land;
- heavily qualified, permissive powers to extinguish by order certain public rights of way and other rights or easements over land purchased by the Conservancy;
- various powers, applied as to a county council drawn from the National Parks and Access to the Countryside Act 1949, including creation of Local Nature Reserves, Country Parks, camping and picnic sites; access agreements; wardens;
- all the usual statutory powers for the control of navigation and operation of the functions of a harbour authority;
- byelaw-making powers, including one for the purposes of conserving flora and fauna.

Other matters provided for include:

- repeals of old harbour legislation;
- powers of local authorities and other statutory parties to enter into agreements etc with the Conservancy;
- protective provisions for statutory undertakers, the National Trust, ancient charter rights, Coast protection legislation, Crown rights;
- arbitration procedures to resolve disputes between constituent local authorities.

In 1966 the West Sussex County Council commissioned a planning study to review development options in and around the harbour. This was undertaken with special reference to the conservation of amenity and natural values and to the provision of recreational facilities. The study concluded that there was a need to rationalise management control of the harbour. Negotiations between local authorities and other interested parties led subsequently to the enactment of the Chichester Harbour Conservancy Act 1971. The effect of the Act was to bring the harbour area under the management of a single body—the Chichester Harbour Conservancy. Pre-existing harbour powers, vested in district local authorities, were extinguished.

Section 5 of the Act specifies membership of the Conservancy as comprising representatives of local authorities, plus members appointed by an Advisory Committee. In effect, the Chichester Harbour Conservancy is a trust port (*vide:* the Ports Act 1991, which empowers such ports to privatise; the Conservancy has expressly turned its face against such a move). Under Section 21 of the Act the Conservancy's general function is defined as:

> *"to take such steps from time to time as to them seem meet for the conservancy, maintenance and improvement of—*
>
> *(a) the harbour, for the use of pleasure craft, and such other vessels as may seek to use the same;*
>
> *(b) the amenity area, for the occupation of leisure and recreation and the conservation of nature;*
>
> *and the facilities. . . . afforded respectively therein or in connection therewith."*

(Note: " amenity area " is expressly defined by way of a plan).

The Advisory Committee—The functions and membership of the Committee (15–17 persons) are set out in section 11 of the Act. An attempt is made to accommodate representatives of the principal recreation and conservation bodies and interests. Interests represented include local residents, nature conservation, recreational boating, commercial fisheries, angling, wildfowling, and the boatbuilding industry. The Conservancy has a qualified duty to consult the Committee on specified matters. The Committee may offer unsolicited advice, which must be taken into consideration by the Conservancy.

The Conservancy is a statutory consultee over town and country planning matters which affect the harbour. The constituent planning authorities are West Sussex County Council, Hampshire County Council, Chichester District Council and Havant Borough Council. Statutory development plans and relevant non-statutory plans and policies are drawn up in close consultation with the Conservancy. A commonality of development plan provisions and other policies has resulted. The system is not perfect but makes for a coherent approach not found elsewhere in the United Kingdom (except to a degree in neighbouring Langstone Harbour).

The harbour has been declared by the Environment Secretary as a Special Protection Area in terms of EC Directive 79/409 on the Conservation of Wild Birds, and as a wetland of international importance under the Ramsar Convention (*vide:* DOE Circular 27/87 " Nature Conservation "). The harbour is notified under section 28 of the Wildlife and Countryside Act 1981 as a Site of Special Interest. It has been a SSSI since 1970.

Since 1982, the RSPB has been nominated to appoint a member to the Advisory Committee to represent the interests of local naturalists. It is an active participant in the work of the Committee, and is currently much involved in its review of planning policies.

Appendix 2

A SELECTION OF CASE STUDIES ILLUSTRATING THE NEED
FOR COASTAL ZONE MANAGEMENT IN THE UNITED KINGDOM

Memorandum by Wildlife Link[1]

The coastal zone has traditionally been seen as a boundary area from both land and sea—the limit of jurisdiction for administration, planning, management and legislation. Yet it is the interface of these two very different but interlinked environments which makes it unique. It is totally inappropriate to treat the meeting point of these environments separately by incorporating the landward part into the land planning procedures and for dealing with the seaward part in a different manner. Such an approach bears no relationship to the way the environment operates nor the uses to which it is put. Activities on the land have direct implications for the marine environment and activities and vice versa. It should come as no surprise therefore that the existing system has led to problems. These problems have been identified by a range of different organisations and interest groups, many of whom are now calling for action on coastal zone management (Annex 1).

Coastal zone management can provide practical solutions to tackling these problems through a system which links planning for both land and sea, developing integrated policies rather than dealing with sectoral interests in isolation, and having initiatives and support at local, regional, and national levels. Support at national level is the key to success yet it is absent in the United Kingdom.

The main issues of concern are listed below and expanded in Annex 2. They are illustrated by 22 case studies which show the unco-ordinated and piecemeal approach that typifies current coastal administrative mechanisms.

- lack of focus and co-ordination in Government to provide a lead, guidance and support in dealing with coastal zone matters
- the absence of a national strategic planning perspective for the coastal zone
- the many agencies who have responsibilities in the coastal zone but who lack a clear framework to integrate their operations
- the absence of a planning regime which covers developments spanning the whole of the coastal zone, ie coastal land, the intertidal zone and inshore waters
- fragmented planning control and management of activities both across and along the coastal zone
- loopholes and anomalies in the law concerned with the coastal zone
- inadequate structure and legislation to fulfil international responsibilities
- poor liaison between the many bodies involved
- unclear, overlapping and, in places, inappropriate responsibilities given to these agencies
- sectoral based approach to decision making in the coastal zone
- public involvement in planning largely restricted to the landward side of the coastal zone
- planning and management arrangements that regard the coastal zone as a boundary area rather than a focus for specific policies.

There is an urgent need for a clear lead from Government at a national level to overcome these shortcomings and to ensure that the coastal zone is given the specific attention it needs. Without this it will continue to be regarded as a boundary area, on the edge of everyone's concern and the responsibility of no one.

Case study 1: Whose responsibility?

The wreck of the *Kimya*, carrying 1,500 tons of sunflower oil, sank in Caernarfon Bay in January 1990. The vessel was moved a short distance by the salvors to a more sheltered position and now lies some hundred yards from a shore which has AONB, Heritage Coast, SSSI and NNR status as well as being a proposed MNR. The salvors subsequently abandoned work on the vessel, and the owners of the wreck and cargo also decided that they had no further interest in the wreck. The Marine Pollution Control Unit (MPCU) did not believe there was a sufficient case to pump out the oil and suggested a "controlled leak" against the advice of the Countryside Council for Wales whose views are supported by the NRA. The Borough Council are also alarmed that an oily mixture might end up on amenity beaches in Anglesey. Meanwhile oily residues have been observed around the wreck and mussels on the shore have died—analysis shows high quantities of sunflower oil derivates in the tissue. The Borough Council cannot intervene as the wreck lies outside its jurisdiction, the salvors and owners of the cargo have been able to relinquish all responsibility, and the MPCU will only support a "controlled leak". The situation is in stalemate despite potential environment impact on a stretch of coast that has virtually every conservation designation afforded by the law.

---

[1] Wildlife Link is the liaison body for all the major voluntary organisations in the United Kingdom concerned with the protection of wildlife. This Report was compiled in recent months by Wildlife Link member organisations working on Coastal Zone Management Policy. The Report was specifically in response to a request from the Secretary of State for the Environment.

Case study 2: Seaward boundaries of local government areas and planning control

There is confusion about the seaward limits of town and country planning in England and Wales and inconsistencies with the situation in Scotland. The maximum extent of planning control currently depends on local government areas. For anachronistic, historical reasons they normally end at mean low-water mark on the open coast although a few happen to have been extended out to sea by ancient charter or private legislation. There has been no legal delimitation of most river mouths or estuaries. Instead, administrative practice relies on an extra-statutory procedure adopted by the Ordnance Survey in the nineteenth century which produces arbitrary and inconsistent closing lines. Moreover, it is uncertain whether planning control is legally exercisable in these subtidal waters within local government boundaries. Ministerial statements that such planning powers do exist in England and Wales are inconsistent with a decision of the Court of Session in Scotland based on the interpretation of identical legislation. Local planning authorities are thus unwilling to assert jurisdiction in estuaries without clarification of the seaward scope of their functions.

Case study 3: An isolated view of environmental impacts

The fitting of Flue Gas Desulphurisation (FGD) at Drax power station (Yorkshire) will increase the loading of the River Ouse with heavy metals, (particularly cadmium and mercury) yet the biological impact on the river system has not been considered in the environmental impact assessment. If the FGD programme proceeds with further retrofits this area could see wastewater loading of the River Ouse (from Drax), the River Aire (from Ferrybridge, and possibly Eggborough) and the River Trent (from Radcliff), all of which empty into the Humber Estuary. This would lead to a significant increase in many heavy metal contaminants in this estuary, and will therefore be in breach of the North Sea Conference agreements to reduce inputs of cadmium and mercury. This type of problem arises from taking a narrow view of the impacts of activities in the coastal zone.

An increase in the metal content of the water will also lead to an increase in the heavy metal contamination of the river and estuarine sediments. The Humber is a primary United Kingdom shipping estuary and to maintain the shipping channels and harbour berths it is necessary to dredge the sediments, the dredged material is subsequently disposed of in the North Sea. Again the North Sea Ministerial Conference declaration agreed to reduce the contamination of dredged materials entering the North Sea from ports and harbours, by reducing inputs at source. However, the current system which deals with effluent input into the coastal zone does not also consider the effects of subsequent disposal of sediment which is contaminated by such effluent.

Case study 4: Development at the coast

A proposal for a marina at Swanage illustrates the sorts of problems which can arise in the absence of a unified system for authorising the placing of " structures " in different parts of the coastal zone. The scheme was for a combined marina and housing development in Swanage Harbour. A Private Bill was required to get permission for the Marina (because of the navigational implications) whereas the housing development was covered by the Town and Country Planning system. Permission was refused for the marina but not the housing. The result has been the partial completion of the work, and bankruptcy of the developer. The project was only viable in combination but there was no mechanism for considering it in this way.

Case study 5: Water quality

Eastbourne beach has qualified for a European Blue Flag for the last three years. Its bathing water passes the mandatory standards of the EC Bathing Waters Directive 76/160. Eastbourne's sewage is discharged from a long sea outfall at Langney Point in the Borough of Eastbourne. The typical water circulation here causes the sewage to affect Pevensey Bay (Wealden District Council) just along the coast whose water quality has been borderline for some years. In effect, Southern Water discharges Eastbourne's sewage from one district/borough council to another, with all the concurrent public health and tourism implications.

Case study 6: Marine fish farming

The marine fish farming industry illustrates many of the problems with current arrangements for planning and administration of activities in the coastal zone. The lack of any planning mechanism spanning the whole of the coastal zone has had significant implications for Highland Regional Council. They have incurred substantial unplanned expenditure in repairing rural roads damaged by the heavy vehicles which service fish farms, yet the siting of these operations is largely outside their control. One such case was the need to upgrade access to Loch a'Choire which initially cost the Council approximately £300,000, and subsequently a further £450,000. In another case (Kylesku-Lochinver) £270,000 was spent on road repairs in 1987 and 1988, largely due to fish farm traffic. Despite the costs to the Council, permission for the siting of fish farms is outside their remit.

Existing legislation is also inappropriate for regulating this coastal zone industry and its use has led to some confusion. Thus marine fish farms appear to be treated as fisheries. Section 9 of the Conservation of Seals Act 1970 permits the killing of seals to prevent " damage to a fishing net or to fishing tackle " and fish farmers have been allowed to rely on this defence even though fish farming equipment does not fall within the definition of net or tackle. Similarly, licences issued under s.10, which may be given to protect "fisheries", are also inappropriate. Freshwater fish farms, including salmon hatcheries, however are regarded under planning law as an agricultural use. We thus have a situation in which an activity, that is neither fishing nor farming in the usual understanding of these words, may be treated as either according to the circumstances.

Case study 7: Passing the problem along

Many of the problems which have arisen over coast protection and flood defence works are a consequence of the lack of a coastal zone management system. On the south coast, groynes erected by Bournemouth Borough Council have been restricting movement of sediment into Christchurch Bay. This has had an incremental effect around the bay and led Christchurch Borough Council to build a series of groynes to trap sediment. They have now reached the easternmost extent of their coastal boundary with a series of groynes. The last groyne in the system has created a classic scour pocket which is eroding away the borough council's own land (a landfill site which has been built on) but they cannot add any more groynes as this would have to be in an area under the jurisdiction of New Forest District Council. In the absence of any coherent strategy for coast protection the problem has gradually moved on from council to council. The solution for each one has been considered in isolation from the possible impacts on the area under the jurisdiction of the adjacent council. The result has been considerable expenditure which has not resolved the problem but passed it along the coast.

In another case on the south coast, East Head spit within the Chichester Harbour SSSI is threatened as a result of coast protection works up channel. Serious loss of level, attributed to heavy offshore dredging, and erosion along the whole coastline, have necessitated the building of breastworks and groynes. These are protecting part of the coast but preventing any build-up of material on the spit. Protection works are being proposed at a cost of £20,000.

Case study 8: Legal status of SSSIs in the coastal zone

The requirement to notify owners, occupiers and local authories of the existence of Sites of Special Scientific Interest (SSSIs) and the absence of such bodies for marine areas means the legal status of SSSIs beyond low water is unclear. This raises doubts about whether the United Kingdom can fulfil its international obligations under the Ramsar Convention and the EC Birds Directive 79/409. In the United Kingdom the SSSI system is used to protect Ramsar sites and Special Protection Areas. Both these designations must cover areas of sea yet the legal status and powers of protection in coastal waters provided by SSSIs do not appear to allow this.

With regard to SSSIs there is also no justification for differences in the legal extent to which SSSIs can extend over the intertidal zone in different parts of the United Kingdom. In England and Wales they can only extend to mean low water, thus excluding and failing to give any protection to those areas falling between this point and low water spring tide even though they may be of equal scientific value. In Scotland, SSSIs can extend to the mean low spring tide mark and in Northern Ireland no interpretation has yet been made as to the lower tidal limit. Situations where this has led to uncertainty and difficulties include trying to control bait digging which has led to damage of eel grass beds in the Portland Harbour SSSI, bait digging in Langstone Harbour SSSI and the cultivation of the Pacific Oyster in the Fleet SSSI.

Case study 9: Development control in the coastal zone

The role of the Crown Estate Commissioners (CEC) is another illustration of the difficulties which arise because of a policy vacuum on coastal zone management. The CEC, who own much of the foreshore and the seabed on behalf of the Crown, have considerable powers over the use to which the seabed is put, but few of the responsibilities for environmental protection which should be associated with such powers. As principal landowner in the marine environment, and in the absence of any planning system which covers sea areas, the CEC, under pressure from the Government have assumed a quasi-planning role by default, despite the fact that they are not a publicly accountable body. At the same time the Government views the CEC as independent and has designated them as the "competent authority" for the purposes of the Environmental Assessment (Salmon Farming in Marine Waters) Regulations 1988. Despite the fact that they have a financial interest in the outcome of any assessment the Government has found it necessary to use the CEC as more than merely a landlord.

Case study 10: Unclear and unresolved responsibilities

In 1986 a semi-submersible oil rig, the *Ocean Kokuei*, was "parked" about half a mile from the shores of the tourist village of Mevagissey. It was, and still remains, unclear how this type of action can be regulated despite the potential threats to navigation, fisheries, amenity and nature conservation. Such actions may also become more common with the decommissioning of oil rigs. The Department of Energy can regulate operating oil rigs through a works licence: when it is underway the structure is considered to be a vessel and therefore the responsibility of the Department of Transport, but it is not clear who is responsible when such structures are "parked". The Crown Estate Commission cannot intervene if the structure is a "vessel at anchor". The Ministry of Agriculture, Fisheries and Food may also have a role because of its responsibilities over dumping at sea. In this particular case the Department of the Environment intervened on the grounds of "visual impact".

Case study 11: The threats to estuaries

The recent NCC estuaries review has detailed the piecemeal damage and destruction of estuaries and their wildlife in Great Britain. Most estuaries fall within the jurisdiction of several local authorities. Hence no one body can act independently to ensure their safeguard as an ecological unit. For example the Dee Estuary marks the boundary of England and Wales. It falls within the jurisdiction of two county councils and six district councils. Despite being designated as an SPA and a Ramsar site, there is no co-ordinated approach to strategic or regional planning. This is reflected in the complete lack of any whole estuary management approach.

Case study 12: Lack of national ports strategy

The successful passage through Parliament of the Felixstowe Dock and Railway Act 1988 by means of a Private Bill allowed the Felixstowe Dock and Railway Company to destroy part of the conservation value of the Orwell Estuary SSSI, a candidate SPA/Ramsar site.

The expansion of the docks was promoted by the Dock Company through Parliament as a scheme of national economic importance and as the clear and obvious place to develop further dock facilities. The expansion of the docks was proposed when Felixstowe Docks had a competitive advantage over other United Kingdom docks but, with the Government decision to abolish the National Dock Labour Scheme and privatisation of Britain's state owned ports, they have since lost this advantage. The complete lack of national guidance for port use and development in this case allowed the pointless and unnecessary destruction of an internationally significant habitat.

Case study 13: No channel for concern?

The Lymington to Yarmouth ferry service has recently been taken over by Wightlink who would like to replace existing ferries with vessels some 20 m longer and with a greater capacity for vehicles. To cater for this dredging will be required to straighten the channel and widen the approaches. The District Council have expressed their concern on three points. The saltmarshes in the area are rapidly eroding and a recent study identified wash from ferries as one of a number of contributing factors. Larger ferries will exacerbate this problem. Increased shore traffic for the ferry, especially at peak times, will lead to congestion and the number of small boats moored along the river give reason for concern on safety grounds. Despite all this the District Council cannot intervene. The Harbour Commissioners who have also voiced concern, believe they cannot deny access and would need to carry out the dredging work to ensure safety.

Case study 14: Unclear and conflicting responsibilities at the coast

The National Rivers Authority has statutory duties upon and powers to carry out works relating to flood defence, environmental quality and pollution control, fisheries, recreation and conservation and water resources management. The NRA is the Sea Fisheries Committee in the Dee estuary, where concern has been expressed by the Dee Estuary Conservation Group (DECG) about cocklers operating on the foreshore disturbing wildlife, damaging the mudflats and leaving unsightly litter. As a Sea Fisheries Committee the NRA can introduce bylaws to conserve fish stocks (eg restricting the taking of small cockles). However it cannot introduce byelaws on conservation grounds alone, eg to reduce disturbance to birds. The NRA is also the navigation authority for the Dee Estuary and the DECG has voiced concern about the effects on wildlife of recreation activities in the estuary, especially in light of recent applications for speed boat and water skiing facilities on the Dee. Whilst the NRA can create byelaws to restrict water recreation where they threaten the safety and freedom of navigation, they cannot restrict this use on conservation grounds alone or restrict freedom of navigation.

Case study 15: Disregard for planning guidance

Morrich More is a stretch of the Dornoch Firth coastline, designated an SSSI. It is within a designated National Scenic Area, and it is part of a candidate SPA/Ramsar site. In 1986 following a public inquiry the Secretary of State granted planning permission for the construction of a pipeline fabrication plant at Morrich More. The development was allowed to proceed in spite of contrary national and local guidance: damage to the SSSI was not sanctioned in the Highland Region Structure Plan. The development was expressly contrary to the Scottish Development Department North Sea Oil and Gas Coastal Planning Guidelines. Furthermore, although clearly not retrospectively applicable, the development was contrary to Government thinking on damage to SPAs later included in Circular 1/1988 which spelt out interpretation of the EC Birds Directive 79/409.

The public inquiry which followed the application did not contest that Morrich More was of national and international importance in terms of wildlife. However, despite viable alternative sites for the development and contrary to policy it was allowed to proceed.

Case study 16: Liaison over coastal matters

The port of Mostyn lies within the Dee Estuary which is an SSSI and is a designated SPA and Ramsar site. The Mostyn Docks Harbour Empowerment Order 1988 has enabled the Mostyn Docks Company to become a statutory harbour authority under the Harbours Act 1964. This allows the Harbour Company to introduce and enforce byelaws, which may include directions concerning harbour use, and dredging for sea-going vessels but does not include nature conservation duties. In spite of the entire seaward part of the Company's operation area being within the Dee Estuary SSSI the Mostyn Docks Company and the Department of Transport failed to consult with the NCC over the Empowerment Order. Only through the tortuous method of a separate non statutory agreement has the NCC managed to secure any consultation over the safeguarding of conservation interests.

The Department of Transport also failed to consult with the Department of the Environment concerning the Empowerment Order and could have compromised the Government's position under the Ramsar Convention and the EC Birds Directive 79/409.

Case study 17: Lack of Government co-ordination

Early in 1990 the Department of Transport produced a consultation document proposing to replace the existing Decca Navigation system with another land-based system known as Loran-C. In the meantime the Ministry of Agriculture, Fisheries and Food, while looking at ways of increasing the monitoring of vessels dumping industrial waste, sewage sludge and dredged sediment through placing marine position recorders (" black boxes ") on disposal vessels, had drawn up a specification for the equipment based on the Decca Navigation system. The Department has recently dropped its proposal to introduce Loran-C much to the relief of the shipping, fishing and port industries which are increasingly relying on the more accurate satellite system, GPS, which has been further enhanced following the Gulf conflict and could potentially outplace both Decca and Loran-C before the end of the century.

Case study 18: Legal complexities and anomalies

The powers of local authorities in England and Wales to regulate public recreational activities on beaches and close inshore are too limited and out of date. District councils are empowered by the Public Health Acts to make byelaws regulating bathing, boating and other public behaviour, but only in order to prevent danger, nuisance or annoyance to people using the shore or bathing in the sea. Consequently, those powers cannot legally be used to control the same activities if they are harmful in other ways. For example, local authority byelaws restricting bait digging, such as those at Southend and Alnwick, could only be made to prevent a public nuisance, but not for reasons of nature conservation. The law is also too narrowly drafted, and fails to take account of modern developments in water sports.

In port or harbour waters different leisure craft, ie windsurfers, powerboats, yachts and high-powered personal water craft (PWCs) (eg jetskis) can be zoned to avoid conflict however this cannot be done on beaches outside the port's jurisdiction. This year Weymouth & Portland Borough Council, in its role as port authority is trying to extend the limits of its jurisdiction through a Harbour Revision Order so enabling it to adequately control the use of PWCs. Weymouth & Portland Borough Council has however now banned the launching of PWCs from council-owned beaches in the area. This is not good management of the coastal zone. The problem and the conflict of use has merely been moved to a different location. The complete lack of guidance on regional coastal management or adequate controlling powers has led to a confused situation.

Case study 19: Lack of Government focus

Currently the provisions of the MARPOL 73/78 Convention and shipping and port related environmental initiatives are implemented by the International Maritime Organization following discussion by its Marine Environment Protection Committee (MEPC). The United Kingdom delegation to the MEPC consists of officials of the Department of Transport and representatives of the shipping industry. The only input on environmental issues from officials of the Department of the Environment has been at pre-meeting briefings for one agenda item only—on the use of tri-butyl tin (TBT) in antifouling paints. This is despite the agenda covering such important issues as designation of particularly sensitive sea areas, adequacy of the provision of waste reception facilities in ports, air emissions from ships and shore installations.

Case study 20: Coastal zone responsibilities at the Department of the Environment

The Directorate of Planning Services of the Department of the Environment (DOE) has recently invited bids for a " Review of Coastal Planning Policy and the Role of Earth Science Information in Coastal Planning " as part of its Geology and Minerals Planning Research yet was unaware that DOE have offered grant aid to a voluntary body to produce a coastal zone plan in north west England. At the same time, the Environmental Protection Central Division has expressed interest in coastal zone management and has commissioned work in this area through the North Sea Task Force under its Marine Branch. At a recent workshop on coastal zone management organised by the Marine Conservation Society, the DOE sent a representative from a third area of its work, namely the Wildlife Branch, based at Bristol. On occasions, funding for research on the coastal zone has come from more than one part of DOE as, for example, in the case of a study of sandeels in Scotland which was funded separately by DOE in London and Bristol.

Case study 21: Environmental assessment

Responsibilities for environmental assessment are artificially fragmented in the coastal zone, and projects may be divided between two authorities at the coastline. Above low-water mark, environmental assessment is incorporated into the planning process, and local planning authorities have jurisdiction. Beyond low-water mark, the Department of Transport or the Crown Estate Commissioners take over, and environmental assessment is included in the application procedures for harbour orders, tidal works licences or seabed leases for marine salmon farms. Thus, the environmental implications of the landward and seaward portions of a development that crosses low-water mark, such as a harbour, marina or fish farm must be separately considered by different authorities, despite their obvious inter-relationship. Moreover, although local planning authorities are responsible for considering the environmental assessment of new sea defence works, the implications of improvements to the same works are determined by the Ministry of Agriculture, Fisheries and Food. On the other hand, the need for environmental assessment of dredging for marine minerals is decided by Government through the Department of the Environment whilst dredging of harbour sediments within ports does not require any environmental impact assessment.

Case study 22: Overlap of responsibilities

Following the establishment of the NRA under the 1989 Water Act it has become apparent that a number of direct overlaps in responsibilities now occur between the NRA and harbour authorities. There is also considerable confusion as to who is responsible or who should take the lead in certain circumstances. Who has the duty and who has the capability?

For example, who should take the lead in the event of an oil or chemical spill on an estuary? Does it vary depending on whether the spill is from a vessel or a shore-based installation? Who is responsible if the spill results from a hose linking a vessel with a shore-based installation? Who is responsible for prosecution of the offending installation or vessel? In some areas it is not only the harbour and the NRA that will be involved. The local authority also has certain responsibilities as may the local fire brigade in some areas. The County Council may also be involved and the Department of Transport's Marine Pollution Control Unit takes the lead when the spill is outside of harbour waters but how does this relate to the NRA's responsibilities up to three miles offshore?

The NRA and harbour authorities are also able, under different pieces of legislation, to make byelaws for the same area of water. The requirements of the different bodies will not necessarily be compatible as the responsibilities are different. The harbour authority has a responsibility to maintain safe navigation within the estuary while the NRA will be concerned with the water quality. These could potentially conflict, particularly with respect to dredging requirements.

Annex 1

ORGANISATIONS THAT HAVE EXPRESSED SUPPORT FOR RATIONALISATION OF COASTAL ZONE PLANNING AND MANAGEMENT.

Association for the Protection of Rural Scotland
Campaign for the Protection of Rural Wales
Centre for Environmental and Economic Development
Council for the Protection of Rural England
County Planning Officers Society
Countryside Commission
Greenpeace
Heritage Coast Forum
Institution of Civil Engineers
Marine Conservation Society
Marine Forum
National Trust
National Trust for Scotland
Nature Conservancy Council
National Rivers Authority
Royal Institute of Chartered Surveyors
Royal Town Planning Institute
Royal Society for the Encouragement of Arts, Manufactures and Commerce
Royal Society for the Protection of Birds
Tidy Britain Group
World Wide Fund for Nature

Annex 2

ISSUES OF CONCERN

1. LACK OF FOCUS AND CO-ORDINATION IN GOVERNMENT TO PROVIDE A LEAD, GUIDANCE AND SUPPORT IN DEALING WITH COASTAL ZONE MATTERS.

The Government provides no lead or co-ordination on coastal zone matters or a forum which draws together the many agencies and authorities currently involved in aspects of planning and management at the coast.

2. THE ABSENCE OF A NATIONAL STRATEGIC PLANNING PERSPECTIVE FOR THE COASTAL ZONE.

The coastal zone covers two very different but interacting environments and has special problems associated with it yet there is no overall view of this resource or support from Government for developing strategic planning for the coastal zone.

3. THE MANY AGENCIES WHO HAVE RESPONSIBILITIES IN THE COASTAL ZONE BUT WHO LACK A CLEAR FRAMEWORK TO INTEGRATE THEIR OPERATIONS.

There are more than 30 Government departments and agencies with responsibilities on the marine side of the coastal zone but very limited links in planning and management with those involved on the landward side. This is despite the fact that the impacts of activities carried out in either part can be far reaching.

4. THE ABSENCE OF A PLANNING REGIME WHICH COVERS DEVELOPMENTS SPANNING THE WHOLE OF THE COASTAL ZONE.

The Town and Country planning system provides a framework and focus for the main organisations involved in resolving conflicts on land and carrying out forward planning. This system does not cover inshore waters and therefore does not span the whole of the coastal zone.

5. FRAGMENTED PLANNING CONTROL AND MANAGEMENT OF ACTIVITIES BOTH ACROSS AND ALONG THE COASTAL ZONE.

Very few of the administrative boundaries at the coast bear any relationship to the physical environment. For example planning and development control around the shores of an estuary rarely falls to a single authority even though the estuarine environment functions as a single unit. The problem also occurs on open coast. Decisions on the need for coast protection works for example can, and are, taken without due consideration of the impact on the coastline of adjacent districts.

6. LOOPHOLES AND ANOMALIES IN THE LAW CONCERNED WITH THE COASTAL ZONE.

There are many examples where legislation relating to the coastal zone is unclear. Legislation has also had to be applied to situations for which it is inappropriate whilst interpretations of legislation have highlighted anomalies.

7. INADEQUATE STRUCTURE AND LEGISLATION TO FULFIL INTERNATIONAL RESPONSIBILITIES IN THE COASTAL ZONE.

The existing structures and legislation used to implement the Ramsar Convention and the EC Birds Directive for example do not appear to be adequate to discharge these responsibilities in the coastal zone.

8. POOR LIAISON BETWEEN THE MANY BODIES INVOLVED.

Liaison between the many organisations involved in planning, management and administration of coastal matters does not work very well. This is true between adjacent districts and counties on the landward side as well as between the many agencies on the marine side as well as across the coastal zone as indicated in point 3.

9. UNCLEAR, OVERLAPPING AND, IN PLACES, INAPPROPRIATE RESPONSIBILITES GIVEN TO THESE AGENCIES.

Conflicting roles within the same agencies is a problem that has come to light on many occasions along with cases where the situation is so confused that there are difficulties in knowing which agency should be responsible for dealing with a particular issue.

10. SECTORAL BASED APPROACH TO DECISION MAKING IN THE COASTAL ZONE.

The scope for integrated planning in the coastal zone is very limited at present. Decision making tends to focus on specific activities rather than having a broader perspective.

11. PUBLIC INVOLVEMENT IN PLANNING LARGELY RESTRICTED TO THE LANDWARD SIDE OF THE COASTAL ZONE.

The planning system provides a mechanism for public participation in preparing development plans yet it does not apply to coastal waters. For coastal waters the Private Bill procedure has been used to sanction developments. In the latter situation concerned individuals must seek a locus to petition directly to Parliament. Realistically this puts it beyond many who would like to participate.

12. PLANNING AND MANAGEMENT ARRANGEMENTS THAT REGARD THE COASTAL ZONE AS A BOUNDARY AREA RATHER THAN A FOCUS FOR SPECIFIC POLICIES.

There are always difficulties in dealing with " boundary areas " however this problem is especially acute in the coastal zone because administrative boundaries do not span the whole of the coastal zone.

Memorandum by the World Wide Fund for Nature

7 October 1991

Dear Dr Phipps

COASTAL ZONE PROTECTION AND PLANNING

I am enclosing the World Wide Fund for Nature's written evidence on Coastal Zone Protection and Planning and also a number of copies of WWF's Marine Update on Coastal Zone Management in the United Kingdom[1].

WWF welcomes the Environment Committee's decision to conduct an enquiry into coastal zone protection and planning. Our submission is in fact somewhat broader than protection and planning in the coastal zone as WWF believes that any use of the coastal zone should be considered within an integrated framework of coastal zone management. It is hoped that this is in keeping with the Committee's enquiry.

Our submission does not contain a great amount of detail on the current land planning system in the United Kingdom. It is recognised that the legislation varies in different parts of the country. Other evidence from bodies with which we work closely, will contain far greater detail.

A number of examples of problems in the coastal zone have been given. It is felt that an integrated system of coastal zone management would solve these difficulties. It must be stressed that these examples are only a selection of problems that currently exist.

Our submission contains references to a number of other documents. If the Committee has not already seen any of the documents we would be very happy to supply copies.

WWF is grateful for this opportunity to provide evidence to the Environment Committee on coastal zone protection and planning and would welcome providing further written or oral evidence if requested.

If there are any problems or if further copies are required, please do contact me.

Yours sincerely

Dr Sian Pullen
Marine Conservation Officer

WRITTEN EVIDENCE SUBMITTED TO THE HOUSE OF COMMONS SELECT COMMITTEE ON THE ENVIRONMENT BY WORLD WIDE FUND FOR NATURE

SUMMARY

1. The lack of a national strategy for the use of the coastal zone has been a matter of concern to WWF for a number of years. WWF's commitment to addressing this problem has resulted in a number of projects and reports addressing the overall management of the coastal zone and some dwelling on a specific activity or issue within the coastal zone.

---

[1] *Not printed.*

2. The lack of a national strategy has resulted in indiscriminate development with subsequent loss of vital wetlands and intertidal areas. Lack of control of effluents entering rivers and coastal waters has resulted in many waters becoming seriously polluted. Poor management of our fisheries has resulted in over-fishing of most commercial fish stocks while increasing demand on resources has resulted in conflict between the wide variety of users of the coastal zone, eg leisure interests and commercial craft competing for space in over-crowded waters, development of new industries such as fish farming and aggregate dredging without satisfactory legislative frameworks. All the problems place overbearing stresses on marine habitats and marine wildlife.

3. In general, planning and control of activities in the coastal zone are considered independently of each other. Vital integration is totally lacking, however, integrated management of coastal and marine areas is increasingly recognised as a necessity for environmental protection, planning and development.

A number of recommendations on coastal zone management are suggested.

INTRODUCTION

1. The World Wide Fund for Nature (WWF) is the major international organisation concerned with the conservation of nature and natural resources. Its conservation work is based upon the central tenets of the World Conservation Strategy: the sustainable and wise use of natural resources, and the conservation of biological diversity.

2. WWF UK has over 1,000,000 active supporters in the United Kingdom and has spent nearly £8·5 million directly on over 2,000 United Kingdom conservation projects since being established in 1961. Further funding for projects all over the world is provided by WWF UK to WWF International, based in Switzerland.

3. The marine environment is one of a number of conservation priorities identified by WWF UK in the current three year work plan. As such it is determined to be an issue of urgent concern and high conservation importance. WWF UK has been involved with work on the marine environment for many years and this resulted in the establishment of a Marine Unit two years ago. The focus for the Unit within the marine theme is integrated management of the coastal zone, including conservation, fisheries and pollution.

4. WWF UK has funded a number of reports addressing the overall management of the coastal zone and many investigating a specific activity which should form a component of management of the coastal zone (see Appendix 1).

Further information on problems and conflicts in the coastal zone is available from WWF UK's Marine Update, a quarterly report which addresses different issues within the marine environment and coastal zone. To date topics covered include:

—Marine Conservation and WWF
—The Crown Estate Commissioners
—Coastal Zone Management
—A Century of Concern: North Sea
—European Fisheries (in preparation)

Copies of the issue addressing Coastal Zone Management are enclosed, all other issues are available on request.

5. WWF welcomes the Environment Committee's decision to conduct an enquiry into coastal zone protection and planning. The marine environment, and in particular management of the coastal zone, has been inadequately addressed by successive Governments. Before addressing in detail the practicalities and problems of protection and planning in the coastal zone it is worth considering that designation of protected marine areas (as Marine Nature Reserves) is possible but is extremely limited in scope and has been agreed by many bodies to be inadequate. Apart from non-statutory designations such as the voluntary Marine Consultation Areas (MCAs) designation in Scotland, which has also been shown not to work (for example, it appears that the conservation importance of sites within voluntary MCAs is not given any greater weight than other sites when fish farm lease applications are considered) no other mechanisms are available to provide protection and management for the coastal zone. This is despite the fact that there are large areas of estuaries, coastline and marine sites, which fall between the most stringent designation of Marine Nature Reserve and the other extreme of derelict industrial areas which are contaminated and of very little conservation or scenic importance, but are still worthy of some protection and urgently require management. It is this lack of adequate protection and management that has existed for many years that first prompted WWF to address marine conservation issues. Pressure on the marine environment has never been so intense. It is imperative that attention is now focused on the coastal zone and it is to be hoped that the recommendations of the Environment Committee will be followed by urgent Government action.

6. WWF UK is grateful for this opportunity to provide evidence to the Environment Committee on Coastal Zone Protection and Planning.

BACKGROUND

7. Coastal Zone Management (CZM) is a term that has been used increasingly in the United Kingdom in the past few years, however, to define coastal zone management is not simple. " A Future for the Coast? " a report by Dr Susan Gubbay (Marine Conservation Society) for the World Wide Fund for Nature identifies a number of definitions and one that is particularly appropriate is from CAMPNET, 1989:

> " A dynamic process in which a co-ordinated strategy is developed and implemented for the allocation of environmental, socio-cultural and institutional resources to achieve the conservation and sustainable multiple use of the coastal zone ".

8. Integrated Coastal Zone Management is now receiving international attention. The United Nations Conference on Environment and Development (UNCED) Secretariat's progress report on ocean-related matters to the second session of PrepCom recognises that CZM is a necessary tool for development and environmental protection and provides guidance as to what integrated CZM should encompass (PC/30 paragraph 17):

> " It relies on modern methods of planning and resources management as well as inter-disciplinary expertise and involves the setting of national policies and goals; inventories of resources and available statistics, identification, selection and implementation of management systems such as guidelines, permits and economic incentives; socio-economic and environmental assessment; conflict resolution techniques; land-use planning; enforced zoning and protection of sensitive areas ".

9. The Environment Committee's inquiry is more specifically into coastal zone protection and planning, however, WWF UK believes that any use of the coastal zone whether addressing engineered sea defences or the establishment of marine nature reserves should be considered within an integrated framework which *takes into account all uses of the coastal zone in a given area and secures the long-term future of the resources of the coastal zone and the conservation of biological diversity.*

10. It is WWF UK's view that integrated CZM should consider all uses of the coastal zone including fisheries; navigation, commercial and leisure; off-shore industries eg, oil and gas exploration; pollution—land based, dumping and accidental, recreation—water sports and land sports in coastal areas, nature conservation, inshore industries—marine fish farming, bait digging, port development, housing and marina development, sand and gravel extraction; and coastal protection. This is not intended to be an exhaustive list, however, each of these will form an important component of integrated CZM.

11. It becomes apparent that it is necessary to attempt to define the coastal zone, particularly in respect of addressing land-based pollution which may originate anywhere within the watershed of a river system and fisheries and shipping-borne pollution (deliberate or accidental) which may be focused many miles offshore.

12. As integrated management will require addressing the interactions between land, air and water, the seawards limit should extend to the exclusive economic zones and into the deep oceans while landwards the entire drainage basin or watershed affecting the coast should be included. In some states it may be that this covers the whole land area as emissions into rivers and the atmosphere that emanate from the farthest point inland have implications for the coastal zone. It is important that a holistic approach is undertaken and that activities in the coastal zone are not separated into landward and seaward components. Activities on land impact on the marine environment and vice versa. Similarly the coast should not be seen as a boundary between the sea and the land but as an interface within an interactive system.

THE PROBLEM

13. The lack of a national strategy for the conservation and sustainable use of the coastal zone and management planning to implement it has resulted in:

(a) loss of wetlands and intertidal areas

(b) inadequate protection for marine species and marine habitats

(c) diminishing fish stocks

(d) serious pollution of coastal waters

(e) conflict between the wide variety of users competing for space and resources.

14. In July 1991, at the Government's request, Wildlife Link produced a paper "A Selection of Case Studies Illustrating the Need for Coastal Zone Management in the United Kingdom ". The case studies have been selected to show the unco-ordinated and piecemeal approach to management of the coastal zone in the United Kingdom. An integrated management system would provide the framework to practically resolve the conflicts and inadequate protection currently afforded to our coastal zone.

15. Examples are given to illustrate the problems identified (la–e) above and an indication of the way in which an integrated national strategy for the coast could hope to resolve these problems. (One of these is taken from the Wildlife Link paper; this is indicated.) It must always be considered that two of the primary aims of CZM should be to secure the long-term future of resources and the conservation of biological diversity.

(a) Loss of wetlands and intertidal areas.

" The successful passage through Parliament of the Felixstowe Dock and Railway Act 1988 by means of a Private Bill allowed the Felixstowe Dock and Railway Company to destroy part of the conservation value of the Orwell Estuary SSSI, a candidate SPA[1]/RAMSAR site ".

(Example taken from Wildlife Link paper)

The expansion of the docks was promoted as a scheme of national economic importance. The expansion took place when Felixstowe docks had a competitive advantage. However, with the Government's decision to abolish the National Dock Labour Scheme and now the privatisation of Trust Ports, this advantage has been lost as other east coast ports with excess capacity once again become commercially attractive. The lack of national guidance for port use and development has allowed the pointless and unnessary destruction of an internationally significant habitat.

A system of integrated coastal zone management including a national strategy for the coast should consider the availability of alternative port facilities in the south east and the strategic importance of the internationally significant habitat for many species of wildlife.

The Cardiff Bay Development Corporation's proposal to build an amenity barrage permanently flooding Cardiff Bay will, if given the go ahead, destroy a Site of Special Scientific Interest and part of a proposed Special Protected Area/Ramsar Site. The original proposal did not consider the nature conservation significance of the site nor the effect that a barrage would have on groundwater levels for the adjacent community. Subsequent private legislation has had to address both of these aspects, however, it still seems likely that a Government Hybrid Bill, to be laid before Parliament this autumn, will not resolve these conflicts of interest. Furthermore the route of Private and Hybrid Bills for major developments such as this restricts local involvement in the decision making process by making progress via a route that is out of reach financially and practically of much of the local community.

Integrated management could be expected to consider the requirement for the barrage, the impact on the internationally significant site and the effect on the local community.

(b) Inadequate Protection of Marine Sites and Marine Species

The designation of Sites of Special Scientific Interest beyond low water is unclear. It is doubtful whether the United Kingdom will fulfil its international nature conservation obligations as the Government has decided that the SSSI system is to be used to protect Ramsar sites and Special Protection Areas. Both these designations cover areas of sea, however, the legal status and powers of protection in coastal waters provided by SSSIs do not appear to allow this. The procedure for designation of Marine Nature Reserves is too cumbersome and inadequate with only two Reserves being identified in the 10 years since the legislation took effect. In any event, it would not be the most appropriate mechanism to use in these instances.

Reports of dolphin and porpoise sitings in United Kingdom waters have decreased dramatically over the past two–three decades yet although these animals are protected under the Wildlife and Countryside Act 1981 it is virtually impossible to protect the habitat on which they are dependent.

A resident population of wild bottlenose dolphins, one of only two currently known in United Kingdom waters, exists in Cardigan Bay. A recent study for WWF identified constant threats to the group from the intense pressures of pollution and disturbance. Competition for food and space also create additional problems. It is not possible to designate Cardigan Bay under the current legislation and what protection does exist is not enforceable.

The island of St Kilda is a site of recognised high conservation importance particularly for its large seabird colonies. The extent of this recognition is such that it has been designated a World Heritage Site. However, the statutory protective mechanism—SSSI and National Nature Reserve—only covers the land area down to low water. This means that the world's largest gannet colony is protected but not the areas in which they and the other seabirds feed. There are no controls on fisheries' activities around St Kilda nor other activities involving boats. As a result the birds may be drowned in coastal gillnets, disturbed by vessels, threatened by pollution or outcompeted by the fishing industry.

A system of integrated coastal zone management would address the problems of pollution, disturbance and over-fishing in a site of significant nature conservation importance.

---

[1]This is a Specially Protected Area under the EC Directive 79/409 on the Conservation of Wild Birds.

(c) Diminishing fish stocks

Fishing is not immediately perceived as an activity that takes place in the coastal zone, however, whether fisheries are based in coastal waters or further offshore, the nursery grounds for commercially exploited fish are generally in inshore waters.

The fishing industry is very concerned that the consequences of sand and gravel dredging in estuaries and in coastal waters will devastate the fragile environment on which juvenile fish and shellfish are dependent.

In July 1991 the Ministry of Agriculture, Fisheries and Food and the Crown Estate Commission rejected an application for a trial dredging project around the Maplin Sands, Thames Estuary following massive representation from local fishermen concerned as the importance of the grounds for sole and cockle fishing.

The Crown Estate Commissioners regulate the use of the seabed, however, they have a dual role as they are also the landlord profiting from any commercial use of the seabed. The lack of an appropriate planning regime and of a national coastal strategy means that new industrial activities, for example as has happened with marine aqua-culture, in the coastal zone tend to develop rapidly without adequate control and the impact on the environment is not considered thoroughly in advance of development.

(d) Pollution of Coastal Waters

Pollution of estuarine and marine waters may result from a number of different sources. Pollution originating from discharges directly into rivers, estuaries and coastal waters and from deliberate dumping has frequently been cited as the cause of elevated contaminant levels in the marine ecosystem. Levels of contaminants in fish has given cause for concern in a number of cases.

Well reported was the case of industrial mercury discharges accumulating in fish eaten by the population of a Japanese fishing village. In the early eighties, levels of mercury in fish from Liverpool Bay approached the limits beyond which it is considered unsafe to consume the fish. The Faroese Islanders are currently recommended not to eat whale meat more than once a week to minimise the risks from high contamination levels present.

Sewage pollution in coastal waters is currently causing considerable concern for bathers and those involved in water sports. While litter from shipping, sewage outfalls and beaches and fishing debris cause further problems for wildlife and other sea users, eg plastic bags have been found in the digestive system of marine turtles presumably mistaken for jelly fish, birds have been throttled after becoming tangled in plastic can holders, abandoned nets continue to " ghost " fish and divers become entangled in fishing line.

To address pollution in the coastal zone each contributing source should be considered within an integrated management plan that ensures provisions are included to require adequate facilities are available to reduce and eliminate pollution without creating further stress to the environment onshore.

(e) Conflict of use

Many estuarine and coastal sites support a wide variety of users and this inevitably leads to a conflict of interests.

Strangford Lough, Northern Ireland, is a complex of an enormous variety of habitats supporting some of the richest and most varied marine wildlife in the British Isles. It is designated an Area of Special Scientific Interest (ASSI), the Northern Ireland equivalent of an SSSI and also qualifies as a Special Protection Area (SPA) under the EC Birds Directive 79/409 and under all three sets of assessment criteria for designation as a Ramsar Wetland under the Ramsar Convention 1971, which the United Kingdom signed in 1973. SPA and Ramsar designation have, however, not yet been applied. The foreshore, the islands and the landward margins for varying distances inland all around the Lough were designated an Area of Outstanding Natural Beauty (AONB) in 1972.

The surrounding land is divided between private ownership and public ownership while the majority of the seabed and much of the foreshore is owned by the Crown. Sporting rights are divided between a number of bodies while many private land holdings hold rights to take seaweed, sand or gravel. The fishing and shellfish industries have historically been significant in the Lough, while sailing, windsurfing, canoeing, water skiing, angling, sub aqua diving, wildfowling, walking, dog training, horse riding, bird watching, tourism and golf are important in the Lough and surrounding land.

With such a variety of users in an area of international importance there is inevitably a conflict of interests. Recent concern has been expressed over large parts of the Lough bed being seriously damaged by dredging for scallops, disturbance of wintering birds reducing the value of the site as a feeding ground and reclamation and dumping. Significant changes in use can occur very quickly with the result that conflicts arise relatively suddenly and there is no framework within which to manage the activities, for example improvements in wetsuits mean that watersports now continue into the winter and cause disturbance to wintering birds. Traditionally this conflict did not arise.

The need for a management plan for Strangford Lough was recognised over 15 years ago yet the current management structure under consideration will have no statutory responsibility to implement the management plan unless new legislation is provided. Integrated coastal zone management would provide a framework on which local and regional management could be structured and implemented.

COASTAL ZONE PROTECTION AND PLANNING

16. Coastal zone protection and planning must be considered within an integrated management framework. It is recognised that the current planning legislation varies in different parts of the United Kingdom.

The Marine Conservation Society report to the World Wide Fund for Nature, " A Future for the Coast? " contains further detail on the areas identified by the Environment Committee for inclusion in its enquiry.

Dynamics of Coastal Change

17. Human impact on the coastal zone is more intense than at any point previously. With increasing population, wealth and leisure time we have demanded more from the coastal resource in terms of space eg for marinas, housing, commercial developments; protein ie fish and shellfish both wild and cultivated, and energy eg oil and gas industry and alternative forms of energy. We have also relied increasingly on the resource to dispose of our waste products—sewage, industrial waste and garbage.

18. Most of these demands have been made and are still being made without due consideration for the impact on the nature conservation importance of the coastal zone or on the sustainability of the coastal zone resource. Many coastal developments are carried out in a piecemeal fashion changing the landscape of the coastal zone without any consideration of other developments in the same adjacent areas or estuaries. The result is that large areas of mudflats or intertidal areas are lost with subsequent loss of wildlife since alternative sites have been destroyed or have reached their wildlife capacity. Similarly fish stocks are exploited without considering the effects on other species competing for the same food resource or considering the effect of removing a major link in a food chain.

Energy sources have been exploited without due consideration of the disposal of the waste products and structures of the disturbance impact on the habitat and ecosystems. Waste disposal at sea has not traditionally addressed the capacity of the environment to absorb the waste or the impacts of contaminants on foodchains.

19. It now seems likely that over the next century we will experience a significant rise in the sea level due to the effects of global warming. Various options are available in the event of a sea level rise, of which the most appropriate for habitat conservation and the maintenance of biological diversity is a phased retreat from the vulnerable coastal margins which would allow valuable mudflats, wetlands and intertidal areas to regenerate in phase with the sea level rise. Within a framework of integrated coastal zone management all options can be sensibly addressed.

Risks to Coastal Settlements and Coastal Ecosystems

20. Risks to coastal settlements are varied and it would be beyond WWF's remit to address these. Risks to coastal ecosystems from pollution have already been discussed in some length. Flooding, coastal erosion and landslides are a part of nature's natural processes and most ecosystems have adapted to be able to cope with these pressures. It is only man's interference that results in these processes irreparably affecting a coastal ecosystem eg if the ecosystem is under additional stress from perhaps pollution or disturbance, it may not be able to respond adequately to the natural pressure of flooding, erosion or increased sediment load.

Coastal Protection Systems

21. Measures proposed to protect coastal settlements and farm land should always consider the impact on coastal ecosystems, through the provisions of an environmental assessment (EA). An EA in the coastal zone should span the landwards and seawards component of a development and not treat each aspect individually. It must not be acceptable for artificial coastal defences to destroy or irreversibly alter the nature of coastal ecosystems of national or local significance.

Planning Policy in Coastal Areas

22. As discussed more fully in the report " A Future for the Coast? " (Gubbay, 1990) in England and Wales planning on the landward side of the Coastal Zone is controlled under the Town and Country Planning legislation (now the Town and Country Planning Act 1990), however, the jurisdiction of local planning authorities generally ends at low water. In Scotland and Northern Ireland the legislation on planning controls is different. It is inappropriate to divide the planning system in this way since activities on land will impact on the marine environment and vice versa, it simply enhances the concept of a boundary between the two interlinked environments. Developments spanning the coastal zone are therefore not fully covered by the planning regime while administrative boundaries, on land, have a tendency to create further boundaries with the result that no consideration is made of a development's impact on the adjacent area. This is seen

particularly on estuaries, at county boundaries and at regional boundaries where there is a tendency for the limits of local authority jurisdiction to be designated. This results in two or more policies impacting on one estuary or county/regional boundary. A national framework for integrated coastal zone management would help to address this problem and ensure that the management strategy and planning for adjacent authorities could be fully integrated.

23. The Private Bill and Hybrid Bill procedure allows many coastal developments to proceed without a comprehensive environmental assessment. This should be addressed and all prospective developers in the coastal zone must be required to complete a full EA before any development or commercial activity is undertaken.

Role of Authorities Responsible for Protection and Planning

24. The role of the many authorities involved in the coastal zone has been discussed in the report " A Future for the Coast? " (S Gubbay, 1990; MCS/WWF).

In many cases there is an overlap in responsibility and it is not always clear which is the lead agency. In some situations different agencies are responsible for different aspects of one activity. The whole situation is extremely complex and frequently confusing. A national strategy for the coastal zone headed by one lead agency would help to resolve much of the current confusion and resultant lack of management.

RECOMMENDATIONS

25. WWF urges the Committee to consider how the present system of coastal protection and planning can be better addressed, and to consider in particular the need for:

—the introduction of national legislation for integrated coastal zone management for the whole of the United Kingdom

—the establishment of a new agency to provide a coastal zone management unit

—the implementation of national strategic planning for the coastal zone

—the preparation and implementation of regional and local coastal zone management plans

—the establishment of formal liaison between all organisations operating within the coastal zone.

WWF recognises the problems associated with the management of the coastal zone and provides funding for a number of projects to address the conflicts experienced. Some projects have been completed and the work has helped to raise conciousness of the conflicts while others are still ongoing.

REFERENCES

CAMPNET (1989) The Status of Integrated Coastal Zone Management: A Global Assessment. Preliminary summary report of a workshop convened at Charleston, South Carolina, July 4–9, 1989.

Gubbay, S (1990) A Future for the Coast? Proposals for a United Kingdom Coastal Zone Management Plan. A report for the World Wide Fund for Nature from the Marine Conservation Society.

Wildlife Link (1991) A Selection of Case Studies Illustrating the Need for Coastal Zone Management in the United Kingdom unpublished document presented to the Secretary of State for the Environment, July 1991.

Appendix I

Coastal Directory for Marine Nature Conservation. Susan Gubbay, 1985. A report to the World Wildlife Fund from the Marine Conservation Society.

Conservation of Marine Sites. A Voluntary Approach. Susan Gubbay, 1986. Marine Conservation Society funded by World Wide Fund for Nature.

Coastal and Sea Use Management. A Review of Approaches and Techniques. Susan Gubbay, 1989. A report to the World Wide Fund for Nature from the Marine Conservation Society.

The Effects of Sea Level Rise on Sites of Conservation Value in Britain and North West Europe. Ted Hollis, David Thomas, Shelagh Heard, Department of Geography, University College London, 1989. A study funded by World Wide Fund for Nature.

Statutory Marine Nature Reserves in Great Britain. A Progress Report. Lynda M Warren, 1989. World Wide Fund for Nature.

Using Sites of Special Scientific Interest to Conserve Seashores for their Marine Biological Interest. Susan Gubbay, 1989. A report for the World Wide Fund for Nature from the Marine Conservation Society.

A Future for the Coast? Proposals for a United Kingdom Coastal Zone Management Plan. Susan Gubbay, 1990. A report for the World Wide Fund for Nature from the Marine Conservation Society.

Appendix II

MARINE LEGISLATION REVIEW

To determine the extent to which existing legislation can be used for the purposes of marine conservation. The review will clarify the present assortment of legislation and identify gaps and overlaps, draw attention to legislation not yet used for conservation purposes, highlight potentially damaging legislation and form the basis for a study aimed at reforming the legislation.

FISHFARMING PROJECT OFFICER

Marine fishfarming is a major, relatively new industry, still in exponential growth and experiencing very rapid change. This project involves the employment of a dedicated specialist in fishfarming to act as a fieldworker, adviser and source of up-to-date information for member bodies of Scottish Wildlife and Countryside Link.

COASTAL AND SEA USE MANAGEMENT PLAN

To further marine conservation by developing and promoting a national strategy for coastal and sea use management. This project is undertaken by the Marine Conservation Society and will tackle a number of key issues and needs for marine nature conservation identified by an earlier WWF funded project, including Marine Protected Areas.

MARINE AND COASTAL PROGRAMME

This project aims to provide a strong science base for the defence of coastal and marine habitats and their species against damaging development or activity. It involves a comprehensive review of the threat to the United Kingdom's estuaries, a review of the demand for marine facilities in the South of England and a project to examine the extent of the use of monofilament nets and the impact on seabirds.

CALF OF MAN NATURE CONSERVATION AREA

This project co-ordinates the necessary scientific work and community education programme for the establishment of the Calf of Man as a Marine Nature Conservation Area.

FISH FARMING CAMPAIGN, SCOTLAND

The project aims to promote the development and management of the fish farming industry on an environmentally aware and ecologically sustainable basis. Work focuses on wildlife interactions, pollution from chemcial and waste inputs, development control and regulation of the industry.

SANDEEL PROJECT

The project, funded together with a consortium of Governmental and other non-Governmental organisations is investigating sandeel distribution, ecology and behaviour in the Shetland area to determine the factors effecting survival of young sandeels and the causes of seabird breeding failures.

CARDIFF BAY BARRAGE

WWF has supported a consortium of conservation organisations opposing the Cardiff Bay Barrage Bill. The work is supported to prevent the precedent being set of the total loss of an SSSI to an amenity development, proposed by a Governmental backed Development Corporation, and to oppose the use of Private Bills as a means of by-passing planning law.

TAWE ESTUARY BARRAGE MONITORING

The undertaking of an ecological assessment of the Tawe Estuary prior to the impoundment phase of barrage construction. The work will be followed up by a further survey to identify change and undertake a comparison pre and post barrage construction.

LAND-BASED MARINE POLLUTION CONTROL

Support is provided for scientific input into a research project identifying the need for a new global and regional regime to regulate land-based marine pollution.

SAND AND GRAVEL EXTRACTION

Funding to support a survey to assess the effects of marine aggregate extraction on environmental aspects in Filey Bay.

Examination of witnesses

MR PHILIP ROTHWELL, Manager of Aquatic Policy Unit, MR STUART HOUSDEN, Head of Conservation Planning, The Royal Society for the Protection of Birds, DR SIAN PULLEN, Marine Conservation Officer and DR CHRIS TYDEMAN, Senior Conservation Officer, The World Wide Fund for Nature, examined.

Chairman

339. For the second part of this Session I welcome representatives from the Royal Society for the Protection of Birds and the World Wide Fund for Nature. Mr Rothwell, will you introduce yourself and your colleague.

(Mr Housden) Perhaps I can do that, Sir Hugh. Mr Philip Rothwell is the Head of our Coastal and Marine Unit at the RSPB. My name is Stuart Housden. I am the Head of the Conservation Planning Department. Dr Chris Tydeman from the World Wide Fund will introduce himself and his colleague.

(Dr Tydeman) Yes, I am concerned with the United Kingdom conservation programme and Dr Sian Pullen is the Head of our Marine Unit.

340. Thank you very much. The first question is one I would like to address to you, if I may, Dr Tydeman. The WWF—I shall say that as it is less of a mouthful—suggests that the coastal zone should include the land landwards for the entire drainage basin or watershed affecting the coast. We have heard about the 12-mile limit, now you want to go landwards. This suggestion would virtually encompass the entire United Kingdom, would it not? Is there some other definition you can use, or are we going to say that we must include the whole of the Thames basin, the whole of the Mersey basin and right the way through?

(Dr Tydeman) We have put this thought together based on looking at the international problems of coastal zone management. Most other countries like America have much bigger areas and you can take into account river basins much more easily. It would indeed encompass most of the landscape of Britain, if you were to do what we suggest. However, it is not quite as stupid as it might seem in that if we include the National Rivers Authority in a more enhanced capacity in terms of coastal management, they are currently putting together management plans for whole river catchments all of which would fit very nicely into the scenario of having a central co-ordinating body for the management of the coast.

341. You have heard the previous witnesses talk about integrated coastal zone management and how it was suggested perhaps that this could be a unit within the Department of the Environment. Are you suggesting that the body responsible for a national coastal strategy, if we were to have such a body, should be the NRA because it would then marry that in with its existing responsibilities for river basin management?

(Dr Tydeman) It is one option that we have looked at and we have not disregarded it. At the moment we join our colleagues in favouring a unit within the Department of the Environment which might then, as Dr Harvey said earlier, evolve into something else. They are not mutually incompatible. At the moment

we see the NRA with their management plans for the river basins as slotting into the DOE unit system.

342. The trouble is that we get tiers and tiers and tiers and it is always difficult to know where we are going in that kind of situation. It can lead to friction.It certainly can lead to things falling between two stools. We have seen that happening on more than one occasion and that is why I was trying to probe to see exactly whether you had a defined structure in your mind in putting forward this suggestion?

(Dr Tydeman) We see whatever becomes the central unit as having a very strategic national overview of the problem. Wherever it sits it must have a national strategic overview of the problem and everything else comes under that umbrella.

343. And this extension of the coastal zone relates to your concern for the lack of a national coastal strategy. Is it this unit that you are talking about that would be responsible for producing the national strategy?

(Dr Tydeman) The unit would be responsible for producing the national overview strategy, yes.

Mr Pendry

344. Lady and Gentlemen, I am sure you will be relieved to know that even though a general election is looming I shall not relate my question to a constituency matter. In fact, it will be much more global, European and international. I must declare an interest also. I am a Fellow of the RSPB, so I shall relate my question to the WWF submission. In that you say that integrated coastal zone management plans should include all uses of the coastal zone. Fisheries and navigation, to name but two, are areas where decisions are often taken at a European or international level. How would the national and regional plans cope with that kind of situation?

(Dr Tydeman) With respect, Mr Pendry, I believe that lots of things on land currently emanate from European legislation and we still have some discretion as to how we deal with them. For example, agricultural holdings are regulated by European legislation, but they do not actually say where you put the sheep. There is some discretion nationally as far as that is concerned, so I do not think it would be impossible to do what we suggest.

345. You have obviously thought this through. Do you have more detail than you have been able to give us in your submission on how you feel you could cope with such a structure?

(Dr Tydeman) Not that I could give you this minute. I am quite happy to provide an additional paper, if that would be helpful to you.

346. Thank you for that.

(Mr Housden) Perhaps we could add one or two points there, first to endorse what Dr Tydeman said

Mr Pendry *Contd]*

in the sense that a great many of the things that we deal with in the United Kingdom are now much influenced by directives or regulations from Europe. For example the Environmental Assessment Directive requires many of the things that affect the coast to be given an environmental impact assessment in order to help decision making in terms of location. We cannot think of occasions where European directives or regulations actually instruct the United Kingdom as to where the location of, say, a road, a port or other infrastructure or other matters should be. We see these things because they impact considerably on the coast. This is where the regional strategy would come into its own to help the wise use and planning within the overall structure and guidance given by this unit within the Department of the Environment, which the previous witnesses this morning outlined and which both WWF and the RSPB endorse as well. It is that national strategy, the linkage across the various interests that is the area of shortfall at the present time.

Mr Steen

347. You mentioned a national central authority in your paper, but this will have an enormous increase in bureaucracy. Would not some other idea be more simple in this complex area?

(Mr Housden) I do not think that we do envisage an enormous increase in bureaucracy, rather we wish to empower all the various agencies that exist at the present time—and there are over 30 statutory bodies that have an interest or have some decision-making process in the coastal zone. We wish to see a system that brings them together and enables them to work to what is a national picture of priorities. A relatively small elite lean unit within the Department of the Environment would do the task, thus enabling the National Rivers Authority, the local authorities and many other bodies to be able to work together and dovetail their interests. We would not wish to take away from them and replace them with something else, but rather empower them and make them more effective together.

Chairman

348. Would you advocate, then, this being a unit within the Department of the Environment having an overall view, or would you accept that the National Rivers Authority, perhaps, following what we have heard about the river basins, should be the authority to be given the responsibility for a national overview of the coastline so that it then integrates two functions?

(Mr Housden) The first thing I would stress is that we would wish to see a national policy or strategy having ministerial endorsement and backing, so that even if the NRA were the body that might help draw that together, they would have to answer to the Secretary of State for the Environment.

349. They are answerable to them already to some extent.

(Mr Housden) But that is a key point to bear in mind because we would wish it to be the Secretary of

State for the Environment, rather than the Minister of Agriculture or one of his other colleagues. That is the first point. We think, in order to gain departmental acceptance for such a strategy—and remember there is the Department of Transport, the Ministry of Agriculture, Department of Trade and Industry and others involved in this—that a small unit within the Government itself would be required to take that policy and deal with it.

350. There is always a strong argument for saying that where there is an overlap of a number of Departments, it is probably happier to have a free standing authority that is separate from all of them.

(Mr Housden) Providing the policy was endorsed and agreed and was seen as a national policy, that would help resolve the conflicts that at present we have with sectoral decision making. The different departments and agencies plough their own furrow, if I may use that language.

351. Whatever happened Parliament would wish to maintain ultimate responsibility and we can only do that by seeing that a Minister is responsible. Whatever form the structure takes, I think Parliament would insist—whatever this Committee may decide upon—on accountability through the normal democratic processes that I have described to you. When we come to structure and administration other questions arise. What we must have in our mind is what is likely to be the most efficient way of doing it, while having regard to the need for accountability. It does not necessarily follow that the best way of doing it is to expand a large department of state even further.

(Mr Housden) Our view is that as with many other fields there is a Directorate of Rural Affairs within the Department of the Environment that deals with nature conservation and landscape matters, but the executive authority is English Nature or the Countryside Commission. Similarly with planning there is a Planning Directorate of some large scale but local planning decisions are dealt with at a local level. We see a shadow within the Department of the Environment as being necessary and for accountability reasons also.

352. I have no doubt that the officials of the Department of the Environment who are listening to this exchange will be gratified to find that they find favour somewhere because they get a lot of stick from other directions.

(Mr Housden) That does not mean to say that they are perfect!

Mr Jones

353. I want to follow that up because I should like to ask a question that is rather contrary to the question that Sir Hugh put. The NRA has a lot of other responsibilities which include recreation, water quality, and also the supply of water. I could envisage a situation, if they acquired yet more responsibilities, where there would be so many internal conflicts that they would not only be the poacher and the gamekeeper, but they would be the beaters, the laird and probably even the game itself. That is something

Mr Jones *Contd]*

that I would find a great deal of worry about. Surely the coast has its own problems and while they obviously interface with problems further inland, it is much better to have a more specialist approach making sure that the input is there from elsewhere?

(Mr Housden) The National Rivers Authority have certain assets which we have to see put into play in terms of overall coastal protection and coastal conservation. To begin with they have jurisdiction out to three miles, which we have already identified this morning local authorities do not; although we wish they did have that jurisdiction. The second thing is that they are organised on a regional basis which allows them to take an overview more effectively than perhaps local authorities have managed hitherto. They are also the agency that carries out coastal defence and coastal management in terms of sea defences against rising sea levels and, as an executive agency, they relate to the Ministry of Agriculture on those matters, something that we find somewhat concerning on occasions as we think the interest is perhaps broader than just agricultural. For all those reasons they have to be involved very much indeed. They have the powers and our small elite unit within Government could not work without them. Nevertheless we are looking for some structure and some ability to bring together the coastal local authorities and empower them so that planning is brought into regional and coastal zone management systems. There are questions of local democracy. They have infrastructure, planning authorities and committees that are accountable to people locally and which can then take some of these balanced decisions. For those reasons we see a welding together of all of the best interests as being the only successful way forward. The NRA are working with coastal cells, as they call them, but they are just looking at coastal defence, flood defence and protection matters, important though they are. We, for example, feel a little left out of that process because they are looking at it strictly in terms of the remit given to them by the Ministry of Agriculture.

354. Can I take you on to this regional coastal zone planning because you are recommending whole estuary plans and I should like to know how that fits in? Either by that you mean a region should cover the whole of an estuary or you mean that there should be something in addition to the regional coastal zone planning units and that is yet another tier. Perhaps you could clarify that.

(Mr Rothwell) The scope of regional plans and estuary plans has to be looked at in terms of the topography of the coast. If you can imagine an area such as the Solway, Morecambe Bay or the Wash, which are huge estuaries covering huge areas that would lend themselves to a coast wide coastal zone plan. In other areas, some of the smaller estuaries in Essex, for example, like the Blackwater, would be a subset of a regional plan and the regional plan would address strategy along that area of coast and would suggest that the local authorities or combinations of local authorities worked together to produce a recreational plan or a small estuary management plan within that wider strategy that is outlined in the broader regional plan. It would depend on scale and

one has to be pragmatic as to the topographic boundaries and the regional and district council structure when looking at whole estuary versus regional management plans.

355. Let us take it as read that where they are relatively small estuaries and fall within an obvious region—and you have given an example of that—that the arrangements will be much more straightforward. What I am concerned about are the big estuaries, the Solway and the Mersey, the Dee, the Wash and so on. Do you see those coming within one of these regional coastal zone planning units, as put forward by the Marine Conservation Society?

(Mr Rothwell) Taking these larger estuaries as a whole is a very sensible way forward. They have their own particular complications. They are often the boundary of many local authorities. The Dee estuary is a classic example of two countries, three counties and probably eight districts. Often they have conflicting interests as local authorities. The individual estuary is designated as a site of special scientific interest or even in the Dee's case a Ramsar site and a special protection area under the EC legislation, yet there is no one bringing together those local authorities to work towards a sensible holistic management plan for the estuary. On the bigger estuaries a very detailed targeted approach by some overviewing body, bringing together the local authorities, is essential.

356. Would this be an *ad hoc* body or a statutory body?

(Mr Rothwell) I think it would have to relate to statutory powers and then, in the case of the Dee, we would see something like the National Rivers Authority as being the co-ordinating body with its fisheries, coast defence, recreational and conservation remit there.

357. We would have very real problems where these happen to be the boundaries between England and Wales or England and Scotland and you would see that as transferring the responsibilities to the NRA, would you?

(Mr Rothwell) Not necessarily.

358. Could you explain that?

(Mr Rothwell) I think the role of co-ordinating the production of a management plan must be done by one body that has an overview. I would not wish to see it ceded to one lead authority. I think an independent regional view needs to be taken.

Mr Field

359. Would it fit the bill, Mr Rothwell, if there were some form of aquatic/maritime national park which took up these very large estuaries?

(Mr Rothwell) With the aquatic/marine national park you always have to address the landward side because the two are indivisible.

360. I mean a type of body similar to a national park, not dealing with the landward side, but dealing with the marine interests.

(Mr Rothwell) I think that our marine protected areas proposals that Dr Earll referred to earlier verge on the marine national park in that they zone

Mr Field *Contd]*

activities and seek management of the sea bed for a variety of uses and the sea surface for a variety of uses. Marine protected areas are slightly different from marine national parks. Marine national parks would require another bureaucratic structure, but we are seeking to reduce the number of structures, or at least rationalise them rather than add to the bureaucracy. Although marine national parks may be something we would like to consider, at the moment we have not addressed that.

Mr Jones

361. I would like to go back to a point about the NRA. There is some prospect, to put it no stronger than that, that the NRA may be subsumed into an environmental protection agency and broadly that idea has been welcomed in the conservation world. But that EPA would then have an enormous range of responsibilities and might the responsibilities for looking at estuarial protection, planning and so on, not be tucked away and given a rather low responsibility, whereas some unit that had specific responsibilities might be better placed to fight for the protection of the environment in those areas?

(Mr Housden) I think the published statements about the EPA and the reaction of the conservation movements initially was one of concern at the break up of the NRA. The RSPB organised a letter which included seven or eight other bodies which went to *The Times* and also the Prime Minister expressing concern at the proposal that breaking up the whole river management and water cycle process was inherently a bad one and that we wished to see that preserved. As a consequence the Government have brought forward a consultation paper with four suggestions, two of which would meet our concerns and one of which is certainly to lump everything together and have one super agency. Our concern, and it goes back very much to what Dr Tydeman said at the outset, is that flood defence, land drainage, management of wetlands and coastal areas, which is a key role of the NRA, is somehow, like the iceberg, and is nine-tenths of the NRA at present and the high profile pollution control area is the one that is being kicked around. We are very anxious indeed that the legislation that governs these matters is brought up to date. It is 40 to 50 years old in its thinking. It empowers the agriculture industry very largely to remove water, drain land, defend it against flooding and that is inimicable to the conservation of wetlands, coastal sites and so on in certain cases. We want to see the balance restored. Whatever we do with the agency itself—and we firmly believe that we have to keep river management and the whole hydrological catchments and basins together—we need also to address this other issue. One of the options in the paper put forward by the Government is, if you like, to put water pollution, water management and hydrological and coastal defence issues in one agency and the use of the land and air, which at the moment is split among two or three authorities, under another. So we would have two agencies where three or four work at present. Both would meet the bill. Administratively it would lump together 10,000 plus staff.

Chairman

362. We are just concluding an inquiry into the possible shape of the environmental agency. I do not want to say anything that may anticipate what the view of the Committee may be. The Committee has not yet met to form a view, so to that extent I am still on very neutral territory. You mentioned the Government consultation paper and the options. One was that mentioned by Mr Jones where the whole of the NRA is subsumed into the environmental agency. The other option is that some of the functions of the NRA be hived off and put into the environmental agency. If the perception is that the environmental agency should in essence be a protection agency, then just the pollution control functions of the NRA would go into this agency. That is how I understand the Government have put their option. The NRA would remain responsible for river basin management, which I think I can say the majority of our witnesses felt should remain in one hand. It is merely the interpolation of that and what you have been saying this morning—assuming that is what the Government decided upon—which would result in the NRA retaining responsibilities for river basin management, but without pollution control responsibilities, and in turn that would make it subject to what it did to the environmental agency. Would the NRA then be the right vehicle, with all its regional functions but without pollution control, to take over the coastal management, still being subject to an environmental protection agency?

(Mr Housden) I clearly see the point you are putting to me. There are two issues as we see it arising from this. The first is in relation to pollution. We are not convinced that separating the management of discharges to rivers and separating its impact in terms of flood defence, potable water supplies, coastal protection and the management of wetlands is necessarily sensible. The quality of rivers at the moment is suffering. We think we need to insert biological criteria connected with the use of land surrounding those rivers. We see some inherent difficulties which we know are shared by Lord Crickhowell, for example, in the evidence he has given you——

363. We have received evidence from him.

(Mr Housden) And also in the discussions that we have held with him on this. The other thing is to come back to the point I made to Mr Jones that at the present time that would leave a body, if it were split in this way, that would deal largely with the Ministry of Agriculture. They fund coast protection and flood defence. Their flood defence committees are, in the main, dominated by landowning and agricultural interests, yet the spend of money is now very much geared towards urban flood defence and protecting the coastline for nature conservation and other purposes. If this option were chosen we would expect to see fundamental reform of the flood defence legislation as well.

364. So really your greatest reservation, if I understand what you just said, is of any responsibility in this area remaining with MAFF?

Chairman Contd]

(Mr Housden) We think that the Department of the Environment would be a better lead authority in that it can take the balanced overview, that we came back to at the beginning. At present we find it difficult to believe that the NRA will be able to push through its executive flood defence committee's decisions on phased retreat as an option for coastal defence and things of that nature, so from that point of view we have our concerns. I agree with you that they are a body that offers an option to help draw together and produce coastal zone plans. They remain as that body. We think that has to be a DOE led matter.

365. These options are all in play and the balance between them is not an easy one to determine. As I understand from what you are now saying, you would not object to the NRA being the lead body, in so far as coastal protection is concerned, provided it was answerable to the DOE rather than to MAFF. Perhaps that might be a trade off if it were to lose other responsibilities because the DOE feel that perhaps these ought to go into another body. Those were the thoughts that we are having to juggle with at the moment.

(Mr Housden) Both WWF and ourselves are doing the same because we have to submit our response to the Government's consultation paper by the end of January, which we are finalising at present. We think there is everything to play for and we are not convinced that the original options put forward by Government were very well thought through.

Chairman: You are producing other options? It will be interesting to see those. We shall not have the advantage of having considered them, though, before we publish our report on the Environment Agency.

Mr Jones

366. I should just like to tidy up one thing as a result of my questioning of MAFF. It emerged that all of the chairmen of the flood defence committees have agricultural interests. I think I ought to ask you for your reaction to that and whether you think that some change ought to be made in the way that they do their appointments?

(Mr Housden) I do not wish to say anything about the abilities of the individuals concerned. The fact is that the Ministry of Agriculture is the appointing body of the chairmen and has a block selection of the people who go on to the flood defence committees. We see their role as crucial as they are at present executive bodies. They can instruct the NRA to do things or not as the case may be. We would like to see reform to make them advisory so that the NRA can act in what we think are the wider and best interests. We would like to see reform covering the whole approach towards flood defence so that it takes into account new ideas about the environment and conservation which is not present in the legislation at the moment.

Mr Field

367. The Royal Society for the Protection of Birds recommends that Chichester Harbour Conservancy is a model for all port and harbour authorities and conservancies. Can you tell the Committee what elements have made it an example worth repeating?

I should be particularly interested to know why you chose Chichester as opposed to, say, Pagham harbour?

(Mr Rothwell) Dealing with the first point, the Chichester harbour model appears to work very well. There are some good reasons for that, not least of which is that the harbour conservancy has environmental duties and very strong environmental duties which are not common to many other harbour conservancies throughout the United Kingdom. It has statutory powers relating to moorings, zonations, to wildlife sanctuaries and it appears to be very effective in its consultation and linkage with local authorities and the land based planning structure. It also has an effective committee structure on which local interests are well represented, particularly the recreational interests and our own concerns, the conservation interests. That combination of an effective committee structure working well with local authorities with effective powers is what makes the Chichester harbour model work particular well. I am not sufficiently well versed in Pagham harbour to answer effectively the second part of your question, but others may be.

(Mr Housden) I am not aware that Pagham harbour has a harbour conservancy. Certainly it is a local nature reserve.

368. It certainly has a rule, I believe, that no internal combustion engines are allowed in the harbour.

(Mr Housden) There are by-laws, certainly, that can achieve that. There are one or two other models. There is Langstone harbour with Chichester and also you mentioned the Poole Harbour Conservancy where there is not representation of the broader interests of nature conservation.

369. So to get it clear in my mind, you are telling the Committee that the committee structure and representation draw in a wide body of diverse interests and that aspect of it works well? It is quite a big piece of water, of course.

(Mr Rothwell) That is right.

370. Does that apply to bait digging?

(Mr Rothwell) I do not know if there are powers to limit bait digging in Chichester harbour. There are certainly other harbours around the country where bait digging is a known problem.

371. Is there a conflict of interest among the bait diggers and conservancy in Chichester?

(Mr Rothwell) I am not sure about Chichester. In other harbours that is certainly the case, where there is a conflict of interest.

Mr Steen

372. Are you satisfied that the sites of special scientific interest, special protection areas and the Ramsar designations afford the necessary protection for coastal sites? And what do you think about the implications of the new Circular 1/92? As you and the Committee knows I have a number of estuaries in my constituency and I know you have produced a report on the marina industry that came to the conclusion

Mr Steen Contd]

that recreational demand would grow. Can you relate that to the fact that the SSSIs are supposed to be protecting all these places? Are they?

(Dr Tydeman) The short answer to the question is no.

373. No?

(Dr Tydeman) No, largely because SSSIs have no presumption against development. They are certainly still being damaged, if not destroyed, at a regular rate. The figures from English Nature and the other bodies that replaced the NCC still show loss and decline of SSSIs. Special protection areas and Ramsar sites offer no further protection than that which an SSSI covers because that is the Government view, that they will not make a site an SPA or a Ramsar site unless it is an SSSI and it confers no real extra advantage. You did not mention marine nature reserves, which is one of the other options. In fact it was one of the options that the Government gave the European Commission as being the way in which special protection area legislation would be implemented in the United Kingdom for areas offshore, that is below low water mark. I do not think anyone now would dare argue that that has worked or could work. Marine nature reserves would be a massive sledge hammer to crack a tiny nut, even if you could persuade the people in the areas concerned that a marine nature reserve should be put into place. The legislation as currently drafted is a bit of a hotch potch. I would throw in one other example which makes it even more ludicrous to our eyes, which is the World Heritage Convention, which is about as strong a convention as you are ever likely to get in terms of site protection. St Kilda is a world heritage site, largely because of its massive sea bird colonies. But the protection ends at low water mark, so you are protecting the birds all the time they are sitting on nests, but when they go offshore to feed the area that they are feeding in has no protection whatsoever, neither can it unless it becomes a marine nature reserve and even then legislation just is not operable. We remain extremely concerned at the level of protection that is currently available to sites below low water mark. In answer to your question about the new circular, we welcome anything that strengthens the SSSI network, but because SSSIs only go down to low water mark there is one boundary around which you cannot draw the consultation zones. It confers nothing on the seaward side at all.

Mr Field

374. Can I add a point about St Kilda? It is rather interesting, and I have been there. There is no low water mark. There is no shore apart from Village Bay. It is vertical. Although I am sympathetic to the RSPB's evidence on this point about the limit, when you quote St Kilda I just cannot see where the conservation conflict is there in terms of bird life, once they get off their stacks.

(Dr Tydeman) There is, for example, no control over fishing. You can put gill nets right underneath the cliffs and those can trap birds. The very reason

you are protecting St Kilda is for the sea birds, or one of the main reasons for it. It is just an example where there is a continuum that is not regarded as a continuum but two separate entities.

(Dr Pullen) In fact, perhaps not in St Kilda because it is such a remote site, but there are other examples where there are very important cliffs designated for the birds that live there. It is not only the impacts of the fishing activities that will affect the birds once they enter the water, but there is the competition between the fishing industry for the food source and pollution——

375. We had that conflict over the sand eel in Scotland last year.

(Dr Pullen) Yes, exactly, and disturbance from recreational craft as well.

376. It was the St Kilda point that puzzled me.

(Mr Housden) Can I just add two brief points? English Nature in their first term report recorded that 200 SSSIs were damaged or, in a few places, destroyed in one year, which is getting on for 5 per cent of the total in England, so they do not provide absolute protection even where they occur. The second point is that SSSIs are linked to the jurisdiction of local authorities. Legislation currently before Parliament, which will change the unitary authorities and so forth, has a proposal in it to move up the shore the limit of local authority jurisdiction by a small amount, but we wish it to be moved down the shore. Everyone says it is low water, but it is in fact mean low water, which means at the very lowest tides vast areas of the Wash, Poole harbour and other sites are not protected as SSSIs. Thus the circular in effect relates to everything to the landward side of SSSI protection and offers nothing one metre offshore below low water. This is a gap concerning which the Government is unable to fulfil their obligations under the EC birds' directive or the Ramsar Convention. We are legislatively unequipped to deal with it.

(Dr Pullen) A classical example of how an SSSI designation is not going to work is being shown at the moment by the promotion of the Government's Bill for the Cardiff Bay Barrage, where the site has been designated as advised by the Government's nature conservation advisers and yet the Government are now promoting a Bill which will destroy the whole of that site.

Mr Cummings

377. What do you believe are the implications of Her Majesty's Government signing a recent agreement known as Natura 2000 which the Committee is led to believe will eventually cover all coastal sites?

(Mr Housden) The first thing to say is that, like all these agreements and directives emanating from Europe, how they are implemented in the United Kingdom is left correctly to national governments. The way we shall implement it in the United Kingdom is through the means of a site of special scientific interest which we have just been discussing. That is under the Wildlife and Countryside Act 1981. In effect we have problems with that legislation in

Mr Cummings *Contd]*

that it does not go far enough out to sea, as we have just discussed, and the protection of those sites is inadequate in that it is a voluntary approach ultimately and there are lots of cases where there has been failure to reach proper management agreements or indeed to prevent damaging developments. This is a very welcome agreement. We are very pleased indeed that the directive has been agreed and in April we trust that it will be formally adopted. It still poses these dilemmas to the British Government. We cannot implement it effectively, particularly in the coastal zone.

378. Do you ever have any opportunity of making your views known to the Department of the Environment on a regular basis?

(Mr Housden) Certainly from the RSPB viewpoint we see Mr Trippier on a fairly regular occasions to put to him our views on matters such as this and on some of the issues that Sir Hugh raised earlier over the proposed EPA. It is not for the want of our trying to persuade the Minister.

(Dr Tydeman) That is true for us as well. I do not think there is any doubt what our concerns are. The problem is that none of them is being met.

Chairman: Getting the structure right is part of the battle. Gentleman, Madam, I thank you for your attendance this morning. I do not think we have any more questions to ask you. We are grateful to you for the assistance you have given us.

WEDNESDAY 29 JANUARY 1992

Members present:

Sir Hugh Rossi, in the Chair.

Mr John Battle	Mr Tom Pendry
Mr Barry Field	Mr Anthony Steen
Mr Robert B Jones	

Memorandum by English Nature

INQUIRY INTO COASTAL ZONE PLANNING AND PROTECTION

Summary of Memorandum

1. ENGLISH NATURE'S OBJECTIVES FOR THE COASTAL ZONE

1.1. English Nature is committed to working with others to

—conserve irreplaceable coastal features of wildlife and earth science interest;

—ensure the natural heritage passed on to future generations is as rich and varied as that which we have inherited;

—encourage the enjoyment and appreciation of our natural heritage by all.

1.2. The coastal zone should be managed to achieve these objectives. We need to ensure that the strengths of present systems are retained and deficiencies remedied. In this memorandum we outline the importance of the coastal zone for nature conservation, its relationship with natural processes and the pressures faced. We identify three policy areas which need to be addressed and make 10 recommendations which, if followed, would help achieve our objectives.

2. IMPORTANCE OF THE COASTAL ZONE FOR NATURE CONSERVATION

2.1. The English coastal zone contains a major share of the national capital of wildlife and earth science features. A measure of its importance is that 38 per cent of the total area notified as Sites of Special Scientific Interest in England occurs on coastal land and foreshore (298,000 hectares). In addition some 286,000 hectares of inshore seabed have been identified as being of equivalent quality for nature conservation. The complex of coastal and inshore marine habitats form part of an interdependent international network of sites supporting important fauna populations, especially waterfowl and seabirds. English estuaries make an outstanding contribution to this network. The coast is also very important for earth science conservation, particularly its geomorphological systems and the many classic geological exposures to be found in coastal cliffs.

3. COASTAL DYNAMICS

3.1. Much of the English coast comprises low-lying land having a gently-sloping intertidal zone. Such areas are net receivers of sediment from various sources, both offshore and from neighbouring eroding coasts. Many of the sand and shingle bars, dune systems and saltmarshes which develop in such areas provide a high degree of coast protection. If these features are starved of their sediment supply they will gradually erode, and their nature conservation and coast protection value diminish.

3.2. The coasts have always been subject to dynamic change; the sea level has fluctuated repeatedly, and the natural processes of coastal erosion and accretion have continued under the influences of wave, wind, freshwater inflows and marine currents. Wild plant and animal communities have accommodated these changes over thousands of years. Such changes, moreover, are central to the value of the coast for the earth sciences. If eroding coasts are protected, their value both for earth science and as a source of sediment supply is lost. If the present coastal and flood defences are retained in their entirety, the likely consequence is that rising sea level will bring low and high water marks closer together and the area of intertidal land will be greatly reduced. Huge areas of habitat may be lost.

4. PRESSURES ON THE COAST

4.1. Since the coastal zone forms such a vital part of our natural heritage, it is important to plan future human utilisation so as to maintain and enhance the natural features. To achieve this, planning mechanisms need to be underpinned by a general policy for pursuing the wise and sustainable use of our coasts. This is consistent with international conservation obligations entered into by the United Kingdom.

4.2. While the coastal zone has not been subject to the same degree of human influence as the English countryside generally, such influence has been considerable. Coastal land and foreshore, particularly in estuaries, has been reclaimed for agriculture, and used for the development of ports, industry, housing and their associated infrastructure. To keep ports operational, maintenance dredging has to be undertaken. The attractiveness of sandy and hard rock coasts has encouraged resort development and associated recreational impacts. These pressures are continuing and can have, not only direct effects in the immediate locality, but indirect effects elsewhere along the coast. If the coastal resource is not to suffer gradual degradation, regulation must be administered effectively and with a full appreciation of all the implications.

5. COASTAL PLANNING AND PROTECTION

5.1. Existing legal and administrative controls have the capacity to achieve the wise use of coastal land and foreshore; problems which do arise result from decisions which have not given sufficient weight to nature conservation and the environmental consequences. The same level of control is not available for the marine environment. This is because such areas are subject to the common rights of fishing and navigation which makes control difficult, because the present regulatory bodies have interests and responsibilities which can conflict with the need to protect the environment and because existing controls are limited in application. Modern methods of fishing, particularly dredges and beam trawls, have the capacity to cause damage to the flora and fauna of the seabed, while activities such as the extraction of marine aggregates, offshore oil and gas drilling, and the possibility of tidal power barrages, all have implications for nature conservation.

6. CRITICAL POLICY AREAS

6.1. If nature conservation objectives are to be achieved, we consider that the following policies need to be adopted:

1. Regulating authorities should ensure that the coastal resource of nature is not diminished and should take decisions after a full assessment of the environmental consequences. [Recommendations 1–5]

2. Further legal and administrative measures should be taken to enable areas of marine nature conservation importance to be conserved effectively. [Recommendations 6–7]

3. Regional and national strategies should be developed to enable planning and other regulatory decisions to be taken within a framework of policies for the coast as a whole. Local strategies should be developed to ensure that coasts are appropriately and sensitively managed. [Recommendations 8–10]

7. RECOMMENDATIONS

Decision-taking

7.1. Coastlines are naturally mobile and there is a need to work with, and not against, natural processes. Hard defences will continue to be essential to protect life and property in many coastal areas, but it is important for nature conservation to limit their further spread and to encourage the use of " soft engineering " alternatives (such as beach feeding, sediment recharge and offshore breakwaters) wherever possible.

7.2. With relative sea level rise already occurring in Southern England, and forecast to increase in the future, there is a need to develop a coastal defence strategy which includes the moving of sea walls inland in areas where the need to protect life and property permits.

7.3. The extraction of aggregates from sand dune systems, shingle bars and the foreshore should be phased out. The present, cautious approach to licensing fresh marine aggregate dredgings should continue and further research undertaken to model the long term effect of such extraction.

7.4. Where navigational dredging has been shown to interfere with longshore drift, schemes to transfer material across the channel (thus bypassing it) should be considered. Alternative uses should be established for dredged materials such as beach feeding or sediment recharge so as to avoid the loss of material to the coastal zone.

7.5. A primary principle for the regulation of activities in the coastal zone should be the wise and sustainable use of the resource so as to ensure that the nature conservation value of this environmental asset is not diminished. The " regulatory " role of regulatory bodies should be clearly separated from any role they may have, additionally, for the development of commercial potential.

Further legal and administrative measures

7.6. Existing mechanisms have not proved capable of conserving the areas of known nature conservation importance in the marine environment below mean low water mark. Other legal and administrative mechanisms need to be established which will result in the effective regulation of activities which are damaging, or likely to cause damage, to such areas.

7.7. The existing statutory site protection framework needs to be amended to enable full compliance with international obligations for nature conservation as regards areas below low water mark.

Developing coastal strategies

7.8. Legal and administrative mechanisms need to be established which will ensure that planning and other regulatory decisions are taken within the framework of a broadly regional strategy. Such regional strategies should be determined on the basis of criteria established by Government. These criteria should include consideration of the relationship between sediment supply and deposition, the influence of the hinterland (including, for estuaries, freshwater inflows), and the relationship between coasts and inshore marine waters. The framework will require cross-linkages to be established between adjacent authorities.

7.9. Those bodies with a responsibility for regulating activities in the inshore marine environment should contribute to the regional strategy (which should cover their areas of responsibility) and should carry out their functions in conformity with the strategy.

7.10. Formal groupings should be established to co-ordinate the preparation of management plans for defined stretches of coast, and individual estuaries, in partnership with the local community and relevant public agencies. Plans could include such matters as recreational zoning, habitat management and the careful control of development.

8. CONCLUSION

8.1. English Nature believes that if these recommendations are implemented, and the needs of nature conservation met within an overall strategy for coastal protection and management, the present coastal resource of nature, illustrated graphically in NCC's Atlas for 1991,[1] will continue for the enjoyment and inspiration of future generations. English Nature is committed to working with others to achieve these objectives.

Memorandum by English Nature

9. ENGLISH NATURE

English Nature is the statutory adviser to Government on nature conservation in England and promotes the conservation of England's wildlife and natural features. Its functions include the provision of advice and information, the designation and management of nature reserves, the notification of Sites of Special Scientific Interest and the conduct and support of research and other work relevant to nature conservation. English Nature shares responsibility for the provision of advice on nature conservation matters affecting Great Britain as a whole, and internationally, with the Countryside Council for Wales and the Nature Conservancy Council for Scotland; this responsibility is discharged through the Joint Nature Conservation Committee established for this purpose.[2]

10. COASTAL ZONE

For the purposes of this submission the coastal zone is defined as comprising land (and inland water) subject to direct marine influences, together with the foreshore and inshore waters extending seawards to three nautical miles.

11. IMPORTANCE OF THE COASTAL ZONE FOR NATURE CONSERVATION

11.1. The English coastal zone includes extensive areas where landforms and vegetation have developed over thousands of years in response to natural processes. These areas contain a major share of the national capital of wildlife and earth science features. They act as reservoirs of wildlife and as ecological laboratories where the effects of dynamic and global change can be studied. They are England's last wilderness and are areas of great and subtle beauty.

11.2. A measure of the importance of the coastal zone is that while in England as a whole some 7,800 sq km of land and intertidal water have been notified as Sites of Special Scientific Interest, 2,980 sq km of this, or 38 per cent, occurs within the narrow coastal zone.[3]

---

[1] " Atlas of nature conservation sites in Great Britain sensitive to coastal oil pollution ". Nature Conservancy Council, 1991.

[2] Part VII. Environmental Protection Act 1990.

[3] Source: "Atlas of nature conservation sites in Great Britain sensitive to coastal oil pollution." Nature Conservancy Council, 1991.

11.3. Coastal land and foreshore contains several habitats not found inland including saltmarshes, intertidal flats, sand dunes, vegetated shingle, saline lagoons and maritime cliff grasslands. Each of these habitats has its own characteristic assemblage of plants and animals. Together they support 48 nationally rare and 65 nationally scarce plants. Intertidal flats contain enormous numbers of invertebrate animals which in turn provide the food supply for important populations of fish and birds. The diversity and size of Britain's estuaries and soft shore coasts, the majority of which occur in England, are unrivalled anywhere else in Europe and make up 28 per cent of the estuarine area of the European Atlantic and North Sea coastal states.

11.4. The mosaic of coastal habitats forms part of a complex international network of sites upon which depend large and diverse migrant and wintering waterfowl populations. In January alone over 1·75 million waterfowl are present on British estuaries, this is 62 per cent of the total British wintering population and over 10 per cent of the relevant international population. The population of British seabirds is also outstanding. Many of these breed in large colonies on rocky cliffs and undisturbed islands, others are characteristic of shingle and sandy shores. About 400,000 pairs of seabirds breed on English coasts.

11.5. The seas around Great Britain support the highest diversity of marine habitats, communities and species of anywhere on the European Atlantic coast. The subtidal area is substantial, covering some 68,000 sq km within the three mile limit (equivalent to 30 per cent of the British land area). In the subtidal environment of rocky coasts, assemblages include such species as sea fans and cup corals and a wide variety of seaweed communities; while in that of soft coasts extensive beds of marine grass and eel grass occur, as do more restricted beds of coral-like algae (maerl). Of the 122 or so species of fish recorded around the coast a number are rare or endangered, while other provide food sources for sea birds and sea mammals. These areas support 5 per cent and 40 per cent of the world population of common and grey seals respectively, together with important populations of harbour porpoise and other small cetaceans. Some 2,860 sq km of seabed have been identified as being of particular importance for nature conservation within three miles of the English coast.

11.6. Coasts are also very important for earth science conservation, particularly in terms of:

(a) geomorphological systems and associated landforms, and

(b) the many classic geological exposures to be found in coastal cliffs.

Many large SSSIs, designated for their geomorphological interest, are located on the coast. This interest may be static geomorphology, such as the relict beach ridges of Dungeness in Kent, or active, such as the coastal process responsible for shaping and grading the cobble ridge at Chesil in Dorset. Conservation of a wide variety of static and active sites is of great importance in interpreting coastal processes and landform evolution and for underpinning the effective management of natural systems elsewhere.

Geological exposures are of importance both nationally and internationally for their representation of sections of geological time. They provide an important research resource, for example using post-glacial sediments to deduce past sea-level changes. The wealth of exposures around our coasts continue to make an essential contribution to global knowledge of geological history and provide a valuable teaching resource for students, for the oil industry and for other areas where the earth sciences are applied commercially.

12. Coastal Dynamics

12.1. Much of the coast of England is occupied by low-lying land with a gently-sloping intertidal zone. These areas are generally net receivers of sediment from various sources, principally offshore but also from neighbouring eroding coastlines. Many of the geomorphological features formed in such areas, be they sand or shingle bars, sand dune complexes or saltmarshes, offer a high degree of natural coast protection. These features exist in a very fine balance with sediment supply and are highly susceptible to its disruption.

12.2. Earth science sites are closely linked to this sediment system and all are dependent in some way on coastal erosion. Geomorphological sites require an input of sediment to maintain the integrity of the feature or to facilitate its development. Geological exposure sites require erosion to keep the face fresh as well as provide a natural beach fronting the cliffs which will in turn provide source material for other down-drift coastal features.

12.3. The study of sediments laid down in low-lying areas shows that in the past coastal habitats migrated long distances in response to gradual changes in sea level. In the historical period some coastal features have been altered dramatically, by natural processes, without adversely affecting the habitats they support. Many coastal ecosystems and individual species are adapted to and depend upon continued instability. Marram grass, which helps maintain sand dunes, can only thrive when being continually buried by fresh sand. Many short-lived dune plants cannot survive without fresh areas of bare sand in which their seeds can grow. Terns breed mainly on the bare sand and shingle of accreting dunes and will desert sites where these areas become too stable.

12.4. Curbing the natural instability of the coast has been an important goal of human society. The storm surge of 1953, with its attendant loss of life, powerfully reinforced this attitude. At present 32 per cent of the coastline of England is held at an artificial line by artificial structures, including a very high proportion of the most naturally dynamic areas.

12.5. This desire for stability can conflict with the long term conservation of the coastal environment. Two examples illustrate this; in Essex, saltmarsh is being rapidly reduced because it cannot migrate landwards in order to adjust to a relative lowering of land levels. On Humberside, Spurn Point moved constantly until fixed by sea defences in the 19th century. Since then it has been subject to a series of measures aimed at keeping it in one place. These have destroyed much of its geomorphological interest, obliterated or damaged substantial areas of dune vegetation and reduced its appeal as a landscape feature. These attempts at fixation are moreover doomed to eventual failure because of the general retreat of the Holderness Coast. Attempts to withstand the inevitable could lead to catastrophic failure following a severe storm; gradual adjustment is to be preferred.

12.6. Sea level rise, to which natural systems could respond without difficulty, will aggravate this conflict. In the South of England sea level rise is a fact. Globally, the sea is rising at an average rate of between 1.0 and 2.0 mm per year and in Southern England this is exacerbated by the sinking of land as a by-product of the last ice age. The rate of sea level rise may accelerate to around 6 mm per year over the next century if the best estimates of the effects of global warming are realised. On soft-rock low coasts with sea walls in place, the nature conservation interest will at best diminish in area and at worst drown or be eroded away. In the long term, habitat conservation and sustenance of natural sedimentary systems in these areas will only occur if the coast is given the opportunity to encroach inland.

13. PRESSURES ON THE COASTAL ZONE

Land take

13.1. The claiming of coastal land for agriculture has taken place since Roman times and continues to the present day. Since the industrial revolution extensive land claims have been undertaken for ports and manufacturing industry, together with housing and associated infrastructure. These claims have treated shores, particularly in estuaries, as wastelands of little intrinsic value. In contrast, the attractiveness of certain coasts resulted in many seaside towns being purpose-built as holiday resorts and in more recent years there has been an explosive spread of major recreational developments. These impacts have beed described for estuaries in a recent NCC publication.[1]

13.2. Land-claim continues. In 1989, 123 land-claims affected 45 estuaries, mainly for rubbish and spoil disposal. Proposals in 1989 affected 36 internationally important British estuaries. While adequate regulatory mechanisms exist to control these developments, effective control depends on good decision-taking. Although some of these claims are extensive, a general characteristic is that land-take is piecemeal and progressive.

Coastal defence and protection

13.3. Concurrent with the developments referred to under paragraphs 13.1 and 13.2 above has been the development of coastal defence and protection systems. One option is the maintenance of the existing line of sea defences. Adoption of such a policy would necessitate the rebuilding and heightening of sea walls, many of which have been in place since the storms of 1953, and could ultimately end with virtually continuous defence along all soft coasts. This option poses a threat to nature conservation and natural coastal systems. Saltmarshes, sand dunes and their associated habitats will be squeezed between a rising sea level and a hard sea defence leading to a loss of intertidal habitats, the potential loss of sediment from the near-shore zone and the consequent undermining of the sea wall.

13.4. The continued use of hard coast protection structures such as concrete sea walls presents the threat of further disturbance to natural coastal processes by locking up cliff sediments and increasing beach loss due to the hydraulic performance of sea walls. Maintenance of natural sediment systems requires that the link between coastal processes and sediment stores are not severed by such defences or by limiting the landward accumulation of sediments. If sand dune or saltmarsh is lost from in front of sea walls, then the full force of the sea will rapidly erode the wall. Moving the sea wall inland, giving the natural system room to respond, may represent the best solution from both cost and nature conservation view points.

Recommendation 1: **Coastlines are naturally mobile and there is a need to work with, and not against natural processes. Hard defences will continue to be essential to protect life and property in many coastal areas, but it is important for nature conservation to limit their further spread and to encourage the use of " soft engineering " alternatives (such as beach feeding, sediment recharge and offshore breakwaters) wherever possible.**

[1] " Nature conservation and estuaries in Britain ". Nature Conservancy Council, 1991.

Recommendation 2: **With relative sea level rise already occurring in Southern England, and forecast to increase in the future, there is a need to develop a coastal defence strategy which includes the moving of sea walls inland in areas where the need to protect life and property permits.**

English Nature will assist the Government, National Rivers Authority and local communities in the development of such strategies.

13.5. In the light of the above, we welcomed the statement by the Minister of Agriculture accepting the need to work with natural processes, that coast protection schemes need to be integrated into an overall coastal policy and that the most natural ways of defending the coast will be used. We have accepted the Minister's invitation to join a periodic review of Government flood and defence policy and look forward to investigating possible approaches with the Ministry. The research presently being carried out by the National Rivers Authority and others on these matters is, potentially, of great value.

Sediment removal

13.6. Removal of sand and aggregate from dunes, shingle features and beaches is almost always detrimental to the conservation of coastal environments and often causes or accelerates coastal erosion. These dangers are widely recognised and led to the passing of Section 18 of the Coast Protection Act 1949, which gave local authorities in England and Wales powers to prevent the excavation of material from the seashore. Despite this, damaging extraction continues at a number of sites.

13.7 A more recent development is the removal of large quantities of sand and aggregate from the sea bed. At least 25 million tonnes of material were taken in 1990/91. This lies outside the land-based planning system and is largely controlled by the Crown Estate Commissioners using the "Government View" procedure where applications for licences to dredge are scrutinised for possible effect on the coastline. There has been no proven link between marine dredging and coastal erosion in recent years. Nevertheless, the rapid and continued increase in the quantities extracted has led to concern over possible long term, cumulative, effects.

Recommendation 3: **The extraction of aggregates from sand dune systems, shingle bars and the foreshore is rarely a satisfactory option and should be phased out. The present, cautious, approach to licensing fresh marine aggregate dredgings should continue and further research undertaken to model the long term effect of such extraction.**

Channel dredging and dumping

13.8. Another major source of sediment removal is the dredging by port and harbour authorities of deep water channels to maintain or enhance opportunities for navigation. Improvement to shipping channels on the Mersey alone required the removal of 400 million tonnes of material. Dredging on this scale, especially when followed up by maintenance dredging, can seriously disrupt patterns of longshore drift. Moerover, most of this material is currently lost to the coastal zone, much being dumped in deep water while some is sold as aggregate or disposed of for landfill. Existing methods of disposal have caused damage to sensitive ecological areas and we are working with MAFF to reduce these adverse environmental impacts.

Recommendation 4: **Where navigational dredging has been shown to interfere with longshore drift, schemes to transfer material across the channel (thus bypassing it) should be considered. Alternative uses should be established for dredged materials such as beach feeding or sediment recharge so as to avoid the loss of material to the coastal zone.**

Recreation

13.9. Recreational activities now affect all types of coastline and increasingly they occur throughout the year. The massive increase in water-borne recreation has reduced the area and the time over which wading birds can feed. Bait digging for angling further increases disturbance and depletes the food supply. Climbers can disrupt nesting sea birds and even walkers can cause significant disturbance, especially when accompanied by unrestrained dogs.

13.10. The development of diving equipment has extended recreation into the underwater marine environment. Recreational diving tends to be concentrated in diverse and attractive areas, which are often of high conservation value. Conflict can result from collection of specimens and from accidental damage to delicate species.

13.11. Recreation can work in harmony with nature. Some coastal reserves are major tourist attractions (eg the Farne Islands) providing income for conservation and stimulating public interest in wildlife. As such they are excellent examples of "Green Tourism" in operation. The challenge is to find ways for people to enjoy the coast without destroying what they come to see.

Water quality

13.12. coastal waters receive inputs of contaminants from rivers and from industrial and sewage discharges. The impact of contaminants on biological communities is dependent on a number of factors including concentration, toxicity, bioaccumulation and persistence. The problem of oiled sea birds has been widely publicised, as has the effect of certain toxic substances, including tributyl tin. English Nature forms part of the Department of Transport's rapid-response procedure for dealing with oil and chemical spills. The factors leading to the occurrence of toxic algal blooms and of anaerobic events in the North Sea are poorly understood, but eutrophication is affecting some inshore waters. Saline lagoons are a rare habitat containing a large number of endangered dependent organisms; they are extremely vulnerable to gross salinity changes and contamination.

13.13. The effects of many contaminants are poorly understood and a precautionary approach is called for. In our view industrial wastes should be treated at source. We were pleased when the Government announced its proposals for the treatment of coastal sewage effluents and for ending the dumping of sewage sludge at sea. We look forward to working closely with the Department of Environment and the National Rivers Authority in the setting of standards for key determinants for estuaries and coastal waters as part of the new statutory water quality classification.

Energy barrages

13.14. In the effort to develop renewable energy resources, there is considerable current interest in developing tidal power, with schemes under study in seven estuaries in England. While tidal power barrages are semi-permeable, the schemes proposed would inevitably reduce tidal amplitude within their impoundments, and substantially alter and reduce intertidal areas upstream. The problem here is that very large areas of present intertidal habitat would be lost as a consequence of a single major scheme. The effect of several such schemes could prove very severe, for bird populations particularly.

Fisheries and shellfisheries

13.15. Dredging for molluscs such as oysters, clams and scallops can be damaging to seabed habitats and non-target species. The method of hydraulic dredging fluidises the sediment surface and this can result in loss of fine particles and release of toxic materials. It efficiently removes both target and non-target infaunal species. Problems caused by crab and lobster fisheries stem from the ecological changes caused by the removal of these carnivores and detritivores, and through damage to the habitat, communities and non-target species caused by the various fishing methods employed. Modern heavy fishing equipment, particularly dredges and beam trawls can also be damaging to sea bed communities.

13.16. The cultivation of molluscan shellfish and fin fish can pose a variety of problems. These include deoxygenation of the water through decomposition of waste-food, eutrophication caused by build up of excretory products, smothering of the benthos and alteration of the substratum, side effects from the use of pesticides, and general disturbance.

14. Control of Activities in the Coastal Zone

14.1. Arrangements for the regulation of activities in the coastal zone are complex. Development falls under the control of the local planning authority (LPA) down to low water mark. Below this, within territorial limits, ownership of the seabed, subject to the common rights of fishery and navigation, is vested in the Crown. Regulatory control over these inshore waters is exercised through this ownership and by statute. The Nature Conservancy Council identified some 18 Government Departments and agencies exercising various regulatory functions under at least 88 Acts of Parliament.[1] Consideration needs to be given to whether a degree of rationalisation is called for. In some instances, for example in the case of the Crown Estate Commissioners, the Department of Energy, and the Ministry of Agriculture, Fisheries and Food, the regulatory role of these bodies is not clearly separated from their role in developing commercial potential. As a consequence a conflict of interest may arise; the separation of " regulatory " from " development " functions is very desirable.

Recommendation 5: **A primary principle for the regulation of activities in the coastal zone should be the wise and sustainable use of the resource to ensure that the nature conservation value of this environmental asset is not diminished. The " regulatory " role of regulatory bodies should be clearly separated from any role they may have, additionally, for the development of commercial potential.**

14.2. The public has a common law right to fish tidal waters, and in the sea within territorial waters (12 nautical miles from baselines), except where a private property right has been obtained which excludes the common right.

14.3. There is a general right of navigation in tidal waters and on the high seas, although within territorial waters individual countries have the right to exclude particular vessels or to restrict their activities. The right of navigation in tidal waters includes some incidental rights with respect to anchoring and mooring.

[1] " Legislative responsibilities in the marine environment—a review ". Nature Conservancy Council, 1989.

15. Conservation Mechanisms in the Coastal ZoneïThe Role of English Nature

Conservation Legislation

Sites of Special Scientific Interest and Nature Conservation Orders

15.1 The cornerstone of nature conservation in Great Britain is the protection afforded to Sites of Special Scientific Interest (SSSIs). English Nature has a duty to notify areas of special nature conservation importance to the relevant local planning authorities, statutory water organisations; owners and occupiers and the Secretary of State. Thereafter, those notified are required to consult English Nature before granting planning consent (in respect of LPAs), or undertaking activities likely to be damaging (with respect to the water organisations and owners and occupiers), on SSSI land. This mechanism is the principal means of achieving safeguard of the 2,980 sq km of coastal land and foreshore of nature conservation interest referred to in paragraph 11.2. Its principal limitations, in the present context, are that it does not apply to the general public (who are, therefore, not required to consult English Nature) and that it is restricted to land above mean low water mark. In certain circumstances, the Secretary of State can make a Nature Conservation Order on the land which can prohibit the general public from carrying out damaging activities. The limitation of mean low water mark still applies.

Statutory Nature Reserves (Land)

15.2. English Nature can establish and manage National Nature Reserves (NNRs) in the national interest and make byelaws for their protection. Local authorities have analogous powers with regard to establishing Local Nature Reserves (LNRs) in the local interest. These powers relate to areas above low water mark. In England there are at present 28 NNRs (180 sq km) and 25 LNRs (38 sq km) in the coastal zone, and plans exist for expansion.

Marine Nature Reserves

15.3 Areas of sea within three nautical miles (this can be extended to 12 miles by an Order in Council) may be designated as Marine Nature Reserves (MNR), for the purpose of nature conservation, and thereafter managed (in England) by English Nature. However, designation is only carried out after any substantive objections have been resolved. In practice, this has resulted in only one area being designated in England (30 sq km) (there is a second in Wales). Moreover, English Nature's powers for protecting MNRs (through byelaws) is very restricted; for example they cannot prohibit commercial fishing (which would require byelaws to be made by a Sea Fisheries Committee), discharges from vessels or some forms of navigation. To date, Sea Fisheries Committees have proved cautious about imposing fishing bye-laws to protect MNRs.

15.4. Areas of Special Protection for Birds can be designated by Order of the Secretary of State and the Order can prohibit access to areas (including marine areas within territorial waters) for the purpose of protecting birds. Such Orders, which cannot be made if there is an unresolved objection by an owner or occupier, can prove locally effective but can only be used for the protection of birds.

Recommendation 6: **Existing mechanisms have not proved capable of conserving the areas of known nature conservation importance in the marine environment below mean low water mark. Other legal and administrative mechanisms need to be established which will result in the effective regulation of activities which are damaging, or likely to cause damage, to such areas.**

15.5. We welcome the Government's recent announcement that they will consult in early 1992 on proposals for establishing a system of Marine Consultation Areas in England and Wales. English Nature will assist the Government in the development of this initiative.[1]

Problems with implementing international obligations

15.6. The limitations referred to in paragraphs 15.1 to 15.3 above create difficulties for the implementation of certain international obligations in marine waters. The listing and safeguard of important wetlands required under the Convention on Wetlands of International Importance especially as Waterfowl Habitat (Ramsar Convention) extends to marine water up to six metres in depth, and to deeper water within such areas. The requirement to designate Special Protection Areas under the EC Directive on the Conservation of Wild Birds extends to territorial waters. Existing legal and administrative mechanisms are not normally sufficient to meet these requirements.

Recommendation 7: **The existing statutory site protection framework needs to be amended to enable full compliance with international obligations for nature conservation as regards areas below low water mark.**

Planning Legislation

15.7. On land and foreshore, development is controlled by the appropriate local authority under Town and Country Planning legislation. Strategic planning is achieved through the preparation of Structure Plans at County level, and more detailed forward planning through Local Plans at District level. These plans are now

---

[1] " This common inheritance—the first year report ". HMSO, 1991.

the first consideration in the process of determining individual planning applications. As stated in paragraph 15.1 above LPAs are required to consult English Nature before granting planning permission on an SSSI. If the LPA is minded to grant permission against our advice English Nature can request that the Secretary of State call-in the planning application for determination; a vital safeguard. However, the threat from development to other coastal areas is less likely to be subject to such close scrutiny.

15.8. Hitherto, strategic coastal planning might have been addressed by a County planning authority through the mechanism of a Subject Local Plan. However, a County authority is now unable to prepare "local" plans (other than for minerals and waste disposal). It will be more difficult for a District Local Plan to address coastal issues satisfactorily unless as a contribution to a wider framework such as a regional plan.

Private and Hybrid Bills

15.9. Where a proposed development would restrict a public right of navigation, a Private or Hybrid Bill is required to obtain the authorisation of Parliament. Such Bills have been used to obtain the necessary consent for large developments, bypassing the planning system with its provision for objection and local inquiries. Parliament can only be advised of the views of English Nature by our petitioning. NCC's experience of the Private and Hybrid Bill procedure leads us to believe that it favours the Bill's promoters, especially when a Hybrid Bill is involved. Moreover, at present, the Private Bill procedure is exempt from the provision of the EC Directive on Environmental Assessment (85/337/EEC) although we understand that there are proposals to remedy this. We consider that the same environmental considerations and constraints should be applied to such Bills as to developments regulated by the planning system. The NCC's detailed recommendations have been submitted previously.[1]

16. Planning Policy in Coastal Areas

16.1. Since the coastal zone forms such a vital part of our natural heritage, it is important to plan future human use of coasts so as to maintain and enhance the natural features. To achieve this, planning mechanisms need to be underpinned by a general policy for pursuing the wise and sustainable use of coasts. This is consistent with the international commitments to safeguard wetlands and bird populations.

16.2. It is possible to quote many examples where the current piecemeal approach to coastal zone management has led to a coastal engineering domino effect; where the down drift ramifications of structures were poorly considered or ignored, sediment trapped and erosion problems passed on to neighbouring authorities. The economic cost of this is high. The present consultation procedures followed by MAFF should help forestall new problems from arising.

16.3. The sustainable use approach presupposes that risks of damage to the environment will be anticipated and avoided. Since many parts of the resource, especially in estuaries, have already been lost or damaged, the minimum standard for natural resource maintenance is very high. The sustainable use approach also facilitates opportunities for enhancing degraded parts of the coastal resource.

16.4. To achieve integration of coastal engineering and the natural coastal process system, it is essential that each engineered structure is viewed within the context of the geomorphological system as a whole. The coastline of England can be divided into a series of large sedimentological process cells, for example from the Humber to the Thames estuaries. This cell provides the basis for the East Anglian Coastal Authorities Group (ACAG), a regional group composed of district authorities from Cleethorpes to Southend. ACAG, as with other similar groups, was set up in recognition of the need to take a more strategic view of coastal planning. Process cells, of which the above is a large example, should be recognised as an essential criterion in the development of a regional planning unit for coastal zone management. If planning authorities were to adhere to a basic framework for management of the coast according to such cells, then engineering solutions would be more compatible with environmental demands and be more cost effective in the long term.

16.5. Similarly, consideration needs to be given, within the planning process, to the effect of development in river catchments behind the coast (and of the consequential effluent discharges and water supply implications of these), and for the implications of development (both coastal and inshore) on neighbouring areas of the coastal zone. The identification of natural cells for planning purposes, therefore, needs to be extended inland and also seawards.

Recommendation 8: **Legal and administrative mechanisms need to be established which will ensure that planning and other regulatory decisions are taken within the framework of a broadly regional strategy. Such regional strategies should be determined on the basis of criteria established by Government. These criteria should include consideration of the relationship between sediment supply and deposition, the influence of the hinterland (including, for estuaries, freshwater inflows), and the relationship between coasts and inshore marine waters. The framework will require cross-linkages to be established between adjacent authorities.**

---

[1] Report of the Joint Committee on Private Bill Procedure; 20 July 1988.

16.6. The development of a " regional " planning mechanism, based on a series of natural process cells, as described in paragraphs 16.4 and 16.5 above, does not necessarily require fundamental changes in the administration of decision-taking. While there would be a requirement to undertake strategic plan formulation on a " regional " basis, day-to-day planning administration could be carried out by the County and District Planning Authorities as at present. The same " regions " could be used by other regulators in the coastal zone, for example the National Rivers Authority, Crown Estate Commissioners, Department of Energy, etc in the development of parallel strategic approaches. What would be required would be a high degree of co-ordination.

16.7. While the " regions " referred to above could be the basis for general strategic planning in the coastal zone, strategies will also need to be developed on different scales to this. For example, it is desirable that detailed local coastal planning be carried out on a smaller scale (see paragraph 16.11 below), while some issues (for example the use of estuaries for tidal power generation and the development/rehabilitation of ports with their linkages to the inland transportation system) need to be determined on a national basis.

16.8. The development of " regional " plans for the coastal zone will require the formulation of national guidelines by Government and their issue to regulatory bodies. Such guidelines should be devised so as to include activities undertaken in inshore waters as well as those on coastal land and the foreshore. The preparation of such regional plans should be carried out within a reasonable time frame.

Recommendation 9: **Those bodies with a responsibility for regulating activities in the inshore marine environment should contribute to the regional strategy (which should cover their areas of responsibility) and should carry out their functions in conformity with the strategy.**

16.9 Regional policies, when formulated should be supported by Government, and any intended departure by a regulatory body from these policies should be referred to the Secretary of State for the Environment.

6.10. Within the " regional " planning mechanism, and national guidelines, referred to above, the importance of maintaining the nature conservation resource needs to be fully recognised, both through policies aimed at safeguarding designated sites, and areas of equivalent value below low water mark, and through the adoption of the wise and sustainable use principle.

16.11. For discrete coastal stretches, and for individual estuaries, there is a need for co-ordinated initiatives which would implement regional strategies at local level. Such initiatives would aim, for example, to zone recreational pressures, enhance coastal habitats through sympathetic management and ensure the careful control of development. The Heritage Coast concept is an example of such a co-ordinated approach which, ideally, would be led by local authorities in partnership with local communities and relevant public agencies. In certain circumstances it may be appropriate for other competent authorities to take the initiative, and the major landowners, particularly the Crown Estate Commissioners and the National Trust, will play a vital role. Schemes, such as Countryside Stewardship, could act in support of such initiatives which need to be coast-wide.

Recommendation 10: **Formal groupings should be established to co-ordinate the preparation of management plans for defined stretches of coast, and individual estuaries, in partnership with the local community and relevant public agencies. Plans could include matters such as recreational zoning and habitat management and the careful control of development.**

17. Critical Policy Areas

17.1. If nature conservation objectives are to be achieved, we consider that the following policies need to be adopted:

1. Regulating authorities should ensure that the coastal resource of nature is not diminished and should take decisions in the full knowledge of the environmental consequences. [Recommendations 1–5]

2. Further legal and administrative measures should be taken to enable areas of marine nature conservation importance to be conserved effectively. [Recommendations 6–7]

3. Regional and national strategies should be developed to enable planning and other regulatory decisions to be taken within a framework of policies for the coast as a whole. Local strategies should be developed to ensure that coasts are appropriately and sensitively managed. [Recommendations 8–10]

17.2. English Nature believes that if these recommendations are implemented, and the needs of nature conservation met within an overall strategy for coastal protection and management, the present coastal resource of nature, illustrated graphically in NCC's Atlas for 1991, will continue for the enjoyment and inspiration of future generations. English Nature is committed to working with others to achieve this objective.

November 1991

Examination of witnesses

DR DEREK LANGSLOW, Chief Executive, DR MALCOLM VINCENT, Head of Freshwater, Marine and Pollution Policy, and MR CHRIS STEVENS, Head of Earth Sciences, English Nature, examined.

Chairman

379. Good morning, Gentlemen, may I welcome you to this session of our inquiry into Coastal Zone Protection and Planning and to thank English Nature for the memorandum you have submitted for our consideration. Perhaps you would like to introduce yourself, Dr Langslow, and your colleagues formally for the purposes of the record.

(*Dr Langslow*) Thank you, Chairman, and thank you for the invitation to come. My name is Derek Langslow and I am the Chief Executive of English Nature. My colleagues are Dr Malcolm Vincent, who is the head of the Freshwater, Marine and Pollution Policy Branch and Mr Chris Stevens, who is the head of our Coastal Task Force.

380. As I understand it, English Nature is a body recently established under the Environmental Protection Act, and you provide advice and research into nature conservation in England and that you are working with the NCC on these matters. That is correct, is it?

(*Dr Langslow*) May I just make a slight correction? The Environment Protection Act set up the three country councils, each of which has a full range of responsibilities extending in our case to England and a wider responsibility nationally to Great Britain and to international nature conservation. Our duties are divided into general and special functions. The special functions component is done jointly with the Councils in Scotland and Wales through the medium of the Joint Nature Conservation Committee, the JNCC.

381. As you are a fairly recently created body, and therefore internally have not built up a body of experience and knowledge, can you tell me whether you have inherited any such body of experience and knowledge from any other source?

(*Dr Langslow*) We have indeed inherited it from several sources. First of all, we are an organisation which has come from the previous Nature Conservancy Council and, before that, the Nature Conservancy extending back to 1949. There is a considerable history of knowledge and experience which is relevant to these matters. In addition to that through the recruitment we have of staff, and particularly over the last 12 months, we have sought to bring in other skills in order to complement those which we inherited from the NCC to enhance our ability to carry out our advisory functions to government.

382. Thank you very much. It is very helpful for us to know, especially with a new body, who we are dealing with. In paragraph 7.10 of your paper you give support for the concept of formal groupings in order to co-ordinate the preparation of management plans for defined stretches of coast. Who should form the basis for these coastal groups and what kind of resources should be made available to them and by whom?

(*Mr Stevens*) We believe that there would be between seven and 10 such cells or groupings around the coastline. The terms of reference for those groupings might be considered at two levels. A core, necessary series of functions would be to bring together flood defence and coast protection and their environmental consequences and other activities that have a major effect on the natural systems around the coast, like channel dredging and coastal aggregate extraction and onshore development that would cause coastal protection or flood defence to take place in the future. We can envisage a second level of terms of reference that took their function more widely to include, for example, the co-ordination of recreation around the coast, inshore fisheries and water quality. To address those terms of reference we believe the groupings should consist of local authorities, the National Rivers Authority, local representatives of the Ministry of Agriculture, Fisheries and Food, of English Nature, the Countryside Commission and English Heritage and, if the broader terms of reference were adopted, bringing in recreation and fisheries, then there would need to be additional members, perhaps including the sea fisheries committees and recreational representatives. The second part of the question concerned the resources that they would need. We see a prime need for a resource in terms of data and information. The sort of material that would be required for that group would be represented well by the NRA/Anglian sea defence management study or the information on coastal cell sedimentary systems collected by the Standing Conference on Problems Associated with the Coastline, SCOPAC, on the south coast. The second requirement would be in terms of expertise. We would expect that to be drawn from the constituent organisations. The third, depending on their exact legal status, might be the resources to undertake monitoring or possible enforcement. A last prerequisite for a group would be a national strategy, some national series of guidelines at a central level which gave a framework within which they addressed their management strategies in the cells.

383. Two further questions occur to me from the answer you have just given. First, why seven to 10? Why not 17 to 20 or why not one or two? What is the reason for splitting up the coastline, as it were, into seven or 10 areas? What is the particular reason for that? Secondly, when you split up the coastline you will get a boundary between two groups. What happens at that boundary? Part of the problem we see from the present arrangement is that we get conflicting policies between different authorities who have adjoining pieces of coastline, especially in an estuarial situation across the estuary. How do you get that co-ordination? Does that come part of the national strategy that you have just spoken about and does that national strategy mean that the separate groups are accountable to a superior body that oversees everything on a national basis? If that is the case, are you talking of a new unitary authority

Chairman Contd]

that would be responsible, lay down the strategy to be followed in practice on the ground by these groupings or will it be the Department of the Environment or the National Rivers Authority? Have you thought that through?

(Mr Stevens) Thank you, perhaps I might address the first part of your question and the first part of the second then perhaps one of my colleagues would like to build on from there. First, the reason for believing that seven to 10 cells around the coast would be the logical unit for planning arises from a series of studies that have been undertaken by a consultancy under commission from the Ministry of Agriculture, which has defined the coastal sedimentary cells, which are the prime determinants of coastal processes within which all of these activities must operate and these coastal cells are the reason for their interdependence and the reason for their impact on nature conservation and the rationale behind much of their economics. There are between seven and 10 such cells which are relatively, but not completely, independent of one another around the coastline of England and Wales. The imprecision over the number arises from the fact that a further report producing a final recommendation is in the process of being commissioned and should be available within a matter, we understand, of months. The majority of the material, which is commonly referred to as the Macro study, is currently available.

384. Before you answer the second question I asked you, could I interpose another there? Do you have, or could you produce for us, a map showing these seven or 10 cells? The reason is that I have invited the NRA to produce a map of the coastline showing the ownership of the various stretches of beach: what is NRA, what is marine local authority, what is private ownership and so on. It would be an interesting exercise to match what you consider to be a unitary cell of responsibility with the way in which it is fragmented at the moment. Would that be asking too much of you, to produce a map of the coastline indicating those seven or 10 areas?

(Mr Stevens) We believe that a map of that kind would be available. I think we would like to stress that it would not be the product of our work, but it would be the product largely of commissioned work undertaken by the Ministry of Agriculture.

385. Perhaps you would then tell the Clerk of the Committee after you have made your enquiries whether you can produce the map or where he can write to obtain the map on behalf of the Committee.

(Mr Stevens) Certainly.

Chairman: I am most grateful.

Mr Jones

386. Perhaps I can follow that first half up before we go on to the second half. Would I be right in assuming that none of these cells would divide any major estuary between two of them?

(Mr Stevens) That probably reflects the second part of the question.

387. In that case I will ask another bit and you can deal with it all at once. What would happen at the Welsh and Scottish borders?

(Mr Stevens) One of the by-products of the way that these cells work around our coasts is that they are in general divided by estuaries. Clearly then there is a potential for the same sort of end effects that we have had among all the various participating bodies all the way within the cells to happen at the end. As I believe you have suggested, Mr Chairman, we see one of the functions of national strategy being to ensure that there was an adequate co-ordination of the ends and the interface between these cells. I mentioned earlier that they are not wholly independent of one another and co-ordination would be required for the same purpose[1].

Chairman

388. And the next question is, do you see that as a supervisory role or a body?

(Dr Vincent) Before we end up with a conclusion to that, Mr Chairman, it would be helpful to look at what such a body, were it formed, might do. The prerequisites that we see for this is that we believe that there need to be clear terms of reference for these cells, so somebody has to determine those. Secondly, we need to consider the question of the inland boundary of the cells: what the hinterland of these cells should be. National guidance is needed for those and so we should look for guidelines being produced from government to the cells to cover both those areas. Within government the clear body, I should have thought, with at least some responsibility for this must be the Department of the Environment, because in part we end up talking about planning and planning strategies. So perhaps we could look to government to decide who to take the lead, but we would have looked to the Department of the Environment to have a major input into that.

389. Not MAFF?

(Dr Vincent) I am sure that the Ministry of Agriculture would be considerably involved in this. Whether it is one department or whether it is some kind of coalition or committee between departments is an open question. But I think both ministries will need to be involved.

390. We have noted a certain amount of reluctance on the part of public bodies to relinquish to some other body whatever they may be doing at the moment. MAFF have a very thick finger in this particular pie. That is why I asked whether you had considered their remaining in the saddle?

(Dr Langslow) I think the crucial thing in making these work, whether they are in statute or whether they are in some sort of voluntary framework, will be

---

[1] Note by witness: "English Nature will, in 1992/93, be using part of its enhanced grant-in-aid from the Department of Environment to institute a programme of management plan preparation for individual estuaries. It is intended that these management plans will be prepared in collaboration with local authorities and other public agencies and in consultation with the local community. The aim of the plans is to provide a policy framework capable of achieving the wise and sustainable use of estuaries. The plans will include proposals aimed at enhancing the value of the estuaries for wildlife; implementation of these proposals to follow rapidly. It is expected that this programme will be continued through the 1990s; the aim being to cover 80 per cent of the estuarial area of England by the year 2000."

Chairman *Contd]*

the commitment of the different groups to work together and the commitment to look beyond their own narrow confines. That would apply just as much to a local authority, to the Ministry of Agriculture or to an organisation like my own. There will be a need within the government framework to take a lead. It seems to us that the lead department is likely to be the Department of the Environment, because somewhere it will have to fit with the existing planning system and be either a development of that or an add-on to it. The impetus will need to come, therefore, from that department. Of course the Ministry of Agriculture is bound to be closely involved, as will other Departments of State as well.

Mr Field

391. Dr Vincent was just alluding to the setting of the boundaries. Who sets the boundaries for the National Environmental Research Council? They have quite separate departments for the marine side and the landward side and so on. Who draws their boundaries up?

(Dr Langslow) Within the National Environmental Research Council, I am afraid I do not know how they draw their boundaries up.

392. Do you know what their boundaries are?

(Dr Langslow) I think I know broadly what the boundaries of NERC are in relation to their duties as set out in their charter.

393. But separate within each of their departments?

(Dr Langslow) I would hesitate to comment about precisely how they draw their boundaries. I certainly know the broad structure of the way they operate, but I think I could not answer a question which was precise about where they drew the boundary between, say, their geological science side and their oceanographic side, as an example. Clearly there are interfaces there in terms of the way they do their work.

Chairman

394. What would be the relationship between these coastal zone management plans to be produced by these cells and the current statutory plan system?

(Dr Vincent) How we envisage the management plans being developed is essentially as a series of strategies based on each of the coastal cells. The area will obviously relate to the area of those cells and the remit of the strategies will be the total of the involvement of the membership of those cells in so far as they are relevant to coastal zone management. What we see is the development of a series of inter-related policies coming out of those strategies, which are agreed to by the members of the regional group, and which form the basis for decision making by the members of the group in the carrying out of their various functions. In the longer term—indeed, as development plans come up for renewal—we would see those policies being integrated into the statutory development planning system in the normal course of events. At the moment that is where our thinking rests.

395. This is really what was behind the question—you envisage the statutory development plan as being the legal mechanism for bringing into place the strategies, because in recommendation 8 you talk about legal and administrative mechanisms needing to be established which will ensure that planning and other regulatory decisions are taken within the framework of a broadly regional strategy?

(Dr Vincent) What we have in mind is that strong encouragement, to say the least, in the form of very strong guidance from government would be given to these cells to develop strategies. The strategies would be implemented by the existing powers of the membership of the group, depending on who had the power in relation to a particular function. As a consequence of evolution those policies, where relevant, would be incorporated into development plans. It would not necessarily be the structure or local plans which would be the framework within which these management plans would be developed, rather the other way round. The management plans would feed into structure and local plans.

Mr Jones

396. Can I have an anwer to my question about Scotland and Wales?

(Dr Vincent) In relation to Wales, when we know clearly the boundaries of the regional cells that might sort itself out, in so far as we are tied to the boundaries of the cells, rather than the administrative boundaries that you are talking about.

397. That is the point I am making. If logically a regional cell covers, for example, the whole of the Dee estuary area, which in geological and general environmental terms makes sense, then it crosses the administrative boundary between England and Wales and then you have problems with the boundary responsibilities between the Department of the Environment and the Welsh Office. What are you suggesting there?

(Dr Vincent) The guidance given would be guidance from government, so it would be signed up to by the Department of the Environment and the Welsh Office. At the moment I am not quite sure that I see any real problem about this since I have explained that the functions are being carried out by the existing bodies. I do not know whether any problem exists.

398. We already have a problem in that the local authorities on the south of the Dee are pursuing totally contradictory policies from the authorities on the north of the Dee. These are approved of respectively by the Welsh Office and the Department of the Environment through the plans, so I think I am right to identify a problem and I am asking you about your solution to it. Can I be absolutely clear, therefore, that you are proposing that these cells should indeed cross national boundaries where these are appropriate?

(Dr Langslow) Given that the definition of the cells is based upon sediment movement etc, then quite clearly they will cross an administrative boundary between England, Scotland and Wales. I

Mr Jones *Contd]*

do not believe that the problems of solving the institutional mechanisms is any greater crossing an England/Scotland/Wales boundary than it may be crossing an individual local authority boundary, say, between Norfolk and Suffolk. The principles are precisely the same. It may be that you have more players, but if government is pursuing, as you seem to be suggesting, different policies in different parts of the Dee then that is clearly something that government has to tackle to resolve.

399. So you are saying there should be a unit crossing that boundary and that it is up to government to sort out the problems on their side? I accept that. Now could I move on to what should happen concerning nature conservation below the low water mark. You have been critical of the present site protection arrangements. We have had some suggestions from other witnesses and I should be grateful if you could let me know what you think should be done to amend the present arrangements.

(Dr Langslow) The major gap is that below low water mark there is no mechanism in statute, other than a marine nature reserve, for delivering any kind of nature conservation policy or programme. This has implications in the sense, for example, that some of the international conventions, to which the Government is a signatory, require a conservation action in areas below the low water mark: special protection areas, for example, under the EC Bird Directive, Ramsar sites under the Ramsar Convention on International Wetlands. Therefore one needs a legal mechanism which allows the conservation of those sites. One possibility is to extend the current SSSI mechanism and amend the legislation necessary to introduce SSSIs. There are other probably simpler legislative mechanisms than the SSSI one where you could define certain minimum requirements in those areas. You could begin, as another alternative, by defining what are in effect special areas on a map and then deciding what activites are permissible within them and do it on a site-by-site basis. There are a number of things, but the major gap is that currently there is no legal framework at all, other than an MNR, which is not appropriate for most of these situations to deal with this legislative gap.

400. So what exactly are you proposing out of that short list of suggestions?

(Dr Langslow) I think the best one would be a proposal to extend an SSSI-type mechanism. I say carefully " SSSI-type " because I do not believe that necessarily simply translating what is presently in the 1981 Act directly into a new statute below the low water mark would be the best way of handling the detail of it. But the principle of setting up a framework which allows negotiation and consultation on the activities going on in a particular area would be the way to proceed.

401. What about the idea of having a zonal planning system for the area between the low water mark and the high water mark?

(Dr Langslow) The situation between the low water mark and the high water mark in terms of nature conservation is already covered by the existing

legislation so that the SSSI mechanism currently extends to that. If you introduced a further stage between low water mark and high water mark then you would either need to change other legislation and set that aside or you would need to incorporate it. But I do not think in principle it would be any different from the kind of idea that we are putting forward for the regional cell, except that we believe that having these very narrow bands on the coast is not the right way to proceed. There is such a close relationship between the intertidal and the inshore tidal area, that these things need to be considered as a group, as a single eco-system.

402. Can I refer to your text in which the words " wise use " and " sustainable use " are used quite often? A bit like motherhood and apple pie, what does it mean? It sounds good and we all support this. Is there any definition of these terms?

(Dr Langslow) As I am sure the Committee is aware, there are dozens of definitions of these things. Those that we find as helpful as any are those that were written more than 20 years ago and are enshrined in the Ramsar Convention, where the wise use of wetlands is defined as their sustainable utilisation for the benefit of human kind in a way compatible with the maintenance of the natural properties of the eco-system. The convention then goes on to define sustainable utilisation of a wetland as a use which meets the needs of the present generation while maintaining its potential to meet the needs and aspirations of the future generations. Those two definitions bring in what I think is the essence of sustainable use, which is that one is looking at the future and at maintaining the natural capital. One is looking to ensure that there are no irreversible changes which damage that long-term future, and that there is an attitude and a commitment towards those natural systems which allows that. The difficult area is distilling that down into practical programmes. But the kinds of proposals that we are putting forward in terms of strategies for the coast and in terms of the regional developments provide mechanisms where this could be done and we are looking at a programme that we are starting in a couple of months' time, what we call our " estuaries initiative " where a major component of that effort will be the practical implementation of sustainable use and how at a local level in a particular place one brings together the different parties to find a framework which allows all of that to be put into practice.

403. You referred to wetlands just now. Do you think we have the necessary skills and understanding to recreate and manage natural wetland habitats on lands encroached upon by the sea?

(Mr Stevens) The main habitat that might be encroached upon by the sea during a managed retreat is the grazing marsh. The main nature conservation value of grazing marsh arises from the birds that use it and also for a proportion of the grazing marshes from the invertebrates and plants that grow on them. We have the techniques to recreate the interest of the majority of the grazing marshes and sustain their bird interest. We do not currently have the techniques in a

Mr Jones *Contd]*

proven way—they are still very much in development and will need rather long time scales to prove that they are going to work—to bring back the floristic assemblages to areas that are wholly transferred. If we recreate adjacent to an area, then the populations can spread into it. If we recreate at a great distance, then they cannot. So for this category of grazing marsh, which is a rather small minority, where the floristic assemblages are particularly important we would not be advocating a general set back across those areas. But for the bulk of them we have the techniques.

Mr Steen

404. Can I ask you about the effects of modern fishing methods upon the sea bed? Can you estimate what proportion of the inshore waters of the coastal zone is affected by these operations and give some indication of how serious this matter is? I represent the coastline of South Devon and I have both scalloping and dredging and I also have lobster potting and crabbers. The shell fishermen have no doubt that over the last 10 years the amount of shellfish has declined dramatically and the sea bed has been destroyed. I have seen some of the evidence they have and I tend to support their view, but I wonder what your feeling is about the controlling of beam trawling and scallop dredging. The further point is the European context and to what extent you feel what we do in this country would in any way be able to control what they do in Europe. The damage seems to be extremely great and I should like to know whether it is something you have majored on or whether it is just a local problem I have in South Devon?

Chairman: Perhaps I could add a footnote to that question. When the Committee were considering the environment in Northern Ireland in one of our previous inquiries we were very disturbed to find that large areas of Strangford Lough had been turned into a moonscape by the hoovering up of scallops along the sea bed by international trawlers. In addition to beam trawling and scallop dredging, have you evidence of hoovering within English coastal waters?

(Dr Vincent) I will try to take the points one at a time and see how far I get. As regards trawling we can say, without too much challenge, that major parts of the North Sea and the Irish Sea are trawled at least once a year, virtually the whole area. The effect of this has essentially been to change the nature of the communities on the sea bed. I am not necessarily saying that they are less valuable, but they are certainly losing diversity. Those species like the heart urchins and some crustaceans, which are relatively slow growing and slow at reproducing, are gradually being replaced by other species, such as polychaete worms. So the effect over a period of time is to change the nature of the communities on the sea bed. The way this happens is that the chains attached to the trawler scoop up the sediment and with it they scoop up the invertebrates. They scoop up to a few centimetres in depth and that is how it works. The mortalities of those invertebrates are considerable and the changes occur as a result. The effects of scallop dredging can also be locally very significant indeed. Strangford Lough has been quite severely

damaged by the pursuit of scallops in recent years. Similar problems used to occur from scallop dredging round Skomer off the coast of Wales and in the Firth of Clyde, so it is quite serious. In the Firth of Clyde the communities damaged were these calcareous algae communities called maerl which are very significant in a European scale. In Devon scallop dredging occurs in Lyme Bay and in that kind of area one should seriously be looking at less damaging ways of collecting the scallops. In the Channel Islands scallop dredging has been banned and has been replaced by diving for scallops. I am not saying that is the answer, but I am saying that perhaps we should look at alternatives. Our general conclusions are that in areas which are highly sensitive and highly important for their marine communities, then this type of heavy beam trawling and scallop dredging are not compatible with the maintenance of that interest. Less severe methods should be employed. As regards what communities are left, in the sense of not being damaged in this way, the areas which are not affected are quite limited. We are talking about areas deliberately set aside in the main either for low intensity fishing or where these methods are excluded. Things like marine nature reserves and protection areas for spawning, nurseries areas and so on, are actually quite small areas in total. I am afraid I do not feel confident in responding to the point about the European Community. That is a question definitely for the Ministry of Agriculture.

405. But you believe that there is sufficient damage being done to ensure that some provision is made to reduce the amount of activity of scalloping and dredging in these areas or in all areas?

(Dr Vincent) In sensitive and important areas where there are marine communities, we have no doubt in saying that there is a need to reduce this type of activity. In other areas of middle importance we should be looking at ways of mitigating or other ways of fishing which will be less damaging.

Mr Field

406. Can I just press you on this? You mentioned earlier the sea fisheries committees. By and large, certainly in my area, they are responsible for licensing and continue to do so in vast numbers. One wonders what fishermen can even be beginning to find. At Hamble Point, for example, you can count as many as 16 inshore vessels just going up and down day after day. One wonders what is left on the sea bed with that sort of number just trawling one after the other.

(Dr Vincent) I forgot to answer the question about lobsters and crabs, but I will come back to that. The concerns which we have are more the community changes which are occurring. I mentioned earlier that this method of fishing does not necessarily reduce the quantity of fish being caught, at least not necessarily for this reason of the disturbance of the sea bed. The communities change and other types of organisms come in which, as fish food, can be just as satisfactory. But what is happening is that the methods employed are destroying the sensitive species like sea fans, sea urchins, communities which

Mr Field *Contd]*

take a long time to colonise themselves. They are being replaced by commonplace communities. Our problem is ensuring that that physical damage does not occur and that those mortalities do not occur. The same thing is occurring in response to modern methods of crabbing and lobster potting. It is not the take of the crustaceans which is damaging to our interests, particularly, but it is the physical damage which the hawsers and lines of these pots cause when they are put down and taken up and drag across the sea bed. Those are the things which are affecting nature conservation, rather than the actual take of the species.

Mr Steen

407. You say it is changing the beds, but then the whole of the world is changing. Can one say it is changing for the worse? Do you have any evidence to suggest there is damage or that it is just being changed?

(Dr Vincent) What is happening is a loss of diversity. A previous community or a present system which was previously highly diverse is being modified. There were lots of different types of communities with many types of organisms and they are being changed into a much more uniform kind of community. There may be as many species in that community for any given hectare or square kilometre of that bed, but overall we are losing species so that the overall diversity is much reduced. That is what is happening. That is the damage to nature conservation.

Mr Pendry

408. I am not sure I got your answer to the Chairman's footnote about hoovering, as distinct from dredging. What we saw in Northern Ireland was this hoovering. They may well have been after scallops, but they were destroying all kinds of species in the process.

(Dr Vincent) I am talking now completely from recollection and I believe that this is occurring in places such as the Solway Firth in relation to cockles. The effect is, as you say, to remove not just the target species but everything else and that causes all the sorts of damage that I have been talking about. Long-lived slow reproducing species tend to be lost as a result.

Mr Field

409. The British Aggregate Construction Materials Industries stated in its evidence that "marine dredging is an environmentally friendly industry". Can you let us have your response to this claim?

(Mr Stevens) I should like to start by drawing the Committee's attention to three specific impacts that marine dredging can have. The first is a direct effect. Clearly the animals and plants living on the bottom will be destroyed in the immediate area of dredging. There is another effect that arises because of all the disturbance that is caused by the dredging: the turbidity of the water increases, the amount of light that gets down into the water decreases over a wider area, and so there is a knock on effect outside the immediate zone. However, having said that, there is impact from any sort of aggregate extraction,

including terrestrial aggregate extraction. Much of the impact from this effect depends on how sensitive the area that is being extracted is, whether it is a very rich area or whether it is a comparatively sparse area. The second impact can arise from the interference of aggregates, the taking out of aggregates from active coastal cells. For the most part licences have been granted rather cautiously in the past. Consequently there probably have not been major impacts for offshore aggregate extraction on cells. There certainly have been historical examples, but under the present system of licensing we know of no examples where there is a demonstrable effect. There is a difficulty, though, because we have been looking at these cells for rather a short time and aggregates that are offshore, lying on the sea bed, may be mobilised to sustain systems; in other words, to replenish beaches naturally, and spits and bars, on long time scales with rather infrequent major storms. A cautious approach is certainly the right approach for the future. The third point is that a lot of the aggregates that lie close to the shore are potential sources for beach recharge and for other rather sensitive ways of managing the coast that we might want to use in the future, so the extraction ought, in our view, to be planned on the basis of the creation of reserves of aggregate for specific purposes in the future, or at least not prejudicing our options to use those in the future in cases where they seem likely to be used. Those are three of the impacts, but we recognise also that there are impacts from every other form of aggregate extraction. I do not think as an organisation that we would be seeking to have marine dredging for aggregates cease as an activity.

410. MAFF told us in their evidence that they were very happy with the research that the Crown Estate does on potential aggregate dredging—I think the Crown Estate gets its research from the Hydraulic Research Organisation—but MAFF had never seen one of those reports. Do you ever see any of the reports?

(Mr Stevens) Yes, there are a series of generic reports which give general guidance over the sort of depth below which you do not need to worry about whether they are going to get involved in coastal cells. So if they are in sufficiently deep water or are sufficiently far off shore, it is generally accepted that they will not cause impacts on the coast. We have seen such a report. There are also specific proposals and we only see a summary of such reports. There is a need for more understanding in this area. I do not think this is a completely understood area, and I do not think that we believe as an organisation that it is sufficiently understood to deviate from the present very cautious approach to licensing.

411. Just taking the fishing, for a moment, the fishermen at Lowestoft and at Great Yarmouth, in the Solent area and Poole, all believe that aggregate dredging does harm to the spawning grounds of local species. Is that your view or not?

(Dr Langslow) The one thing that you can say is there is not any reason to think that aggregate extraction will be significantly more damaging than some of the trawling or some of the other activities.

Mr Field *Contd]*

Any activity where you take an area of sea bed and in one way of another plough it up, has a potential effect on the animals that are there and therefore you can have that. What we have no immediate or direct evidence for—and I can certainly undertake to take it away to check it—is that aggregate dredging in itself has affected fish spawning grounds. It is worth adding here that in terms of the Crown Estate Commissioners' role, they operate a very correct and cautious approach. Their attitude is very responsible and they evaluate very carefully the various views. If they believe there is any evidence that damage is to be caused to another interest, then they do not grant the licence.

412. I am sure that Mr Parrish, who is sitting right behind you, will be pleased to hear that. Can I ask you about this business of being an exact science in terms of the littoral drift and so on? I have to apologise to Tom Pendry about this because otherwise he will be at me for a constituency point. Many of my constituents have photographs where they have frontages to the coast showing very much better beach nourishment in the 1940s and the 1950s to that which exists today. Can they be so very wrong in their views that this constant taking of aggregate in, say, the Solent area has depleted their foreshores?

(Mr Stevens) Your constituents are undoubtedly right in recognising that there has been a deterioration in the extent, depth and width of their beaches since the 1940s. We believe that the underlying causes of that are largely the starvation of sediment as a result of coast defence, which has stopped cliff erosion and stopped the input of sediments into systems and has also reduced the extent of spreading along the cells. In each cell there is a source of sediments, there is a transport zone and perhaps a sink. There were also in the past many well-known examples where aggregate extraction from the sea bed has caused loss of the beach, sometimes very dramatically and there is an area in South Devon, which is probably the leading case history. It is not an exact science. We believe at present on the balance of evidence that the prime issue is the coast defence works that have been undertaken rather piecemeal around the coast and that we are now having this as a knock-on effect from those and that there is not very much evidence that dredging, as it is currently licensed and carried out is having an impact. I return to the original point that it is not an exact science and that particularly long term events that we have not had a chance to witness yet may mean that there is more of a connection between the areas that are being dredged and the health of the coastal systems than we currently have evidence for.

413. Do you have a view on marine archaeology as an organisation and the destruction of archaeological sites by aggregate dredging?

(Dr Langslow) I do not think we have a view on it in the sense that we have a knowledge of the scale of the issue.

414. Do you have any knowledge?

(Dr Langslow) The only knowledge that we have on the archaeological side relates to our more general duties to integrate with other activities. Anything which disturbs the sediments, for example, will clearly affect the archaeological interest. It affects the historical record and a whole lot of other things, so the activities that we describe which affect the nature conservation interest are quite likely, therefore, to affect some of the archaeological interests, but we have no specialist knowledge on that.

415. I am told—I do not know whether it is anecdotal or not because I have not seen it for myself—that it was a regular feature of aggregate dredging in the Solent for the ships to unload cannon balls in vast quantities after dredging in certain areas. That would seem to me to be the destruction of some quite good historical sites.

(Dr Langslow) There is nothing I can add.

Mr Steen

416. Can I talk to you for a moment about your channel dredging and dumping which flows naturally from what we have been discussing about sea bed dredging? From what you were saying there is little evidence that dredging in itself, although you are looking into it, has done great damage to the sea bed or to the area. Obviously there is one area which we have been rather weak on in the channel dredging and dumping and that is that there is a commercial element to it, not just for the company that dredges, but for those people who are affected by there not being dredging done. I am thinking particularly of the estuaries which need regularly to be dredged so that the fishing vessels and the yachts can get in and out. I have in mind particularly the Salcombe estuary, which has now been declared a site of special scientific interest. Salcombe is a good example of this, where it has been so painful to get the necessary licences and so delayed and so expensive because English Nature and others have suggested that dumping silt in places where it has been dumped for the last 50 years will cause various damage to the ecology that they have had to take the silt miles and miles out to sea at enormous cost. The delay was such that the bottoms of the fishing vessels were scraping along the ground. Now there are further problems, not just in the Salcombe estuary, but in other estuaries, where the yachts cannot get out from the commercial yards where they are being repaired because the silt has built up so much. They cannot get the extraction licences and they cannot dump the silt anywhere, so that is affecting the trade of small enterprises. I feel that there is a lot in this report which is purely academic but does it not relate to the commercial dealings that people have to have in their ordinary way of life? I just wondered whether you could comment on that.

(Mr Stevens) May I say at the outset that there is a major distinction in the impact between channel dredging and aggregate dredging. Our previous comments related to aggregate dredging. Also, as a general comment, we fully recognise that there are many and conflicting uses of the marine environment and what is required is a balanced compromise to achieve the best national good. Channel dredging itself directly can lead to difficulties with areas adjacent to the channel which slump into it and there

Mr Steen *Contd]*

are many examples round the Thames estuary and up the Essex coast. The disposal of fines can obliterate environments, of course, and we believe that they are sometimes major impacts. However, the effect that channel dredging has on coastal systems where it cuts them is one that is subject to a sustaining option whereby the material is simply taken out of the channel—it would flow into the channel because the channel is being filled up naturally by material moving down the coast—into the area where it can naturally carry on and move down the coast. By that means the impact can be minimised. The material that is taken out of the channel may find great use in habitat recreation because one of the issues that we face in managed retreat over low-lying land is that because it has been drained it has gone down considerably below sea level. That is a natural shrinkage of the soil that has taken place as a result of the drainage. That would need to be built up again. In many cases it seems likely that the fine dredging material could be used for that purpose. There are a lot of environmentally friendly—if I may use that expression—end uses for that material. We believe that there is in most cases that we have met room for a solution that does not reasonably preclude other coastal uses.

417. Do you think that in your report and in your thinking you should have regard to the commercial impact of some of the things you are recommending and the effect on jobs and wealth creation in the locality by suggesting, for example, that silt has to be carried out to various places, which puts an enormous cost on a very small shipyard or other small enterprise? I think that your organisation ought to have some regard for that. Not just in the recommendation for channel dredging but for all recommendations, it should have some idea of the cost to an organisation and the rules and regulations which would reduce their effectiveness if you insisted on various things.

(Dr Langslow) Let me reassure you that we certainly have a very sharp eye for that kind of thing. But we also have a sharp eye to our parliamentary duties as set out in the Act to give advice. In very few of these instances are we an authority which tells people finally what to do. We give advice. Where you have a special resource, we are talking here about our sustainable utilisation principles, where what is potentially a very short term issue in commercial terms overrides the long term issue and the future generational issues. It is a very difficult balance to strike and we do not underestimate that. Also the need, particularly if the cost is falling on a very small user, must be dealt with as part of the institutional mechanism.

Chairman

418. Listening to this exchange on what to do with the product of channel dredging, it occurred to me that this is part of the overall problem of what to do with waste generally which we have met on other occasions. The immediate reaction of those who produce waste is to get rid of it at the cheapest possible price. I would imaging the same thing occurs with the results of dredging out the middle of a

channel. What is needed is for someone to sit down and to decide whether to treat this silt as waste and just to dump it at the most convenient and nearest place where we can get rid of it. At the same time, they should also ask whether there is some use to which this waste should be put. The small businessman engaged in this type of operation does not have the means, the facilities, or the knowledge to work that kind of problem out for himself. I should have thought, with respect, that this is where an advisory body could sit down and do some work and produce advice to those who wish to engage in this operation.

(Dr Langslow) I certainly share that, Chairman. It is also precisely the kind of issue that we would see addressed in what we have termed these regional cells. There you do have situations where there may be a shortage of sediment in some areas and an excess of it in others, because the cells are defined based on these movements. That is precisely where we would see that kind of thing, by taking it out of the very narrow confines and looking at it on a wider scale, produce real benefits, I would certainly share that and we will note that point carefully to ensure that our advice feeds in strongly there.

Mr Field

419. Can I then take you on to marine pollution? How much is marine pollution, particularly oil, a problem for coastal habitats and wildlife and are clean-up and preventive arrangements working to your satisfaction? I believe that you collaborated with the RSPB in producing a survey of all the coastline which is now on the Marine Pollution Control Unit's computer so that they can turn it up in the event of a major spill.

(Dr Vincent) One of the tools to help in this which we have produced working with the RSPB and from our own information, is a coastal atlas. This indicates in colour the various habitats which occur round the coast of Great Britain. It was an initiative started by the NCC which we are taking forward. Each sensitive site is earmarked and references on this coastal atlas are referenced to a data sheet which explains the value of that site, what is there and the preferred action to take in the event of an oil spill. Copies of those atlases are available to the MPCU and part of their database and they are also circulated to local authorities for them to use. The information available to help the MPCU and local authorities with oil spill clean-up on the nature conservation side is quite good now. In answering your question about how damaging is it, the most sensitive areas seem to be areas of calm water and sediments where the oil can percolate into the sediment and then is released slowly over a longer period of time. The habitats which seem to suffer most badly from this kind of spill are salt marshes, or mud flats, calm water sediments, whereas habitats which seem to be able to recover most quickly are rocky shores and where there is wave action, for example. How serious is it? Over a period of years the communities seem to recover. Colonisation of the marine environment is better, generally speaking, than on a terrestrial environment. If you wait long enough the communities seem to come back. There are apparent exceptions to this. For

Mr Field *Contd]*

example, if oil and dispersants are combined then the effect on some communities like eel grass beds, which are valuable both to us for the community of animals and plants which are associated with them and also as a food for wild fowl, seem to be quite susceptible. So some habitats seem to be hit in certain circumstances. The coastal dynamic of populations is very complicated and sometimes it is very difficult to understand whether what appears to be long-term damage is due to one cause rather than another. At the end of all this I have to say that recovery in the main seems to take place, but that some damage probably is occurring, but it is very difficult to attribute it to oil pollution, if that is the cause.

420. Do you support the Department of Transport's reliance on dispersants?

(Dr Vincent) The use of dispersants in certain circumstances, such as at sea, is probably reasonable. In general we advise against the use of dispersants on salt marshes and mud flats. We tend to think that they are often a first line of defence taken by local authorities to demonstrate that they are doing something when oil threatens to come ashore on amenity beaches. Our advice would be, where the beach is an important one for nature conservation, to scoop off the oily sand in some instances or just to let it naturally biodegrade. When it comes ashore we tend to advise against dispersants more often than we advise the use of them, but offshore they are often a reasonable option.

Mr Steen

421. Have you seen the report by the South Hams District Council on the effect of the oil pollution resulting from the *Rose Bay* disaster a couple of years ago when a tanker was holed by a fishing vessel that was not looking where it was going and the oil came out and affected the whole of the coastline, similar to the *Torrey Canyon*, but not on the same scale of course? Would you be seeing those reports and the criticisms made by the council as to the way in which the Department of Transport responded? If you have not, and you did, would you then make a view known to the Marine Pollution Control Unit? Is that your role, or would you be advising them at the time the accident took place?

(Dr Vincent) We would be advising them at the time the accident took place. In fact, one of our officers would be monitoring—not all the time but from time to time—what was happening and whether it was being efficient in keeping the oil offshore, if the booms were working and that sort of thing. In relation to the South Hams report, we as an organisation have seen it. I personally have not. It would be within our function to draw attention to the MPCU those sorts of things within such a report with which we agreed. We would not hesitate to do that.

422. Has that been done?

(Dr Vincent) I believe that on several occasions, such as on the Mersey, we have drawn quite strong attention to the situation which we were dissatisfied about.

423. But you do not know about the *Rose Bay*?

(Dr Vincent) I honestly cannot remember about the *Rose Bay*.

(Dr Langslow) We could happily write to you, Mr Chairman, if that would be helpful.

Mr Steen: Thank you very much.

Mr Field

424. Can I take you away from oil pollution and talk about marine pollution generally? We still seem to get a phenomenal amount of plastic waste coming ashore on our beaches wherever they are exposed to the prevailing wind and have a catching effect. Do you have a view on this and the fact that there does not seem to have been a marked reduction despite the new legislation?

(Dr Vincent) Plastic waste is a concern to us in the sense that it is the sort of thing which is ingested by sea turtles and by dolphins and could well lead to mortalities, so we are in agreement that it should be strictly controlled.

425. Have you seen any improvement, do you think?

(Dr Vincent) I have no evidence to say that the new regulations have resulted in actual improvement on the beaches one way or the other, I am afraid.

426. There has been an absence of plastic cups since the cross-Channel ferries were prevented from dumping them over the side. You can physically check that, but the squeezy bottles and the drinking water bottles from France still seem to turn up in vast quantities on our beaches.

(Dr Vincent) I am afraid I do not think I have anything to say on that. Obviously we would prefer it to be cut down.

(Mr Stevens) There appears to be a wider consideration here, and that is that we are very concerned that some of our less visually attractive habitats—salt marshes and mud flats are a good example—are seen as more valuable and seen to have their true value. If they are covered in waste and litter it detracts from that and the value of the coast is an integrated value of nature conservation, scenery and a wide range of other heritage values. One of the values that suffers first when that kind of pollution occurs is its scenic value and its attractiveness. That has an impact on the way in which it is managed and the support that there is for the conservation issues. It concerns us in that way as well.

Chairman

427. Time is marching on and we have the Countryside Commission to talk to now. Perhaps we could at this point release you, Dr Langslow, Dr Vincent and Mr Stevens. Thank you so much for attending this morning and helping us in this inquiry.

(Dr Langslow) Thank you very much indeed, Mr Chairman. May I add one comment, that if you wish us to illustrate in the field any of the points that we have put to you about coastal processes, we would be very pleased to arrange something near London for you.

Chairman: Thank you so much. I would be grateful.

Memorandum by the Countryside Commission

COASTAL ZONE PROTECTION AND PLANNING

1. This submission:

—sets out the Countryside Commission's remit;

—defines the coastal zone for the purpose of this submission;

—explains the Commission's concerns with coastal matters, and outlines recent public statements and initiatives on coastal policy by the Commission;

—identifies pressures on the coastal zone;

—indicates areas where the Commission believes new policy approaches are required.

THE COMMISSION'S REMIT

2. Established under the Countryside Act 1968, the Countryside Commission is an independent public agency with the duty to promote the conservation and enhancement of the natural beauty of the English countryside and to give the public better opportunities to enjoy and appreciate it. As part of a wide-ranging remit, we have a special responsibility for advising government and Parliament on countryside issues.

3. Prior to 1 April 1991, we also had responsibilities in Wales. These responsibilities, alongside those of the former Nature Conservancy Council, have been assumed by the Countryside Council for Wales.

THE COASTAL ZONE

4. For the purposes of this submission, the definition of the coastal zone adopted is that agreed by delegates from 11 member countries of the European Community, together with Sweden, at a European Workshop on Coastal Zone Management held at Poole in the United Kingdom from 21–27 April 1991 (see also paragraphs 19 and 20).

5. The Communication from the Workshop says:

" The coastal zone is a dynamic human and natural system which extends to seawards and landwards of the coastline. Its limits are determined by the geographical extent of the natural processes and human activities which take place there. Coastal zone management should extend as far inland and seaward as is required by the management objectives. "

It should be noted that this definition does not employ precise geographical limits. Indeed, the Workshop concluded that the range of socio-economic issues and natural pressures affecting the coastlines of Europe prohibit a rigid definition that can be applied to any coastal zone.

THE COMMISSION'S CONCERNS WITH COASTAL MATTERS

6. The coast has a special place in the history and life of Britain and is a focus for tourism and recreation. The Commission has always had a concern with the undeveloped coastline but has had a special interest in its finest scenic stretches where it pioneered the concept of Heritage Coasts in the early 1970s. Outlined below is an explanation of its concerns with Heritage Coasts and with the wider coastline.

HERITAGE COASTS

7. Since 1972, in partnership with local authorities, 44 stretches of coastline have been defined as Heritage Coast, some 1,496 km, and in most cases special arrangements put in hand for their planning and management.

8. In January 1991 the Commission published a policy statement on Heritage Coasts—" Heritage Coasts: policies and priorities 1991 ", a copy of which is attached at Annex A. Prepared after extensive consultation with all those with an interest in these areas it:

—sets out Heritage Coast objectives;

—explains the status of Heritage Coast definition;

—discusses the possibility of defining further heritage coasts, including on estuaries or on previously degraded coasts which have been restored;

—sets out policies for the control of development;

—describes the management measures needed for Heritage Coasts to secure their conservation and public enjoyment and proposes management targets;

—sets out the role the Commission believes local authorities and other interests should play in Heritage Coast matters;

—identifies resources requirements;

—sets the policies in the context of the wider coastline.

9. The policy statement has been warmly welcomed, organisations with Heritage Coast management responsibilities have responded with statements of support for Heritage Coast objectives, and a statement of support from government has been sought.

10. The Commission is encouraged by the achievements which have resulted from Heritage Coast definition and from the establishment of Heritage Coast management services (described further in paragraph 15 and in the policy statement) although it recognises that such services can only tackle small scale management tasks. The achievement of the Heritage Coast management targets proposed will require action by many organisations and deployment of increased resources. The Commission urges the Committee to commend the approach and to recommend that adequate resources are made available to achieve the targets proposed.

THE WIDER COASTLINE

11. In preparing its policy statement for Heritage Coasts, the Commission recognised that Heritage Coasts cannot be protected by action within their boundaries alone. The statement goes on to say:

" At a national level this policy statement sets the framework at least for the undeveloped coast. What is missing is a national policy framework for the coast as a whole. Many organisations have coastal responsibilities but co-ordination between them is not well developed, and the role of local authorities in particular is limited below the high-water mark. Yet the inshore zone faces many development pressures and problems, to which must now be added the prospect of a significant rise in sea level. This brings with it difficult decisions on the nature, extent, and cost of flood protection and coastal defence measures, made more complex still by divided administrative responsibilities. A wider review of coastal policy is necessary for Heritage Coast protection to have a proper context. "

12. On 25 April 1991 the Commission formally wrote to Mr David Trippier, Minister for Environment and Countryside, seeking government support for a review of coastal policy and suggesting the scope for such a review.

13. We suggested that it should deal with the land, the foreshore and inshore waters and should extend to all of Britain, and possibly include Northern Ireland as well. Most importantly, it should be led by government and be seen to embrace the full range of government interests in the coast; it should therefore be allowed to question the current institutional arrangement and responsibilities of all the organisations concerned. Specifically, it might do three things:

(i) identify the current and foreseeable development pressures and other trends affecting our coastline, and the threats these pose to the long term conservation of its environmental quality;

(ii) consider the adequacy of current policy and institutional arrangements to deal with those trends;

(iii) recommend a national strategy for the conservation and sustainable development of the coast, this to include the roles and duties of the public agencies involved and the arrangements needed for effective co-ordination between those agencies.

14. On 1 October 1991 the Commission published " Caring for the Countryside: a policy agenda for England in the nineties. " In this policy paper we reiterate the points made above. At page 7 the statement says:

" The Commission believes that an integrated approach to conservation and development at the coast is urgently needed, with arrangements to ensure co-ordination among the many organisations responsible for managing the resources of inshore waters, shoreline and coastal fringe. A number of issues require urgent attention: the quality of inshore waters, the cleanliness of beaches, the need for a strategy to respond to rising sea levels, the need for environmentally sensitive sea defences on soft coasts, provision of marinas and other coastal developments, the management of water-based activities, and the exploitation of resources on the sea bed.

The Commission will:

—advocate a national coastal initiative to achieve more effective and better integrated coastal zone management;

—recommend that government should indicate a presumption against further development on the undeveloped coast.

PRACTICAL ACTION TO PROMOTE COASTAL CONSERVATION AND PUBLIC ENJOYMENT

15. The Commission has traditionally supported action to conserve coastal scenery and to promote public enjoyment on coastal lands, particularly through the development of local authority countryside management services and by supporting voluntary organisations, especially the National Trust, to acquire and manage coastal land. These programmes are well developed in Heritage Coasts where most such coasts now have a Heritage Coast Officer and ranger service (the Heritage Coast management service) to undertake management tasks to conserve and enhance the coastline, to provide information services and generally to assist visitors, although their brief has often been limited and the resources available small. These initiatives are explained in more detail in the Heritage Coast Policy Statement. Action outside Heritage Coasts has also been supported.

16. In June 1991 Countryside Stewardship was launched. This is a joint initiative with English Nature and English Heritage, but led by the Commission. It is a national pilot scheme to demonstrate how conservation and public enjoyment of the countryside can be combined with farming and land management through a system of incentives and agreements. Among several options, the scheme specifically targets coastal areas and aims to:

—sustain and restore existing areas of coastal vegetation and the wildlife it supports;

—restore and protect characteristic landscape features;

—create and improve opportunities for people to enjoy the landscape and its wildlife.

17. Agreements will be at the Commission's discretion and we wish to attract land into the scheme which has most potential for environmental improvement and public benefit. We believe, if properly resourced, Countryside Stewardship has the potential for major beneficial change on the coast through enhancing landscape character and wildlife. It could also impact on the need for specific coastal protection measures by encouraging, for example, the conservation of dune systems and the development of reed beds. Full details of Countryside Stewardship are attached at Annex B[1].

18. The Commission urges the Committee to commend the approach of Countryside Stewardship and to recommend that adequate resources are made available so that the scheme can achieve its full potential.

INTERNATIONAL INITIATIVES

19. The Commission and the National Trust, with advice from the European Commission and support from the Department of the Environment, were instrumental in organising the European Workshop on Coastal Zone Management referred to in paragraph 4. The specific objective of the event was to prepare and agree upon a proposal of priorities and an agenda for action within the European Community and its neighbouring countries. The Workshop achieved its objective and resulted in an agreed "Communication" from the participants. The Committee's attention is drawn to this communication (copy attached at Annex C)[2] and in particular to the recommendations (Section 4) with their proposals for action at the European and National levels.

20. At its meeting in June 1991 the Commission endorsed the Communication. A copy was sent subsequently by the Commission's Chairman to Mr David Trippier, Minister of State for the Environment and Countryside, who responded that it had been widely distributed within the Department and that he had asked that due regard be paid to its principles in the Department's work.

PRESSURES ON THE COASTAL ZONE

21. The coastal zone is subject to increasing pressures of all kinds from land reclamation of intertidal areas, from development and from a greater intensity of activity in many uses of the coastal zone. New issues are also emerging, notably concern about possible sea level rise and exploitation of renewable energy resources.

22. Land take, particularly from estuaries, continues for industrial uses, recreation, waste disposal and for agricultural purposes with the loss of characteristic areas of upper tidal flats and saltmarshes.

23. Onshore development pressures for industry, housing and recreation uses continue, while there is increasing demand for sea dredged sand and aggregates and pressure to exploit offshore oil and gas reserves. Interest in renewable energy resources is leading to proposals for tidal barrages and for wind farms at coastal locations.

---

[1] *Not printed.*

[2] *Not printed.*

24. With rising affluence, greater mobility and more leisure time there has been a great increase in waterborne leisure activities of all kinds. The result has been increasing conflicts between different marine uses, for example between conservation and recreation, between different recreation activities, between fisheries and recreation and between commercial and leisure craft. Leisure activities and increasing public concern for the environment has also raised public awareness about water quality and pollution issues and has led to demands for higher standards.

25. Concern over sea level rise is leading to a reappraisal of coastal defence needs, including the strengthening of existing sea defences along low-lying coasts. Hard engineering solutions to this issue, with concrete sea walls for example, would have major consequences for natural coastal landforms and habitats.

AREAS WHERE NEW POLICY APPROACHES ARE REQUIRED

26. The Commission's main concern remains to ensure the effective conservation of the undeveloped coastline and that it can be enjoyed by the public. In the light of the pressures on the coastal zone illustrated in paragraphs 21 to 25, the Commission believes new approaches would be helpful in three areas. These are:

—the provision of a national strategic policy framework;

—improved procedures for the control of development in the coastal zone;

—a review of responsibilities among the many organisations with an interest in the coastal zone to reduce any unnecessary duplication of responsibilities and with a view to improving co-ordination.

A NATIONAL STRATEGIC POLICY FRAMEWORK

27. There is no national strategic policy framework within which issues affecting the coastal zone can be considered. Such a national perspective would provide a sense of direction and a long term view and help inform regional and local decision making. National guidance would be especially helpful in relation to the recently emerging coastal issues such as rising sea levels, exploitation of renewable energy resources, increasing exploitation of aggregates and oil and gas offshore, and proposals for reclaiming estuaries.

28. The Commission would wish to see such guidance providing a strong framework for protecting the undeveloped coast. Indeed, as already noted at paragraph 14, in "Caring for the Countryside" the Commission is recommending that government should indicate a presumption against further development on the undeveloped coast. The Commission has also endorsed the overall aims for the coastal zone agreed at the European Workshop (paragraphs 19 and 20). These are:

—to conserve remaining stretches of undeveloped natural coastline;

—to restore natural coastline wherever possible;

—to ensure that the development and uses which take place in the whole coastal zone are sustainable and respect the precautionary principle.

29. The Commission suggests that a strategic policy framework should be prepared by government, possibly through an interdepartmental working party, should be subject to extensive consultation and should be promulgated widely, for example through a planning circular. We urge the Committee to recommend accordingly.

LAND USE PLANNING

30. The present statutory planning system does not deal adequately with issues in the coastal zone and we argue it requires substantial modification to enable it to regulate activities in the coastal zone effectively. In particular:

—there is no statutory requirement for planning at the strategic level covering both land and sea together within a coastal zone, and the present development plan and development control systems do not apply below the low water mark except where local acts extend the planning authorities' jurisdiction. Furthermore, planning policies in structure plans and individual local plans do not add up to a comprehensive planning framework for a coastal zone which will often cover more than one district council area;

—some key uses and activities are outside the scope of planning control, for example coastal engineering works, land drainage, and the exploitation of minerals, aggregates and oil and gas offshore;

—harbour authorities are exempt from planning control in carrying out most of their own operations;

—most of the United Kingdom's seashore and the sea bed is owned by the Crown Estates Commissioners (CEC) on behalf of the Crown. The CEC lease areas of the sea bed for the mooring of boats and fish farm cages for example and licence activities such as cable laying and minerals extraction. They are under a general duty to enhance the value of the Crown Estate but also to show due regard to good management but they are not publicly accountable for their decisions;

—existing requirements for Environmental Assessment are not wide enough. For example, projects subject to private bill procedures and the development of offshore oil are not covered. In relation to the Crown Estate, the Crown Estates Commissioners themselves, after consultation, take the decision on whether an EA is required for proposals for dredging for minerals offshore and for marine salmon farming projects;

—use of private bills to promote coastal development projects has circumvented the planning system and reduced the scope for public debate, for example the development of Felixstowe Docks and in relation to the Cardiff Bay barrage.

31. In briefing Members of the House of Lords, and Commons, on aspects of the Planning and Compensation Bill, the Commission gave its support to an amendment which would have provided for the preparation of coastal zone local plans which would be prepared jointly by all the local authorities in the coastal zone. Government was not prepared to accept the amendment which it regarded as in conflict with its proposals for simplifying the development plans system and which provided only for county-wide structure plans and district-wide local plans, relying instead on co-operation between adjacent authorities to ensure that individual plans for a coastal zone are consistent. It remains to be seen whether such procedure will be adequate.

32. A number of organisations have made a case for extending planning controls below the low water mark. Hampshire County Council, for example, in "A Strategy for Hampshire's Coast" argues for the extension of planning controls to cover coastal waters up to three miles from the shore while the Marine Conservation Society in a report for the World Wide Fund for Nature "A future for the coast? Proposals for a United Kingdom Coastal Zone Management Plan" proposes an extension of planning jurisdiction to the 12 nautical mile limit of territorial waters. Northumberland County Council, in their consultation draft Northumberland Coast Management Plan, are suggesting an appropriate offshore boundary of the Plan area would be the 10 fathom contour.

33. The special position of the Crown Estates Commissioners has been raised by many organisations, including this Commission in its evidence to the House of Commons Agriculture Committee Inquiry into Fish Farming where it argued that marine fish farming should be brought within the scope of planning control in the long term. A copy of this evidence is attached at Annex D.

34. We are aware that the Department of the Environment is preparing a Planning Policy Guidance Note on the Coast and will be responding to consultations on draft guidance in due course. However, the Commission argues that a much wider review of planning legislation and procedures is required for the reasons illustrated above and urges the Committee to recommend accordingly.

RESPONSIBILITIES FOR PLANNING AND MANAGEMENT IN THE COASTAL ZONE

35. A great many central and local government agencies and organisations are involved in aspects of planning and managing the coastal zone. They include, in England alone, in addition to ourselves, the Department of the Environment, Ministry of Agriculture, National Rivers Authority, Department of Transport, Department of Energy, Harbour and Port Authorities, Sea Fisheries Committees, Crown Estates Commissioners, Ministry of Defence, The Home Office, English Nature, English Heritage, English Tourist Board, Sports Council, Rural Development Commission, County Councils, District Councils and several National Park Authorities.

36. Such a multiplicity of statutory organisations with responsibilities in the coastal zone makes for difficulties in co-ordination. In the case of coastal protection and sea defence discussed below there is also a division of responsibilities in this key area.

37. The Commission suggests there are two needs:

—an examination of the current allocation of responsibilities of statutory organisations involved in the coastal zone to see if a reallocation of responsibilities might be appropriate in some cases;

—an examination of methods of achieving closer collaboration between the statutory organisations, voluntary bodies and local interests to ensure that there is proper co-ordination in securing planning and management objectives at the national, regional, coastal unit (a length of coast which it is sensible to plan together) and local level.

38. In relation to collaboration at the England and Wales level, the Commission has had direct experience in its initiative in establishing the Heritage Coast Forum to provide an effective mechanism for liaison and co-operation between organisations with responsibilities for Heritage Coasts. At the national level the Commission is also aware of the Marine Forum, which serves as a point of contact between a number of organisations and individuals which have a particular interest in the marine environment such as the Marine Conservation Society, the RSPB, and several government departments.

39. At the level of a coastal unit a number of maritime local authorities have come together informally to co-ordinate planning and management policies for a particular area. There are groups, for example, concerned with the Mersey and Dee estuaries. There are examples of local initiatives too. Attached at Annex E is a copy of the Cardigan Bay Charter which has been drawn up by the Cardigan Bay Forum to promote the conservation of this important area among as wide a range of organisations as possible.

40. The Commission commends the concept of fora to promote co-ordination for particular coastal issues (for example the Heritage Coast Forum) and fora for individual coastal units where it believes these have a valuable role to play in resolving conflicts of interest through improved management arrangements. However, it also believes that there is need for a formal mechanism for liaison on coastal zone issues at the United Kingdom level.

41. The Commission urges the Committee to recommend the establishment of local fora for each coastal unit, and a national level committee of all key interests to ensure effective co-ordination at the United Kingdom level.

COASTAL PROTECTION AND SEA DEFENCE

42. A particular issue of co-ordination arises in relation to coastal protection and sea defence. With possible rising sea levels as a result of climate change, together with sinking land in the south east, this is a major issue, yet responsibilities are at present split between District Councils in the case of coastal protection works and the National Rivers Authority for general flood protection through sea defence works. In both cases schemes are substantially funded by the Ministry of Agriculture.

43. A green paper in 1983 proposed that a future Coastal Protection Act should operate only through MAFF and the Water Authorities (now NRA). The Commission supports a simplification of the present arrangements giving all responsibility for coastal protection and defence to the National Rivers Authority so that investment in protection and defence works can be considered together for whole lengths of coast and urges the Committee to recommend accordingly.

44. The Commission has welcomed the separate inquiry recently announced by the Ministry of Agriculture which is to examine, through consultation with conservation organisations in England, its sea defence and coastal protection policies to ensure that they are sensitive to conservation and environmental concerns. The Commission's policy is to achieve the minimal use of obtrusive built structures and maximum use of natural defence mechanisms compatible with other constraints (principally human safety).

SUMMARY OF KEY RECOMMENDATIONS

The Commission urges the Committee to commend the Heritage Coast approach and to recommend that adequate resources are made available to achieve the management targets (paragraph 10).

The Commission urges the Committee to commend the approach of Countryside Stewardship and to recommend that adequate resources are available so that the scheme can achieve its full potential (paragraph 18).

The Commission suggests that a strategic policy framework should be prepared by government, possibly through an interdepartmental working party, should be subject to extensive consultation and should be promulgated widely . . . we urge the Committee to recommend accordingly (paragraph 29).

. . . the Commission argues that a much wider review of planning legislation and procedures is required . . . and urges the Committee to recommend accordingly (paragraph 34).

The Commission urges the Committee to recommend the establishment of local fora for each coastal unit, and a national level committee of all key interests to ensure effective co-ordination at the United Kingdom level (paragraph 41).

The Commission supports a simplification of the present arrangements giving all responsibility for coastal protection and sea defence to the National Rivers Authority so that investment in protection and defence works can be considered together for whole lengths of coast and urges the Committee to recommend accordingly (paragraph 43).

September 1991

Examination of Witnesses

Dr Roger Clarke, Director, Policy and Mr Richard Lloyd, Head, National Parks and Planning Branch, Countryside Commission, examined.

Chairman

428. Dr Clarke, I welcome you and Mr Lloyd to this part of our inquiry. I do not know whether there is anything you want to say generally in opening. You have submitted a very full report, and there is a great deal of evidence before us.

(Dr Clarke) Thank you very much for the invitation to come along this morning. As you will have gathered, my name is Roger Clarke. I am the Policy Director at the Countryside Commission. Richard Lloyd, Head of our National Parks and Planning Branch, has specific responsibility for our work in relation to coasts. Having listened to the discussion with English Nature, I might make a couple of points about where we are coming from which may help the Committee. Our expertise in this area is in coastal management, particularly focusing on the landward side of the coast. We have had what we regard as a very successful heritage coast programme, defining the finest stretches of coast and arranging for local management arrangements to care for them, to plan for them and to try to improve the quality of environmental management and their enjoyment by the public. As I think you will know, heritage coasts now cover about 1,000 km of the English coastline, a little over 30 per cent, and about 100 or so staff are working at a local level, not our staff, but the staff of the local heritage coast services, doing a lot of good work on the ground. We come to this recognising the success of that programme, but also its limitations. We have become aware of how some of the wider coastal issues cannot be tackled solely through local management action by the heritage coast services and it is almost in a sense of frustration about what we cannot do that brings us to comment not from a position of great scientific expertise, perhaps, but from a position of wanting to see improved coastal management on some of the wider issues which are the wider concern of your Committee and which I imagine we shall be discussing this morning.

429. Thank you for that introduction. That touches on my first question. In paragraph 26 of your memorandum you draw attention to the multiplicity of organisations with an interest in the coastal zones and you suggest a review of responsibilities among these many organisations to reduce unnecessary duplication and to improve co-ordination. Which particular organisations do you have in mind, and what suggestions do you have for simplifying these arrangements?

(Dr Clarke) Institutional arrangements are always very tricky. Just to step back from that for a moment, I would say that our primary concern at a national level is that there should be on the part of the nation and therefore of government a strategic view of coastal policy. At the moment one cannot turn to any single place or document and be able to say that that is the Government's view about our coastline, yet the coastline is immensely important in national and international terms for all sorts of reasons, including our own particular interests. There is no unified approach. Perhaps before speaking about which specific institutions might or might not do what, I think that a national strategy for the coast is one of the things that would help us most and help the coast in general. We would like to see such a strategy. Creating such a strategy implies the getting together of the different institutions involved with coastal issues and it implies a political lead to invite the different agencies to collaborate, or ask them to collaborate, in devising such a strategy through some form of review process or some form of initiative. The key players in relation to that will be as evident to you as they are to us: the National Rivers Authority, the Department of the Environment, the Ministry of Agriculture, the Crown Estates, and local authorities I suppose would be the front runners. Other organisations such as ours perhaps have a narrower focus of concern and responsibility.

430. You mention the NRA. In paragraph 43 you say that you would put all the responsibility in the NRA as a national supervisory body, whereas English Nature were suggesting that role might perhaps be one for the Department of the Environment. Let us take your scenario. If we had the NRA as the overall national body responsible for coastal protection and defence, where would the maritime local authorities fit into that?

(Dr Clarke) I would distinguish between overall responsibility for coastal policy and responsibility specifically for coastal protection/coastal defence issues. Coastal policy obviously is more wide ranging as a concept and there the responsibility is a political one and ultimately, therefore, a departmental one. I would guess, for reasons which we may come on to, that the Department of the Environment would be the most logical home for responsibility for coastal policy as a whole. In relation to coastal defence/coastal protection, as institutions are arranged at the moment, it would seem appropriate for the National Rivers Authority to have a strategic responsibility for our coastal defence/coastal protection programme for very obvious reasons. Individual local authorities responsible for a particular bit of coast may do well to construct groynes or sea walls on their bit of coast, but they do not necessarily have to have regard for the impact on the neighbouring coastline. So somebody needs a strategic view. The appropriate national agency, as things are organised at the moment, is the NRA. Local authorities might implement bits of that strategy and be responsible for individual bits of sea defence, but within a strategy prepared by the NRA.

431. Where would the Crown Estate come into all that?

(Dr Clarke) The Crown Estate, as I understand it, has responsibility for the sea bed and is primarily a licensing organisation regulating what activities may or may not take place in relation to the sea bed. We

Chairman *Contd]*

do not have a great deal of first hand experience of their work, but from a distance they would not appear as publicly accountable as a local authority. We think, although we do not have a definitive view on this subject, there is merit in considering whether some aspects of local authority powers, particularly planning powers, might extend offshore to inshore waters because some of the activities that take place in inshore waters relating to dredging or to the use that is made of the water space are clearly very closely linked to what happens on land. Since local authorities through the statutory planning process have responsibility for the landward side and there are very close land and sea links, we think there is merit in looking at the issue about local authority responsibilities in the immediate offshore area. That would not be to take away all Crown Estate responsibilities for, say, licensing, but that might take place within a framework set by the local authority to which the Crown Estates would contribute.

Mr Field

432. You have urged us to recommend the establishment of a national coastal committee of all key interests to ensure effective co-ordination at a United Kingdom level. Can you explain how this committee would be structured, how it would operate and what powers it would have? How would such a committee be related to a national coastal zone unit which has been recommended to us by other witnesses?

433. As I have already indicated to the Chairman, I do not think our view of institutional arrangements is clear. Our view that there is a need for a strategic approach is clear. Before a decision is taken about any particular continuing committee, government should institute a review of existing issues and arrangements. We think there are a number of strategic issues which require attention at a national level in an integrated way: the management of inshore waters, the implications of sea level rise, policy for coastal protection and coastal defence, how we can achieve improved water quality, beach cleanliness, the appropriate scope of development at the coast. All those issues might be addressed in some form of government review. So that would be a time-limited committee or group which would be needed to undertake such a review. Out of that might well come recommendations for some continuing institutional arrangements. I do not think it is our view that there should be a new super coastal agency subsuming within itself the responsibilities of other agencies, but clearly in our judgment existing co-ordinating arrangements are not fully effective, so that might well imply some continuing committee. But we do not have a precise view about how it might be structured at present. We think the review and the strategy are the first things.

434. You commissioned a report on the environment from Posford Duvivier[1]. How would you seek to act on the recommendations in that report?

(Mr Lloyd) We jointly commissioned that work

---
[1] *Not printed.*

with English Nature, who you were discussing matters with previously, with the National Rivers Authority and with the Department of the Environment, so we are one sponsor out of four. We feel that this is a fascinating document and some of the recommendations and conclusions in relation to the idea of managed retreat may have application in a wider strategy for coastal defence. Our first concern is to get the knowledge in this report more widely known and promoted in some way. We hope to run a joint seminar with English Nature and the Joint Nature Conservation Committee later on this year to debate some of the ideas coming out of that work and to make them more widely available. We think there are some fascinating ideas in there which have a part to play in coastal defence strategies in the future. That is basically what we want to see done with it. It has only very recently been published and we have made copies available to your Committee. We have yet fully to digest its implications.

Mr Jones

435. I want to follow up that point about the Duvivier report because it says that you do not support, in general, the principle of monetary valuation for landscape assets. How on earth do you carry out a cost benefit analysis without attributing some notional monetary value to them?

(Mr Lloyd) That comment is not entirely correct. It is an area that we are concerned about. It is a very difficult area and we are currently doing some work on just this subject. We certainly do not have any definitive answers at the moment and at the end of the day it might be possible to attribute monetary values to things like landscape and amenity. It will be very rough and ready and at the end of the day there is no substitute for informed judgment about these things. It is an incredibly difficult area and we are commissioning work on it, but we have no answers as of now.

436. Any cost benefit analysis—involving things that most of us do not——

(Mr Lloyd) Intangibles, yes.

437. Intangibles—must be rather unscientific, but it is a useful tool, would you accept that, in trying to reach conclusions?

(Mr Lloyd) Indeed it has to be a contribution to the decision-making process.

438. So this was rather overstating your position in the report?

(Mr Lloyd) It is a three-line comment in a very large report.

439. I am with you. Can I now ask you about the countryside stewardship scheme? Have you ever considered expanding that? Do you think it would be desirable to expand it to allow for the payment of landowners for management agreements?

(Mr Lloyd) The countryside stewardship scheme is a government initiative, basically buying environmental goods from the farming community. There are five target landscapes of which the coast is one and we have provided details of this scheme to the Committee. It is a very recent initiative. It has

Mr Jones *Contd]*

only been up and running over the past six months and clearly it is very early days. In principle what one would be doing through countryside stewardship, if we applied it to the coastal retreat situation, would be using government money to pay for the creation of new coastal habitats. In principle I do not see why stewardship prescriptions could not be devised to do that job, if that job is what we collectively want to do. In principle there is no reason why the scheme could not be modified to do that.

440. What sort of money does that imply would have to be spent and what about staff implications for you?

(Mr Lloyd) We have devoted between three-quarters of a million and one million pounds to the coastal element of it. Most of that money has gone into recreating grasslands on arable land and on cliff tops, that kind of thing. Although it is coastal, it is the landward frontage of the coast. Again, it is very difficult to put a figure on it at this stage. Perhaps we are talking about the same amount of money again. I imagine the coastal retreat areas would be relatively small geographically, so we are not talking about vast sums of money, but we would be looking for an incentive to make it worth while for landowners to buy into the process.

Chairman

441. You seem to be working so closely with English Nature on so many of these things and you said in reply to the first questions that I was asking that you were looking for simplification and review of responsibilities. The thought has occurred, what about English Nature and Countryside Commission becoming merged? How would you view that?

(Dr Clarke) That is a much wider issue. If you were starting from scratch you probably would not create the same institutional arrangements as we have today. They are a legacy of the thinking in the 1940s, but our present view is that the disbenefits of merger would outweigh the benefits. The two organisations have evolved with quite different histories and styles of work and different locations. I suppose the Countryside Commission has primarily a planning and management, whole of the countryside orientation. English Nature has a scientific, tending to be site focused orientation. The process of creating an integrated agency in our judgment would not be worth it in terms of the disruption that might be caused. The Government have recently endorsed that view in commenting on the recent report on national parks, the Edwards Report. However, we have integrated agencies in Scotland and Wales, and I am sure we shall all want to follow their progress with interest and see how they get on.

442. So you have an open mind about it. What we find fascinating emerging from this inquiry is that everybody has a criticism to make about the present system, and thinks that simplification can take place, until it affects them directly. It is always somebody else who has to be reorganised, but " we are working well and please do not disrupt what we are trying to do ". This is quite a fascinating exercise.

(Dr Clarke) Can I just respond to that? Our major programme in this area is the heritage coast

programme. We think that within its own terms that has been successful, but we would be at pains to say that it could be more successful if it was part of something bigger, to use the point you have just been making, if that were subsumed within or related to a grander strategy involving other organisations. That in terms of the discussion we are having today is saying that our interest could be merged with or linked to other interests. I have said that our work would be greatly helped by a national strategy for the coast. The other substantive point we want to put to you is that our work at a local level through the heritage coast programme would be helped by a more strategic view of the coast at a local or sub-regional level and the development of some wider management framework for the coast at that level within which our programme could fit. That is the real issue about integration affecting this particular inquiry. The people on the ground, the heritage coast staff, with whom we are in touch, are very keen to see the wider dimension of some of the things that they are engaged in and feel a little frustrated that that particular organisational backing is primarily planning oriented dealing with land. Some of the issues that we are very much aware of concern visitors. For example: how clean is the beach, how clean is the water, how are the conflicts about the use of inshore waters to be regulated. They want to be part of some structure which can deal comprehensively with those programmes and problems.

Mr Steen

443. I should like to take up that one point about merger. We are doing another inquiry on an environmental agency and there are a lot of agencies there as well. Can I get one thing clear just for the record? I know it is available, but can you please repeat it. What kind of staffing and costs do you have in relation to English Nature?

(Dr Clarke) The Countryside Commission has a budget of about £35 million a year and a staff of about 220. Most of our budget is disbursed through grant aid to other projects and organisations at a local level. About 75 per cent of it goes in that way. The countryside stewardship scheme, which Richard Lloyd mentioned, is a substantial single programme within our overall budget and when it is fully operational will have a budget of up to about £9 million and about 40 staff working specifically on that scheme.

444. Compared with English Nature, you are very much smaller then?

(Dr Clarke) In terms of budget we are about the same. In terms of staffing they have about four times as many staff. Their style of operation, as I mentioned, is different. Their grant-giving role is more limited. They have a larger number of specialist staff in house, whereas we tend to operate at arm's length, both in terms of our own research and in terms of staff on the ground. Their equivalent to heritage coast staff would be employed within their own organisation, whereas the heritage coast staff in South Devon are employed through the county council.

Mr Steen *Contd]*

445. That helps me on to the questions about the heritage coast, of which I have a large chunk in my constituency. Do you think the heritage coast focuses too much on areas which already have a measure of protection? Can the existing system of designations be simplified in any way? At the same time perhaps I can ask you how would you see the heritage coast management plans relating to the wider regional coastal zone management plans advocated by organisations such as the RSPB? Perhaps I can throw in on those points a personal one, that I am very troubled about the fact that the whole of this heritage coast business is dependent upon the local planning authority. It does not matter what you say about areas of outstanding natural beauty and coastal preservation, it all depends on the local planners. The whole thing depends on local district councillors, so it does not matter how much work you do someone other than you has to make a decision at the end. How does that all work and should you be taking the heritage coast out of planning or should you be giving it certain sacrosanct planning controls? How would you relate that to the national parks? The national park is going to be weakened in some ways if the plans that the Government are proposing happen, that the national park comes out of the county council and becomes a separate organisation. That all has to be fitted together. I am sorry to ask you so many questions all at once.

(Dr Clarke) There is a cluster of very interesting issues within all that. The heritage coast programme dating from 1970 is saying that we need to conserve our finest stretches of undeveloped coast. In European terms we are fortunate, largely through the achievements of the town and country planning system, to have retained undeveloped coastline in a very crowded island. I guess that by the 1970s we recognised that we needed not just to have planning policies, but also to have management policies to care for these areas and to look after things on the ground, not just to say " no " to certain kinds of development. We think a focus on conserving the finest bits of coast has been a good focus. It is relatively easy to grasp, has public appeal because these areas are important holiday areas. We do not think it is appropriate at this stage to go for another form of official designation called " heritage coast designation ". We are aware of the comment that there are a lot of forms of designation and many people see that as a negative thing. However we would like to see greater government recognition of the importance of the heritage coast. Last year we published a policy statement on heritage coast, of which I think Committee Members have a copy.

446. But to what end would the Government recognise it is a good thing? I mean, so what?

(Dr Clarke) The Government, through planning guidance and in other ways, shape policy. The Department of the Environment is working at the moment on a statement which would form a foreword to the reissue of this document saying that the Government think that heritage coasts are a great idea. That contributes to our overall programme of coastal management and we would like a positive strong Government endorsement of the heritage coast programme as one of the building blocks of coastal policy. How are things managed locally and what is the role of the local authorities? I suppose we see this issue very acutely in relation to national parks or any other nationally important area of countryside. There is a national interest. These areas are nationally important, but there is a local interest too and local people have strong feelings about how their local area is managed. We think by and large it is right that the initiative should be with local people and with local authorities to provide for the planning and management of these areas, usually through some form of special committee which can include local authority people, county and district and, increasingly perhaps, outside organisations such as NRA, tourist boards and the National Trust where appropriate because coastal management requires participation by a consortium of interests. We think at present that the local authority base for that is probably the right base, rather than taking out of local authority hands into some national agency which immediately attracts criticism for being remote, insensitive and all these kinds of things.

447. Having something called the heritage coast—this is my problem personally—I do not understand what it means because at the moment in my constituency there is a massive new plan for some huge building right on the heritage coast at Bigbury-on-Sea. It is to replace an appalling amusement arcade which I hope will be demolished but it is a massive structure. I do not know who was consulted about it. I do not know who is involved because it is heritage coast. It is just part of the local plans. Of course, it is against the local plans because the local plans do not have buildings, but there is a trade off about knocking down the amusement arcade and putting up a massive building there. What does the category "heritage coast" mean to the local planners? Do they have much greater concern? They may have concern, but here you are talking about preserving an undeveloped part of the coastline and here is this new development plan coming up. Then, to the contrary, you will find applications from the owner of a bungalow to put an extra floor on the house and it is immediately turned down as being against the heritage coastline.

(Mr Lloyd) Most of the heritage coasts represent the coastal frontages of either areas of outstanding natural beauty or national parks, so most of them are already subsumed in a wider landscape statutory designation which government have endorsed. The point about heritage coast definition is that it represents something special and we would hope to see appropriate policies in the local plans. Indeed the heritage coast boundary should be marked on the local plan.

448. It is.

(Mr Lloyd) We would hope the planning policy guidance note on the coast which the Department of the Environment is shortly to issue will have something very positive to say about heritage coasts.

449. Do you know how shortly?

(Mr Lloyd) It is on its way. I could not put a time scale on it. We would hope that that will firmly endorse the need to protect these areas for all time in the national interest. It is a very systematic process to

Mr Steen *Contd]*

identify the lengths of heritage coast based on scenic qualities. They are there. We want to make the most of them and the right vehicle is through getting the proper policies in the local plan. Government is moving towards a plan-led system and those plans will be more important than ever before. They are the blueprints against which local politicians should be taking their decisions on development control cases and if planning applicants are not happy with the decisions and they go to appeal they will be the blueprints upon which planning inspectors will have to decide appeal decisions. We argue and we hope the Government will endorse that heritage coasts are very special as part of the nation's landscape heritage. The right policies should appear in the planning framework. There is no reason why that cannot happen as of now. Your Bigbury case worries me and I want to have a word with my regional colleagues about it.

450. Do find out about it. The point I am getting is that you do not believe there should be any national protection, other than through a PPG?
(Mr Lloyd) If PPGs mean what they should, that ought to be sufficient. As I was saying, most of the heritage coast are either the coastal frontages of national parks or they are the coastal frontages of AONBs. At one stage it was suggested that the remaining heritage coast—there are about half a dozen—should have some statutory protection perhaps by making mini little AONBs on the coastal frontage. We do not believe that is really necessary if the planning policy guidance and now the new raft of district wide local plans can get the right policies built into them, so it should not be necessary.
(Dr Clarke) We would hope that in planning policy guidance or any other statement the Government take a strong view of the need to conserve our remaining undeveloped coast. We are very conscious of it as a national and as a European asset. We have a lot of coastline, and, comparatively speaking, we have a lot of relatively undeveloped coastline. It is a scarce resource. I am not suggesting that there should be complete embargo on development at the coast, but the Government need to give a very strong lead that we wish to see such coast protected.

Mr Jones

451. Have you any reason to assume that that is not the case? All the comments I have ever heard from Ministers or, for that matter officials speaking for the DOE have actually made that point.
(Dr Clarke) The point Mr Steen has just been making illustrates some of the difficulties that arise. Clearly there is constant pressure for marina development or other developments at the coast, because the coast is very popular. A lot of people want to do things there.

452. The point Mr Steen was making will arise in any planning system because it is a planning gain point. All these planning gain points are contrary to plans, whether they are structure plans or county- or district-wide plans, but local politicians think it is worth the trade off. Sometimes I strongly disagree with them. But that will arise, whatever the system, will it not?
(Dr Clarke) The Government can do their part by indicating the sort of policies they wish to see adopted.
(Mr Lloyd) There is a point to add to that. The ultimate form of safeguard is protective ownership in some way. We have encouraged the National Trust over the years to concentrate their coastal acquisition programme on heritage coasts. We put quite a lot of grant aid their way and presumably that has come up in their evidence. But the National Trust are key players now on a lot of heritage coasts, owning for the public benefit substantial tracts of countryside and all power to their elbow.

Chairman

453. Just one final question, from your experience of working with multi-agency groups and fora, how effective do you find them in reaching agreements on policy and management issues?
(Dr Clarke) Fora are not effective as executive groups because that is not what they are. They are effective as a means of exchanging information reaching agreed or understood positions on topics and encouraging organisations to be a little less parochial in the way they think about things. That is particularly important in any given situation where there will be lots of players involved, whether there are one or two more or less. They need to be able to talk to each other and a lot depends on the skill with which the forum is constructed and chaired as to its effectiveness. A lot of our work depends on persuading other people. We have very little in the way of coercive powers and we would not want such powers. If you are going to persuade people there have to be fora in which that persuasion and discussion takes place.

454. Thank you so much for your attendance and for your help this morning.
(Dr Clarke) Thank you very much. We wish you well with your inquiry and we hope to see your report before the election.

455. We have to agree it before the election, whenever that may be.
(Dr Clarke) Thank you.

WEDNESDAY 5 FEBRUARY 1992

Members present:

Sir Hugh Rossi, in the Chair

Mr John Cummings   Mr Tom Pendry
Mr Barry Field

Supplementary Memorandum and Statement by The Crown Estate Commissioners

INQUIRY INTO COASTAL ZONE PROTECTION AND PLANNING

INTRODUCTION

1.   The role of the Crown Estate Commissioners in authorising the use of those parts of the foreshore and the sea bed which it owns was described mainly in Paragraphs 45 to 50 of the Government's Evidence to the Committee. At its hearing on 4 November, the Committee were further informed that the Crown Estate Commissioners did not take decisions on the environmental aspects of applications for marine aggregate extraction; that was a matter for Government. Nevertheless the Government, with the full support of the Commissioners, was prepared to review the procedures involved in determining applications, with particular reference to the Crown Estate's role.

2.   At that hearing various references were made to the Crown Estate's ownership of the sea bed, its exemption from planning permission and its policy on moorings. In view of these references, the Commissioners have set out below some further information on the situation in England and Wales which they hope will clarify the Crown Estate's position and be helpful to the Committee in its important deliberations. In particular, Commissioners would like to emphasise their willingness, as landowners, to work within any planning framework for the coastal zone that is agreed by Government. They have no particular wish to act in any quasi-planning role, provided that an appropriate planning authority can be established. Indeed the Commissioners are anxious that they should be allowed to concentrate on their responsibilities as landowner under the provisions of the Crown Estate Act 1961.

THE CROWN ESTATE

3.   The Crown Estate is an estate in land which, as well as the marine portfolio, includes substantial blocks of commercial and residential property, primarily located in London, and some 250,000 acres of agricultural land in Great Britain. Since 1760 revenue from the Crown Estate has been surrendered to the Government by the Sovereign at the beginning of each reign, as part of the arrangement for the provision of a Civil List. However, the Estate itself remains the property of the reigning Sovereign in right of the Crown. It is not Government property but neither is it part of the Sovereign's private estates. In 1990–91 the revenue surplus paid to the Government, after deducting management expenses, was £61 million.

4.   The Crown Estate Commissioners manage the Crown Estate under the provisions of the 1961 Act. The general duty under Section 1(3) of the Act to maintain and enhance the value of the estate, and the return obtained from it, is subject to the responsibility to have due regard to the requirements of good management. This means that Commissioners take account of environmental and conservation considerations when fulfilling their general duty. Indeed Commissioners have set themselves environmental objectives across the wide range of their diverse activities.

5.   Further details of the Crown Estate's activities are included in the attached Report by the Commissioners for the year ended 31 March 1991. The Commissioners are about to publish a policy statement on their stance on environmental issues and a copy is also attached for the Committee's reference. It may enable the Committee to view the Commissioners' attitude on marine issues in the context of their policy on the portfolio as a whole.

OWNERSHIP OF FORESHORE AND SEA BED

6.   The position is stated succinctly in Halsbury's *Laws of England*, Volume 8, paragraph 1418:

"By prerogative right the Crown is prima facie the owner of all land covered by the narrow seas adjoining the coast, or by arms of the sea or public navigable rivers, and also of the foreshore, or land between High and Low Water Mark. . . . There is a presumption of ownership in favour of the Crown, and the burden of proof to the contrary is on a claimant. In Scotland the right to the soil extends, it seems, from the coast outwards to the territorial limit."

Authority for this stems from a classic text on this subject, Sir Matthew Hale's *De Jure Maris* of about 1640 in which he said: " The narrow sea, adjoining to the coast of England, is part of the waste and demesnes and dominions of the King of England, whether it lie within the body of any County or not. . . . In this sea the King of England hath a double right, *viz.* a right of jurisdiction which he ordinarily exerciseth by his Admiral, and a right of propriety or ownership."

7.   Thus from early times it seems to have been settled that the extent of the Realm at least coincided with the extent of the Counties, and the seaward boundary of the Counties was at low water mark. The foreshore is, therefore, clearly in the Crown's ownership except where, over the years, it has been disposed of. The Appendix to this Memorandum sets out the more significant parts of the foreshore and sea bed in England and Wales which do not fall within the Crown Estate's jurisdiction as landowner. The Committee will note that there are some very sensitive areas that are not owned by the Commissioners.

8.   As for the sea bed, the United Kingdom has claimed sovereignty since the 1830s over a belt of sea out to one marine league (now three international nautical miles). There is considerable case law both in England and Scotland examining the Crown's proprietary rights within that area and accepting that it has such rights in full. The Territorial Sea Act 1987 extended the territorial limit to 12 miles and with it the Crown's rights. The Commissioners have recently completed a conveyance to the Government of the Isle of Man of the additional nine-mile band of sea bed around that Island.

9.   Management of the foreshore and sea bed was given to the Commissioners of Woods by the Crown Lands Act 1829. The Crown Lands Act 1866 transferred that management to the Board of Trade where it remained until transferred back to the then Commissioners of Crown Lands by the Coast Protection Act 1949.

10.   Any rights exercisable by the United Kingdom outside territorial waters with respect to the sea bed and subsoil and their natural resources (except hydrocarbons) are vested in " Her Majesty" by virtue of Section 1(1) of the Continental Shelf Act 1964. These rights fall under the management of the Commissioners who exercise them by grants of licences for such activities as aggregate dredging and the laying of cables and pipes of various kinds.

The Commissioners' Policy on Moorings Charges

11.   During the course of the last 12 months or so, there has been much misinformed and unfounded speculation by interested parties and the Press about the policy of the Crown Estate Commissioners on mooring leases and about the level of rent increases that might prevail on their renewal. Reference was made to the issue by the First Commissioner in his Preface to the 1991 Annual Report (Page 3) and the Commissioners' Policy was outlined in Pages 26 and 27 of the Report. Members of Parliament whose constituencies included significant moorings were also advised of the policy through a letter from the First Commissioner in June 1991. In short, the salient features of the issue are:

(i) Rental values in old leases have become virtually meaningless over the years;

(ii) Block leases are granted to Harbour Commissioners, Local Authorities and Fairways Committees so that local interests can participate in the management of moorings and charges can be allocated in the light of local circumstances;

(iii) While Commissioners have statutory obligations to obtain a reasonable return in the light of market circumstances, the Valuation Office of the Inland Revenue acts as an independent expert should disputes arise and determines the appropriate level of rent;

(iv) When a significant increase in return is justified, the new rent will be phased in over a suitable period.

12.   Two recent decisions agreed with lease-holders, without recourse to the Valuation Office, amply illustrate the responsible attitude taken by Commissioners. At Lymington, the Harbour Commissioners believe that the new rents, which will be phased in over a six year period, can be met while continuing to offer reasonable rates to moorers. On the River Hamble, the Commissioners and Hampshire County Council have agreed a number of measures which will ensure continuity of management over the next few years, without any effect on the charges for most moorers. In particular, the Council has been granted a Conservation Lease at a nominal rent for the stretch of river without moorings, on the basis that no moorings development will be allowed there. For the rest of the river, the Council are committed to its long term conservation, while ensuring a balance between the needs of public access and recreation; their continued role as lesees will help to safeguard these policies.

13.   There is no doubt that demand for moorings will remain at a high level and the environmental and financial aspects will need to be addressed collectively by all concerned in or with the industry. The Commissioners are committed to playing their part in ensuring that a healthy market in moorings is maintained in a socially and environmentally responsible manner.

PLANNING PERMISSION

14.   There is no statutory requirement for the Crown Estate Commissioners to apply for planning permission for development on land which they own. However, like all Crown authorities, the Commissioners have been subject to the requirements of Departmental Circular 18/84, and its predecessor Circular 7/77; these circulars established administrative procedures shadowing the requirements of planning legislation. In addition, Part XIII of the Town and Country Planning Act 1990 sets out provisions relating, *inter alia*, to application for planning permission etc in anticipation of the disposal of Crown land.

15.   The Commissioners wish to reassure the Committee of their commitment to adhere to planning rules and any environmental assessments required under them. One of the main difficulties is that no planning authority exists beyond the low water mark and, until such an authority is established, problems will continue to arise, particularly in relation to the role of the landowner.

Appendix

EXAMPLES OF NON-CROWN ESTATE TIDAL LAND IN ENGLAND AND WALES

    All the harbours, rivers and estuaries in Cornwall

    Salcombe and Kingsbridge, and most of Plymouth Harbour in Devon

    The River Dart

    Part of Poole Harbour

    Parts of Portsmouth Harbour

    Much of Christchurch Harbour

    The Beaulieu River

    Much of Langstone Harbour

    Much of Chichester Harbour

    Littlehampton and the River Arun

    Shoreham and the River Adur

    Newhaven and the River Ouse

    The Swale, Whitstable

    The Medway

    Most of the Thames

    Burnham and the River Crouch

    The River Colne

    The River Orwell and Lowestoft

    Great Yarmouth and much of the foreshore of Norfolk

    All the foreshore in Lancashire

    The Dee Estuary

    Bangor

    Much of the Severn Estuary

STEWARDSHIP IN ACTION

A Crown Estate Policy Statement

The Crown Estate portfolio covers a wide range of properties in the built, rural and marine environments. The Crown Estate Act 1961 charges the Commissioners with a duty " To maintain and enhance the value of the Estate and the return obtained from it, but with due regard to the requirements of good management ".

Historically, the Crown Estate has an enviable record of stewardship, particularly in the fields of urban renewal, renovation of listed buildings and agricultural management. However, an Environmental Audit carried out recently showed that there is much that can still be accomplished. Following the Audit, a number of policies have been refined and consolidated to form a policy statement for the environment with an overall objective:

    To balance environmental considerations with commercial and other conflicting needs, in order to achieve effective stewardship of the Crown Estate.

1. THE URBAN ESTATES

To preserve and enhance the built environment, the Crown Estate will:

(a) maintain the highest conservation ideals for and significant levels of investment in the long term maintenance of buildings;

(b) set the highest standards of architecture and landscaping of new developments and refurbishments and promote the incorporation of works of art;

(c) ensure that appropriate environmental assessments are carried out for development proposals and that measures are taken to address any issues which such assessments bring to light, including early identification of archaeological potential.

2.   THE RURAL ESTATES

To maintain the highest standards of stewardship of the rural environment, the Crown Estate will:

(a) encourage conservation and environmentally sensitive practices on Crown Estate farms and continue the annual Conservation Award for the tenant farmer who has made the greatest contribution to conservation;

(b) continue looking for opportunities to invest in conservation projects on tenanted farms;

(c) continue the programme of investment in fixed equipment to reduce risk of pollution incidents on farms;

(d) encourage public access where this can be achieved without danger to wildlife, the ecology of the area and the viability of agricultural businesses;

(e) instruct tenants to maintain statutory paths and bridleways;

(f) preserve and restore traditional farm buildings which are of environmental or architectural significance;

(g) draw up management plans in partnership with tenants, supported by advice from conservation bodies and statutory organisations;

(h) keep forestry policy under review in the light of changes in environmental and conservation concepts.

3.   MINERAL EXTRACTION

To ensure that land-based mineral resources on the Crown Estate are exploited in the most acceptable way, the Crown Estate will:

(a) consider requests for mineral extraction only when satisfied that there is a genuine economic need for the materials to be extracted from a particular location and that extraction can be carried out without unacceptable disturbance to the local community and irreversible damage to the environment;

(b) require pre-extraction studies on flora, fauna, noise, traffic generation and other relevant issues;

(c) insist on the adoption of environmental quality objectives and defined conditions for land restoration.

4.   THE MARINE ESTATE

To safeguard the foreshore and sea bed, helping to balance the demands of conservation, development and recreation, the Crown Estate will:

(a) continue to refuse to grant dredging licences unless approval is given by the Department of the Environment (or the Welsh Office or Scottish Office where appropriate);

(b) continue to provide financial support for research into dredging and related issues;

(c) co-operate with and support marine conservation bodies, complying with requests to designate areas as SSSIs and Marine Nature Reserves;

(d) extend the amount of foreshore under lease to conservation bodies for long term management at nominal rents, particularly when these areas are part of the Heritage Coast;

(e) carefully consider proposals for marine developments and require measures for enhancement of the environment or provision of wildlife habitats where appropriate.

5.   THE SCOTTISH ESTATES

To continue to act as a responsible landlord in Scotland, the Crown Estate will follow the policies set out in earlier sections where they are appropriate to their estates in Scotland. Moreover, they will:

(a) remain committed to balancing all the views expressed during the public consultation process for granting fish farming leases and insist on Environmental Assessments where appropriate;

(b) maintain a significant level of investment in fish farming related research;

(c) encourage leaseholders of fish farms to adopt sound environmental policies in relation to good management practices;

(d) continue with a specialised management system for control of river fishing with full regard to the need for effective conservation measures;

(e) maintain the impetus already gained in developing the Glenlivet Estate with due regard to diversification, tourism and recreational resources, without materially detracting from the rich natural heritage of the area.

6. THE CROWN ESTATE OFFICE

To manage an efficient and environmentally aware office the Crown Estate will:

(a) encourage energy saving practices;

(b) recycle as much waste as possible;

(c) use products which are not harmful to the environment;

(d) specify products which come from sustainable resources.

The Crown Estate is in an unique position. It is an important contributor to the Treasury, but it also forms part of the national heritage. The Commissioners have a statutory duty to exercise good management and this concept forms the bedrock of their decision making. " Stewardship in Action " is an extension of this duty and will help to preserve and enhance the Estate for future generations.

Examination of witnesses

THE EARL OF MANSFIELD, a Member of the House of Lords, attending by leave of that House, First Commissioner, examined.

MR CHRISTOPHER HOWES, Second Commissioner, MR FRANK PARRISH, Business Manager, Marine Estates and MR MARTIN GRAVESTOCK, Business Manager, Scottish Estate; of the Crown Estate, examined.

Chairman

459. May I welcome you, Lord Mansfield. You are the First Commissioner of the Crown Estates Commissioners. I wonder if you would like to introduce those you have brought with you?

(Lord Mansfield) As you have said, I am the First Crown Estate Commissioner and Chairman of Commissioners. On my right is Mr Christopher Howes, who is the Second Commissioner, Accounting Officer and, more especially, Chief Executive of the Crown Estates. On his right is Mr Frank Parrish, who is the Manager of our Marine Estate in England and Wales but not in Scotland, where the vast bulk of our fish-farming interests lie. On my left is Mr Martin Gravestock. His title is Scottish Receiver but, in fact, he is the Manager of all our Scottish Estates both on land and on the marine side; and that, of course, includes fish farming.

460. May I welcome you here, gentlemen, to this session of our Inquiry into Coastal Zone Protection and Planning and thank you for the memorandum that you have submitted to us in advance. I believe, Lord Mansfield, that you would like to make an opening statement. Before you do that, Mr Barry Field has indicated that he has an interest he wishes to declare.

Mr Field: Thank you, Chairman. May I declare an interest in that I am in negotiation with the Crown Estate for the revision of the lease which I hold from them. This is not an interest which appears in the Members' interests because it is appertaining to my own dwelling in my constituency.

Chairman: Thank you very much, indeed. We will not comment on the negotiations but we hope they go well. Now, Lord Mansfield, the floor is yours.

(Lord Mansfield) Thank you, Chairman, for this opportunity to make an opening statement. I promise to be brief and to the point, bearing in mind the diversity of the Crown Estate's interests. The underlying aim of the 1961 Crown Estate Act is that the management of Crown lands should be separated from the function of Government. In place of the Government Departments which had administered the Crown lands, the Commissioners are appointed by Statute with defined powers and duties as if, in effect, they were trustees. The Commissioners have a large measure of executive responsibility but imposed on them is a degree of accountability to the Chancellor of the Exchequer, the Secretary of State for Scotland, the National Audit Office and ultimately the Public Accounts Committee of the House of Commons. The Commissioners have always sought to manage the Crown Estate as exemplary landlords. We exercise the general duty under section 1(3) of the Act to maintain and enhance the value of the Estate and the return obtained from it but with the important additional responsibility to have due regard to the requirements of good management. This is a very wide duty which we take seriously. It takes account of both environmental and conservation factors and the general welfare of those who live and work upon our various estates. As the fish farming and aggregate industries developed, the lacuna in planning controls beyond low water mark became more evident and the Commissioners, as responsible landowners, felt a need to extend their role with the full encouragement of Government. But a misconception has arisen that the Commissioners act as a planning authority over a wide range of other activities below low water mark; and I should like to take this opportunity to point out that they do not. Even within the Government View Procedure on the extraction of marine aggregate, the Commissioners' planning role is limited to the collation of the views of the interested parties. It is not generally realised that the Department of the Environment consider the application and the views on it and, after consultation with other Government Departments, take the final decision on whether to approve it or not. The Crown Estate plays no part in such decision making. Thank you, Chairman.

Chairman *Contd]*

461. Thank you for that helpful statement. We will come to planning matters in a moment or two; but I should like to ask some wider and more general questions before we start. Can you indicate to us the size and value of the Commissioners' marine estate in relation to the inland estate and what proportion of your staff (and how many) work in the marine estate?

(Lord Mansfield) Perhaps I may start in (shall I say?) cash terms. The capital value at 31st March, 1991, of our marine dealings was worth £150 million, of which £89 million came from dredging, £10 million from fish farming and £51 million from other coastal activities; and that amounted to approximately seven per cent of the total Crown Estate property valuation of £2,085 million. Size is difficult to quantify. The Marine Estates comprise 55 per cent of the foreshore, almost all the sea bed out to the 12 mile limit and it includes the right to exploit natural resources on the United Kingdom Continental Shelf, excluding the rights to oil, coal and gas which have been (if I may coin the term) "nationalised". This, comparing it with our land-based interests, compares with 250,000 acres approximately of agricultural land and substantial blocks of commercial and residential property mainly in London. I think you wanted to know about the staff numbers, Chairman.

462. Yes, the staff.

(Lord Mansfield) Sir, the Crown Estate is a small organisation in staff number terms. It operates mainly through a network of local agents. The Marine Estate has 23 posts out of a total of around 270; that is to say, 9 per cent—although I have to say that consideration has been given to diverting more resources to that department in view of the increase of applications for production licences in connection with marine aggregate. We also employ a large number of consultants and agents. In Scotland, our fish-farming interests employ, I think, 10 in the office at Charlotte Square, Edinburgh; and Mr Gravestock will correct me if I am wrong. We also employ a number of local agents in that part of the world.

463. Do I take it that the main concern of the Commissioners, as such, is to maintain the capital value of these assets and also to maximise the income that could be derived from them? That is your function. Is that correct?

(Lord Mansfield) I would respectfully go back to the Act. Section 1(3) says that we should maintain and enhance the Estate but with due regard for good management. So that I dislike the expression, "maximise the return", because—and I know that certain parts of the general public feel that, too, if not honourable members of this House—because "maximise" is to be equated with a certain (shall I say?) ruthlessness. And that is very foreign to our policy.

464. But "good management" implies, does it not, maintaining to ensure that the value is held?

(Lord Mansfield) It goes much wider than that. We have found over many years that, unless the property is running smoothly, unless the people who are tenants (and it does not really matter whether they are on land or sea) and the people who have commercial dealings with us are satisfied with the deal which they are given, then the estate is that much more difficult to run, it costs us more money to do it, we become less efficient and eventually we return a lower revenue surplus to the Treasury. So that it is not a simple matter.

Chairman: I accept that. Can you tell me what is the rationale for the Crown Estate to own the sea bed in this day and age?

Mr Field

465. Before we go to that, Chairman, may I just refer the First Commissioner to page 48 of his Annual Report?[1] On the Marine Estate, it says in regard to new lettings:

> The valuation of Marine Estates at 31st March, 1991, includes £721,039 relating to areas of land and water which have not in prior years been the subject of any lease/licence. The comparative figure in 1989–90 is £171,642.

That is an increase of some 600 per cent. Just before we leave this position of the size and value of the Marine Estate, I wonder whether you could tell us why, if there was such a massive increase in areas which hitherto had not been subject to leases——

(Lord Mansfield) If I may, sir, I shall ask the Second Commissioner to answer that.

(Mr Howes) Sir, It may be that Mr Field would want to discuss moorings in more detail a little later; but if I may answer it in rather general terms, the substantial increase which you quoted arises primarily as a result of a number of long-standing leases falling due—leases which were granted over 25 or 30 years ago, often at very, very nominal rents.

466. You will appreciate that that is what I thought you would tell me. But, in fact, what your report says, if I may say so, is this:

> . . . relating to areas of land and water which have not in prior years been the subject of any lease or licence. . . .

(Mr Howes) Yes. Also, there were areas where the occupiers had paid no rents to us. They had acknowledged our proprietary right and, as part of the general policy of trying to improve the management of the foreshore, they were therefore brought in. Often they were adjacent to new areas which we were re-negotiating in any case. It is part of the gradual process of good estate management of the foreshore.

Mr Pendry

467. Can it not be said that there was a certain amount of ruthlessness in this?—and Lord Mansfield was objecting to that word earlier.

(Mr Howes) No. I think it is striving to achieve a degree of equity between adjacent occupiers.

---
[1] *Not printed.*

Mr Pendry *Contd]*

468. It seems like maximisation to me.

(Mr Parrish) May I make the point that these new dealings are in response to the demand we receive for new uses as well as a question of our going out and, so to speak, regularising uses that are already there; ie bringing the landowner's interest into being where it was not in being before. That is a comparatively small part. The new dealings that are referred to in the annual report comprise every type of dealing on the Marine Estate. And those include such things as oil pipelines, cables, *et cetera*, right across England and Wales and Northern Ireland. So that all we are doing really is responding to new dealings and we are documenting the figure that we have received for these new dealings each year. And that is what we have done in the annual report for many years.

Mr Field: Nevertheless, it is quite a sizeable sum.

Chairman

469. Perhaps we may move from there. What is the rationale for the Crown Estate to own the sea bed in this modern age?

(Lord Mansfield) Sir, I think it is important to have regard to the historical perspective which was outlined in paragraphs 6 to 10 of our Written Evidence. Over the centuries, the monarch has granted land to the Royal Family (and that includes the Duchies of Cornwall and Lancaster) and to the aristocracy. An example is the manorial grant of the Beaulieu River to Lord Montague and, more recently, the Board of Trade managed the foreshore and sea bed from 1866 to the late 1940s. And that body made a significant number of sales, mainly of the foreshore. When responsibility was transferred back in 1949, it was on the basis that it was not necessary for the same authority to be responsible both for protection of public rights and the management of the Crown's proprietary interests. The Trustram Eve Report in 1955 indicated the need to sever connections with Government by not having a Government Minister as chairman; in other words, splitting the executive function of Government from the management of the Crown lands. Those conclusions were accepted by Government and the 1961 Act, in effect, preserved our position on the foreshore and sea bed. I could go on to say—and perhaps it is to the good that I should do so—that the present position is that 55 per cent of the foreshore is owned by the Crown Estate and all the sea bed, apart from some estuaries and tidal rivers. I would respectfully point out that there does need to be a landlord in the sense that the sea bed has got to be in the ownership of some body; and the Royal Prerogative decrees that in the main it should be the Crown Estate and the Commissioners appointed as if in effect they were trustees. We are very nearly administering this Estate on behalf of the State in the shape of Her Majesty the Queen.

470. What you are saying is that, for historical reasons, there have been modifications over the years as to the benefits to be derived from the ownership. Now the State, through the Department of the Environment, seems (from your opening statement) to have the major control over what happens; although nominally the ownership is in the Crown or the Crown Estate Commission. You are responsible and answerable both to the Treasury for the money you produce and to the Department of the Environment for the way in which you undertake your good management. I believe that that is the situation now.

(Lord Mansfield) I must not leave out the Secretary of State for Scotland to whom, by statute, we are also responsible.

471. So that really what we are going back to is the statement in paragraph 6 of your memorandum that the fount of this seems to be Sir Matthew Hale's works of 1640. I cannot recall whether that was just before or during the Civil War between Parliament and the King. It sounds to me to be a Royalist statement that was made just at the outbreak of war to determine the Crown's claims in that particular dispute; and that was allowed to remain as it was without any change as a consequence of the Civil War. Is that historical perception correct, or not?

(Mr Parrish) I am afraid, sir, that none of us knows.

(Lord Mansfield) I have a feeling that Charles I raised his standard a little after 1640.

472. So that it was part of the divine rights of kings.

(Lord Mansfield) That is what it comes to. Bringing the thing up to date, it is clearly tidy if there is only one owner, whoever it is. And it makes for proper administration by——

473. Well, an owner of 55 per cent we are talking about.

(Lord Mansfield) That is the foreshore. I am talking about the sea bed.

Chairman: I shall come to that in a moment; but I think Mr Field is catching my eye.

Mr Field

474. Thank you, Chairman. Through Dr Dexter, who died so sadly and very suddenly, you very kindly arranged for me to come and actually to have a look at the nature of the Marine Estate that you occupy and service and rent. You showed me (not you, personally, First Commissioner, but your colleagues) what I would term in my layman's language a sort of abstract of title which had bits of paper stuck on it referring to agreements and sub-agreements. It was (without wishing to sound disrespectful) of a Dickensian nature. We have heard advice in this inquiry from MAFF and from the Department of the Environment and so on as to their data base. The Department of Transport has a data base about the foreshore. Could you tell the Committee what stage you are at in terms of the record keeping and the quality and nature of the administration of the Marine Estate and how much of it is computerised?

(Lord Mansfield) I shall ask Mr Parrish to deal with that.

Mr Field *Contd]*

(Mr Parrish) I think that what Mr Field is referring to, Chairman, is perhaps our terrier of the estates, the maps that we hold which show how the ownership has progressed over time as we sell, lease and dispose of property as claims come against the Crown Estate and we document them. Obviously, this map is very much a working record. It is an extremely good record, in fact. It may not look impressive but it is an extremely accurate and living record of the Crown Estate's ownership. We are certainly looking at new ways of holding that information; and perhaps the Second Commissioner may want to talk about computerisation in the Crown Estate more generally. Perhaps I could mention two things in particular which show how we are moving. First of all, on the offshore side, the Marine Estate dealing with aggregate dredging, in the last two years the Commissioners have approved and implemented a very sophisticated Geographic Information System computer base which is fair state of the art. It is a very good system which actually relates all the data from the marine aggregate dredging, both its reserves and the commercial information about it on a spatial basis; so that it is referred to by maps, it can create maps. So that we are moving in that direction and in Scotland (if I may move on to Mr Gravestock's territory a little) we are at present implementing a further Geographic Information System for the Scottish Marine Estates following the same type of technology. So that we are moving very closely into electronic technology to hold our maps; and we are transferring it over. But at the moment I should say that our current paper maps are good, accurate records and that there will not be much of a problem when it comes to making them electronic and using them in a better way.

(Mr Howes) If, perhaps, I may add something on a rather broader front, the Crown Estate on its urban and on its rural estates has introduced over the last two years what is generally regarded to be the most comprehensive and sophisticated computer-based data system that exists for any property-owning organisation; so much so that Government Departments, public bodies and major leading property-owning organisations regularly come and visit and look at our data base which is a fully-integrated data base dealing not only with the day-to-day management but has on file all the relevant details for all of the thousands and thousands of tenancies that we hold. And that process is now being extended to include the Marine Estate. Meanwhile, the Marine Estate, through its consultants, has also produced a very sophisticated Geographic Information System. So that it is merely a question of just rounding the circle and we are nearly there. But I think that one needs to look at other parts of the Estate to see how we may actually achieve the objective that we have set ourselves.

475. There is just one small stage further. You very kindly took up my suggestion of standard agreements in terms of some of these smaller lettings and leasings; so that you are very receptive to points of constructive concern. But would I be totally unjustified in concluding that the figure that I see in page 48 of your annual report which shows this difference between £171,000 and £721,000 in one year for areas of water and land that previously had not been the subject of leases, or agreements or licences was actually a reflection on that data base or lack of it?

(Mr Howes) No. I think it would have no relationship to it at all.

Chairman

476. I wonder if you could clarify something that puzzles me a little. Probably I am not reading the document correctly. Returning to your Paper, the Supplementary Evidence, in paragraph 6 you quote Halsbury as stating the legal position. As I read that, what is being said here is that in so far as England, Wales and Northern Ireland are concerned, the Crown is *prima facie* the owner of all land in estuaries, bays (in between the arms of the bay) and also the foreshore between the high and low water marks. Then a distinction is made with Scotland where it is suggested that the right to the soil extends from the coast outwards to the territorial limit which was originally three miles but is now 12 miles. In other words, the inference of that is that in so far as England, Wales and Northern Ireland are concerned, there is no Crown ownership in the soil from the low water mark out to the 12 mile limit; although that is the case in Scotland. That is how I read. I may be corrected on that because, later on, in paragraph 8, you talk of the exercise of rights over the sea bed everywhere up to the 12 mile limit and then talk of the various rights of management and extraction and all the rest of it. I do not quite understand the point in Halsbury making that distinction in the light of your paragraphs 8, 9 and 10. I wonder if you could explain that to me.

(Lord Mansfield) Sir, I think it is the way in which the editor of Halsbury has drafted his paragraph. It was not concerned with Scotland; it was the *Laws of England.* And there is possibly a faint curl of the lip as he comes to the last few lines of that paragraph. But, so far as both countries are concerned, we treat it all the same.

477. So that up to the 12 mile limit, you regard the soil as yours to do with as you think fit, subject to the various constraints that there are upon you through Government assertions which have grown up.

(Lord Mansfield) Yes.

478. Well, that clarifies that and I am grateful to you. Now, you set out in your Appendix to your memorandum (the document to which I was just referring) a list of land that is not in the ownership of the Crown Estate. Who owns that if the Crown Estate does not? And how did the Crown Estate become divested of it. Is it historical again or commercial deals?

(Lord Mansfield) As you have said, sir, we give in our appendix a substantial number of examples, both of the foreshore and the sea bed below mean water mark, which are no longer in the ownership of the

Chairman *Contd]*

Crown Estate. In estuaries and tidal rivers, the picture is rather difficult. Disposals of the Crown Estates, tidal and submerged land, have taken place over centuries—some of the earliest examples being the Duchies of Cornwall and Lancaster. But other examples of historic grants are manorial grants such as the River Beaulieu which went to Lord Montague and the upper part of the River Severn, which went to the Duke of Beaufort of the day, town grants such as substantial areas of the foreshore at Southampton, the foreshore and river bed of the River Blackwater at Maldon; and during the 19th Century when, as I have said, the Board of Trade had control of what are now our foreshore marine interests, they made a large number of sales (mainly of foreshore) to local authorities. Examples of these are at Bournemouth and Brighton. The Crown Estate, itself, at moments has sold foreshore and sea bed to port authorities; and this practice continued on a lesser scale until comparatively recently. Some of the largest exceptions from Crown ownership of the River Thames were those which were sold to the conservators of that river in the 19th Century. Ownership is now divided between the National Rivers Authority above Teddington Lock and the Port of London Authority below the lock. Also on the River Dee, from a line between the Point of Air and Hilbre Point, the Crown Estate's ownership was displaced in favour of the River Dee Company and the Dee Conservancy Board by an Act of 1732. This interest appears to have devolved to the Dee and Clwyd River Authority and thence, we believe, to the National Rivers Authority but we do not have details of that. In 1869, a 999 year lease of the whole of the River Humber was made to the then Humber Conservancy Board which is now Associated British Ports; and there are numerous examples of sales of lesser amounts of tidal land to port authorities. During the same period a small number of sales were made to private individuals or companies; and in the time since they were sold out of the Crown Estate, of course, many areas have been sold again in whole or in part; so that there is a very complex pattern of ownership of foreshore and the beds of estuaries and tidal rivers. The biggest owners of estuarial beds and river beds are the harbour authorities. They have wide responsibilities and policies under their individual Acts so far as management is concerned. Where areas are in private ownership, the situation will depend upon whether the planning boundaries extend into the area. For example, the River Severn, which is in the ownership of the Duke of Beaufort, is within the planning of the County of Gwent and, to our knowledge, planning consents have been granted for the dredging of sand. So, sir, it is a very complicated pattern.

479. This is a departure from the principle that it is desirable to have all this land within one ownership or, rather, the sea bed in one ownership. We are talking of 45 per cent that has become fragmented—in fragmented ownership—over the years. One can look at it in two ways. One can either say that that was not desirable, that it was a pity and that perhaps

one has to consider reversing the situation. Or, on the other hand, one can argue, "Well, it does prove that one ownership is not absolutely necessary; we can manage just as well in having it in a number of different hands."

(Lord Mansfield) If I may correct you, sir, I am talking about the foreshore and not the sea bed. The sea bed remains ours.

480. It remains yours. Then let us deal with the foreshore. You were arguing before (were you not?) that one ownership is a matter of great usefulness and convenience; that somebody would have to own it and that it might as well be—

(Lord Mansfield) Certainly, when it comes to the sea bed, it would be tidier and probably much more efficient in resource terms if the foreshore was also in one ownership. But, unfortunately, successive monarchs took the view that they wished (mostly by way of "Thank you" presents) to get rid of bits of the foreshore. If it is not improper, my own family was in receipt of some from the King of Scotland and the present position has therefore grown up. I do not think that it would be profitable to try to get back to 100 per cent in ownership of the Crown and neither can I envisage some sort of privatisation of all of the rest of our foreshore interests which would make sense.

Chairman: To bring it back into the Crown Estate would imply compensation; and that is another dimension altogether.

Mr Field

481. One could put that in a specific sense in terms of the Duchy of Lancaster. You have here on your list as not in your ownership, "All the foreshore in Lancashire". Presumably that is because the Duchy own it. Would that be correct?

(Lord Mansfield) That is right. It was given to the Duchy by the King.

482. And some of us consider the Duchy as something of an anachronism today; and using your own word, "tidiness", would it not be right for those particular interests of the Duchy to be vested back in with the Crown Estate?

(Lord Mansfield) I do not think I am going to enter into that field of controversy, sir, if only for the fact that I do not know the full range of activities on the foreshore of the Duchy of Lancaster; but I think they are pretty large.

Chairman

483. You mentioned a little while ago the question of planning and the care that the Crown Estate takes over these matters. May I once again revert to your Supplementary Evidence? In paragraph 2 of that document you state that the Crown Commissioners do not wish to have a quasi-planning role and that you would apparently approve that an appropriate planning authority be established. What do you see as the most appropriate arrangement that could be

Chairman *Contd]*

made? And, until the arrangements are made, are there any interim measures that could be taken to improve the current uncertain situation?

(Lord Mansfield) I shall ask Mr Howes to deal with that, Chairman.

(Mr Howes) Paragraph 2 is drafted because the Crown Estate were conscious of the criticism that as landowners (and, as landowners, recipients of revenues from the assets) perhaps we were inappropriate to be seen exercising regulatory or quasi-planning functions. Commissioners are very aware of the various options that have been put before this Committee and discussed elsewhere— whether a single-purpose marine authority or, indeed, Ministry or extension of local authority powers. The Commissioners are not convinced that there is a single, simple solution, due primarily to the complexity and the range of marine activities. The Commissioners feel that perhaps marine planning is best handled on a functional basis. They hope that better integration can be achieved to overcome the diversity of responsibilities and they are pleased that the Department of the Environment has announced that a Planning Policy Guidance Note is being prepared on coastal planning. That is a move in the right direction. If I may, sir, I shall turn to two more specific areas where we are involved in a quasi-planning role. We are involved with fish farming and other of my colleagues will answer to that; but in Scotland we have seen an evolutionary and adaptive process emerging which we have welcomed. But another area in which we are involved is that of marine aggregate extraction. We are involved as a participant (a leading participant in the Government View Procedure) which leads to the grant of a licence for the extraction of marine aggregate. We feel that this is an area where the existing system is capable of improvement. Although the present procedures were reviewed only three years ago, we feel that the process involves the Crown Estate in being perceived to be operating as a quasi-planning department; and, of course, this is not the case because the ultimate decision is taken by the Department of the Environment and we would never grant a dredging licence unless the application received a favourable Government View. So that, in terms of marine aggregate extraction, we would prefer a simplified system which could be statutorily based, perhaps similar to onshore planning procedures and, perhaps, with a public inquiry process; and with the ultimate decision being seen to be taken by the Secretary of State for the Environment. When the Department of the Environment gave evidence to this Committee on 4th November last, they said that the Government was willing to give further thought to the role of the Crown Estate in relationship to the extraction of aggregates and we look forward to this review being progressed quickly with the Department in the hope that it will produce a simpler and quicker decision-making process. But one would also have due regard to the importance of environmental considerations. Sir, if I may, I shall finally just touch on the role of the local authority in the planning process. It has been argued that local

planning authorities' jurisdiction could be extended seaward for marine activity. The Marine Conservation Society suggested 12 miles while the National Trust and others suggested three miles. But the Commissioners are doubtful whether a local planning authority's powers would be appropriate for all (and I stress, "all") marine activities. Extending local authorities' jurisdiction is not without its problems over the multiplicity of local authorities' and consequent difficulties of demarcation. And there are serious doubts as to whether local authorities necessarily always have sufficient expertise to handle oil and gas and marine aggregate, to say nothing of the complications of the public rights of navigation and fishing. But there could be specific forms of development beyond the low water mark which could be included within local authorities' planning jurisdiction. It could include the extra bit of marinas which go out into below low water; it could include the laying up of oil rigs. It is more difficult when you start to look at their powers in regard to moorings because, in a sense, harbour commissioners do exercise a degree of responsibility at the moment. But that is not to say that it would be impossible for local authorities in some areas to deal with intensive applications for moorings. So, essentially, the Crown Estate Commissioners and the Crown Estate are very happy to co-operate with whatever planning regime Government determines is appropriate in all the circumstances.

484. May I put the thing more specifically? I can understand what you are saying regarding the local planning authorities because we have received other evidence of where problems can arise between one local planning authority and another planning authority up and down the coastline. They will not have regard perhaps to the effect of their own decisions upon the coastline further along. We also have had evidence of estuarial difficulties where one arm of the estuary falls within one authority and the other arm within another. Therefore, if that needs correction, we are almost driven to saying. "Should we not have one planning authority that is responsible for the entire coastline of the British Isles?" And then we start saying, "Who? Which authority? Should it be entirely the Department of the Environment; or should it be outside the Department of the Environment because they do not like to get involved in planning matters except on appeal? Should it be the NRA that has a responsibility for conservation of the coastline in many areas; should their jurisdiction be extended so that it becomes the planning authority having regard to all the matters of sea bed shift and littoral waters and all the rest of it that need to be taken into account when any new activity on the coastline is contemplated?" How does that kind of thought strike you?

(Mr Howes) I always feel that, in a sense, these are almost too radical a solution at this stage; because I feel that essentially what is needed is better co-ordination and better integration between the existing bodies. It may be that Planning Policy

Chairman *Contd]*

Guidance Notes from the Department of the Environment will help clarify and lay down guidelines for coastal zone management; but whenever the Commissioners have looked at this, we have never found that there is really one body which would be able to exercise what I call " total planning control " throughout the whole of the coastal zone, particularly if you then extend further than the coastal zone and start going out towards territorial limits. And this is why Commissioners tend to feel that it has to fall back upon a function base.

485. We are driven to this by the fact that you say you have no wish to exercise a planning role and to be a regulatory authority, as such. Historically, you are; because, as landowners working through a leasehold system or a licensing system, you do act as planners. One need only take, for example, (forgetting the sea bed and the coastline for a moment) the Nash Terraces which you own around Regent's Park. By the covenants you impose upon your leaseholders requiring them not to affect the facade in any way, to paint them a uniform colour at regular intervals, you are doing now what some local authorities are trying to do through conservation areas. The great landlords of the past were, in fact, the preservers of the urban environment and, no doubt, you (through your licensing and leasing system) were the preservers of the marine environment. But, inevitably, there arises a clash or a conflict between your needs and aspirations as owners of the land and what the community as a whole may require as regards the use of that land. And it is this area, perhaps, where a blurring has occurred; and, because you act as landowners controlling and regulating, they think you are acting as planners as well—which you are not! What I am trying to suggest to you is this. Yes, carry on as you do as landowners and exercise your rights in insisting upon your tenants regulating their affairs in a particular way; but, in turn, you should be subject to a planning authority that is not concerned so much with the same matters as you are concerned about but with wider issues that affect the community at large. Therefore, if one looks at that aspect, should there be perhaps another body to which you have to refer rather than the sort of *ad hoc* arrangements that you have at the moment?

(Mr Howes) I am not sure that they are exactly *ad hoc* because, as landowner, we rely upon a potential user of the sea bed and foreshore to obtain the necessary consents that they require under statute from existing bodies that have statutory powers. So, we have no statutory powers to act as a regulator— with the one exception of fish farming in Scotland. So that what has tended to have evolved has been that somebody who wishes to create a pipeline requires his statutory consents to do that. Once he has acquired his statutory approvals, then he talks to us as a landowner and we would grant him a licence or a lease on the assumption that he had received a valid approval from the relevant statutory body. It is rather different from land and it is also rather difficult to impose covenants over certain aspects. But what

we try to do and where we go close to being a quasi-regulator is this: where there are environmental aspects in connection with marine aggregate extraction then we incorporate within our " licence under private law ", as it were, conditions and covenants to ensure that dredging companies adhere to the terms of their licence and adhere to any environmental aspects that the Government has seen as appropriate. To that extent, that is the only example I can think of, sir, where the link is made much more explicit, where a licence will reflect the desires of either Government or a statutory body.

Mr Field

486. The code of practice for the extraction of marine aggregates came into force on 1 January, 1982. In the establishment of that code you, as the Crown Estate, had been involved. Is that the code under which you are still operating?

(Mr Parrish) That code of practice is the one between the Ministry of Agriculture, Fisheries and Food and the trade federations for the dredging industry. As Mr Field says, the Crown Estate was, indeed, involved but is not a signatory to that code of practice. We understand that MAFF has proposed a revision of that code of practice but at present it is still in force.

487. It actually says under the signature of the then Minister of State, Alick Buchanan-Smith: " . . . and I recognise also the encouragement provided by the Crown Estate Commissioners in the establishment of the code." In prospecting terms, paragraph 2 says that MAFF headquarters will consult with its fishery laboratory at Burnham on Crouch and the District Inspector of Fisheries but no outside interests. If we turn to Scotland, under " Extraction ", paragraph 3.11 goes on to say that this is an inter-Departmental procedure and does not represent a basis for public consultation. In both those terms, this code of practice in which the Crown Estate was involved makes it quite clear that there will be no outside representation whatsoever on these matters in terms of public consultation.

(Mr Parrish) May I take first of all the question of prospecting? There is no general consultation before a prospecting licence is issued. What we do is to go to MAFF and seek their advice as to whether the prospecting itself would be likely to cause, mainly, interference with fishing or, most unlikely, would possibly have effects upon marine life of commercial interest. The prospecting itself is, in essence, no different from that carried out on a very wide scale by academic bodies and all sorts of bodies. It is non-intrusive; it uses side-scan sonar (which does not even touch the bed) and seismic techniques which have minimum interference. In addition, there are grab samples which take very small amounts of sediment and a limited number (and it is a comparatively limited number) of core samples from the bed. So that, in principle, we do not actually consult about the issue of a licence except about the effects of the prospecting on fishing.

Mr Field *Contd]*

488. That is in prospecting; but, under the code of practice for Scotland, paragraph 3.11 says—and I will read it in full:

> Under the Government View procedure, the Department of the Environment will consult DAFS headquarters on an application for an extraction licence. DAFS HQ will consult Aberdeen and the Inspectorate.

It goes on, as I have already said:

> This is a Departmental procedure and does not represent a basis of public consultation.

It makes it quite clear there that the public—or any other body—cannot play any part. And that is under " Extraction ". That is not under " Prospecting ".

(Mr Parrish) The procedures followed in Scotland are identical to those followed now in England and Wales. That particular code of practice deals with the practice of MAFF and the dredging industry. It is completely overtaken by, and is subservient to, the general Government View Procedure which is, indeed, a very wide-ranging consultation procedure run by the Crown Estate and subject to the decision of the Department of the Environment. So it has been totally overtaken by the review by the Department of the Environment of the Government View procedure, itself, in 1988–89.

489. And do the Department of the Environment invite representations in those terms?

(Mr Parrish) Under the Government View Procedure, the Crown Estate carries out the informal stage during which there is wide-ranging consultation leading, eventually, to a public advertisement of the application in the local press and the fishery press. All of that is followed by referral to the Department of the Environment which then carries out a further stage of consultation with the statutory bodies and Government Departments involved.

490. Could you take us very briefly through that advertisement stage? Is it the *London Gazette* or some well-known periodical like that? Is it a local newspaper?

(Mr Parrish) We seek to make it accessible.

491. And for how many days?

(Mr Parrish) We normally place a single advertisement at the moment. But the advertisement is at the final stage. We place it certainly in *Fishing News* and the fishery press invariably; and we try to place it in appropriate, local papers. The aim is to make it available, not to conceal it.

492. And for how many days?

(Mr Parrish) At the moment we place a single advertisement, I believe. If that is wrong, may I come back to it?

493. How many days have people to make representations?

(Mr Parrish) There is normally a six weeks period. I think that is so, but I would have to come back if that is wrong. I think it is a six weeks period but, in practice, we allow people to make comments until the matter has been resolved and until it is put forward formally to the Department of the Environment. We do not exercise a guillotine on it.

Mr Cummings

494. You state to the Committee:

> " There is no statutory requirement for the Crown Estate Commissioners to apply for planning permission for development on land which they own."

Do you think that that is an appropriate arrangement at the present time in the light of your newly-drafted, " environmental objectives "?

(Lord Mansfield) Well, there may not be a statutory duty upon us to seek planning permission for land-based operations; but I can tell you that, by convention, we submit ourselves to the planning procedures just as if the Act applied to us. In fact, the Crown Estate, in everything that it does on land, now behaves in exactly the same way as any other individual or corporation would behave.

495. Why then do you not have it enshrined in legislation? Why do you object to that?

(Lord Mansfield) We do not object to it. There is very little point in it. Some of us believe that there is far too much legislation. In this particular instance, it is an important but, nevertheless, a narrow point. There is no need to compel the Crown Estate by statute to conform to a procedure to which everybody knows they are already very happy to conform. It may be untidy but untidiness is sometimes, if I may say so, the British way of regulating itself.

Chairman

496. Just to make this perfectly clear, you do not make a formal application for planning permission because there is no basis for you to do so. What you do, I suppose, is to consult with the local planning authority—with the local authorities which we were discussing a few moments ago. You consult with them. And do you do so on the strict basis of: " Assuming that we are applying for a planning permission, how would you deal with that normally?"

(Mr Howes) We apply for planning permission as if the Town and Country Planning Act 1990 applied to us. We fill in the form and if, on rare occasions, we do not accept the local authority decision then we appeal and we will then stand by the determination of the Secretary of State. To my knowledge, sir, there have been no cases where we have developed without a valid planning permission. If I may touch very briefly on just one aspect, the early Planning Acts (the 1947 Planning Act, the 1968 and the 1971 Planning Acts) were totally silent with regard to the Crown. This was interpreted by the Law Officers as meaning that we were not able, as a Crown body, even to apply. As a result of that, there were the Circulars in 1977 and a further Circular in 1984 which gave

Chairman *Contd*]

Crown bodies a shadow system under the Circular. But we have always applied direct, certainly since the mid-70s. Before then, I think it was a rather more *ad hoc* arrangement based primarily on the Circular and earlier Statements. But we operate in exactly the same way as a private property owner when dealing with the local planning authority or when dealing with the Secretary of State for the Environment on appeal.

Mr Cummings

497. That, I am sure, is because you have enlightened officers; but let us assume that sometimes you might not have enlightened officers. You are enlightened officers but what would happen if you were not?

(Lord Mansfield) I think that the situation has now become that it would be regarded in law as a convention that we do go through these procedures. I cannot see, even if the Commissioners were unenlightened, that they would seek to go back on this. And, if they did, I think it could be challenged. But that is my own private opinion.

Chairman

498. And you are talking in respect of planning applications for the whole of the United Kingdom where you have land. Does this mean that you apply for the equivalent of planning permission in Scotland for fish farms, for example?

(Lord Mansfield) No. A fish farm does not come under planning at all; unless part of it is land-based. But we do not ever develop a fish farm. On land, if a fish farmer wishes to put up some sort of a shed or building or an erection of some kind, then he has to go to the local authority. But that has nothing to do with us.

(Mr Howes) In Scotland, if there were to be an application for a fish farm, the application would be made by the prospective fish farmer and he would go through the process that applied in Scotland. Part of that would mean that he would apply to us for approval.

499. It is suggested that fish farms can cause pollution of the areas, the coastal waters, in which they are situate. Or am I wrong on that?

(Lord Mansfield) Certainly, in certain circumstances, they do cause pollution.

500. Then should they not become subject to planning applications in that case; and this is an argument then in favour of extending the jurisdiction of planning authorities outwards from the coastline?

(Lord Mansfield) This is very much of a separate and quite bigger question as to what sort of procedures should be adopted for fish farm applications. This is obviously going off land and on to the sea; but quite different considerations have applied to it than would apply to a land-based operation.

501. You have drawn a distinction in so far as fish farms are concerned. Those that use land for part of the operation require planning permission; but those which, perhaps, do not require land you are saying that there is no need for any planning permission at all. That does bring into play the discussion we were having earlier: the extent to which the Crown Estate regard themselves as within the planning system. Are we drawing a distinction between your land-based operations and those seaward? In other words, if it is seaward you do not feel that there is any need to make any application for planning permission or to discuss with the local authorities because no other landowner would be required in those circumstances because the planning authority's jurisdiction does not extend outwards?

(Lord Mansfield) No, sir. I think we are getting at cross purposes.

502. Please clarify these, if you would.

(Lord Mansfield) In England and Wales (or, indeed, in Scotland) we do not (so far as I know) undertake any marine-based activities which would need planning permission. Everything which is done is done by tenants or licensees; so that the question does not arise. I have no doubt that if in England and Wales (or, indeed, in Scotland) we ourselves undertook an operation which would require planning permission from a land-based authority, we would go through the normal channels—which we do on land, in any event. I have no doubt about that. In the fish-farming context, the situation is entirely different. We do not own any fish farms; we do not operate any fish farms. The situation has been an evolutionary one. The fish-farm industry which slowly developed during the '70s came on stream with the first production of marketable fish in about 1979; and during the next five years (I suppose up to 1985) the procedure was that the would-be fish farmer would go simply to the office in Edinburgh and apply for a licence, which became a lease, for a portion of the sea bed where he wanted to farm his salmon. And, without being facetious, I have to say that in the very early days it was almost like going to a post office counter and obtaining a television licence. It rapidly became apparent as the whole industry took off that regulation was needed and, at the request of the Scottish Office, (and with some reluctance) we have evolved a regulatory procedure which since 1985 has gradually evolved into the system in which it finds itself today; and which was examined by the Agricultural Select Committee in 1990. Your question, sir, actually originated on environmental damage but I think that, if I may, I had better go on with the planning aspect of this. There has been dissatisfaction in the past at the seeming role in which we acted as the planning agent and authority although we also were the owner of the sea bed. And it has been asked before now whether that dual role really is proper and compatible with our status. Some years ago, we worked out a system of guidelines and a consultation procedure which, in effect, brought together the public at large and all the voluntary and statutory bodies who were interested in the subject in Scotland; so that for the last few years there has been a procedure by which every

Chairman *Contd]*

application for a fish-farm licence is advertised in the press, the adjoining riparian owners are informed, as are the Department of Agriculture for Scotland, the local authorities, the River Purification Boards (because we do not have in Scotland anything which approximates to the NRA), the District Salmon Boards, if applicable, and one or two other bodies the names of which, I am afraid, are not on the tips of my fingers, so to speak. I should also say that the NCC, which is the Government's advisor—although I think that we must now call it "Scottish Natural Heritage"—is also informed. All of these are given a chance to make representations and they do so. If any objections are made which cannot be reconciled by some sort of remedial action such as resiting of the fish farm or limiting its size; if there cannot be something like that and the objection is sustained by one of the statutory bodies such as a regional council or the NCC, then the Secretary of State has set up an appellate committee which considers the appeal (so to speak) and all the circumstances of the case. We have undertaken to (and we do) take the advice of that committee so that if it happened that at the end of the day the committee came down against the application, then it would not be granted.

Chairman: Thank you. That is very clear. Mr Cummings has a further question.

Mr Cummings

503. Do you, Lord Mansfield, actually subscribe to the idea that the jurisdiction of planning authorities should extend out to sea?

(Lord Mansfield) Chairman, that is a very wide question. I think that it depends entirely on the activity which is contemplated. For instance, aggregate extraction which is taking many miles off the Humber coast has really no part to play in the lives of the citizens on land. It is difficult in some cases even to decide into which local authority area it falls; the local authority probably would not have the resources with which to process what can be a long and complicated procedure which, as I say, would have little to do with it. So, at that end of the extreme, I can see no case for extending the jurisdiction of the local authority. If I come right back to the coastline, then I think there are a number of options which could be considered. They could include marinas and they could include also a procedure which, to the English, would appear a rather esoteric one, which is the laying up of oil rigs; although, in fact, one took place in the West Country. And where there is an intensive use of moorings that is something which could be brought under the aegis of the local planning authorities. At present, for instance, in waters where they have control (and I am thinking now of, say, Chichester) the planning consent is given by the Chichester Harbour Authority; and that works quite well. So that the answer to your question is, Yes, there are certain aspects where I think it would be perfectly possible and an option which could be pursued.

504. And do you believe that there should be Environmental Impact Assessments for developments proposed in areas owned by the Crown?

(Lord Mansfield) We are going back to the whole planning scene and not just fish farming. Well, yes. We in the Crown Estate think that Environmental Impact Assessments have a very important role to play. They would need to be within the European Community requirements under Directive 85/337 but it is essentially for Government to decide or to determine whether an Environmental Impact Assessment is appropriate. The Government has limited environmental assessment to control frameworks even if there is no statutory planning system; but, of course, sir, your question breaks down really into two main activities with which we are concerned today; namely, marine aggregate extraction and fish farming.

505. But, of course, the Government does not agree that an Environmental Impact Assessment should be carried out where there is no overall planning control. I also understand that the European Community disagrees with this particular stance. Do you have any observations on that, Lord Mansfield?

(Lord Mansfield) We sit in the middle (if that is the correct term) and we will loyally abide by whatever is the outcome.

506. The Government's memorandum (paragraph 43) states that Ministers have agreed to a review of the scope for rationalising the existing system of consents for development below the low water mark. What has been your involvement with this review; and have you an idea of how it is progressing?

(Lord Mansfield) Perhaps Mr Howes may deal with that.

(Mr Howes) This study has been undertaken by the Deregulation Unit of the Department of Trade and Industry in co-operation with the Department of Transport. It is really for the Department of Trade and Industry to respond on how it is progressing. From the Crown Estate's point of view, we have participated but we have pointed out to the Deregulation Unit that, irrespective of how the systems of statutory control are streamlined, there will always be the legal necessity for there to be a separate contract between ourselves, as landowner, and a tenant so as to preserve what I call the "Landlord and Tenant" relationship. It is not an area where we have been directly involved. We have merely pointed out to the Department of Trade and Industry that in our case our relationship is that of the landowner in private law, which is rather different from the consents granted by others under statute.

Mr Pendry

507. May I ask you about the effectiveness of the Government View Procedure and, in particular, this question? How many applications year by year were there in the last, say five years for marine aggregate extraction? How many of those received a negative Government View and how many licences were granted?

Mr Pendry *Contd]*

(Lord Mansfield) Perhaps I may ask Mr Parrish to deal with this question, Chairman.

(Mr Parrish) You asked about the effectiveness of the Government View procedure. I think your starting question was on the effectiveness of the proceedings?

508. It was on the general nature of it; but it depends upon your answer.

(Mr Parrish) You asked for the number of production applications in the last five years. To begin with, there were six applications outstanding at the end of 1986. Then, in 1987, there was one; in 1988, there were five; in 1989, there were six; in 1990, there were eight; and, in 1991, there were six. That makes a total of 32. Of those, three were rejected before proceeding to the formal stage of the Government View Procedure (ie reference to the Department of the Environment for decision); one was given an unfavourable Government View by the Department of the Environment; nine have been given a favourable Government View; and there are 19 outstanding at present.

509. How many licences were granted?

(Mr Parrish) I think, Mr Pendry, that I am going to have to come back with the exact figure of how many were granted.

Mr Pendry: That is fair enough. Can I get from you some specific examples of where you feel that the Government has not provided a clear Government View? Before you answer, Mr Parrish, I think Mr Field would like to come in with a question.

Mr Field

510. Can you tell us how many licences there are actually in existence? The last question pertains to the issue of licences over a period of time but there will be others already in existence.

(Mr Parrish) There is a total of 80 areas licensed and, in total, about 100 licences because some of those areas have more than one licensee or operator on them.

511. Can you tell us what that total tonnage represents in terms of extraction?

(Mr Parrish) In terms of extraction, I can put it in two ways. In any given year, the total number of licences would permit up to about $43\frac{1}{2}$ million tonnes to be taken from those licences. In terms of reserves in those licences, we estimate that there are currently something like 415 million tonnes in those licences.

512. When we were talking about the Duchy of Lancaster, the First Commissioner said that, although you do not own the foreshore, you own the sea bed everywhere. You have 100 per cent ownership of the sea bed.

(Mr Parrish) If I may clarify it, we said it was foreshore and sea bed within estuarial areas. What we have is about 55 per cent of the foreshore—that is, mean high to mean low water. Below mean low water in estuarial areas, we have roughly about the same proportion, not always exactly the same areas but roughly the same proportion. Just to give an

example. You asked specifically about the Duchy of Lancaster. They own the tidal bed of the River Ribble and they own half of the tidal bed of the River Mersey—to the median line between Cheshire and Lancashire. And that goes out basically to the point between the two outmost headlands.

513. I was really wanting to know about the point Mr Pendry was making about extraction, which is taking place rather further offshore. And in those terms, you own 100 per cent of the sea bed. Could you therefore explain to the Committee how it is that some local authorities retain the right to license extraction in their own right?

(Mr Parrish) I shall try. Under the Coast Protection Acts (the latest one being 1949) there is a section 18 of the Act which permits local councils to take out what is called a "Section 18 Order". Under such orders, councils may control extraction of sea bed sediments within three miles of their coastline. However, in most cases dredging is taking place a good deal further out than that—and, certainly, the sort of dredging that is licensed by the Crown Estate. It is only in very exceptional cases where it comes within that three mile band. One case in point is that of South Wight Borough Council which has held a Section 18 Order since, I think, the early '60s. In fact, to my knowledge they are the only coast protection authority which is actually using the Section 18 powers to control licensing within the Solent—that is to say, the Solent between Hampshire and the Isle of Wight. They set the extraction limits each year under their Section 18 powers for the licences that fall in that area; and we co-operate with them in doing so.

514. Can we be definite about this, Chairman? I have asked all the other witnesses if they could help on this particular point of how many local authorities actually license extraction in their own right; and all of them said that they were unaware of this. Can we be quite certain that it is only South Wight that still does this?

(Mr Parrish) According to a list that we have received from the Ministry of Agriculture and Fisheries (and we understand that it is up to date) there are 30 local councils which hold Section 18 Orders. As I say, the only one of which I am aware which actively uses its powers in respect of licensed dredging is South Wight in respect of those licences in the Solent.

515. But you own all the sea bed; they exercise their right under that Act, the 1949 Act. Are they subject to the same requirements to evaluate the resources in terms of hydraulic research that you are?

(Mr Parrish) No, sir.

516. Or do they just get on and issue a licence, willy-nilly?

(Mr Parrish) As far as I can see, it is entirely a matter for their discretion how they do it. They have a statutory base to do so; ie the Coast Protection Act 1949; and the Government View Procedure is an entirely informal non-statutory system which must give way before it.

Mr Field *Contd]*

517. Finally, Chairman, may I direct a question to the First Commissioner? You laid great emphasis in your opening statement on this question of tidiness. I do not believe that you have brought any evidence before this Committee to suggest that this particular anomaly should be addressed. Is there a reason for that?

(Lord Mansfield) Save that it is no part of our function. This is a statutory right which is exercised. They do not have to get permission from us, they do not have to consult us or even to inform us.

518. But if they were acting irresponsibly—and I am not for a moment suggesting that they are; in fact, South Wight I think have just announced that they are not going to increase or to allow any dredging licence for the current period—they actually would be denying you your foreshore rights, would they not? So that the Crown Estate has a very real interest in this matter, does it not?

(Lord Mansfield) I do not see how they would deny us our foreshore rights; because they are operating below low water mark.

519. It is well established that irresponsible extraction will lead to erosion of the foreshore. There is no disputing that, is there?

(Mr Parrish) In practice, Mr Field, we have not actually had a particular problem over this. As I say, South Wight is the only authority which is active in this field and we co-operate very closely with them. We consult each other about it and they tell us what they are going to do. We have not had a case of irresponsible use of these powers. As I said to you earlier, dredging licences (nowadays, certainly) are a great deal further off the coast than the area in which these particular powers exist.

520. With respect, Mr Parrish, when I put down a Parliamentary Question on the rationale of maintaining dredging for aggregate in the Solent, given its very heavy use in terms of recreation and shipping I was told that this was a requirement so that in heavy weather the licence holders could continue to dredge within sheltered waters. The fact that we always seem to find them at midnight on a flat calm in the middle of August is a matter of a lot of local conjecture.

(Mr Parrish) There are some very small ships which tend to use those licences. The quantity—and, if you wish, I can provide full details to you—has declined significantly over the last few years. One company, ARC Marine, has already announced the abandonment of those licences and——

521. That was——

(Mr Parrish) —— and, without prejudging the situation, I think you can expect further developments about those licences being ended very soon.

Mr Field: That would be very good news. Thank you.

Mr Pendry: On the line of questioning——

Chairman

522. May I ask one question on that? The dredging may be for aggregates but, on the other hand, it may be just to clear a channel. What happens to the stuff that is not then required? Where is it dumped? Is it dumped somewhere else on your Estate; and do you have any say as to where it is dumped?

(Mr Parrish) Navigational dredging, whether it is to maintain an existing channel or whether it is to deepen a channel for larger vessels, is subject to the powers of harbour authorities under their harbour powers. There are statutory provisions in each of the Harbours Commissioners Acts to cover those situations. The Crown Estate would not license maintenance or capital port dredging in the normal way at all. If, exceptionally, a situation arose where there was a proposal to do some maintenance or harbour dredging in an area which was not subject to harbour authority powers then I believe we would go through the Government View Procedure, or some version of it, before giving the Crown Estate proprietary consent. As far as dumping is concerned, yes, it takes place under licences issued by MAFF under the Food and Environment Protection Act. It takes place on Crown Estate sea bed and generally, as long as it is within an existing licensed area—and these are fairly well established by now; and have been so for several decades, if not more—then the Crown Estate charges a proprietary consent fee and gives its consent as landowner essentially to preserve the Crown Estate's proprietary interest.

Mr Pendry

523. Perhaps I may repeat the question that I began to put a few minutes ago. Can you provide us with specific examples of where you feel the Government has not provided a clear Government View?

(Mr Parrish) I cannot think of any cases where the Government View that has been given has been unclear. If necessary, we would go back and discuss it with the Department of the Environment before we framed the appropriate clauses in our licence; but we do not actually have many cases (or any cases that I can bring to mind) where we need to do that.

524. So say you all! That is the view that is shared by you all?

(Lord Mansfield) If the oracle of Delphi were not clear in its answer, then we would go back until it was; because we are not going to give a licence unless we get a clear Government View that it would be proper to do so.

525. So that it does not happen very often that you go back and carry out a healthy " threshing"?

(Mr Parrish) Not very often. It is an exceptional situation. It is where one gets perhaps a string of very difficult conditions occasionally before we actually write the licence. As the First Commissioner has said, we would want to be absolutely clear that we had

Mr Pendry Contd]

understood what it was that the Government was telling us should be incorporated into the licence. That is all.

526. Can I come to marine aggregates?—and it may be that you will be repeating yourselves a little. In view of the fact that marine aggregates are not a sustainable resource, what is your estimate of the total available reserves and of the annual rate of extraction both now and in the future?
(Mr Parrish) Before giving (and, perhaps, repeating) the reserve assessment, may I put a little caveat on it? It is only comparatively recently that we and the Department of the Environment have carried out the regional assessments of the geological background to the aggregate reserves around the United Kingdom. It should be borne in mind that the United Kingdom is perhaps very unusual in Europe (and perhaps even in the world) in having such aggregate reserves around its coastline. So the studies by the British Geological Survey, tasked by us and DOE, started in 1986 and they are still going on. And over that same period, prospecting techniques also have been improving; so that our assessments are very much a moving situation as better information becomes available. We collate them on our Geographic Information System that I mentioned earlier. So, with those warnings that it can change, I will give you our assessment. I have already said that we estimate that there are currently 415 million tonnes in currently-licensed reserves. We estimate that there are a further 152 million tonnes in areas which have already been prospected and are subject to the current production-licence applications that I mentioned earlier. We also estimate—and it all gets vaguer as we go further into these estimations—a further 250 million tonnes in current areas which have been prospected and which may become production applications. Those are the figures that we estimate at present in total. Obviously, the latter two figures are entirely subject to the decision of the Government about whether they should be worked or not; and the conditions under which they should be worked, so that one may not always be able to extract all of that reserve. Finally, in making estimates, we cannot take account of future prospecting technology or developments in dredging or even in such things as treatment of sediments at the wharf and the way in which companies can access what are regarded at the moment as unworkable, uncommercial reserves. So, with those caveats, we believe that that is as good an estimate as can be given.

527. Tell us about your advisors, Hydraulics Research. Are you confident that they fully understand the complex sediment régime in licensed dredged areas? What studies have been undertaken in this respect?
(Mr Parrish) I have no doubt that Hydraulics Research understand what is required. It is laid down in the Government View Procedure itself, and it has been laid down and reiterated comparatively recently in 1989 that the basic criterion is "nil effect" on the adjacent coastline. Otherwise, the dredging will not

be permitted. Hydraulics Research are quite clear about that. They have carried out both for us, for MAFF and for the Department of the Environment in the past a series of studies; and, obviously, they—

528. Are those studies in the public domain?
(Mr Parrish) They are a privatised company now. They used to be the Hydraulics Research Laboratory, I think. But I can provide to the Committee, if it would be helpful, a full list of all the studies they have carried out. They range from generic studies into the way that sediments move in dredging areas, whether they move shoreward, and how one can predict when sediments will move through to site-specific studies on dredging application areas. And they also include such things as mathematical modelling techniques and studies on how to predict wave diffraction and the effects of changes in wave patterns caused by changes in sea bed configuration which might be caused by dredging. There is a full list of our reports from Hydraulics Research which are available; and both we and they make those reports generally available as required.

Mr Field

529. May I put a question to Mr Parrish? Could you, Mr Parrish, tell the Committee how you actually audit the amount that is extracted and landed by the dredgers, given that they come at all times and tides and at night. how does the Estate monitor how much is actually extracted?
(Mr Parrish) First of all, it is a condition of our licences that every document of relevance to the dredging operation is available to the Crown Estate and its agents. A full audit is now carried out on a cyclical basis. It has been carried out annually over the last few years and we satisfy ourselves that every cargo, every tonne of gravel, that is sold has been properly accounted for. The wharves themselves are pretty sophisticated now and their documentation is sophisticated; so that we do not have much problem in that regard. It would take fairly extraordinary cunning and duplicity to avoid our audit and we are quite satisfied that that is not happening.

530. How many of your licences require ships to carry " black boxes "?
(Mr Parrish) The licences require the " navigational fit " on the ship and that it should be as required by the Crown Estate. At present, we do not have " black boxes " as such. What we have undertaken is that an electronic monitoring system which will record the passage and the dredging operations of every vessel on Crown Estate licences will be fitted by 1 January, 1993. The basic aim of the electronic monitoring system is to have a permanent record of the time, the date and the position of a dredging vessel whenever its dredging equipment is deployed. And that should be automatic and tamper-proof, basically. The information is stored on discs and it will be security encoded so that it will come back to us in a secure form so as to be decoded, plotted and monitored on our Geographic

Mr Field Contd]

Information System. That will be in place by 1 January next year on every vessel that is operating on a Crown Estates licence.

531. You said earlier that you were happy with the arrangements in relation to the Government, in terms of its Government View Procedure and the way in which it signalled the acceptance of the dredging licence. When the further licence at Great Yarmouth was granted there was considerable correspondence between the Department of the Environment and the Chief Executive of Great Yarmouth as to that licence. Mr Bide of the Minerals Division of the Department of the Environment says in a letter:

"As you will have noticed, this area lies within a much larger area that is already licensed. I am sure that you will agree it would be unreasonable to hold up this dredging proposal without good reason."

In that particular case Posford Duvivier had actually contradicted the advice of the Hydraulics Research; and, in a letter to your Mr Purkis, Hydraulics Research acknowleged that the coastline of Great Yarmouth is sensitive and is subject to erosion; but they go on to suggest that the reason they are quite happy with licensing the larger area is because it is beyond the 18-metre depth contour, which is the point that has been made continuously this morning—that this is very much further offshore. But the point is that the Hydraulic Research actually base their view on the 18-metre depth contour on the basis of the fact that no movement of shingle was detected in an extensive field investigation by Hydraulics Research off Worthing. You will know that many residents along the coast and more than a few local authorities are utterly convinced that aggregate extraction does have an effect on the foreshore. Is it not a remarkable statement from Hydraulics Research that they are quite happy with this very large licence—I think it is a million tonnes— off Great Yarmouth; but there is no effect to it because the study off Worthing proved that there was no shingle movement below the 18-metre contour?

(Mr Parrish) May I say, first of all, that I am unaware of the particular correspondence. That is obviously DOE correspondence with Great Yarmouth and that is where they are fulfilling their role under the formal stage of the Government View Procedure before telling the Crown Estate what is their decision. As far as Hydraulics Research is concerned, I think the Crown Estate Commissioners must take the view that what they are seeking to do, in all these cases, is to provide Government with the proper and full information on which that decision can be taken and do not seek to evaluate it or to assess it themselves. We do not seek to do that. What we do is to make sure that the basic criteria for the assessment are properly understood, that it is done on a consistent basis by Hydraulics Research and that that information is, first of all, made available to coast protection authorities and the National Rivers Authority which can then, themselves, raise questions with Hydraulics Research. For example, we arrange meetings between Hydraulics Research and the experts of these bodies and, in some cases, require further studies to be carried out before we pass it to the DOE. So that what we are trying to do in every case is to make sure that it has been fully tested and that the arguments are properly laid before the decision-taking body. Now, Hydraulics Research, themselves, are internationally renowned in the field of coast protection and in sediment movement. They are not going to make rash statements about issues of this sort and, while I am not going to take their part (for I think they can defend their own methods and their own expertise better than I can) they have stated publicly on several occasions that their whole methodology is designedly cautious because of the unknown factors in this. For example, if coast protection schemes were subjected to the same criteria, then most of them would fail. So they are taking an extremely cautious approach and that is what we encourage them to do. Their report is only the first stage of the whole Government View Procedure; it is commented upon and tested by other bodies involved and the decision is taken by the Department of the Environment at the end of the day.

532. But we have asked several of the witnesses we have had about this question of foreshore erosion; and one of the points they have made is that very little is known about the total concept of the whole of the marine structure, if I may use that expression, and that the erection of sea defences by one authority may actually deprive another part of the coast of littoral drift and so on. I just wonder how it is that you can be so certain—as a result of this particular study commissioned by you, yourselves, off Worthing— that this whole question of the impact of marine aggregate extraction on processes like littoral drift is not having serious impact; when there is so little geological information available for inshore waters. And I raise this point particularly because some constituents of mine have photographs going from the 1940s, 1950s, 1960s and 1980s which all show their foreshores being very severely eroded.

(Mr Parrish) I think there is no doubt that the foreshore has been eroded in a number of cases. The Isle of Wight is an eroding system fundamentally and has been so over a geological time scale. That is why (come to mention it) it is an island. But the view of Hydraulics Research (which, as I say, is designedly cautious; and they are well aware of the Government requirement of "nil effect") is that dredging is not contributing to that process of erosion. Our understanding and our advice are that erosion is certainly being exacerbated by such things as concrete coast defences and so forth which stop circulation within the coastal zone itself. But, as far as can be scientifically determined, most of the dredging that we are talking about is outside the area of influence of that coastal region. What we are seeking to do at the moment, as you may be aware, is to look better at the regional science; and particularly so along the south coast. And we and SCOPAC have recently decided that we will commission Hydraulics Research to do a full regional assessment of sediment

Mr Field *Contd]*

transport patterns along the south coast to add to the science.

Chairman: I do not think that we can really pursue this point very much further. You have expressed what we have asked of you: the degree of confidence that you have in Hydraulics Research. Clearly, that is very high; and you place entire reliance upon them. We cannot really ask you to comment upon their methodology. These are really questions that we, ourselves, would have to ask them as to whether or not a study off Worthing is valid for what happens outside Yarmouth. We would have to find out what are the scientific and technological base for those conclusions. Perhaps, Mr Field, we might consider a visit to Hydraulics Research and be given a presentation of their work. That would then give the Committee a better idea of the sort of matters that are giving rise to your concern. So that perhaps we could leave that particular matter there.

533. May I please ask you about public accountability? You state in paragraph 4 of your Supplementary Evidence that you have set yourselves environmental objectives across the range of your activities. Can you tell the Committee who carried out the initial environmental audit? Can you tell us how you intend to monitor your performance in achieving your environmental goals; and to whom you will answer in this process?

(Mr Howes) Thank you very much. If I may answer that, the environmental audit was carried out by the Crown Estate. It was carried out by the Crown Estate in our capacity as landowners. But, in view of the size and diversity of the Estate, we involved all our consultants—architects, planners, surveyors; we talked to conservation groups, to English Nature, RSPB and others with whom we maintain regular contact in any case. So that our environmental statement has now been written into each of the separate business plans within the Crown Estate. We set up a monitoring system which will continuously monitor internally but with each business giving a formal report to the Commissioners as to how they have managed to move forward in achieving those objectives. As new environmental issues arise in the future, we shall assess those and add and adapt to those.

534. You might recall I suggested that perhaps the Crown Estate could have a travelling roadshow to try to take some of the mystique out of the Crown Estate and perhaps make it more (shall I say?) user friendly in terms of its local perception in coastal areas *et cetera* so far as its main estate is concerned. Can you tell the Committee how far that might have progressed?

(Mr Howes) We engaged a number of consultants who give us advice on public relations and so forth. This is a matter they are actively looking at. As you will appreciate, with an estate as wide and as diverse as the Crown Estate and my concerns for public accountability, we would have to have hundreds of road shows on the road every day so as really to cover.

535. Well, you say that, Second Commissioner, but the NRA have now embarked on a very extensive programme on a sort of blow-by-blow basis of having public meetings to which they invite all and sundry. They had one in Arundel; they had one on the Isle of Wight recently and I think they have had several others to try to raise the level of accountability and public information. If they can do it—and they cover the whole of the United Kingdom—why is it so difficult for the Crown Estate?

(Mr Howes) We are going to Cowes, we are going to Portsmouth. Regularly, Commissioners and senior officers go down and talk at public meetings. It is the concept of a road show. Perhaps I have misinterpreted you. If the road show was a matter of individuals going and talking to those who are concerned about the activities of the Crown Estate, then that road show is taking place all the time.

536. May I move on to fish farming? I think that we very extensively covered this earlier, Chairman; but may I just ask this. The Government's evidence in paragraph 49 states that there has been no evidence of serious environmental damage resulting from the manner in which the fish-farming industry has been controlled. What damage has occurred? How do you assess how serious this is; and what steps do you take to ensure that the conditions of leases are adhered to once they have been approved?

(Lord Mansfield) The most obvious effect of having a set of cages, for instance, is the limited effect which occurs through the deposit on the sea bed of solid waste which may consist of food and faeces from the fish in the cages. Research which has been partly funded by the Crown Estate indicates that this area which is affected extends no more than 50 metres from a group of cages and it is often less—particularly where you get a strong tide or other hydrographical factors. This waste is biodegradable and the sea bed recovers extraordinarily quickly if the cages are moved. As a result of this research and the quite startling facts both as to the recovery rate and the general lack of what I might call permanent damage by the presence of sea cages on the sea bottom, we have encouraged our tenants—and, in fact, in some cases we make it a condition of the lease—that what we call "fallowing" now takes place. In other words, they rotate the cages perhaps year by year and we encourage the farmers to take these measures and any other preventative measures. The Department of Agriculture for Scotland pay particular attention at the application stage to what I might call "the flushing rate" so that in any particular part of a sea loch (or, indeed, the sea) they would, as I say, monitor the situation. Also the River Purification Boards— and, if I may correct the honourable Member, the NRA has no remit in Scotland—who take that particular function to themselves in Scotland have a statutory role. If they, as our consultees, are concerned at any possible adverse environmental effects by a proposed fish farm, they in fact, by their statute, can refuse a licence or they can lay down conditions both as to density of the fish farm or its output. So far as monitoring is

Mr Field *Contd]*

concerned, all our fish farms are inspected on a regular basis to ensure that the conditions written into the lease are being met. The River Purification Boards monitor the position from their point of view and the Department of Agriculture for Scotland, where it is appropriate, monitor the impact on the marine environment generally—and disease levels, in particular.

Chairman

537. Well, I think that concludes our session for this morning, Lord Mansfield. We are most grateful to you and your colleagues for your attendance and for the assistance you have given us this morning. Thank you so much.

(Lord Mansfield) Thank you, sir.

WEDNESDAY 12 FEBRUARY 1992

Members present:

Sir Hugh Rossi, in the Chair

Mr Barry Field   Mr Anthony Steen
Mr Ralph Howell

Memorandum by the Joint Marine Panel of the British Aggregate Construction Industries and the Sand and Gravel Association

COASTAL ZONE PROTECTION AND PLANNING

A. INTRODUCTION

1. BACMI and SAGA are the two United Kingdom trade associations which represent the aggregate industry. For marine dredging of sand and gravel they have established a joint committee—the BACMI/SAGA Joint Marine Panel. This Panel represents all major sand and gravel dredging companies in the United Kingdom, covering, perhaps, 98 per cent of all sand and gravel dredged for aggregate purposes in United Kingdom waters.

2. The marine aggregates mining industry does not normally operate in the area commonly recognised as the "Coastal Zone" ie the inshore area. With very few exceptions the industry works in offshore licensed areas well outside the area being considered.

3. Although we do not have details of the quantities, we understand that the total of material dredged and dumped by various port and harbour authorities, far exceeds the quantity of material extracted by the marine aggregate industry.

4. The harbour dredging is, of course, taking place in the Coastal Zone and historically has been, for example, outside any coastal erosion control by virtue of long-standing rights under Act of Parliament.

5. The question of why the marine mining industry has been included is interesting. We operate outside the Coastal Zone—any question of erosion is dealt with as a part of the Government View Procedure by HRS as advisers to the Crown. If erosion is a potential threat no licence is issued.

B. BACKGROUND

6. Sand and gravel of usable quality is found in a number of locations around the coastal waters of Great Britain on the continental shelf. They are of similar geological origin to many adjacent land based deposits often having been deposited in river valleys long before the sea covered them.

7. In most respects they are the same, mineralogically, as their land based equivalents. Such differences as exist are derived from the action of the sea after water levels rose to cover them following the last Ice Age.

8. Because of wave action, they are usually rounded compared with the more angular deposits on land. Being under water they are not usually contaminated with clay and often the softer parts have been worn away.

9. The main demand for sand and gravel in the United Kingdom is in South East England and thus much of the present activities of the marine industry are concentrated in the Southern North Sea and the English Channel. In fact there is at present only occasional dredging north of Liverpool on the West Coast and little north of the Humber on the East Coast.

10. In 1989 sand and gravel production in Great Britain amounted to 131 mt of which 21 mt [16 per cent] was dredged from the sea. Most of this marine dredging [15 mt] was landed in South East England and was dredged from licensed areas offshore from the Thames Estuary, off the Norfolk and Suffolk Coasts and offshore of the Isle of Wight.

11. There are currently three other important dredging areas which supply their more local markets—the Bristol Channel, off the Humber Estuary and Liverpool Bay.

12. Further reserves of sand and gravel exist outside the current licensed dredging areas. There are, however, two fundamental restrictions which limit the amount of sand and gravel which can be usefully extracted:

(i) the depth of the deposit—current technology limits the maximum depth of dredging to 46 metres but in fact 25 metres is the present-day norm.

(ii) the distance of the deposit from the port. The normal operating cycle is the 25 hours tidal cycle, including dredging, steaming to port, discharge of cargo and steaming back to the bank. Some dredgers in the Bristol Channel can operate on a 12·5 hour cycle whilst some of the larger dredgers working in the North Sea can manage to work on a 37·5 hour cycle.

13. Marine dredging is an environmentally friendly industry, the dredgers operate often out of sight of land, they extract the sand and gravel without noise or blasting, dust or movement of heavy bulldozers and dump trucks. The vessels include many modern ships, much like any other merchant vessels and are British registered. They land their cargo unobtrusively at wharfs in our coastal towns and cities close to the main points of need thus minimising lorry delivery mileage.

C. A Strategic Resource

14. Aggregates are essential for building houses, schools, hospitals and shops. The country's roads, railways and airports absorb huge volumes of aggregates and these are also used in water and sewage systems, sea defences, agriculture and leisure facilities.

15. Aggregates, quite simply, provide the foundations for the maintenance and growth of our society. Ever improving standards of living mean that use of these materials is at a higher level than ever before and is predicted by the DOE to continue to increase.

16. Each house built requires between 50 and 60 tonnes of aggregate. Each mile of motorway can use up to 200,000 tonnes of aggregates. Each mile of new main line railway will use 70,000 tonnes. Although there has been a slow trend of increasing use of crushed rock for some years, sand and gravel will remain the favoured minerals for concrete and mortar production. In the South East of England, where there is virtually no indigenous rock, sand and gravel will almost certainly remain the main mineral used.

17. Land won sand and gravel production—gravel pits—has long been a controversial activity in lowland England. Gravel pits are unpopular neighbours and the level of concern arising from the latest Government estimates of aggregate demand until 2011 and what this might mean in terms of an increased number of gravel pits, has been enormous.

18. Government policy currently explained in Minerals Planning Guidance Note 6 (MPG 6) has always been to encourage marine dredged sand and gravel extraction as a way to lower the pressure for more land based gravel digging. This policy remains and is likely to be strengthened.

19. Put quite simply 20·7 mt of marine dredged sand and gravel—the 1989 production—means about 400 hectares of land saved from sand and gravel digging each year.

20. Put another way it means one million lorry loads not being moved into towns and cities from gravel pits perhaps 20–30 miles away. It also means perhaps 40 fewer pits each year requiring a further vast round of lorry movements to backfill and restore them.

21. When marine dredged sand and gravel is landed at the wharf this is usually close to the town or city centre—close to the building site and thus involving far, far less lorry delivery mileage than the land based industry. Increasingly, some marine wharves are also being linked to the rail system to further reduce traffic generation.

D. The Government View Procedure

22. The current Government View Procedure for licensing marine dredging was revised in 1989 following extensive consultations with all interested parties, including the industry. The marine dredging industry had many reservations about the new system in particular:

(i) the lack of a strict timetable to apply to the entire procedure.

(ii) the automatic ability of any government department, Such as MAFF, to veto applications.

Neither of these reservations were accommodated in the agreed revised procedure.

23. Also included in that original industry response was a caveat that if the new procedures had been tried for say, three years and the hoped for improvements had not materialised then the industry believed that a statutory local inquiry system should be considered with the Secretary of State for the Environment becoming the final and overriding arbiter.

24. The current industry position is that the current Government View Procedure has failed but that it is too soon to scrap it.

25. None of the 15 or so applications to dredge made to the Crown Estate under the new system for England has reached the DOE and thus the bulk of the new procedure has yet to be properly tested.

26. All of these current applications are stalled at the Crown Estate due to the complications and apparent delays occasioned by what the Crown calls the—" pre-consultation process ". The Crown Estate are required to carry out these consultations which will then have to be repeated again by DOE under the Government View Procedure. This pre-consultation seems to the industry to be unnecessarily complicated and open ended and is an important concern in the operation of the current system.

27. There is a further problem with the current procedure and that relates to the need or otherwise to accompany applications with an environmental assessment to comply with the EC Directive. Whether this should be submitted as a matter of routine remains unclear, as does which is the competent authority to decide whether such an assessment should be required and whether when submitted it is satisfactory. An application which actually reached the Welsh Office for a Government View has left unresolved the question of who is the competent authority.

28. The industry therefore believes that the current procedure should for the present not be replaced but rather amended on the following lines:

(i) the pre-consultation procedure carried out by the Crown Estate should be carried out in partnership with the applicant, greatly simplified and severely time-constrained.

(ii) the total procedure should be strictly timetabled, ideally a six month maximum deadline.

(iii) the DOE role should be enhanced with power to override other dissenting government departments.

(iv) the coastal planning authorities should not have a role to play in this stage of the process.

E. Conclusions

29. Government policy is, as explained, quite rightly to encourage marine dredging of sand and gravel to relieve environmental problems of land based extraction. The industry has made very substantial investments in ships and wharves in support of this policy. The industry does not believe their investment on the basis of that policy has been honoured—the Government View Procedure is not working for the reasons explained.

30. Notwithstanding all the problems with the actual procedure the simple additional fact is it is getting harder and harder to get licences. Objections from fishermen, marine and land based conservation groups get ever more vociferous and coastal planning authorities are raising irrational fears about coastal erosion. Concerns about the effect of marine dredging on coastal erosion are common but ill founded. All dredging licence applications are thoroughly and scientifically scrutinised by the Hydraulics Research Station (HRS) and only recommended for approval if there are no doubts. Fears of the fishermen at the effects of marine dredging on the benthos are also raised. However, it is fair to say that fishermen cause far more disturbance to the sea bed than the dredging industry and their operations should be subject to the same degree of scrutiny if a balanced assessment is to be made.

31. This evidence has been necessarily brief. The industry would be more than happy to support it with appearances before the Committee if that would be helpful.

Examination of witnesses

MR BRIAN WHEELER, Chairman of BACMI Marine Panel; MR MICHAEL BROWN, Chairman of SAGA Marine Section and MR BOB SAMUEL, Member of BACMI/SAGA Marine Panel, examined.

Chairman

538. I welcome to this session of our Inquiry into Coastal Zone Protection and Planning representatives from the British Aggregate Construction Materials Industries and the Sand and Gravel Association. Mr Wheeler, as you are sitting strategically in the middle, perhaps you could introduce yourself and those you have brought with you.

(Mr Wheeler) Thank you, Mr Chairman. I am Managing Director of ARC Marine and Chairman of the BACMI Marine Panel. With me I have Mr Bob Samuel who is Director and General Manager of South Coast and East Coast Aggregates, a member of SAGA, and Mr Michael Brown, who is Chief Executive of British Dredging and NorWest Sand and Gravel and Chairman of the SAGA Marine Panel.

539. Thank you also for the helpful memorandum that you have submitted to us. If I may I should like to start immediately by going to that and asking you one or two questions concerning it. First I draw your attention to paragraph A.2. You define " coastal zone " as the inshore area and then make the assertion that it is commonly recognised as such.

From the evidence that we have been taking so far witnesses tend to regard the coastal zone as somewhat wider than that and, in fact, they cover the area which you say is not part of the coastal zone; that is the area offshore where you carry out your operations. I wonder what basis you have for that rather sweeping assertion that the coastal zone is " commonly recognised ", to quote you, as the inshore zone?

(Mr Wheeler) On reflection perhaps we should not have used those words. The perception of coastal zone varies considerably. I think it is fair to say that we in the industry have referred to the coastal zone in that way over the years, in that the newer licences have been moving further and further offshore and generally we have taken three miles as being the coastal zone and the area of particular concern and the new licences are outside that area.

540. I am glad to have made that correction, because as our inquiry concerns the coastal zone and if you tell us in terms that you do not operate in the coastal zone, you are, in fact saying to us that the inquiry does not really concern you at all?

(Mr Wheeler) I have to concede that point, Mr Chairman, on re-reading it.

Chairman *Contd]*

541. Thank you very much. I now turn to paragraph 30 where you tell us that the coastal planning authorities are raising irrational fears about coastal erosion. We have received evidence that the extraction of gravel has an effect upon sediment movement and that is something we shall be looking into, particularly with the Hydraulic Research people[1]. I wonder if you could help us by telling us what research you have funded that justifies your making the statement that extraction does not affect sediment movement?

(Mr Weeler) In evidence from other parties you have already seen details of the procedure that any licence application must go through. If the Hydraulics Research investigation is found to be negative in that they say it will have an effect on coast erosion, then the licence procedure stops. That is funded by the industry. Although the Crown Estate is responsible for it, the research is funded by the industry. So on each licence application Hydraulics Research is asked to give an opinion on the possible effects on the coastal zone and coastline.

542. So you rely upon Hydraulics Research to tell you in each and every case whether an intended operation is likely to affect sediment movement?

(Mr Wheeler) That is correct. If the report by the Crown says there is likely to be some danger or if there is a possibility, then the licence application does not proceed.

543. It is members of your two associations that pay for this research to be done?

(Mr Wheeler) Correct.

544. That is specific research into specific cases. Do you know whether any general research has been taken into this particular matter?

(Mr Samuel) I believe that the Crown commissions Hydraulics Research to carry out research in specific areas where there is a predominance of dredging. Another point that perhaps should be considered is that by and large the majority of our dredging is done in water depths exceeding 20 metres. I believe that the action of the sea on the sea bed in depths greater than 18 metres has little or no effect.

545. You say you believe that, is that the result of published research documents, or just a general belief within the industry?

(Mr Samuel) I believe it was an area that was considered many years ago by Hydraulics Research.

546. So really we have to talk to Hydraulics Research to test that. One of the points that was made to us the other day was that the effect of works outside Yarmouth were determined by research work carried out off Worthing. Therefore we have to consider whether or not the results outside Worthing are valid for Yarmouth. We are laymen in this and we have to be told by people who know.

(Mr Samuel) The research that Mr Wheeler was referring that the industry pays for is not as a body, but as individual companies, in respect of a specific area that they are looking to obtain a licence in. It is not general research, it is specific research.

(Mr Wheeler) Can I elaborate a little on that, Mr Chairman? I take your point about Worthing and Yarmouth. I noticed reference to that in some of the evidence that has been given. It is important that these sites are investigated on a site specific basis, because it is true that the conditions in one area will vary greatly from those in other areas, but if you are meeting Hydraulics Research anyway, you will hear from the expert body.

Mr Steen

547. You talk about the irrational fears of coastal erosion. I represent the South Devon coastline. Are you saying that the dredging that has taken place off the coast of Devon has had no impact on the devastation that took place in places like Torcross, where the sea came in and destroyed the village, Hallsands, which is crumbling into the sea, another village Beesands, and all along that coast, as a result of massive dredging out to sea in the 1920s, 1930s and 1940s?

(Mr Wheeler) In direct answer to your question, that was not aggregate dredging anyway. It certainly was not under the current licence procedure. It was carried out for the construction of Devonport dockyard, so the Ministry of Defence has——

548. You know about that, do you?

(Mr Wheeler) Surely, yes. I suggest that is something which just is absolutely impossible now. It just would not happen.

549. Today?

(Mr Wheeler) Correct, today, and long before the procedures which are now very firmly in place.

550. Would you have a view about whether that level of dredging and extraction could have an effect?

(Mr Wheeler) I am not totally familiar with the details, but from memory that was carried out on shore. It was taken away by ship, but it was extracted from the shoreline.

551. And actually out to sea as well.

(Mr Wheeler) Yes, but very close inshore. Obviously I expect there would be some effect from an operation like that.

Chairman

552. Can you tell me what quantitative assessment you make of the effect of marine aggregate extraction, first on fishing and, secondly, on the flora and fauna of the sea bed?

(Mr Wheeler) At the moment there is no environmental assessment procedure as such. It is all part of the Government view procedure with the Department of the Environment, who will then decide whether an environmental assessment is required involving the questions you have just raised.

553. The Commission in Brussels will also have something to say about future environmental assessment of these matters, but so far you have not applied your minds to that problem?

(Mr Wheeler) The environmental considerations are taken into the procedure and MAFF are part of

---

[1] *See* Appendix 29.

Chairman *Contd]*

the consultation procedure, of course, as far as the Crown is concerned.

Mr Field

554. In paragraph 12(i) of the memorandum you submitted to us you say: " the depth of the deposit— current technology limits the maximum depth of dredging to 45 metres but in fact 25 metres is the present day norm ". Have I correctly interpreted that paragraph to mean that that is the depth of aggregate that you are actually dredging in the licensed area?
(Mr Wheeler) No, that is the water depth.

555. It says " the depth of the deposit ".
(Mr Wheeler) Yes, correct, I apologise for that.

556. Is it the depth of deposit?
(Mr Wheeler) It is the depth of deposit, it is 25 metres below the surface to the sea bed. We would describe it as the thickness of the deposit if we had been talking about thickness, but we were talking about the water depth.

557. Can you help the Committee by telling us what thickness of deposit you normally extract in terms of vertical measurement?
(Mr Wheeler) It would vary from licence to licence.

558. But what would be a fair figure?
(Mr Wheeler) Two to three metres maximum.

559. Can I take you to " Conclusions " on page 5, paragraph 30 and your last sentence in the penultimate paragraph says: " However, it is fair to say that fishermen cause far more disturbance to the sea bed than the dredging industry and their operations should be subject to the same degree of scrutiny if a balanced assessment is to be made ". I do not know of any fishing method that disturbs the sea bed to the depth of three metres.
(Mr Wheeler) I would accept that on the specific licensed area. But we are referring there to the surface of the sea bed related possibly to the Chairman's question in terms of benthos and in terms of disturbing the surface of the sea bed. The fishing industry obviously have no restrictions and their disturbance of the sea bed is far greater than ours.

560. When you are involved in extraction, a large amount of sediment goes over the side of the ship as the water runs off from the action of suction. Would you agree with that?
(Mr Wheeler) What do you mean by large? There is a wash out in the fines.

561. There is a wash out in the fines as it comes up the suction tube. The heavy materials are deposited in the hold and the rest runs out in some ships through a scuttle and in others just over the side in a washing effect. Very often if the tide is running there is quite a streak downstream of that suction arrangement where the sediment is falling out upon the sea bed. That could well be outside the licensed area, could it not?
(Mr Wheeler) Certainly, but it relates to exactly the effect that beam trawling would have when they are moving across the sea bed.

Mr Howell

562. You say in paragraph 24 that the current Government view procedure has failed, but that it is too soon to scrap it. Can you elaborate on that a little?
(Mr Wheeler) Yes, in 1987 and 1988 there was considerable debate among the Crown Estate, Department of the Environment and the industry about quite major changes in licensing procedures and the Government procedure, because the industry was about to embark on a reinvestment programme of replacing much older vessels—in fact in line with what I have just said, moving further offshore. We were very concerned that since the new licensing procedure came in on 1 January 1989 there seems to have been a blockage in the procedure. The meeting was held on 6 December with the Department of the Environment and the Crown Estate to talk about what had happened and what had gone wrong. As an industry we were concerned that applications had been made and had gone into the system and appeared to be making little if any progress. The meeting on 6 December was very useful. It is minuted by the Department of the Environment and has helped to sort out some of the problems that have arisen with the new procedure in terms of consultation and various other aspects.

563. How do you explain your expression " it is too soon to scrap it "?
(Mr Wheeler) We were concerned that although as companies we made applications, it was not a question of applications being turned down or there being apparent problems with the applications, but just that they had disappeared and we got no reaction about whether there were problems or what should happen. The meeting was to clarify why this had happened and the Department of the Environment said at that time that they had not had any applications from the Crown to put through the Government view procedure to test whether the mechanisms were adequate.

564. So now you are proposing that there should be some form of partnership?
(Mr Wheeler) What we were saying as the industry was that we would like to be involved at the informal stage when the Crown were carrying out the consultations with interested parties, so hopefully we can deal with things rather more quickly than waiting until this informal consultation has been carried out and then dealing with the problems all at once several months later. We suggest that that is a more constructive way of working.

565. In paragraph 27 you refer specifically to the question of environmental assessment of applications. How do you think the present unclear situation would be resolved and how often have environmental assessments been asked for with applications and by whom?
(Mr Wheeler) I think I answered a question from the Chairman earlier that I am not aware of an environmental assessment on an individual licence having been asked for. We certainly think the

Mr Howell *Contd]*

Department of the Environment must make that decision and we would appreciate some guidance on what information will be required for licence applications. That is one of the points of clarification we were also pursuing at the meeting on 6 December.

566. Your criticism of the Government view procedure in paragraphs 22 to 28 seems to contradict the opinion of the Crown Estate Commissioners who appeared to be quite content with the arrangements as they stand.

(Mr Wheeler) I should like to refer to the meeting that was called on 6 December which was attended by the Crown Estate Commissioners and the meeting was chaired by the DOE to address this very problem. There are minutes available which I am sure could be provided if they were considered of use to the Committee.

567. You were quite satisfied with that meeting?

(Mr Wheeler) I am satisfied that we seemed to have made progress. We still need to see the results of statements made, but I think we made some constructive progress.

Chairman

568. The Crown Estate Commissioners therefore are aware of your views?

(Mr Wheeler) Indeed.

569. Did they react to them in any way?

(Mr Wheeler) I suppose they reacted by attending the meeting with the Department and ourselves.

570. They made no comments on the views you were expressing because, as Mr Howell has said, it appears to us that the views you have expressed in these paragraphs is contradictory to the views expressed to us by the Crown Estate Commission. Therefore it is of some interest to us to know that now you have had this meeting and exchange of views whether they have modified their attitude in any way and what kind of reaction there has been, whether they accept what you are saying or whether they reject it?

(Mr Wheeler) We have not been party to them, but we understand that meetings have taken place between the Crown and Department of the Environment as a result of the 6 December meeting.

571. Before you make a licence application, do you ever think it worth while making your own investigations, or do you just identify an area where the right kind of aggregate happens to be and then make your application? What kind of procedures do you follow before you make your application?

(Mr Wheeler) It is important to understand that there are two parts to the application procedure. The first is prospecting. We have to apply to the Crown for the prospecting licence. Clearly the initial application at that stage is made on the basis of past knowledge, examination of charts and geological records and information from people such as BGS. If we do not get a prospecting licence we proceed no further. The prospecting licence then lays down certain terms and conditions. It is for a period of two years and within those two years we must carry out largely seismic work coupled with some very small grab sampling and the information then must be provided to the Crown, whether we proceed with the application or not. Then we would apply for a full production licence if we found suitable material.

572. What is the seismic procedure testing? Depth?

(Mr Wheeler) We are checking the presence of material, depth.

573. Whether there is enough to make it worth while?

(Mr Wheeler) Yes.

574. That is really what you are concerned with. Let us suppose—although you are not required to make an environmental assessment nevertheless it must occur—that while you are carrying out your prospecting, or even after you have done that, you suddenly find that you are on an ecologically rich area. There may be some unique marine specimens that are not found elsewhere. If you find that do you back off, or do you carry on?

(Mr Samuel) It is fair to say that when we commence considering an area, it is normally a much larger area than that for which we would ultimately expect to receive a production licence and there is prior consultation with the Crown and there are records available, for example, where an area may be a significant spawning ground or where there is a particular type of fish prevalent. The Crown would not allow even a prospecting licence to be issued if the present knowledge on that area suggests that ultimately, because of the fishing interest, a production licence would not be approved. There is quite a bit of consultation with the Crown before we do any physical prospecting.

Mr Field

575. In view of the fact that marine aggregates are a finite resource, what alternative materials or sources are you investigating as possible substitutes?

(Mr Wheeler) We sit here as the BACMI Marine Panel and most of us are associated with companies which also have land resources, so our responsibility as far as I am concerned—the other two may wish to add to this—is that we are concentrating solely on the marine side. As groups we have interests on land and alternative materials already.

576. You mention crushed rock in your evidence? What are you saying to the Committee is that you have no knowledge of this because you are entirely marine orientated?

(Mr Wheeler) No, that is not correct. I do have knowledge of that.

577. Are you able to tell us about it?

(Mr Wheeler) About crushed rock?

578. Yes.

(Mr Wheeler) Surely, yes.

579. Is it something that you are investigating? Will it provide a suitable alternative?

Mr Field *Contd*]

(*Mr Wheeler*) My company is the biggest producer of crushed rock in the United Kingdom, so the information is available if it is required. I did not realise that it would form part of the evidence this morning. I am quite happy to make it available.

580. What the Committee is trying to get at is to see whether there has been endeavour to avoid the need for land quarrying in sensitive areas and sea dredging for marine aggregates.

(*Mr Wheeler*) To follow that exactly, I would suggest that the people of Somerset would be very enthusiastic to stop rock quarrying and to take everything from the sea. There has to be a balance here.

581. Indeed, but I notice in your evidence that there is no mention of something we see quite often in the city here of portable mobile crushers producing on site clean hardcore. No 3 Marsham Street, providing we get the civil servants out first, would make some excellent hardcore which would save a lot of marine aggregate, but there is no evidence from you.

(*Mr Wheeler*) I apologise if we have misunderstood what was required, but we are quite happy to extend this to cover other materials. We are talking about approximately 20 million tonnes of material out of an overall market of something over 300 million tonnes and the two companies here are certainly involved in alternative materials.

582. Is it something this Committee in its environmental hat should be looking at? Should there be encouragement to save building materials? For example when Marsham Street is demolished, would it yield a suitable supply of material which would lead to a reduction in the amount of marine dredged aggregate that would have to go into the building that replaces it?

(*Mr Wheeler*) With great respect, there is a Government review on expected increased demand under way at present. We have been asked by the Department of the Environment whether we can increase our production because of the expected shortfall in land production and recycled materials through into the next century. I accept that everything must be looked at. If you are asking for a personal opinion, the more recycling we can do the better. I am all in favour of that, but I do not think that will take away the demand for primary aggregates. It has to be a balance.

Chairman

583. Can I put the question another way, for example, comparing this with other areas of environmental policy that we have looked at? Let us take energy and the burning of fossil fuel and all the problems that are associated environmentally with the use of fossil fuel as an energy source. There is a great demand now and a great need for research for alternative fuels. Perhaps this has not yet arisen in your industry. You have not considered whether or not your operations damage the environment in the sense of the marine environment. You may or may not. I am not passing a judgment. But have you carried out any research into whether there are alternative supplies of materials, not only the one mentioned by Mr Field, but other possible sources of material that might replace aggregate so that you can leave the marine development undisturbed. For example, I believe the Japanese—I may be incorrect in this—who incinerate vast amounts of their waste because they do not have land fills, use the by-products of incineration. They consolidate it and then use that as a building material. Are you as an industry carrying out any research into these other possibilities or is it a case of " We are in the business of recovering aggregates, whether it is from the land or the sea, that is our business and therefore we are not interested in finding other materials as a subsitute "?

(*Mr Wheeler*) The industry uses recycled materials. It has been a growing part of the industry for, for example, road pavements and concrete production. Taking Mr Field's point, when it is easier to collect demolition material—rather than in country areas where it might involve trucks running all over the place which environmentally may not be such a good thing—in places such as major cities the percentage of recycled material is growing. I make the point that because the market seems to be growing the demand from society for aggregates is growing faster. It still will not avoid the demand for primary aggregates. The Japanese, for example, dredge over 60 million tonnes of marine sand. It is the largest marine aggregate operation in the world. I suggest that environmentally their controls are totally non-existent.

584. That is very interesting.

(*Mr Samuel*) A point worth noting is that recycled material has its limitations. you cannot use recycled concrete or recycled brickwork to produce quality concrete. The extent to which recycled materials can be used in strength materials I would not like to say, but I would suggest it would be small. It could never be a replacement for the prime material, even if it is used for recoating for road surfacing. You are limited to the quantity of recycled materials you can use relative to new material.

Mr Field

585. I was aware of that, Mr Samuel. That is what I was hoping to get Mr Wheeler to do, to try to indicate to the Committee the limits that there are in recycling existing material.

(*Mr Wheeler*) Perhaps I can add to that. It is further complicated by specifications and technical requirements becoming tighter and tighter over the years for construction which has made the use of primary materials even more important.

586. Can you tell us what research you undertake to estimate the available reserves of United Kingdom marine aggregates and what account is taken of the demand for aggregate within the Government view procedure?

(*Mr Wheeler*) The estimating is from the Department of the Environment. They carry out that exercise. We come in after that. We do not actually do the investigation as far as the Government

Mr Field *Contd]*

publication of expected demand. The MPG6 process which deals with that is currently under review at the moment. All the papers are with regional working parties consolidating the assessment of demand into the next century[1].

587. So the Department of the Environment do not approach you at all?
(Mr Wheeler) Yes, they do.

588. In terms of you are the suppliers. They only approach the customers. Is that right?
(Mr Wheeler) I apologise for that. They provide the indication of demand and we are then involved in discussions with them. I have referred to that earlier about what quantities would be available within that demand expectation.

589. So the demand figures are entirely theirs, are they? You play no part in the demand figures? What about a company such as Tarmac that has both quarries and construction arms? Do they straddle the fence, so to speak?
(Mr Wheeler) There is consultation on the figures produced and they may well be amended before they become official Government policy, but the initial demand figures, as far as I am aware, are from the Department of the Environment.
(Mr Samuel) I would imagine that companies would assist the Department of the Environment in the preparation of the figures, but they cannot take an overall view in the same way as the DOE can.

590. What is the quality of them? Is it like which one is going to win the Derby? Is it a Treasury forecast? What is it in your opinion?
(Mr Brown) In my opinion it is from the aggregate working parties around the country who are normally attended by the DOE anyway, local planners and people from our industry, to try to get an idea of what the local structure plan is and what the requirement will be locally and therefore where the minerals will come from. That goes back to the DOE.

591. You are digging the stuff out. How often are they right?
(Mr Wheeler) Demand has exceeded forecast in the last few years.

592. Have they always underplanned?
(Mr Wheeler) Yes, that is an interesting thing.
(Mr Brown) At present we are talking about a 40 per cent increase to the year 2010.

592a. How long have you been in the industry?
(Mr Wheeler) Over 20 years.

593. And they have always underplanned?
(Mr Wheeler) No, I do not think it is fair to say that they have always underplanned because there are regional variations and there is a policy laid down

by the Department of the Environment that there should be a minimum 10-year reserve on land planning.

594. Yes, that would be nationally, but regionally you said that it comes out of it by a regional——
(Mr Wheeler) With respect it is not the DOE that is doing the planning. If the local authority refused to plan the land, you cannot say, surely, that the DOE have underplanned.

595. Mr Brown just indicated that it was subject to local plans.
(Mr Wheeler) I have just confirmed that.

596. Can you put a figure on the reduction in sea dredged aggregate as a result of the recession in the building industry? What is the percentage decline?
(Mr Wheeler) I have not seen the figures for 1991.
(Mr Brown) I do not think they are out yet.
(Mr Wheeler) I think the Crown are the only people that would have the overall figures of dredged materials.

597. Your members do not have the figures?
(Mr Wheeler) We do not collect those figures.
(Mr Brown) Only individual companies.
(Mr Wheeler) Individual companies would have their own figures.
(Mr Brown) I could tell you 21 per cent.

598. 21 per cent of what?
(Mr Wheeler) I would think it is something like 15 to 20 per cent overall, yes.

599. Can you tell the Committee what the cost differences are in terms of imperial ton between a sea dredged aggregate and a land won material?
(Mr Wheeler) I apologise. I do not wish to sound facetious, but unfortunately, if we are not competitive we do not sell anything, so it has to be the same or cheaper in the location in which it is landed. In other words, there is no premium for marine materials.

600. So it is always more competitive than land won material?
(Mr Wheeler) It must be more competitive in the area where it is landed, yes. Marine material is a very small part, some 15 or 16 per cent of the overall sand and gravel and only about 8 to 9 per cent of total aggregates, so it tends to be very site specific, Southampton, London, the sea ports that can accept the ships and the material tends to be marketed in about 20 or 30 kilometres.

Chairman

601. Transport presumably then comes in?
(Mr Wheeler) Plus the cost of transport, yes.

Mr Field

602. There has been quite a consolidation in the number of companies owning land, aggregates and materials in the last 20 years. It used to be a very diversified industry. Do you agree with that?
(Mr Wheeler) Yes, that is true.

[1] *Note by witness:* We would like to clarify that the industry provides the DOE and the CEC with estimates of reserves within their licensed areas based on full resource assessment surveys carried out at their expense taking into account actual production. The DOE and the Crown Commissioners collate this information to provide the total industry reserve assessment.

Mr Field *Contd]*

603. It has really been consolidated, so how do the Government avoid companies speculating in their land reserves by not extracting them and increasing the extraction of marine aggregate?

(Mr Wheeler) I am not aware in a democracy that there is any way you can avoid it.

604. Do you think it is happening?

(Mr Wheeler) Speculation?

605. A desire to extract from the sea in preference to the landward base because the price, with scarcity, will go up over the years and it costs nothing to keep it there?

(Mr Wheeler) I doubt that very much indeed, because the marine industry is so long term relatively that with the heavy capital investment in ships, certainly with a minimum 20 year life, I suggest the money would be better speculated on the stock market than it would in aggregate prices over a 20-years period.

(Mr Samuel) If we are talking about sand and gravel, I think it is fair to say that the capital that a company would tie up in land would be so great that it would not be a realistic proposition.

606. A lot of them are historic acquisitions, are they not?

(Mr Wheeler) On sand and gravel I think that is very doubtful quite honestly.

(Mr Brown) There are one or two places in the country where people have paid so much for a deposit in the ground that they are not in a position to use it until prices alter. I have heard of those, but I have not heard of anybody deliberately holding on to a deposit, other than for that reason.

607. Can we now talk about licensing? You state that your investment in ships and wharves on the basis of Government policy to encourage marine dredging of sand and gravel to relieve the environmental problems on land-based extraction has not been honoured. You say this in paragraph 29 of your memorandum. Where have the Government stated that that is their policy, and what evidence do you have to prove that licences are becoming harder to obtain?

(Mr Wheeler) Can I take the second part first, that licences are hard to obtain? Our concern really is what has been touched on earlier. The Government view procedure was changed and the new licensing procedure came in in 1989. But there then seemed to have been a block on the applications. Again I refer to the meeting that was held on 6 December, specifically to address that problem, which we hope will unblock the problems that we have had there in terms of obtaining licences. Again I believe the figures have been referred to in evidence given to you already, in the numbers of licences involved in that procedure. That is why the industry requested the meeting and that is why we were extremely concerned after the considerable debate between the industry and the Crown and others in 1987–88 which led to this new procedure. During that period there was considerable investment, something in excess of £100 million in new shipping in this section of the industry by the British industry. We were very concerned that

the procedure did not seem to be working. As you said earlier, it is a finite resource and as material is worked out we need to be replacing this material to ensure that the continuity is there.

608. Can you direct the Committee to the Government's policy on this point?

(Mr Wheeler) MPG6, I think, the then Minister for the Environment, Christopher Chope in 1988, perhaps I could let you have a copy of the announcement. I am not sure I have one with me.

609. Can you remind us of what he said?

(Mr Wheeler) "To encourage the use of marine aggregates", comes to mind, but perhaps I could let you have a copy.

(Mr Samuel) If it is in order to refer to the evidence that was given by the DOE, section 46 of their evidence specifically spells that out.

610. From what you have already said, Mr Wheeler, you are not going to be able to answer my next question because you said you have come equipped only to deal with marine aggregate today. But on the landward side aggregate quarries and pits are pre-licensing or post-licensing, depending on when they were first extracted from. Would you be able to supply the Committee with advice on the number of licensed pits and unlicensed pits?

(Mr Wheeler) I will ask. I am not sure whether it is all that easy to split as you have suggested there, but perhaps I can take notice of that question and try to come back[1]. Are you talking pre-1948 planning?

611. I am. There are quite a few of those, I believe.

(Mr Wheeler) I will come back on that. That is a matter that is being addressed at the moment, of course.

612. Is it true to say that in some of those pre-1948 pits the companies are now going back and seeking new levels of extraction from them?

(Mr Wheeler) I am not aware of that. I am aware that the Government are changing them. I think these were under industrial development orders, IDOs, and I think there is a closing date that is going through the consultation procedure with the Government at the moment that that will be discontinued[2].

(Mr Brown) I think the closing date is the end of March. If you have an old deposit you must register it by then, or you lose any right whatsoever.

(Mr Wheeler) Which then must completely comply with the Town and Country Planning Act.

(Mr Brown) So anybody who has any doubts I would presume is trying to register.

---

[1] *See* DOE News Release No. 265 dated 8 May 1989 (not printed).

[2] *Note by witness:* The estimated number of working pits and quarries during 1990 was 1,400, but analysis is not available as to whether these were operating under the Town and Country Planning Act or initiated by the IDO procedure. The DOE is currently engaged in an investigation, where all IDOs (covering many aspects) must be registered by 25 March 1992 via local authorities. Only when this is completed will it be possible to establish precise details of IDOs relating solely to mineral operations.

Mr Field *Contd]*

613. For example if you take the Sussex greensand belt, many of those are extracted down to 90 or 100 feet. Now companies, Redlands and so on, are coming back with new technology and saying that they are prepared to go down another 100 feet, subject to the aquifer. Those were pits that were closed some time ago and possibly were going to be subject to infilling using refuse. Suddenly they seem to have a new life, as the price of aggregate moves up.
(Mr Brown) And perhaps increased technology in going deeper and creating a greater future for the tipping of refuse.
(Mr Wheeler) May I ask a question through the Chair? It is the number of licensed and unlicensed pre-1948 pits that you are interested in?

614. I am interested to know how many of those are still active that were unlicensed, if it is possible to get a figure.

Chairman

615. Perhaps you could write to the Committee if you have the figures, if you do not have them off the top of your head.
(Mr Wheeler) I believe they will have been set out during part of the joint exercise with the IDOs.
Chairman: Perhaps the Clerk will be in touch with you about further information for the Committee that you may have.

Mr Howell

616. Despite ministerial reassurances, many of my constituents think that coast erosion is aggravated by the extraction of aggregates from off Happisburgh, Eccles and Sea Palling. I wonder if you can say anything which will convince me that your activities are not causing problems. What distance out is the extraction taking place?
(Mr Brown) We were talking earlier about the coastal zone. Certainly one can say that the largest proportion of dredging takes place beyond the three-mile limit and most of it within the three to 12-mile limit. I do not have the precise figures, but I would suggest that it is in excess of 90 or 95 per cent.

617. Can you let me know what is happening in that area?
(Mr Wheeler) Perhaps I can come back to you on that, Happisburgh and Eccles?

Mr Field

618. Is it still a fact that because a quarry is a wasting asset you get special tax treatment on it, as opposed to your capital investment in ships which you have to write over the normal capital allowance period?
(Mr Wheeler) The second part I confirm, the first part I think you are correct, but perhaps I can come back to you on that.

619. So it is a fact that the Government's taxation policy——

(Mr Wheeler) A depletion allowance[1].

620. ——actually favours landward extraction as opposed to seaward extraction?
(Mr Wheeler) On that basis that is quite correct.

621. If I am correct?
(Mr Wheeler) Yes.

Chairman

622. Arising from your answer to an earlier question of mine when I was asking about the effect of dredging on the sea bed, you told me that the Crown Commissioners were very assiduous in making sure that you did not carry out operations where you might disturb the breeding ground of a particular kind of fish. But the Crown Commissioners are responsible for only some 50 per cent of the coastline overall. Do you confine your dredging operations to only Crown Commissioners' land or do you go outside that to where it is in private ownership, for example? What are the discussions that take place with the private owner over the questions I have been addressing to you?
(Mr Wheeler) I am not aware of any of the members of SAGA/BACMI working at the moment other than with Crown Estate licences. I have been corrected, there is one, my company—I am sorry about that—the Duchy of Cornwall, although it is a very small one. There are other licences with the Duke of Beaufort in the Severn Estuary, but those are operations carried out not by our members.

623. If you are relying upon the person who allows you to dredge to take all the necessary precautions and say that you can or cannot dredge, but do not take any yourself, much depends on how interested the licensor is in these matters as to whether or not it is taken into account, because you do not take it into account—if I understand your evidence correctly.
(Mr Wheeler) If I may I should like to expand on that. That is part of the Government view procedure. The Crown are our landlords, but there is a consultation procedure which involves all interested parties. We are back to the environmental question. MAFF and all others will comment on the application before it goes back to the Crown and before they even consider granting us a licence. We were saying at the prospecting stage that there was not as much detail.

624. On land we have all kinds of restrictions, areas of special scientific interest, SSSIs and all the rest of it where everybody is up in arms to defend, if only to protect one rare breed of butterfly. When it comes to the marine environment we have not had, except for one or two exceptions, a marine SSSI, as it were. Where the interest is taken by the Crown Estates or by MAFF, it is over fish breeding grounds and they are concerned with the commercial impact on that in so far as the fishing industry is concerned.

[1] *Note by witness:* We have checked with the Crown Estate Commissioners who have confirmed that marine aggregate extraction in the Happisburgh and Sea Palling area is taking place at least six miles offshore. The licence concerned provides a very small quantity of material annually.

Chairman *Contd]*

But I do not know, but there may be some rare kind of sea urchin or whatever that is to be found only in one particular spot and that nobody has troubled with so far. When we get into the world of environmental impact assessment studies, as you are afraid you will be moving into, then I am afraid the rare sea urchin or the marine equivalent of the natterjack toad and all these things that are defended so vigorously on land you may find come into your thinking in a way that they have not done up to now. That is a consideration that may affect your work, your industry and your profitability. I do not know to what extent you are giving thought to these matters and incorporating those into your corporate plan thinking over the next five to 10 years.

(Mr Wheeler) Perhaps I might correct something that you have suggested there, with respect, Mr Chairman. We are entirely aware of the environmental interests. That is why we have asked the DOE to give us guidance on the subject, what they want and whether this will become a normal part of every application or whether there will be regional environmental considerations of locations as distinct from individual——

625. I think you ought to be talking direct to DGXI in Brussels as well, if I might with respect give you some gratuitous advice.

(Mr Wheeler) Thank you, that has helped to concentrate the mind. I come back to the point that has been made in the paper, if we are talking about the environment and the sea bed and damage to the sea bed, then the fishing industry will come into that assessment as well. Our use of the sea bed is about 0.02 per cent of the North Sea compared with the fishing industry which covers about 50 per cent of the North Sea every year. So if damage to the benthos is to be judged everything must be in there and the assessment must be done objectively. There is no way we are trying to stop that. We are aware of it, but we would appreciate some guidance about how this should be gone about.

Mr Field

626. Perhaps I may take you back to the point that the Chairman was making about the fact that you leave it to the licence issuer to do the investigation concerning the environmental situation. Your own company holds licences from South Wight Borough Council.

(Mr Wheeler) No.

627. You do not. Members of your association hold such licences?

(Mr Wheeler) That is correct.

628. The Crown Estate told this Committee that licensed aggregate dredging by local authorities does not have to have any environmental investigation whatsoever and does not come within the Department of the Environment's remit.

(Mr Wheeler) I should like to attempt to answer that. Others may wish to add to my answer. What you are referring to are the powers that the South Wight Borough Council take under the Coast

Protection Act. As far as I am aware there is no question of them giving a licence without the Crown Estate being involved. In other words the Crown Estate could give a licence, but the South Wight Borough Council, under the Coast Protection Act, could stop that licence operating.

629. I am not sure that is what the Crown Estate's evidence was.

(Mr Wheeler) I may be wrong because I am not totally familiar with the procedure, but I am not aware that the South Wight Council can give a licence in its own right, which I hope answers your point. In other words, if there is no way the South Wight Council can grant a licence without the Crown involvement that will mean that the whole procedure will involve some assessment.

630. I have not had an opportunity of reading the evidence yet because I do not believe we have received it yet. I am fairly certain my recollection was that the Crown Estate indicated that there are a number of coastal authorities who have the ability to issue dredging aggregate licences under the Coast Protection Act 1949, as you rightly say, and that there is no reference to the Crown Estate or to it being within the Government view procedure.

(Mr Wheeler) I apologise if I have that wrong, but I believe it is the case that all they can say is that under the Coast Protection Act they will control the operation of that licence. I do not think they can issue a licence in their own right. Perhaps we could contact the Crown Estate to obtain clarification on that and let you know. But that is my understanding.

Chairman

631. There is just one other matter I should like to raise with you. Throughout your paper you talk about the tonnes per annum that are dredged from the sea and how those compare percentagewise with other sources. You talk about the maximum and normal depths of dredging, but you do not indicate in terms what area of the sea bed is disturbed annually by your operations. The only indication that I seem to have is in paragraph 19 where you say that the dredging from marine sources " means about 400 hectares of land saved from sand and gravel digging each year ". I am not sure I would be safe in saying that that means 400 hectares of marine beds are disturbed each year because a lot depends upon the depth of your digging and the depth of your dredging. It may well be that you have a thinner layer at sea than you do in gravel pits. Therefore we are talking of a larger area than 400 hectares in so far as the sea bed is concerned. Can you put a figure on the sea bed in terms of hectares or acres?

(Mr Wheeler) Can we come back to you on that?

632. By all means.

(Mr Wheeler) Some figures have been published as part of a SAGA bulletin and also in the report of the International Council—which the Department of the Environment is about to publish in March— which has been looking at marine aggregate dredging. I suggest that would be quite a useful

Chairman *Contd]*

document as well. The SAGA bulletin suggest 0.2 and in the ICES report it says 0.3 of the sea bed of one fishery square is used each year[1].

633. Perhaps you would like to check the figure and drop a note when you are writing to us giving the approximate average area each year.

(Mr Wheeler) I should like to expand on that. Our ship operation is transient. The licence for half a million tonnes a year would see a ship there only twice a week for about three or four hours. At other times it is open to fishing interests and other sea users[2].

634. Will there be any problems in dredging for marine aggregates from greater water depths than the normal at the moment?

(Mr Wheeler) The new generation of dredgers are equipped to dredge down to 45 metres, which Mr Field mentioned earlier. That very crudely would cover all the southern North Sea area and most of the English Channel as currently operated.

635. Currently you do not operate up to 45 metres, do you?

(Mr Wheeler) Yes, some of us do, now we have moved further offshore.

Mr Field

636. Could you supply the figure to the Committee of how many of your members' ships fit a black box?

(Mr Wheeler) At the moment the answer is none because the requirement is coming on 1st January 1993. There is an ongoing debate on the type of equipment to be fitted and the information to be collected. It will be part of our licence procedure.

---

[1] *Note by witness:* It is estimated that approximately 20 square kilometres per year of sea bed is affected by aggregate extraction. This is based on 20 million tonnes production, at less than 0.75 metre depth. It compares with a total sea bed area on the English and Welsh Continental Shelf of 350,000 square kilometres, of which 1,652 square kilometres is licensed for aggregate extraction (effectively only disturbing 1 per cent of the total licensed area annually). The ICES report states that marine aggregate extraction disturbs 0.03 per cent of the sea bed annually in the North Sea whilst fishing disturbs of the order of 55 per cent.

[2] *Note by witness:* The CEC have confirmed that local authorities cannot issue licenses for extraction of sand and gravel under the Coast Protection Act without the Crown Estate's consent as landlords.

637. Do you know how many aggregate dredging vessels there are on the British Register?

(Mr Wheeler) Approximately 50.

638. Are there any that operate in United Kingdom waters that are not on the British Register?

(Mr Wheeler) Yes.

639. How many of the total number of ships operating are members of your association?

(Mr Wheeler) We say about 97 or 98 per cent in tonnage terms.

640. And that includes the non-British flag ships, does it?

(Mr Wheeler) Yes, there are only one or two. We could supply a list of those ships in the British fleet if that would be useful and thought appropriate.

Chairman

641. Thank you very much. I do not think there are any other questions that we need trouble you with this morning. I am very grateful to you for your frank answers to some rather penetrating questions, but we have to ask these and then arrive ultimately, I hope, at a balance between the evidence we receive from the industry and the evidence we receive from the environmentalists and others. That is our task. We have to evaluate it all and we hope to hear from Hydraulics Research and other people about the impact upon the matters that we have been talking about with you this morning. We are grateful to you for your help. I can foresee problems for your industry with the rising profile of these green issues. When you are looking at your corporate plans I think that looking for alternative sources of basic materials you want to see would be something that would not be a total waste of time for you. I am not saying that that is what we ought to do, but I think that is what is going to happen.

(Mr Wheeler) We thank you for the opportunity. I would like to say in closing that we are aware. There is a North Sea Task Force under the auspices of the European Commission which is also interested in our activities and many others. They have just published another document, so we are very much aware of that.

Chairman: Thank you very much indeed for your help. We are most grateful.

Memorandum by the British Ports Federation

1. INTRODUCTION

The British Ports Federation is the only representative voice for the United Kingdom's ports, and includes in its membership over 130 of Britain's commercially important ports. In the course of 1990, the United Kingdom handled 494 million tonnes of cargo and 30 million passengers. An estimated 32,000 people are employed within the industry with a further 200,000 indirectly employed.

As an industry which by its very nature is situated at the interface between land and sea, ports occupy a special place within the coastal zone environment: the abundance of statutory controls combined with government and industry-led initiatives aimed at protecting the coastal zone environment demonstrate that the industry is fully aware of its responsibilities and indeed is already strictly controlled in many of its activities. It should be remembered that port authorities have statutory requirements placed on them to provide for the safe navigation of shipping. Their role, therefore, is unique and distinct and requires special consideration when examining current and future coastal policy.

1.2   The Committee's inquiry focuses on five issues: the Federation's comments will concentrate on two of these: planning policy, and the authorities responsible for coastal zone protection and planning. In particular, we would like to focus the Committee's attention on the following fundamental points concerning the port industry and the environment:—

— that regulatory and planning authorities give full recognition to the development needs of ports, their importance to the United Kingdom's commercial success and the major contribution they have already made to the coastal environment and its management

— that legislation on coastal management issues, whether current or proposed, operates efficiently, clearly and fairly.

2.   CURRENT REGULATORY AND PLANNING PROCEDURES

This section is intended to highlight the measures which ports already take, often at great expense, to protect the coastal environment. The ports industry is controlled by a variety of regulations and statutes which have their origin in international agreements and Community and United Kingdom law.

2.1   INTERNATIONAL CONTROLS

The International Maritime Organisation (IMO), an arm of the United Nations which deals with maritime matters, originates regulations in the form of conventions, protocols and agreements. These are in effect international treaties: Governments which ratify or accept them must make their own national laws to conform with their provisions. The IMO has a committee which deals with the marine environment (Marine Environment Protection Committee) which submits recommendations to a conference of all members of the United Nations. Any treaties that relate to or affect ports which are ratified or accepted by the United Kingdom Government must ultimately be adopted by British ports.

An example of an IMO Convention adopted by British ports is the Marpol 73/78 Convention, otherwise referred to as the International Convention for the Prevention of Pollution from Ships. This directs that ports should ensure the provision of waste reception facilities for ships, thus preventing the discharge of oily wastes, noxious substances, litter (including food waste), and eventually sewage while at sea. As one can recognise, the provision of such facilities is costly to the ports in terms of both manpower and money: ports cannot always pass these costs on to shipping companies for fear of driving them away to other ports—especially those on the continent where such facilities are generally provided with state financial assistance. The Convention has been implemented in the United Kingdom by the Prevention of Pollution (Reception Facilities) Order 1984, the Control of Pollution (Landed Ships' Waste) Regulations 1987 and the Control of Pollution (Landed Ships' Waste) (Amendment) Regulations 1989.

The European Commission also issues directives on environmental matters with which the United Kingdom ports industry must comply. For example, regulations have been made by the United Kingdom Government implementing Council Directive No 85/337/EEC which requires certain public and private projects—including harbour works—to be accompanied by an environmental impact assessment.

British ports will also be affected by a further Council Directive currently under discussion which seeks to establish a voluntary environmental protection system on European businesses. This would entail the use of environmental auditing techniques to set environmental targets for individual sites; this might include, for example, pollution control measures to further reduce emissions. Guidelines for this draft directive suggest reducing pollution levels below those currently permitted.

Other internationally established environmental controls and standards imposed on the ports industry include those which regulate dredging and the disposal of dredged material at sea and on land. Dredging is vital for ports which are under a statutory obligation to provide and maintain safe navigation and adequate depth of water at berths. The main international bodies for dredging and disposal of dredged material are the Oslo and Paris Commissions and the London Dumping Convention (LDC). Their aim is to limit contaminants held within dredged material from being disposed of in the sea and on land. The LDC, for example, has a list of chemicals and chemical compounds which are thought to be hazardous or potentially hazardous and worthy of regulation. The United Nations' Environmental Programme (UNEP) with its Regional Seas Programme has also put forward conventions and guidelines for pollution controls in the marine environment and deals with disposal of dredged material at sea.

2.2   NATIONAL CONTROLS AND ENVIRONMENTAL INITIATIVES

This paper has already provided some indication of how the United Kingdom Government has implemented internationally agreed treaties and protocols. The following provides more detailed examples of United Kingdom measures.

ENVIRONMENTAL ASSESSMENT OF HARBOUR WORKS

In cases where environmental assessments are not required under normal planning procedures because, for example, the development has been specifically authorised by a Harbour Revision Order, regulations have been made which establish that harbour works are potentially subject to environmental assessment. In cases where private Bills have been submitted for harbour works, the Standing Orders of each House have been amended so as to provide for environmental assessment.

FOOD AND ENVIRONMENTAL PROTECTION ACT 1985

This Act provides a potential environmental safeguard on works below the high water mark; in such cases permission has to be sought from the Minister of Agriculture, Fisheries and Food, or from the Secretary of State for Scotland. Permission must also be sought from the Secretary of State for Transport, under the Coast Protection Act 1949, to ensure that such works do not impede the safety of navigation.

THE ENVIRONMENTAL PROTECTION ACT 1990

The Environmental Protection Act 1990 covers a comprehensive list of regulations to protect the environment, and is enforced by local authorities, port health authorities, Her Majesty's Inspectorate of Pollution (HMIP), the National Rivers Authority (NRA) and their regional counterparts. All processes prescribed by the Act for pollution control (which includes treating, handling, sorting and storage of particular cargoes) are currently being registered and licensed provided that the authorities concerned are satisfied that it does not pose a threat to the environment. There are significant compliance costs involved in this legislation which must be borne by United Kingdom ports. In addition there are costs associated with the register of certain processes and renewing licences after a given period.

The Act also covers the handling, storing, transfer and transportation of waste. This incorporates the Control of Pollution (Amendment) Act 1989 requiring that anyone carrying, collecting or transporting waste on a " professional basis " must be registered with a waste regulation authority. This complies, therefore, with Article 12 of Council Directive No 91/156/EEC. Ports, like other industries, are soon to be legally bound by the " Duty of Care " which puts responsibility on to the company managers to ensure that only registered carriers of waste are used, that all waste handling in their jurisdiction is done in a manner which will not harm the environment, and to ensure that those handling their waste beyond their property/land area is done without harm to the environment. Firms will be required to hold records of all transfers of waste within the United Kingdom which can be made available to local authorities whenever required to do so.

The Act also strengthens a local authority's power to act on complaints from the public concerning statutory nuisances, such as noise pollution; the industry is becoming increasingly aware of this potential restriction on the operations of ports.

COUNTER POLLUTION PLANS

In addition to controls on ports imposed by Parliament and international legislation, the Government has initiated a number of environmental protection measures in conjunction with industry. One such initiative involving the ports industry concerns the establishment of a framework for counter pollution plans.

In recognition of the potential consequences for the marine environment from marine accidents, the British ports industry working in conjunction with the Department of Transport's Marine Pollution Control Unit (MPCU), have agreed a framework for port action and preparedness for spillages of oil or chemicals. This builds on the development by individual ports of their own contingency plans.

3.   AREAS OF MAJOR CONCERN TO THE PORTS INDUSTRY

British ports are well aware of their environmental responsibilities and work within an ever increasing amount of legislation designed to protect the environment. However, our comments in this section are concerned with the operation of planning policies and ways in which they might be improved.

3.1 The process by which planning applications are considered is unnecessarily complex and time consuming. Some ports fall within the areas of more than one local or district authority with the result that up to 11 separate consents for the construction of a quay and adjoining berth have had to be required.

Furthermore, if a port wishes to work between the mean high and mean low-water tidal marks or on the seabed (to a distance of 12 miles from the mean low-water mark) it is likely that unless arrangements have already been agreed, or ownership of the foreshore and seabed is with the port, the port will need to gain permission from the Crown Estate Commissioners; this may involve buying land from the Commissioners or, more commonly, leasing. The process is both slow and increasingly expensive: in some cases reports have suggested that it can take in excess of 12 months to receive details of cost and planning consent from the Crown Estate Commissioners.

All these factors may hold back crucial harbour works, sea defences and other important development. The concern is that this situation may worsen as a consequence of the growing environmental awareness shown by the public and local planning authorities and, potentially, their subsequent increased involvement in the planning process. This situation is likely to become accentuated if, as a result of legislation (such as the Environmental Protection Act) local authority responsibility for the environment is increased.

Therefore, serious consequences could result if this delayed essential works being carried out by port authorities with responsibilities for coastal protection systems.

3.2   Matters are further complicated by the establishment of large areas immediately adjacent to ports designated as environmentally sensitive areas which are often accompanied by planning and development restrictions; these include, for example, Areas of Outstanding Natural Beauty (AONB) and Sites of Special Scientific Interest (SSSIs). Ports have developed in these areas because of their geographical proximity to urban centres, or sources of raw materials for industry, or simply because they formed natural harbours.

There appears to be little or no consultation between those who are already situated within these areas and the Nature Conservancy Council who are responsible for designating areas for protection. As a result, ports experience difficulty in establishing the precise boundaries of such areas and the development rights within them. The sheer number of organisations dealing with planning and environmental controls emphasises the need for proper co-ordination and communication between them. We have noted, for example, with great interest the proposals for the future reorganisation of the HMIP and the NRA. We will be commenting to the Department of the Environment on these in detail separately, but the proposals only serve to underline the importance of creating efficient regulatory and inspection bodies which can cope with the pressures placed upon them, bearing in mind the fact that their decisions have not only environmental consequences but far-reaching economic consequences as well.

The British Ports Federation would like also to recommend that developments do not take place in areas where an obvious incompatibility exists with existing land use and industrial activity. A major area of concern for ports in this respect is the encroachment of residential areas around working ports. Such development is both unreasonable for prospective residents and also for the ports themselves when a conflict of interest develops. The Environmental Protection Act 1990 gives added impetus to public complaints to local authorities on matters such as air quality and noise. Local authorities have increased powers under the Act to obtain abatement orders and stop any activity which constitutes a statutory nuisance. Such problems would not arise if residential development had been prevented from being built around the ports. Planning policy guidance should be formulated which advises against incompatible development.

4.   CONCLUSIONS

In summary, the British Ports Federation commends to the Committee the following points:—

1.   The designation of specially protected areas and the development, particularly in recent years, of residential areas around ports represent new pressures on port activity; there must be full consultation with port interests and full recognition of their vital economic role before far-reaching decisions are taken.

2.   The current system of obtaining planning consents can be time consuming and bureaucratic; improvements are required to remove the need for the duplication of applications.

3.   Coastal management and protection is the function of a number of organisations and interests, within which the ports are one factor; the ports industry supports the need for co-operative effort in tackling planning and environmental issues.

4.   The United Kingdom ports industry would support wherever possible new research into the marine environment and the dynamics of coastal change.

Examination of witnesses

MR DAVID JEFFERY, Chief Executive, River, Port of London Authority, MR PHILIP LACEY, General Manager, Shoreham Port Authority, MR JOHN SHARPLES, Managing Director and MR DAVID WHITEHEAD, Director of Policy, British Ports Federation, examined.

Chairman

642. Good morning, in this part of this morning's session we have before us the British Ports Federation. Mr Sharples, do I take it that you are leading the team? If so, would you be kind enough to introduce yourself and your colleagues?

(*Mr Sharples*) Thank you, Mr Chairman. I am John Sharples, Managing Director of the British Ports Federation. I have with me Mr David Whitehead, who is Director of Policy with the Federation; Mr David Jeffery who is a member of the Federation and Chief Executive of the Port of London Authority and Mr Philip Lacey, also a member of the Federation, and General Manager of the port of Shoreham.

643. I thank you also for the memorandum that you have submitted to us. As before I come directly to that memorandum. In paragraph 1 in the introduction you tell us that you represent over 130 of Britain's commercially important ports. What proportion of British ports are members of the BPF? Are there any major ports not represented? Is there any particular reason for that?

(*Mr Sharples*) We represent about 75 or 80 per cent of the ports in the country, if you measure it by the amount of tonnage of cargo that the ports handle, which is a normal measure. There is one group who are not in membership, those are the ports owned by Associated British Ports, which is a company. They left the Federation back in 1986 over issues related to dock labour.

644. So that is the only distinction between BPF ports and others, a historic schism, as it were, over a particular issue?

Chairman *Contd]*

(Mr Sharples) Yes, that is right.

644a. From your paper you seem to pay close attention to environmental matters and you support a responsible environmental attitude. Do you believe that bringing your activities beyond the low water within town and country planning legislation would enhance your environmental credentials?

(Mr Sharples) We do not really feel that there is a need to extend the powers of local authorities below the low water mark in port areas. The types of activities that go on in port areas below that water mark have historically been undertaken by port authorities and are principally related to navigation and dredging, which is the business of port authorities and one in which they are expert. We do not see a need. We could look at the types of development that might go on in those areas, particularly where ports are concerned. We feel that the existing regime and controls and procedures for such things as environmental assessments are adequate in the present situation.

645. At the moment for your activities you have to go to the Crown Estate Commission for any work you do between the mean high and low water marks or on the sea bed beyond those. We have had a certain amount of criticism from witnesses that that regime means that you are exempt under the Town and Country Planning General Development Order 1988. Can you tell me which of your activities constitutes permitted development under that Order?

(Mr Sharples) There are basically three. The first is developments which relate to operational land of a harbour, generally the land on which it carries out its statutory undertaking to maintain and manage the harbour and keep it open for navigation; secondly, dredging activities in the harbour area and, thirdly, lighthouses and other navigational aids that are required for safe navigation.

646. With regard to those exempt areas, do you have any consultations with local authorities and adjacent land owners before you carry out the development?

(Mr Sharples) Generally that has been the trend that has been developing in the last few years. Ports, as a management tool, as well as their recognition of environmental responsibilities, have increasingly where significant developments are involved taken the step of consulting local authorities and other interested parties, residents directly and businesses and so on. There is an example I could give you in the Port of Blyth where they were planning to build quite a large warehouse for storing newsprint and so on. They had the powers under their general development order to go ahead with that development on their own responsibility, but they took the step of consulting and it resulted in painting the warehouse a different colour so that it was less visually intrusive to the residents surrounding the harbour. That kind of activity does go on. Now, particularly, ports are doing that as a matter of course.

647. One can see exemptions existing when the ports were essentially a state activity. There is the general responsibility of the Crown and I assume that, because you always did things in a certain way, when you became commercial ports, free ports, those rights remained. But can you give me any reason why a commercial port, which is very much a business, should have any rights or exemptions not enjoyed by industry generally, in so far as the planning laws are concerned?

(Mr Sharples) It is important to understand the fundamental basis of a port, which is that it is operated as a statutory undertaking regardless of who the owner of that port is, whether it is a private company, the state, a public trust or a local authority. There are examples of the ownership of ports in each of those four categories. But each port of any significance has a local Act of Parliament which determines its powers. In addition there is general legislation such as the Harbours Act 1964, which determines what that statutory undertaking can do. It is under that regime that there are general development powers for the maintenance of a harbour undertaking. One of the significances of developments in the last 10 years, and perhaps the Government's present port policy, is that by bringing more ports under private sector ownership it is giving those businesses, which start off as port businesses, the ability to expand in other directions, such as other transport businesses. If they do that they will be subject, as would any other transport business, to the normal planning controls. It is only in the relatively narrowly defined harbour undertaking and the land required to carry on that undertaking that ports have these general development order powers outside the other planning controls.

Mr Howell

648. It is possible that port-related activities permitted under the general development order could still require environmental assessment under EC EIA Directive. What is your view on whether those activities should be subject to an environmental assessment by the EC?

(Mr Sharples) The present position is that all developments below the low water mark which are significant require environmental assessment—significant in the view of the Secretary of State. An environmental assessment is required. That also applies to works where one needs to apply for a harbour revision order or harbour empowerment order where you as a port do not currently have the powers to carry out those works. A large class of the activities is already covered by environmental assessment. You also will need an environmental assessment for operations on the operational land of a port where hazardous substances are concerned. That, I believe, would require formal planning consent, so there would be inquiry procedures and so on. It is a relatively small class of activities on the land side of a port, warehousing and that sort of activity, where the general development order powers apply and environmental assessments are not always required. But, as I say, I think the ports increasingly nowadays are looking to environmental assessments before they undertake these developments, particularly if they are large ones and

Mr Howell *Contd]*

they are very close to residential communities or other business communities.

649. Have I understood you correctly? It is just the area between high water mark and low water mark?
(Mr Sharples) The port has designated operational land which is generally above its low water mark where it carries on the quay operations, the warehousing and so on. Below the low water mark the port also has jurisdiction for navigation and there any developments will require an environmental assessment.

650. So perhaps everything does anyway?
(Mr Sharples) Most things, yes.

651. Why should not everything?
(Mr Sharples) We feel that on the land side the port needs a flexibility to organise the activities of warehousing and storage of goods and that kind of thing. It needs a commercial flexibility at relatively short notice very often, because markets change, to organise its business without seeking planning consent in all cases. It is also true to say that in recent years the Environmental Protection Act has provided extra control for local authorities over the handling of certain types of cargoes—those that are particularly dusty, for example—and there are restrictions on noise in operations in ports and so on. We feel that already there is a wide variety of controls on what takes place in ports and we do not see a need to extend that.

652. You are opposed to residential development around ports because this encourages public complaints to local authorities on matters of air quality and noise. What consultation is there between ports and local authorities and local residents on environmental problems arising from day-to-day port-related activity?
(Mr Sharples) This is an area where I can give a brief response and ask one of my colleagues to give a specific example. It is something that has emerged relatively recently, I suspect, given development trends in the last decade or so. Ports have to think very long term sometimes and maintain activities in certain areas for long periods of time. What we have seen occasionally in the last few years is that residential developments have taken place rather too close to some of those port activities. We feel that there has not always been effective consultation on behalf of local authorities in issuing planning consents with the port to take into account the port activity and the needs of that activity to continue. That subsequently, once the developments have been completed and people are living in them, has given rise to problems. Perhaps Mr Jeffery would like to give a couple of examples.
(Mr Jeffery) Yes, we have a number of examples on the river Thames as you might imagine with all the development that has taken place in Docklands. On the Greenwich peninsula there are aggregates berths. They have existed for many, many years. Their operations become increasingly difficult, not just from residential accommodation that is built nearby, but from a whole range of residential development that has taken place in Docklands on the other side

of the river. There is a constant flow of difficulty of noise and light, not so much dust, but those sorts of complaints tend to squeeze the industry very tightly. Another example is the Wandsworth refuse system. The transfer station puts refuse into containers, so it is environmentally very helpful, and into lighters to take down river to a landfill site, taking something like 150,000 lorries off the roads and 3·5 million miles off the roads. The process requires staging of barges at moorings down the river which are keyed to the time and distance. Those moorings have been there since the last century. In 1930 they began to be used in Nine Elms for staging barges. Just the very fact that they are there creates criticism, not just the environmental criticism about health and so on, but unwarranted criticism about riverside noises. All those things, because of development close to the river, tend to squeeze these industrial and essential operations, which are themselves generally environmentally very good, tighter and tighter. We will find that they are required to be moved. Some local authorities are producing plans to resite some operations and we are obviously in very close co-operation with those local authorities to help to do that. But there are many other examples on the river Thames.

653. Are you satisfied that there is enough consultation or that steps are being taken to improve this situation?
(Mr Sharples) As Mr Jeffery says, there are consultation procedures in, for example, Greenwich where a local authority is adjacent to a port development. We would feel that given recent examples there ought to be a little more regard paid to the port activity by those local authorities. There are some particular difficulties in a river situation where you have local authorities, or development corporations, on the other side of the river with rather different powers. Again it should be pointed out to them that there is an onus on them to take a little more care in future to avoid problems occurring further on.

Chairman

654. What endeavours do you make? Assume that you had a port operation or some kind of operation going back 30, 50 or 100 years and you have been doing things in the same way. The local authority is there and may be more recent historically than the operation you have been carrying out. What do you say to the chief executive or to the chief planning officer. Do you ever take them out to lunch and say, " Now, look here, as you know we have been doing it this way. There is no other way of doing it economically. It is bound to generate noise, sand, smoke and smell and the rest of it. We thought we ought to tell you of all this so you bear this in mind before you give anybody planning permission because we are going to be pretty unsympathetic if you bring people to the nuisance and then they complain about it ". It is one thing to bring a nuisance to people and another thing to bring people to the nuisance. I would have thought that most intelligent chief officers, executives of local

Chairman *Contd]*

authorities or planning officers would thank you for pointing that out to them. They might not necessarily have taken that into account. This is a human dialogue.

(Mr Jeffery) Absolutely and we do that. Of course you have talked very sensibly about the officials, the chief executive and the officials dealing with planning. But ultimately it is the planning committee which is politically motivated in local authorities that delivers the planning permission or not. You will understand that in my area stretching through 17 riparian boroughs down through the Thames that the persuasions of those groups of people are very different. They have different motives. I do not challenge their motives, but the problems they create differ from place to place. You get a different answer from Lambeth than you do from Greenwich, or you get a different answer from the City than you do from Wandsworth. We get alongside them all, but at the end of the day, if local councillors have a view expressed by their constituents about whether an apartment building will be nice for that stretch of the river, they will take account of that rather than the industrial enterprise which we try to make environmentally very successful.

655. That is the best argument I have heard so far for not having elected local authorities. I think you ought to put that in a submission to Michael Heseltine because he is thinking of all the alternative ways of having local affairs administered.

(Mr Jeffery) Mr Chairman, I did not challenge their motives.

(Mr Lacey) Mr Chairman, may I help a little with an example from Shoreham, a much smaller port than London, but a river port and historic. We go back to 1760 and the urban community, which is all around us—and this happens with a lot of ports in the United Kingdom—grew up because of the port. Its industry was there and so on. We cannot move elsewhere as many businesses in the country can. When it comes to planning we all know the intensity of concern and the conditions are becoming ever more tight. We have to respond to that. We have four local authorities within the small port of Shoreham: West Sussex, East Sussex, Adur District and Hove. We find that we can consult local plans, structure plans, all these proper processes, but we will find, for example, a housing scheme was introduced within the port in contradiction to the local plan. We have tried to have this called in by the Secretary of State but it was not. We find when we try to use the site of a redundant power station within the port—a 25 acre site, a gift in terms of urban development potential and ports lead the country in this, being an island nation—blow me, our plans are called in by one of the local authorities which fears the environmental impact of traffic, a legitimate concern, on the residents and we get stalled in a process which has been going on for four years. It has taken us through a Secretary of State-called inquiry. We challenge the Secretary of State and go to the High Court. We get two of the decisions quashed. We go back with further representations and they are still with the Secretary of State. We have issues of pollution, noise. I have been rung up in the middle of the night by a lady who said there was a ship in the port making a noise and her baby was disturbed. I am very sympathetic to that. Ports cannot go away. We are land-locked. We have all these planning concerns and we probably need a little bit more help and sympathy, but we have this dialogue and are trying to move as far as we can to assist.

Mr Howell

656. It sounds a very unsatisfactory situation with the various local authorities.

(Mr Lacey) It is desperate. Developments are stalled, we are in recession on the south coast of England. We cannot regenerate jobs in the local economy. There is an environment price that we appear to be paying.

(Mr Sharples) There are clearly difficulties in some situations more than others, particularly, as both my colleagues have, where there are a number of local authorities concerned. To come back to the Chairman's point, it is clearly in the port's interest and the community's interest for a proper dialogue between the port and the community because at the end of the day one depends on the other and vice versa. It is everyone's prosperity that is at stake, but within reasonable environmental safeguards. That is what the ports nowadays are committed to achieving.

657. The Government's memorandum states that it is considering the delegation to harbour authorities of consents for proposed works in tidal waters. What steps would those authorities take to ensure that the environmental effects of any works are assessed, and whom would they consult before reaching a decision?

(Mr Sharples) We read that in previous evidence to the Committee. We do not know what particular regulations the Government are referring to in this case, so we cannot comment in detail. We would welcome an amendment to procedures if that simplified them. It may be that consents under the Coast Protection Act are being referred to. Typically if port authorities are undertaking a development in a harbour area they would need to apply for a harbour revision order which would involve an environmental assessment. But separately they would need to apply for a consent under the Coast Protection Act if there was any effect on navigational interest in the harbour while the works were being carried out. It seems rather heavy handed for one body to have to apply for two approvals. Or there may be other bodies in the port who will have an effect on navigation through their activities. It seems sensible for the port to administer that rather than the Department. We are not sure that those are the regulations referred to.

Mr Field

658. In their evidence BACMI and SAGA state that the total amount of material dredged and dumped by port and harbour authorities far exceeds the quantity of material extracted by the marine aggregate industry. Can you provide the Committee with figures of the amount of material dredged by port and harbour authorities?

Mr Field *Contd]*

(*Mr Sharples*) We have been helped by Ministry of Agriculture in this. The latest figures available are for 1989. For England and Wales the total dredged and deposited at sea was almost 41 million tonnes based on 138 separate licences and for the United Kingdom the figure is just over 44 million tonnes based on 171 licences.

659. The Small Ports Conference are not members of the BPF are they?
(*Mr Sharples*) Yes, they are.

660. So you cover all. How many of the small ports are members of the Small Ports Conference? Do we know how many are not members?
(*Mr Sharples*) We have about 60 Small Port members of the Federation. Each year they hold a conference and I would say that probably 35 or 40 of the small ports would attend that conference.

661. Is all their maintenance dredging subject to licensing by MAFF?
(*Mr Sharples*) Yes.

662. So that is all dredging?
(*Mr Sharples*) As I said before, the act of dredging is covered under the general development order, but the vast majority of dredging is deposited in the sea, for which you need a licence from MAFF for England and Wales and DAFS in Scotland.

663. That takes me on to my second question. What do port and harbour authorities do with the dredged material? You have almost answered that already. Is it treated as waste and dumped, and if so where? Or is it put to alternative uses? How much co-ordination is there between the dredging carried out by ports and harbours, local authorities and the aggregate extraction industry?
(*Mr Sharples*) There are a number of examples of the beneficial use of dredgings. I would have to say that they are a relatively small proportion of the material that is dredged. Some of it is used for landfill reclamation. The port itself will reclaim an area of land for port development and use the necessary dredging material to do the infill. That is one possibility. I think the Glasgow garden festival a few years ago recovered soil from dredging in the river Clyde quite successfully. That is another occasion. There have been occasions where bricks have been made from dredged material and there are others, particularly for capital dredging, where it has been recovered for aggregate use because normally then you are not dealing with fine silty material, but harder materials from lower down. That does happen. By and large, as the ports need to stay in business and make a profit, will be very actively looking for ways of capitalising on this dredging activity which, as you can see, at 40 million tonnes a year is quite substantial. It is an expensive operation. We are not aware that there are ways of recovering that on a mass scale. There are some isolated examples.

664. Would you know how much of the maintenance dredging—leaving capital aside because you may be digging into aggregate or clay or something that is useful—is sludge or mud which is pretty powerful stuff?
(*Mr Sharples*) I cannot give you an answer to that. It varies in different parts of the country depending on the harbour.

665. So what you are saying to the Committee in general terms is that where the material is of some use it is used, so if it is sand and gravel washings that come down and create banks in the estuary you extract that and use it as if it is a normal aggregate extraction in terms of maintenance dredging, but where it is the slime and the black mud the only method you see of disposing of that is at sea?
(*Mr Sharples*) I think that is generally the case, but my colleagues might want to add something.
(*Mr Jeffery*) London is perhaps a little peculiar. The dredgings from the river, from the up river berths and the inner channels, of which there is some done, and the docks, is not usable material. It is siltation-like mud which cannot be recovered. We have investigated whether it might be made into bricks and all those sorts of things and it is not of a quality where that can be done. We pump that ashore into lagoons on land that we have used for the past 80 years. We have another piece of land which we hope one day we shall be able to use for the same purpose. That is the inner dredging. In the outer areas in the channels we direct the aggregates companies, who are searching for material they can use, to the areas where the dredging that is done for maintenance dredging is being carried out. That way we try to kill three birds with one stone by, first, restricting the amount of dredging that is done, secondly, using the material which needs to be taken to deepen the channels and, thirdly, getting it done free, which is virtually the way it goes. That is the way it is in London, but it is different elsewhere.
(*Mr Lacey*) May I amplify a little? The port of Shoreham is not quite unique, but we have to dump our materials by statute and perhaps other ports, because of their local Acts, are required to carry out their dredging and dump in certain zones in a controlled way. So when the harbour was enlarged as a result of the Shoreham Harbour Act 1949—which was going through Parliament at the same time as the Coast Protection Act—the controls tended to overlap, such that the dredgings for deepening the harbour at that time had to be placed in a certain location and subsequent maintenance dredgings had to be placed in a certain location. MAFF controlled that with an overlay and there were the fishing grounds where local fishing interests said, "Not in that area". The materials are not beneficial and we have to dump them.

666. That is very interesting. How are MAFF able to control that, which is what you said, when the dumping areas are laid down by statute?
(*Mr Lacey*) We are obliged to dump the materials and it is stated in what general zone. MAFF specifically will control the dumping and will set up their own zone within the port limits and controls, so we have a double overlay of controls.

667. Have they varied those over the years in your experience?

Mr Field *Contd]*

(Mr Lacey) No, because the port of Shoreham has not enlarged itself since that major Act and we dredge approximately 100,000 cubic metres a year. It is a kind of steady state system to keep the harbour clean and dredged to its designated position.

668. Is it always dumped in the same place?

(Mr Lacey) Yes, in the same general area. MAFF have the opportunity if they wish, to ask for sediment samples to be taken, so that if there is concern at the nature of the material they can satisfy themselves.

Chairman

669. I can understand that you are taking care to see what is usable and not usable, then of what is not usable you say you take samples to ensure what the nature of it is. I assume that you have sediment in some port areas that may be heavily contaminated with heavy metals, for example. When you find that what precautions are taken to ensure that that particular material does not damage the environment? If you pump it inshore into a lagoon, do you ensure that it is a lagoon from which there can be no leachate? What kind of precautions do you take? Or if you dump it at sea, what care do you take to make sure it is not an area where it can damage the marine environment?

(Mr Lacey) Chairman, I will have to pass on that because the catchment area of the river Adur which comes through Shoreham is largely agricultural. Apart from the industry within the port, which is controlled by the pollution discharges requirements, it has not been seen as an issue, although I have to say that no samples have been taken of our sediments.

670. With the great interest there is now on contaminated land and the interest that is being taken in the environmental assessment studies, you may find that it becomes an issue.

(Mr Lacey) I think inevitably so.

(Mr Sharples) MAFF and DAFS operate a very closely controlled licensing system so all dumpings of dredged material in the sea require a licence which is specific in relation to the area in which you can dump certain material. Those areas are chosen on a number of criteria, not least the characteristics of the area itself and the proximity to fishing grounds and so on, but also on the composition of the dredged material that will be put there. That is systematically analysed. Clearly in an area such as Shoreham where there is no history of industrial development further up the river MAFF might take a more relaxed approach to the frequency of sampling, but in other areas it is very active and, in my experience, each licence requires analysis of the sediment to ensure that there are not concentrations of heavy metals which ought not to be dumped in the sea. As far as I know no dredged sediments have been found by MAFF as unsuitable for dumping in the sea.

Mr Field

671. I find that very interesting because it is my experience that the reverse is true. It may be one of scale and the small amounts of maintenance dredging carried on by a diversity of small operators who come in on a contracted basis, but I cannot think of one occasion, including dredging for myself, when the material has been analysed by MAFF. But in your large ports where you are talking of very substantial cubic metres per annum, that is perhaps their criteria. We can ask MAFF that. It is true to say, is it not, that this substance is deoxygenated and has a very suffocating effect when it is dumped on the sea bed? Nobody disagrees with that, do they?

(Mr Sharples) I am not a scientific expert and cannot answer that with total confidence. When it is dumped in the sea not all of it will land on the sea bed anyway. The act of dumping it will result in some dispersion inevitably. That is one of the characteristics or our part of the North Sea. Its dispersal qualities, because the currents are very strong, are actually quite good compared with other parts of the North Sea, off the German coast for example. I could not answer about the oxygen content.

672. I have been on a dredging vessel in the North Sea and I cannot remember that it was required to let its load go in terms of the tidal stream, but let us talk of one that is very near to home as far as I am concerned. The Hurst narrows is a licensed dump, but you are only allowed to dump there on the ebb, one hour after the ebb has commenced. There have been tremendous problems in getting those who hold a dumping licence to adhere to that restriction. Clearly if you do, you plug the tide all the way down to the Hurst narrows and then you have to plug the tide all the way back. This is a considerable on cost for the contractor. How, in your experience, is MAFF enforcing those licensing requirements?

(Mr Sharples) They have been considering recently—as cropped up with the previous witnesses—the issue of marine position recorders. In our view it is in the industry's interest and all responsible ports that I represent would take steps to ensure that their dumping grounds are used for that purpose and that corners are not cut. I cannot speak for others who may be dredging outside the Federation. Clearly we would like to see a properly policed system of dumping. We have discussed the issue of position recorders with MAFF and that is something they are still considering. We would support that.

673. Will you go so far as to say that your Federation would be prepared to subscribe to a code of conduct that only dredging contractors who had position recorders and time recorders installed in a tamper-proof operation would be capable of contracting with your Federation members?

(Mr Sharples) That does not seem to me to be unreasonable. It is the port authority that is responsible for engaging the contractors and they would want to ensure that contractors are fulfilling the terms of the licence which is issued to the port.

(Mr Lacey) In practice we dredge by contract dredging substantially now, whereas we used to do our own dredging which was little and often and we controlled precisely where we dump. In the historic pattern there was no incremental increase in the sea bed offshore. The contractors have to put up

Mr Field *Contd]*

transponders, position recorders and dump precisely where our harbour master—who controls dredging—requires it. What I am trying to say is that we are responsible organisations as statutory port authorities. We have a historic code of conduct which goes back many years and we are a seafaring nation—all these things. In all you are saying, it is going to be very hard to refute that we should comply with that code of standard and practice.

(Mr Lacey) For some small ports you might find the cost of installing the position recorders might be prohibitive.

674. That was the point I was about to come to. In order not to spoil your lunch let me make it clear that all the transgressions that I know about have never been members of your Federation. Let me just return to the cost. First of all, would you know how many of your members still own their own dredging equipment and do it as direct labour?

(Mr Sharples) I cannot give you an answer to that today. We could make enquiries. As Mr Lacey said, by and large nowadays it is contracted.

675. I think an avenue worthy of exploration for the Committee is that the area from where marine aggregate is dredged would be a suitable place to dump dredged material, on the basis that the sea bed has already been substantially disturbed. What we have heard from the Crown Estate is that this is quite a long way out from the shore today. Would that be such a substantial on cost that it would make your port operations uneconomical?

(Mr Lacey) You would risk shutting down the port of Shoreham and the employment livelihood if we had to go out to the Owers, for example, where a lot of aggregate is dredged. There may be other issues concerning cross-contaminating the resources of those licensed areas. It would have to be very precisely placed and controlled. We have looked at this on capital development schemes and if we were forced to go out and dump it in a hole off the Nab or something like that it is 20 miles. It adds a lot of cost and in practice we have found that what has been done historically, which we have disturbed very little, has worked to satisfaction. Our consulting engineers on a master plan development recently recommended that the historic practice of putting the material into the coastline for the benefit of the coast in terms of this soft sea defence idea, which we have heard about earlier, is desirable.

676. Mr Jeffery made the point that they have successfully used it for land reclamation purposes or land building purposes. I appreciate that there is nowhere in Shoreham where that would be appropriate, but have you considered that?

(Mr Lacey) We have looked and we looked at the statutory powers. I mentioned earlier that we are compelled to dump, except for various areas and beyond the limits of those controls. We could put into reclamation, but we need the reclamation to put it into. We do not have the GDO rights of more historic ports, so we have become subject to planning and the planning system might then inhibit us if we have, for example, in part of the port an antagonistic local authority.

677. I have your point on the statutory constraints, but, if you have looked at it, it would be helpful to the Committee to know whether you felt that, given all the constraints and the aggravation of the local authority and planning etc, whether it would be economic, given that you would have to take it some distance perhaps for reclamation purposes, and what it looked like in terms of what you do with it at the moment?

(Mr Lacey) I understand your point exactly, and we looked at a scheme whereby, when we dredged the channel, we could economically reclaim land. We found at that stage it was cheaper—because the land reclamation was not immediately needed—to dump it into the system and rely at some future date on importing the material to reclaim the land because of the cost.

678. Sea dredged aggregates perhaps?

(Mr Lacey) That is what is done at Shoreham in our port in our soft sea defences. We are a major port in sea dredged aggregate, but the cost of money and the discounted cash flow analysis means you throw it into the system in as beneficial way as you can or in compliance with statutes. It is a question of timing.

679. Would anyone else like to add to that?

(Mr Jeffery) May I make two observations in relation to dumping at sea? We have done quite a bit of work to establish whether that is a more sensible arrangement than pumping ashore. I will come to that shortly. The question of taking dredgers far out to sea to dump brings in the European dimension and our competitiveness with European ports. The proscriptions placed upon us are very tight as they exist now and are greater than in the continental European ports. I believe there is likely to be a problem in placing on United Kingdom ports a very much tighter regime requiring them to do much more costly activities than their competitors on the continent. But setting that aside, I could go to the pumping ashore area. I have the opportunity to say that that is extremely difficult as well. We have been pumping ashore at Rainham in Essex for the past 50 years. The life of that site is practically at its end. There is about five year's worth of opportunity there. We bought a site in 1931 on the Kent marshes at Cliffe, a completely desolate area. About a third of the site in the area that we were interested in is an old munitions factory. That can be seen from an aerial photograph. Trying to get permission to pump ashore there on that piece of land upon which the port is likely to depend is extremely difficult. We spent a quarter of a million pounds on an environmental impact study just two years ago, which one has to do in these circumstances. We voluntarily did it and voluntarily came to some sort of management plan for the ecology of the area. That is then thrown into the melting pot by the declarations of sites of special scientific interest, of Ramsar sites in Europe, of declaring the whole of the Kent marshes right into the Medway as an area of special interest under European designation. So the restrictions placed upon us are increasingly making life extremely difficult for the continuation of doing what we already do.

Mr Field *Contd]*

680. You tempt me into asking you the difference between, as an overhead, light dues and dredging, but I know the Chairman would never forgive me if I went down that route. You ask that regulatory and planning authorities give full recognition to the development needs of ports. Can you say what work is now being done on an overall view of the best use of port facilities following the privatisation of state owned ports? Can you say whether there was any work done when they were still in state ownership?

(Mr Sharples) First, to clear the terms that we are dealing with, the Government have in the last decade privatised some ports which were state owned, the British Transport Docks Boards—which is now Associated British Ports—and also the Sealink ports which are now either part of Sealink Stena as a company or Sea Containers. The most recent round of privatisation has affected the trust ports which are not state owned in the sense of a nationalised industry, but they are public trust bodies and have always existed as independent trusts in that sense, although some of their trustees and in some cases quite a large number, are appointed by the Secretary of State. However, day-to-day control and independence has rested in the port. In this country we have quite a large number of relatively independent operations managing and maintaining ports. We have not had for quite a long time a Government policy which aims at developing those ports in a particular way. The Government have generally taken the view that the decisions to develop a particular port are best taken locally by the port concerned. That is one which we in the Federation would support. That is the background to your question. The Federation does not have a policy as regards the development of the ports in particular locations as opposed to others. It is best done by the ports themselves.

Mr Field: Thank you very much.

Chairman: I do not think we have any further questions for you this morning. I thank you, Mr Sharples, and your colleagues for your attendance and help this morning. We are most grateful to you.

WEDNESDAY 19 FEBRUARY 1992

Members present:

Sir Hugh Rossi, in the Chair

Mr John Battle   Mr Robert B Jones
Mr John Cummings   Mr Tom Pendry
Mr Barry Field   Mr Anthony Steen
Mr Ralph Howell

Memorandum by the Institution of Civil Engineers

COASTAL ZONE PROTECTION AND PLANNING

The Institution of Civil Engineers welcomes the invitation to present evidence to the inquiry on issues connected with coastal zone protection and planning. The Institution's interest and experience in coastal and other aspects of maritime engineering goes back to its earliest days, and it views the present inquiry as timely and of prime importance.

In the present context it is relevant to mention that the ICE has in recent years set up a United Kingdom Coastal Engineering Database, and produced reports on Research Requirements in this field as well as on the Storm Damage which occurred to coastal defences during the winter of 1989–90. The Maritime Board has produced a report on Future Sea Level Rise, and a paper on this subject was presented and discussed at the Institution. References to these reports are given at the end of this document.

The Maritime Board of the Institution has a continuing interest in coastal engineering, and is working actively to further knowledge, promote technical advances, and express views on issues in which civil engineers are interested or involved. This response to the invitation to submit evidence has been co-ordinated by the Maritime Board on behalf of the Institution.

The problems faced by engineers in protecting the coast are well illustrated by the results of the Institution's Statistical Survey of Storm Damage to Coastal Defences—Winter 1989–1990. The four major storms during this period caused damage to over 110 km of beaches, more than 50 per cent of which proved to be inadequate to resist the attack. In addition damage occurred to over 18 km of sea walls, 12 km of revetments, nine km of sea banks, 22 km of dunes and 16 km of cliffs. Considerable damage was also caused to roads, land, property and possessions.

Further evidence of the difficulties facing coastal engineers is contained in the recent Sea Defence Survey 1990–1991 (Phase I) published by the National Rivers Authority. Here it is stated that some 29 per cent of their defences are fronted by an eroding foreshore.

Although there is a great deal of activity evident in the United Kingdom in the field of coastal planning and management at the present time, the Institution is of the opinion that measures need to be taken to develop these efforts against the background of a national strategy. In addition steps should be taken to co-ordinate the many diverse interests in the coastal zone, and to rectify a number of anomalies which operate against the development of comprehensive coastal plans.

SUMMARY

For the purposes of this submission the coastal zone is considered as the foreshore and the nearshore area, together with areas where leisure activities, construction or industry could affect the regime of the shoreline. Additionally, on the landward side, the zone includes areas likely to be affected either directly or indirectly by coast erosion or inundation. Coastal management needs to be exercised over the whole of this area.

Even without the possibility of global warming there is a need for long term planning in the coastal zone. Liaison between engineers, planners, developers, conservationists and environmentalists is necessary if conflicts of interest are to be avoided or resolved. The recent reports and papers by the Institution referred to in the References, together with " Strategies for Adaptation to Sea Level Rise. Nov 1990 " published by the Intergovernmental Panel on Climate Change (IPCC), highlight the additional reasons for strategic planning to take account of possible changes in sea level due to global warming. The IPCC report is referred to later in the submission.

A central government lead authority should take the initiative in developing a national coastal strategy, and provide guidance, information and advice to regional authorities.

The powers and responsibilities of government departments concerned with the coastal zone should be positively co-ordinated. This is especially important in the light of the numerous Acts relating to the coastal zone.

The Regional Coastal Groups or coastal cells which are already established for most of the coastline of England and Wales would appear to form the best basis for regional coastal management organisations. Recognition and support of these groups as the official agencies for coastal zone management would provide a regional structure responsive to a national strategy. The areas of coast not already covered by the present groups would need to be integrated into the system. The constitution of these groups is such that they would offer the benefit of co-operation between the numerous bodies involved.

If the Regional Coastal Groups are made the official agencies for coastal management, appropriate funding arrangements would need to be made.

Many authorities such as ports, harbours, highway and railway authorities, Crown Estate Commissioners, the PSA (Public Services Authority), and other bodies are often excluded from the provisions of the 1949 Act, although they are required to consult with the coast protection authorities.

In this respect the powers and duties of the Crown Estate Commissioners in relation to the management of the foreshore and the abstraction of sand and gravel offshore need examining in order to ensure that they fit into an overall national strategy, are executed in conformity with regional coastal management plans, and that licences include appropriate conditions.

An essential element in the development of coastal management plans is the understanding of the physical processes acting in each coastal zone. Coastal studies should be carried out along the lines of the Anglian Sea Defence Management Study in areas where this has not already been done.

Conservation and environmental impact are a matter of public concern. Coastal management has an important part to play in ensuring that such issues are integrated into the planning process.

Long term monitoring of the coast and of coastal defences should be instituted on a national basis, with stated objectives and with guidelines to ensure that data is collected and collated in a uniform way. It is understood that the NRA is proposing to institute some monitoring in areas under its control, but there is a need for a systematic national approach.

Databases should be available to planners and designers. This is especially important in relation to emergency protective works when there would be little time to collect background information.

A national wave recording network should be established to provide a reliable picture of the wave climate around the coast, and to assist the development of an inshore wave forecasting service. Wave information is vital to the design of coastal defences, the understanding of coastal processes and the provision of early warning when storm damage and flooding can be expected.

The recommendations on research and the dissemination of information contained in the report of the Co-ordinating Committee of Marine Science and Technology Working Group on Coastal Engineering R and D, and the report on Coastal Engineering Research Requirements of the Institution of Civil Engineers, both published in 1991, should be acted on. Their main recommendations put forward as of national importance should receive urgent consideration.

The establishment of a national co-ordinating committee to develop a research strategy, to encourage a collaborative approach so that programmes are complementary, and to ensure that the research potential of the United Kingdom is being fully utilised, was strongly recommended in both the above reports. It is understood that this recommendation has been accepted and that the Institution of Civil Engineers has been invited to play a leading role in the proposed committee in relation to coastal engineering.

RESPONSE TO THE HOUSE OF COMMONS ENVIRONMENT COMMITTEE INVITATION TO PROVIDE EVIDENCE ON COASTAL ZONE PROTECTION AND PLANNING

1. INTRODUCTION

The need for integrated coastal management plans has been recognised in a number of countries for many years. In the United Kingdom the increasing demands and pressures on the coast have in the main been met by responding to needs as they arise, rather than by the formulation of plans on the basis of a comprehensive forward look.

A number of reports by the Institution of Civil Engineers in the past five years have drawn attention to the need for a co-ordinated approach to coastal engineering and to the pressing need for research and the dissemination of information in this area. These reports are itemised in the list of references attached to this submission.

A general review of the field of coastal engineering with particular reference to coastal defence is given in the excerpt from the Institution Report on Pollution which is attached as Appendix A, which forms part of this response.

The five areas of interest set out in the Environment Committee's letter of 26 July 1991 encompass a wide range of topics many of which are interrelated. In responding to the invitation to submit evidence, the Institution has concentrated on issues which it feels relate to the overall objective of the inquiry. After submitting brief notes on the five original topic areas, the Institution has, therefore, commented under the following headings:

> Current national activity.
>
> Notes on current legislation.
>
> The need for a single central advisory body and lead authority.
>
> Planning in the coastal zone.
>
> Exploitation of offshore mineral resources.
>
> National and regional data collection.
>
> Studies of regional coastal processes.
>
> Research and engineering design.

2. Brief Notes on the Five Topic Areas Set Out in the Environment Committee Letter of 26 July 1991.

A. *The dynamics of coastal change, both human and physical.*

The coastline is in a never ending process of change brought about by the actions both of man and of nature. If sea level rise does occur the process of change will accelerate. Maintenance of the present coastline and defence of low lying areas from flooding would require an increased programme of investment.

Some of the physical man made changes relate to the rise or fall in importance of ports and harbours, the development of marinas, the siting of industry including refineries and terminals, power stations, and the demands of recreation.

B. *The risks to coastal settlements and coastal ecosystems from, for example, flooding, coastal erosion, landslides and pollution.*

A general review of these areas may be found in Appendix A. Flooding and coast erosion depend predominantly on sea level and wave action. Both are subject to short and long term uncertainties in the form of meteorologically induced surges and global warming in one case, and storm action and future increase in storminess on the other. Long term research and monitoring is a continuing need, in order to predict the probability of extreme events with greater confidence.

The House of Commons letter refers to coastal zone protection, but taking, for instance, the Anglian Region as a whole the flood defence of estuary towns is probably of equal importance to the protection of the open coast.

The National Rivers Authority (NRA) considers that the Anglian Region is one of the most vulnerable areas in the United Kingdom, over one quarter of the Region being below flood risk level. Many thousands of people live in this flood risk area and many billions of pounds of investment in property, high grade agricultural land and infrastructure are dependent on protection from flooding.

Current action in relation to data collection, surveys and risk assessment are referred to later in this document.

C. *Coastal protection systems, engineered sea defences and their environmental consequences.*

The choice and design of the most appropriate coastal defence for a given situation depend on a knowledge of the environmental factors including sea level and wave action, together with an understanding of the coastal processes, geological characteristics and assets being defended. Continuing research and field studies and investigations on a regional basis are needed. In addition much research is required on the design of specific works.

Consideration is currently being given to the use of Asset Management to allow more rational decisions to be made about maintenance and capital expenditure, based on the standard of service required, the condition of the existing defence, and the risks and consequences of failure.

The emphasis in recent years has been to use soft defence systems when this is possible. This usually involves beach replenishment using shingle or sand. Beaches are by far the best way of absorbing wave energy, and are environmentally acceptable. Attention is drawn later in this evidence to the problem of the limited resources of shingle, especially in the light of the requirements of the construction industry as a whole.

D. *Planning policy in coastal areas.*

Planning has been a very weak link in the coastal management area. Conflicts of interest inevitably occur. Limited guidance and failure in many cases to take account of engineering advice or to use existing powers effectively have proved to be detrimental to good coastal management. There is some evidence that steps are being taken to improve the situation through the medium of Local Plans.

E. *The roles of the many authorities responsible for protection and planning in the coastal zone.*

Control and responsibility in the coastal zone is frequently seen as being fragmented and confused. There are often as many as 10 different bodies within one coastal area with responsibilities for one or more aspects of coastal planning or management.

3. CURRENT NATIONAL ACTIVITY

This section draws attention to current activity in the United Kingdom which is proceeding in the absence of a national strategy.

It has been clear for many years that coastal planning and management must be based on regional coastal cells involving many authorities. Although provision was made in the Coast Protection Act 1949 for the Minister or for Coast Protection Authorities to form collaborative groups and joint committees, these provisions have been little used. The coastal authorities have, however, in the last five years formed voluntary regional groups to pool information, and address problems of mutual interest in relation to the planning and execution of coastal measures. This year the Ministry of Agriculture, Fisheries and Food (MAFF) has recognised the benefit of an association between these groups and the Ministry, and has arranged a meeting of representatives. In addition, meetings are held at the Institution of Civil Engineers once or twice a year, at which representatives of the groups can exchange ideas and have informal discussions.

The recently formed National Rivers Authority has a regional structure, and has powers to carry out surveys and investigations. Progress has been made in collecting information which is relevant to the development of management plans. The most comprehensive study so far has been made in the Anglian Region, namely the Anglian Sea Defence Management Study. This has been supplemented by a survey of the condition of the sea defence structures in the United Kingdom, and related measures, such as contour maps of low lying areas, and flood risk evaluation.

Nevertheless, although much is being done by the NRA, MAFF, Welsh Office and the maritime authorities in general, it is not being carried out against the background of an overall perception of what the national strategy should be. MAFF has drawn the attention of the Coast Protection Authorities to the need for Coastal Management Plans but has not provided specific guidance on their preparation. Bearing in mind that there are 88 such authorities in England, together with the corresponding authorities in Wales, there would appear to be an overwhelming need for expert advice from Govenment on the scope, content and structure of these plans, in order to optimise their efforts and ensure that topics relevant to each area have been covered. The Scottish Regions have a similar need.

In the area of planning policy it is understood that the Department of the Environment has commissioned a report which is intended to lead to the provision of planning policy guidance. It is to be hoped that this action is being taken in collaboration with other departments and organisations concerned with the need for an overall coastal strategy.

The action outlined in the recent report of the Intergovernmental Panel on Climate Change (IPCC)—Response Strategies Working Group—does not appear to have been matched by a positive declaration of a United Kingdom national timetable.

4. NOTES ON CURRENT LEGISLATION

Coastal defences in England and Wales are governed by two main Acts of Parliament. These are the Coast Protection Act 1949 and the Land Drainage Act 1976. The Water Act 1989 transfers powers to the National Rivers Authority.

Appendix A of this document summarises the powers and responsibilities under these Acts and those applicable to Scotland and Northern Ireland. In addition there are numerous other Acts relating to food and environment, wildlife and countryside, town and country planning, fisheries, pollution, dumping at sea, mining, dredging and numerous private bills and railway acts affecting the coastal zone.

Many authorities are excluded from the provisions of the 1949 Act, although they are required to consult with the coast protection authorities, and there is the possibility of referring disagreements to Ministers.

Many port and harbour operations both above and beyond low water, also fall outside the remit of local government planning.

Crown land cannot be compulsorily acquired, and the Crown cannot be obliged to pay any coast protection charges. The powers and duties of the Crown Commissioners in relation to the management of the foreshore need examining in relation to overall management and planning.

Although it may be said that there are implicit powers for central Government to take national initiatives in coastal strategy and management, it is not a requirement that Government should exercise this role.

Control and responsibility in the coastal zone is therefore often seen as being fragmented and confused. The same applies to funding, and the policies and objectives of the various bodies concerned. This often means that the objectives of the various bodies with responsibilities for planning, environment, conservation, flood defence, harbours, fisheries, pollution, etc are in conflict with one another.

It is therefore absolutely necessary that all these bodies, statutory or otherwise, should operate within a coastal zone management plan.

It may be possible to use existing powers more effectively, without recourse to new legislation. However, the following section sets out the case for a forward looking, pro-active lead authority, with a dynamic co-ordinating role in relation to the fragmented legislation at present governing action in this zone.

5. THE NEED FOR A CENTRAL ADVISORY BODY AND LEAD AUTHORITY

There is sufficient evidence and experience of successful national strategic planning by other countries to form the basis for a coastal strategy for the United Kingdom which, having been set down, could be appreciated by Government, local authorities, the public and the press. At the moment it is difficult to assess the policies which are being pursued by the many agencies involved.

In comparison, in Australia the New South Wales Coastline Hazard Policy 1989 is but one of a range of Government policies and initiatives relating to the coastal zone. As part of this policy, approved by Cabinet in June 1988, a manual was prepared to provide guidance to local government and others involved in coastline development and the management of coastal hazards.

Again, the powers and duties of the Beach Protection Authority, Queensland, indicate that it is a positive lead authority which among other things has a duty to:

—give advice.
—carry out investigations.
—disseminate information.
—conduct national resource data collection programmes.
—implement long term wave recording programmes.
—require to be consulted on any proposed planning scheme in an erosion prone area.
—define buffer zones in erosion prone areas.

The Netherlands Government has carried out an extensive policy analysis and has adopted a national strategy in relation to the maintenance of its coastline. Modified legislation will lay down national standards for coastal defence systems, and set up a technical advisory board. It also places responsibilities on the Minister in relation to the monitoring, planning, financing and execution of measures.

On the other hand present United Kingdom Government policy appears to require coast protection authorities to submit coastal management plans in collaboration with other authorities, without providing guidance as to their structure or content.

Attention has already been drawn to the need for a United Kingdom reaction to the IPCC Report on Response Strategies, with special reference to target dates.

The need for a co-ordinated national research programme with regular dissemination of research results and general guidance notes is referred to later under section 10—research and engineering design.

6. PLANNING IN THE COASTAL ZONE

Planning has been a very weak link in the coastal management area. There are reports of planning authorities granting development applications without taking full account of the coastal implications. The existence of so many authorities with statutory powers relating to the coastal zone can be a source of friction. In the somewhat confused situation at the moment, there are instances of county councils preparing management plans, even though their interest and responsibility for coastal defence is marginal.

The current activity of the Department of the Environment in relation to planning policy has already been referred to in section 3 above.

Even without the possibility of global warming there is a need for long term planning in the coastal zone. Liaison between engineers, planners, conservationists, environmentalists and developers is necessary if conflicts of interest are to be avoided or resolved. Conservation and environmental impact are a matter of public concern. Coastal management has an important part to play in ensuring that such issues are integrated into the planning process.

The subject of climate change and sea level rise has received considerable attention and publicity in recent years. Attention is drawn to the report " Strategies for Adaptation to Sea Level Rise. Nov 1990 ", prepared by the Coastal Zone Management Subgroup of the Intergovernmental Panel on Climate Change. (IPCC). The report defines the broad categories of possible response to sea level rise as those of retreat, accommodation and protection, and discusses their environmental, economic, social, legal and institutional implications.

Sea level rise response strategy is likely to vary from region to region, and an element of retreat for coastal defences may have to be considered in any regional plan.

It is not the purpose of this submission to detail the deliberations and conclusions of the report other than to record that it defines coastal management planning as of the highest priority, recognising that decisions today on planning for coastal development will greatly influence costs for later adaptation to impacts of sea level rise.

On the international scale it proposes a 10 year timetable for the implementation of comprehensive coastal zone management plans, starting in 1991. Although some aspects are already being addressed by various authorities in the United Kingdom, the Institution is not aware of an integrated national strategic plan or timetable.

An essential element in the development of coastal planning and management is the understanding of processes acting along the coast as a whole. Such investigations should proceed in parallel with the evaluation of areas at risk leading to decisions as to whether set-back lines, buffer zones, or other forms of planning controls are appropriate. Although some measures of this kind can be introduced without further legislation, there appears to be a need for coastal hazards to figure more prominently in the minds of the planning authorities as an essential part of the overall coastal strategy.

Some coastal authorities have taken steps in this direction by using the local plan process to restrict development in vulnerable areas. In Norfolk and Suffolk recent Structure and Draft Local Plans have referred to control of development within defined set-back lines.

As has already been mentioned, however, certain developments fall outside normal planning control. Better ways must be found for dealing with these since the sea does not recognise either statutory boundaries or powers.

The formation in recent years of the coastal cell groups, or regional coastal groups, for most of the coastline, comprising representatives of maritime district councils, NRA regional staff, MAFF regional engineers, Welsh Office staff in Wales, and in some groups representatives of harbour authorities, conservation organisations, British Rail and county councils, holds out the promise of coastal zone management plans based upon coastal cells. These groups, however, have no statutory powers or central financial support.

Planning policies are likely to be influenced by the development of comprehensive management plans, especially if these are based on regional groups of authorities.

7. EXPLOITATION OF OFFSHORE MINERAL RESOURCES

Most of the foreshore and the seabed is vested in the Crown. The Crown Estate is a landowner and the Crown Estate Commissioners have the duty to maintain and enhance the capital value of the estate and the return made from it. It is Government policy to encourage the use of marine aggregates, without introducing the risk of erosion.

Although there is a consultation procedure before granting licences to dredge, this is not statutory, and is not made against the background of a coastal zone management plan. There is no statutory planning for dredging material from the sea bed. Coast protection authorities frequently express concern about the long term effects, and about the effectiveness of monitoring systems.

To many authorities it would seem that the Crown Estate Commissioners (CEC) are taking decisions which should fall to a marine planning authority.

There have been recent promises of improvements in relation to the conditions attached to licences granted for this work, and to the control of the operators. These have again been of a responsive nature rather than the result of an integrated coastal strategy.

It is understood that an evaluation is currently being made of the resources of sand and gravel. Crown Estates have already indicated that there is likely to be a shortage of marine dredged shingle around the coast of South-east England, because of the demands of the construction industry. This has an effect on the use of natural materials for the replenishment of eroding beaches, ie for what is known as soft defence measures.

The close co-operation between all bodies with an interest in dredging and its effects on the coastline could be better achieved through a regional coastal zone planning and protection authority.

8. NATIONAL AND REGIONAL DATA COLLECTION

Clearly the establishment of regional and national databases is essential so that all the relevant information in each area is available to coastal managers, planners, engineers and designers when decisions are being made. This information would also be valuable to research scientists. This is especially important in relation to emergency protective works when there is too little time to collect enough background information if it does not already exist.

The establishment of a national system of wave data collection has long been recommended as of the highest priority. This is something which must be set up and controlled by a central authority.

Monitoring of beach variations and of the behaviour of protective works is needed to provide long term information on their behaviour. It is necessary to have guidelines for coastal authorities in order that national data is collected and collated in a uniform way.

9. STUDIES OF REGIONAL COASTAL PROCESSES

An essential element in the development of coastal management is the understanding of processes acting along the coast.

It is now accepted that the justification for protection should not be made on a local basis, but within the context of comprehensive coastal zone management plans based on coastal cells, taking account of all the factors such as wind and wave climate, tidal currents, sediment budget and transport, geology, bathymetry, conservation of natural beauty, geological features, wildlife, buildings and sites of archaeological and historic interest, as well as tourism, coastal amenities, and the basic economic and social factors.

The petrochemical industry is interested in the effects of artificial islands for production purposes, and is investigating the effects on the local shoreline. Seabed movements and shoreline configurations need to be forecast, and a knowledge of the overall coastal processes is of vital concern. It would be of advantage to industry in general if regional coastal management plans existed which had common elements and well defined strategies.

There is, therefore, a need for comprehensive regional coastal studies similar to the Anglian regional study, in order to provide this essential information so that effective and economical plans based on predicted coastal changes can be made.

Sea level rise is likely to be relatively slow acting. In the meantime it is necessary to improve our knowledge and develop strategies that can accommodate change.

10. RESEARCH AND ENGINEERING DESIGN

The establishment of a national co-ordinating committee to develop a research strategy, to encourage a collaborative approach so that programmes are complementary, and to ensure that the research potential of the United Kingdom is being fully utilised, was strongly recommended in the report of the Co-ordinating Committee for Marine Science and Technology on Coastal Engineering R and D and in the Institution of Civil Engineers Report on Coastal Engineering Research Requirements, both published in 1991. It is understood that this recommendation has been accepted and that the Institution of Civil Engineers has been invited to play a leading role in the proposed committee in relation to coastal engineering.

At the moment there are nearly 20 major laboratories and organisations concerned with various aspects of coastal research, together with nearly 50 university and polytechnic departments involving over 80 full time academic staff with expertise in this field. They are severely underfunded.

The recommendations on research in this area made by the Co-ordinating Committee for Marine Science and Technology and by the Institution of Civil Engineers, mentioned above, should be implemented.

The Ministry of Agriculture, Fisheries and Food funds an extensive programme of research aiming for a better understanding of coastal processes, changes in sea level, socio-economic effects of sea level rise, and associated topics. Their annual report gives a general review of their activity, and references are made to some aspects of the research at their annual Conference of River and Coastal Engineers. However, specific design-oriented summaries are not disseminated to coastal engineers and consultants who would benefit from information presented in this form.

In the applied field, consideration is being given to post engineering impact, design codes, standards of service, probabilistic design and risk assessment.

The lead time for major research projects is lengthy, especially where extensive data collection is involved. A co-ordinated and well funded research and data collection programme is urgently required. The type of weather conditions which occurred in 1953 and 1978 will occur again, and in the meantime research and investigation will prove a good investment.

INSTITUTION OF CIVIL ENGINEERS REFERENCES

1. Future Sea Level Rise. Maritime Board Report. Institution of Civil Engineers. 1990.

2. Is there a sea level problem? D T Pugh. Proc ICE. June 1990.

3. A statistical survey of storm damage to coastal defences. Institution of Civil Engineers. 1990.

4. Coastal Engineering database. Institution of Civil Engineers. Thomas Telford. 1991.

5. Coastal Engineering Research Requirements. Institution of Civil Engineers. 1991.

6. Pollution and its containment. Institution of Civil Engineers. Thomas Telford. 1990. (See extract—Appendix A on coastal defences.)

September 1991

Examination of witnesses

MR BRIAN WATERS, a member of ICE's Council and of the Maritime Board, representing the Chairman of the Maritime Board; DR PAT KEMP, a member of the Institution's Coastal Engineering Advisory Panel (CEAP); MR GEOFF COWIE, a representative at the informal meetings of the Regional Coastal Groups, which are hosted by ICE; and MR GEOFF FORD, a member of CEAP and chairman of the informal meetings of the Regional Coastal Groups; the Institution of Civil Engineers, examined.

Chairman

681. Good morning, gentlemen. May I welcome you to this session of the Committee's Inquiry into Coastal Zone Protection and Planning. Mr Waters, you are a member of the Institution of Civil Engineers Council and I wonder if you would be kind enough, as leading the witnesses this morning, to introduce yourself and your colleagues?

(Mr Waters) Thank you, Sir. My name is Brian Waters, I am a partner and a director of an international firm of consulting engineers which has a substantial coastal and maritime department. I represent the Maritime Board of the Institution of Civil Engineers. My colleagues, leading the discussion, Dr Pat Kemp here, Fellow of University College, London and a member of the Coastal Engineering Advisory Panel. Mr Geoffrey Ford is Director of Technical Services, Suffolk Coastal District Council, and also a member of the Coastal Engineering Advisory Panel of the Maritime Board. Mr Geoff Cowie is the Director of Engineering of Sefton Metropolitan Borough Council. We all represent the Maritime Board of the Institution.

682. Thank you very much indeed. May I start by thanking you for the extremely helpful and interesting paper that you have submitted to us about which we will wish to ask you a number of questions. We are particularly interested in seeing your definition of the coastal zone, about which we have had some conflicting advice from other witnesses; but this does seem to be, as one would expect from civil engineers, more practical than some of the other suggestions that we have had. What I would like to start off by asking you concerns the legislative framework. In your paper you make references to the Coast Protection Act of 1949 and the Land Drainage Act of 1976, both of which may be considered by some to have now served their purpose and there might be a need to update or bring in new legislation. Have you any thoughts on that, the need to update the legislation and how it might be usefully done?

(Mr Cowie) Chairman, I think we would take the view that there is probably enough legislation but it actually needs consolidation in general terms. There is a large amount, most of which is enabling rather than mandatory, and I am sure you do not need me to tell you that there are Acts, Regulations, Orders, By-Laws and even European legislation. We would contend that the existing Acts are not obstructive at all but that authorities and other bodies are doing a lot of work on coastal zone management, whether it be monitoring, research, data gathering or consulting together but at different speeds on a different range

Chairman *Contd]*

of issues, different scale and without co-ordination. We believe that it would be advantageous if there were a consolidation, in that coastal zone management, in our view, also ought to be a duty of, say, the structure plan authority or the unitary development plan authority to prepare the plans and this would accelerate the process and co-ordinate the effort. We do recognise that there is a requirement to change or extend legislation to some extent, in planning terms, below low-water mark and that would actually require a different bit of legislation, but in general terms we would say there was probably enough provided if it were consolidated.

683. Talking of this question of management, the general theme running right through your paper is the fragmentation of responsibility and the consequent confusion which seems to exist with conflicting policies being followed by different bodies; and you underline the need of some central direction being required and look for inspiration to what is being done in Australia, for example, in that particular matter. Assume that there were a body which became responsible for coast management plans and strategies; and which body that would be is a separate matter, but if there were a body responsible in a position to give central direction, advice, help, what would be the range of issues which should be dealt with in coastal management?
(Mr Ford) Chairman, I think coastal management is an umbrella term covering, I think, both shoreline management and coastal zone management and the Institution differentiates between the two in this way, that shoreline management is essentially the management of the natural processes causing erosion or accretion along the shoreline, those processes being tides and currents, wind and wave climate, bathymetry, geology, sediment transport and sediment budgets and I think the first issue is that it is essential to fully understand those processes at every location before considering any measures. Comprehensive surveys are needed and, I believe, based on an area sufficiently wide to ensure that all the responses of the shoreline to any measures can be confidently predicted and coastal cells represent the logical area for such surveys. Coastal zone management, on the other hand, is the management of the use of the coastal zone, for example, the use of beaches as an amenity, so it follows there are a number of issues involved there: the managing of the impact of adjacent land uses and the development of that land; any infrastructure features adjacent to the shoreline; the impact of tourism and recreation; conservation of natural beauty; wildlife; and natural features, such as sites of geological, ecological and archaeological and historical importance; heritage coast conservation; water sports; the impact on the coastline of fisheries; mineral extraction; oil and gas extraction; environmental health matters, such as pollution and litter; and, of course, the economic impact of all of these activities. There are many bodies involved and the coastline varies greatly, the pressures are not the same everywhere. Coastal zone management plans aim to manage those pressures and resolve any conflicts and shoreline management is the driving force in the preparation of coastal zone management plans.

684. I will come in a moment to the regional responsibility but what you have just outlined is very, very far reaching, as compared with what we have at the moment, and it would be a very important body which took on all those responsibilities and produced a national strategic plan for our coast. Have you had any thoughts as to which would be the most suitable body, of the existing range of bodies that there are, which might be expanded to take on all those various matters?
(Dr Kemp) In relation to not the national body but the regional ones, I think that——

685. I am going to come to regional coastal groups in one moment, to which you refer in your paper, but you also say there should be some body which gives overall direction. You criticise the fact that MAFF does not give any direction at the moment. There are powers under the Coast Protection Act for the Minister to form groups and also to give advice, but the advice is not coming through and people are floundering around and one area probably has a different approach to that of another area. What I am focusing on at the moment is who should be co-ordinating the work of these groups by giving some kind of advice or direction how the approach should be?
(Dr Kemp) On coastal zone management plans, I think the Institution view has been, especially in relation to the point, Chairman, that you have raised, that perhaps MAFF have encouraged maritime authorities to produce coastal zone management plans but without providing some sort of guidance on it. It may well be, and I think the Institution is standing back from any political view on this because we have not discussed it, that the Department of the Environment should have been the one which encouraged the maritime authorities to do this. MAFF did it and they did say, "We ought to have coastal management plans" and then did not follow this up with guidance, possibly because they were conscious of the fact that there was someone else who was also involved because of the planning factor. I think the engineering response is rather a practical engineering one, that the engineers do not mind which authority is the guiding authority providing they know who it is. There seems, therefore, to be a lack of decision at higher levels as to who is actually responsible to lead this particular operation by being ahead of the game.

686. If you have not really discussed it between yourselves you are not really in a position to help us with conclusions, but one of the matters which this Committee will have to consider is whether or not there should be a national body with overall responsibility, whether that should be a Government Department and the choice would be, if it were a Government Department, between the Department of the Environment and MAFF. The question then arises, if it is a Government Department, what, in practice, does that mean: two or three officials within a vast Department being given specific responsibility for this area and advising the Minister and that kind of thing, or should we go outside a Government Department and have a body which is free-standing,

Chairman *Contd]*

has responsibility, in the same way as the National Rivers Authority has been created and is given its responsibilities? The first question is should this body be the National Rivers Authority, should it be given this additional responsibility, or should some other body be created in order to look after this matter specifically. Then that would be the sole focus of its attention, that would be its job in life and, therefore, you might get people recruited who would then have the time and possibly the knowledge to be able to discharge these functions and advise Ministers? Obviously, the Chairman and the Board of this body, if it were that kind of body, would be responsible to a Minister who would then be answerable to Parliament. We have got these options which the Committee will have to consider and reach a conclusion on, if we can, in our report and we are looking for assistance as to the most practical way to proceed?

(Dr Kemp) This is not a matter, Chairman, that we have not discussed, but we are in some difficulty here and I think the feeling is that we should build the structure, in a typical engineering way, from the foundations upwards, so we start at the coast. Mr Cowie and Mr Ford and Mr Waters are, and I am not, intimately concerned with the way the existing system operates at the moment and the problems which are being faced. If, in fact, it is perceived that this is a good way to proceed, from the bottom up, by building a structure whereby the shoreline management aspects which Mr Ford has described can be addressed then these can then be fed in to what would, I suppose, under the present system be a structure plan which already has extremely wide remits in relation to planning but should take forward the specific problems which are associated with the shoreline protection and how that affects habitats, and so on. Dominantly it is an engineering problem of protecting the coast or not protecting it, as the case may be, then one gets to the stage, Chairman, where, I think, as you have indicated to me, that one has a plan which is a coastal zone management plan, which is now, shall we say, at structure plan level. I think, as far as we could go, speaking personally, as far as the central government is concerned, one would then write a job description for that particular Department which would ensure that whatever the structure plan authority, whether a unitary authority or not, wanted to achieve, it would provide that backing. It would, therefore, define the national issues on which it could pronounce in the most cost-effective way. In other words, if there were regulations or if there were any matter which would be dealt with better nationally than regionally but, at the same time, would serve the regional authorities by providing them with the necessary information, whether it were on a database basis, whether it were on an international basis, then at that stage I think we would say, "Which authority, at the moment, fits into this job description?" If there were not one, and I must confess we have not got to the stage and I do not think we are qualified to get to the stage of doing an in-house examination of what DOE and MAFF and the NRA do at the moment, what their internal structure is, what their resources are, what their

motivation is, to decide or propose which authority could at the moment——

687. Then you cannot really help us. I was rather hoping that, as you have practical, day-to-day experience of these matters and you have had dealings with these various bodies, you might have formed a view of those who might be the most appropriate, but if you find that an embarrassing question, because you have day-to-day dealings with these different bodies, I will not press you on it.

(Dr Kemp) It is embarrassing, Sir; I do not know whether my colleagues would like to comment.

(Mr Cowie) I think, Chairman, there are differences of view; different areas of the country have different responses from different Government Departments. I could speak personally, but on behalf of the Institution we have not got a corporate view.

688. Can I turn then to the other question that I wanted to raise with you about the regional coastal groups, which you tell us, in the Summary of your paper, are already established for most of the coastline of England and Wales and would appear to form the best basis for regional coastal management organisations. These groups, as I understand, are voluntary groupings, mostly concerned with coastal protection; that is correct, is it not?

(Mr Cowie) Yes.

689. They are mostly civil engineering orientated and do not involve themselves in the wider issues of planning and leisure and that kind of thing?

(Mr Ford) They can do, Chairman.

690. Would you tell me?

(Mr Ford) The groups do not have the same constitution, there are a variety of arrangements, but they do all include engineers and they represent the Maritime District Councils and Boroughs and the Councils of the National Rivers Authority and MAFF. As such, they are well-placed, I think, to prepare the shoreline management plans, but in addition I think the arrangement can work well and be integrated into the planning process. I think, with regard to coastal zone management plans, the groups have got access to a wide range of expertise in their constituent authorities. The NRA has its expert advice on environmental matters, for example. The maritime districts have planners, environmentalists, tourism and leisure managers and access to advisory groups dealing with matters such as heritage, coasts and countryside management and all of these can contribute, I think, to the work of the groups as required. I think it is inevitable that groups will look outside for advice as well; there are so many bodies involved, I think they are bound to look, they would be well-advised to look, to the RSPB, Marine Conservation Society, English Nature, Countryside Commission, local fisheries committees, and so on, and I think these bodies would, in any case, be consultees of any planning process and I think early involvement of them would be helpful. I think, as well, the way the groups are emerging is that they are quite sensibly and reasonably realising that the resource involved is quite considerable for a period during data collection analysis and modelling, and so

Chairman *Contd]*

forth, and consultants are being involved, again with a wide range of expertise. So the groups, it seems to me, do offer the possibility of being able to achieve coastal zone management plans and I think it would be greatly helped if the expense of the considerable studies which I have mentioned could, in fact, be subject to special credit approval and grant at the time when they are being carried out rather than associated with any works which might be carried out at some future time. Indeed, I think if the emphasis in studies is shifting from schemes to management of the coastline it seems to me all the more appropriate that it is the studies which should be funded, initially, and I think that would represent better value for money.

691. You see, if we are going to recommend, as a Committee, that some new formal structure should be created to oversee the problems relating to the coast, it is far easier and probably more sensible to build on something which is already there than to try to create something entirely new and sweep away everything which is there and perhaps working reasonably well within its limitations. Therefore, would you say that these regional coastal groups, which are organised on a practical coastline basis, could be turned from voluntary into statutory groups; and if they were turned into statutory groups, with all that implied insofar as resources and funding are concerned, which seemed to be the main problem that you were indicating to us, if that were to happen, what would you say should be the remit of these bodies and should the membership be changed in any way in order to bring in people who were competent to deal with any extended remit they may be given, over and above what they are doing at the moment?

(Mr Ford) I think, again, that this question of whether there should be a statutory responsibility has been aired and there are a variety of views about that. I think most would favour the question of the preparation of coastal zone management plans becoming a duty rather than merely a permissive matter. I think, in terms of the remit and whether that should be widened, as I indicated a moment ago, the groups, in fact, already have access and can call in and do call in wider expertise.

692. We are talking of giving them a responsibility now; at the moment they are doing what they feel ought to be done but this would be giving them a responsibility for which they will be accountable to somebody and that I am going to come to in a moment. How do you answer that?

(Mr Waters) It would need a change in their present constitution, because most of these regional bodies have set up on an *ad hoc* basis and some have elected representatives and their technical officers some have them separately but still reporting to a main group at the end, some bring in experts from the general environmental bodies, so if it became a strategy then probably some common denominator would have to be established as to their membership and responsibilities. It is at that point, I think, that one comes to the difficulties, and to a certain extent the regional coastal groups are addressing them. In

the very extreme difficulties, between a seaside resort which wishes to preserve its promenade and its beaches for purely recreational use and, say, an area which is largely agricultural use or perhaps has cliffs or something like that and they are site specific, then difficulties arise as to what are required in the remit for its protection. Of course, there are even environmental arguments that protection is not necessarily the way that one should be going, so there are some extreme opposites in requirements but depending upon the individual bodies concerned.

(Mr Cowie) Could I perhaps amplify that to say that you talked about giving responsibility; most of the constituent members of the coastal groups originally already have that responsibility as do the engineers of the authorities responsible. It was the coming together of those engineers to share information, to share experiences in a climate which involved all of them, and so that they individually have the responsibility you were talking about, making a collective responsibility, and that is no great problem, as far as I would be concerned.

693. Assume we had this establishment of regional coastal groups on a more formal basis but taking the geographical areas and the kind of people that they have got there at the moment, how would they feed back to the local authorities or to national agencies and would you say that SCOPAC is a useful model in this respect?

(Dr Kemp) I can answer or refer to part of that question, Chairman, because, undoubtedly, the present climate and your Committee in general have focused attention beneficially by all the groups on their role and I do know, for instance, that SCOPAC, which has been mentioned, has taken this aboard and is meeting tomorrow, and I do not think I am speaking out of turn in saying that I know they will consider two possible scenarios. One is that they should be predominantly a shoreline management group and the other one is that they should, to consider the extremes, there will be something in the middle perhaps, consider whether they would be responsible for coastal zone planning. In one way, the answer is that the situation is not crystallised, that the groups themselves, as Mr Cowie has said, came together because they could perceive it was in their interests to do so, and therefore they are very highly motivated. I cannot help feeling that this is something which must be exploited, because if they are getting on together on a voluntary basis and doing the job they want to then this is something to build on, but the groups themselves are actively considering what role they could play and I am sure that they could be adapted, or they would adapt themselves, in order to fit in to this new concept of having a coastal zone management plan. I do not think it has crystallised and, as Mr Cowie has said, SCOPAC, for instance, have a two-tier system, in the sense that the officers meet, they have sub-officer groups, they present their recommendations, and so on, to the main Council, which consists of elected members as well.

Mr Field

694. Thank you, Dr Kemp. As you might imagine, I have had quite a bit to do with SCOPAC, because it is actually run from my constituency. Could you tell us how you would see the regional boundaries intermeshing? It seems to me that if one were to start with a clean piece of paper and look at the United Kingdom the county boundaries may have great history to them but there are no actual natural boundaries in many cases, there are in some, obviously, and, in the terms of this Committee's interest, the way in which the dynamics of the shoreline work might not be the same as the political equations of regional boundaries. How would you match that particular point?

(Dr Kemp) I think it has been matched. Perhaps Mr Cowie can give you a specific example.

(Mr Cowie) I think, quite clearly. Chairman, they do not match, they do not have to match. I am Chairman of the Liverpool Bay Group, which covers part of England and part of Wales, so it is different national boundaries in that sense and different systems of administration of Government Departments. There is no reason at all why the two should be co-terminal; you can report through whatever systems there are to your own local authorities or regional authorities, whatever it may be. Once again, taking the Welsh situation, the chairmen of all Welsh groups all meet with the representatives of the Welsh Office and that works perfectly well, as far as I am concerned and my colleagues are concerned. Reporting back to local authorities is no difficulty at the present time; if it were a regional one, yes, SCOPAC is a model, it is not the only one, it is one which does appear to work in some circumstances. The co-terminality of boundaries is, in my view, Chairman, irrelevant, it is the geographical and dynamic, to use you word, of the coast which is more important. You may have a foot in several camps at the same time but the boundary would be perhaps a particular feature which limited the effects of one river on one area and another river on another area.

695. Yes, that is a point I very much appreciate, that it is the mechanics of the shoreline that matter more than the human boundaries, but who is to be the arbiter in that?

(Mr Cowie) SCOPAC, I think, was the first.

696. Oh, indeed, it was?

(Mr Waters) Yes, I think SCOPAC was the first of these regional coastal groups by a short margin, I think, and, in fact, one could identify several coastal cells within the SCOPAC frontage. That has not made it any less successful from the point of view of engendering a point of view over what should happen on the central southern coastline.

697. Quite so, Mr Waters, but the point is, of course, that is because it is in its infancy, but if you take it on to connect it round the coast you want to make certain that the mechanics of that piece of shoreline are within that region and it is not overlapping and then you get into all the questions of boundaries. That is the point I am trying to make.

(Mr Waters) You are quite right, SCOPAC is based on administrative areas, but it has developed elsewhere and some of the later ones have not; I can think, in Wales again, of Cardigan Bay, where the actual area which is intended to be considered in the Cardigan Bay study actually ends halfway across two local authorities, Dwyfor and Ceredigian, so people are developing these as appears to be right on the ground.

698. Dr Kemp put his finger on a particularly pertinent point, if I may say so, and that is that while SCOPAC is a bigger organisation than existed previously, in terms of engineers coming together, the engineers were already, certainly in terms of county boundaries, working regularly amongst themselves at a district borough level and that continues. It has been suggested to me that the involvement of the elected members has actually introduced a slowing in the process in deliberations, particularly with the regular changing of councillors. I do not know whether you wish to step on that particular minefield but I would rather hope that you would because I think it would be helpful to do so?

(Dr Kemp) I cannot step on it as accurately as you can, Sir, because you, I am quite sure, know more about SCOPAC in general. I have attended one or two of their meetings and I was quite impressed with the way they went, but I cannot really, in terms of the political field, because I do not know how these things do interact, but I can perceive that there will always be a dichotomy between these two, that the elected members may have a different perception of their own particular authority and they want to preserve this, and they do not want to be, perhaps, beholden to anybody else in telling them what they might do. I think through that Mr Cowie and Mr Ford, who must deal with this sort of problem on a day-by-day basis in relation to the elected members against the officers and the inter-relationship between these, could comment further. I would think, personally, that something has to be accepted as it is in a national political field as well.

699. Yes, I understand that point, Dr Kemp. If I have understood the evidence of Mr Ford and Mr Cowie correctly, they do not actually have elected members directly involved in the coastal management matters?

(Mr Cowie) That is quite correct, Chairman.

(Mr Ford) No, we do not.

700. I was really asking the independent-minded spirit of an engineer which was the shortest route to action and which was the preferred model?

(Mr Ford) The Anglian Coastal Authorities Group formed really by engineers getting together and it works quite effectively in the sense that the engineers report back to their constituent authorities where the decisions are made and, of course, where the powers lie to carry out any of the recommendations which come from the groups. That works extremely well and the councils have been supportive of this arrangement, not encountered any problems or slowing-up of the process.

(Mr Cowie) Certainly, my group, Chairman, we have always reported back to our own authorities

Mr Field *Contd]*

because we felt the authorities were sovereign. We have not found it necessary at any stage to directly involve a corporate group of members. I would not necessarily see that as being any disadvantage in the long term of involving members on a formal basis, in that on Merseyside we have recently gone through an exercise of devising a working system for members which will address the transportation problems of that sub-region and that has worked perfectly well with a two-tier system member and officer.

701. Is that a one-off, Mr Cowie, for that particular problem, or will it be a standing system?

(Mr Cowie) It will be a standing of the future, it was formed for that particular purpose, but there is no doubt that the issues which are around will mean it has got to continue. It is not coincident with the coastal planning, of course, but it is a similar system which can work.

Mr Jones

702. It is the sovereignty point I wanted to pick up really, because we have had some evidence that projects have been embarked upon in some local authority areas which have a spin-off effect on others without any thinking through, really, what the effects would be. In your paper you support the idea of coastal cells but I am not sure to what extent the existing regional coastal groups are based on that cell structure and to what extent they are based on these sovereign administrative areas?

(Mr Cowie) It is a collection; certainly some of them are directly linked to geographical features. A boundary which is terminal by a headland or a river estuary, like Liverpool Bay, forms one area which is not co-terminal with administrative boundaries, it works; as we have said before, on Cardigan Bay, that is not co-terminal, others are. It is a real mixture but they all really will work because they are mostly based upon the geographical changes rather than the administrative changes and, as I said before, you can have a foot in both camps, I do for England and Wales and that does not present a real problem.

703. But it does not seem to prevent these developments going ahead of various kinds which have spin-off effects on other areas, and that, presumably, is because the elected members in some cases want to do something for their area and be seen to be doing something for their area, irrespective of whether that is good for everybody else's area?

(Mr Cowie) That is clearly quite possible, Chairman.

704. It is not just possible, it happens?

(Mr Cowie) I am being careful with my choice of words. Yes, it does happen. I think there is a tendency these days to recognise that there are greater geographical and greater environmental impacts on matters than there were some years ago and I think it may well be, my experience anyway indicates, that it may have been so in the past but is less likely to be in the future. You cannot rule it out because there is a sovereignty issue but if sufficient information is available then really that ought to sway, even in my view, members making a decision.

705. The triumph of hope over experience there, I think. Mr Ford, you are obviously expert on the Anglian Coastal Authorities Group and I think they have suggested to us that these coastal zone plans should be based on a "bottom up" approach, which I think is what Dr Kemp said was what good engineers did anyway. Do you share this view that local plans should develop into regional plans rather than regional plans being translated into local plans?

(Mr Ford) Yes, I can say on that point we all do, actually, we have talked about that. I think there are a number of advantages in the way it has evolved. The question of shared ownership of the plans I think is important; if there is to be a commitment in the future to those the "bottom-up" approach is likely to be very helpful in that regard and it is tapping existing enthusiasm and interest and experience.

706. But some of these issues are very broad strategic issues, are they not? How on earth can you possibly promote a cohesive approach to, for example, an area like the Wash, or the Dee, or whatever, if you are starting with these very basic building blocks, which may be inconsistent building blocks, and trying to bring them together and reconcile those differences?

(Mr Ford) It seems that the way forward is, in fact, to reconcile those; it seems to me that shoreline management plans and coastal zone management plans are about resolving all of those conflicts. First of all having a database, having a full understanding of all the processes must lead logically to the right conclusions, it seems to me, and if all the parties involved are parties to the process then the chances of resolution of those conflicts are much greater.

707. Obviously, up to now we have been largely talking about coastal protection in the physical sense; let me put to you a perfectly likely scenario over a planning matter. Supposing there is widely perceived to be a need for a couple of extra marinas in a particular stretch of coastline, a particular estuary, every local authority in that area wants one because they see some potential job impact in their area, but the sum total of all of these people having their way would be an over-use of that particular stretch of the coastline, perhaps in conflict with other environmental priorities. By trying to just reconcile the irreconcilable you will end up with everybody, in effect, saying, "Well, if you let us have ours you can have yours", and so on, and nobody is taking a strategic view of that issue. Is not that the real danger?

(Mr Ford) There might be a danger. If I could sort of draw a parallel with the structure planning process, where similar conflicts have to be resolved, it is a process which is well-established for dealing with that and something similar, I think, could exist in relation to coastal zone management plans.

708. But that is where we have started, with the structure plan, and worked down; that is where you have the strategy and you block in the detail. Here, we start with detail, as it were, and hope that it comes together with the strategy?

(Mr Ford) I do not think it necessarily follows that the fact that one is dealing with the basic

Mr Jones *Contd]*

building blocks means that we are not also talking about strategies, because the groups are definitely talking about overall strategies for their coastline, the elements of which must be these basic building blocks.

(Mr Waters) I think, Chairman, there is, of course, as you have clearly recognised, a dichotomy between engineering solutions to deal with lengths of coastline, whether accreting or whatever the particular problem is, and the future strategic use on development and it is certainly true that the planning strategy has sometimes said, " No, we do not wish to have any further development on the coastline ". I think this is the case in point in Hampshire County Council, where the planners have said, " No, we do not wish that " whereas the local coastal authorities might well wish to do what you have said. That is really something really outside the remit of what we can advise on.

709. It is a problem we are having to address during the course of this inquiry, which is why I was sounding you out. Can I refer to the map which you very kindly sent us, the shoreline map[1]; I expect you have got a copy with you. MAFF have told us that very shortly 98 per cent of the coastline in England and Wales will be covered by coastal groups, but I cannot see how that matches up with your map?

(Mr Waters) No, that is not up-to-date, Chairman.

(Dr Kemp) With respect, Chairman, we did not submit the map as part of our evidence and I must commend this Committee for getting on to the fact that there is a map in a subsequent publication by the Infrastructure Policy Group of the Institution. That has the date 1992 on it, which seems to make it the latest thing, but as we all know papers and presentations for reports take a long time to come through and the draft for that particular document was probably a year ago, when that map was put in. On the other hand, it is possible, Chairman, that the presentation, which I am conversant with, in fact, Mr Ford here wrote the background to that particular chapter, so he is more conversant with it than I am, but the presentation in their document does have some full lines and some double lines and that is simply to differentiate betwen adjacent groups, so I do not know whether that was a confusing point.

710. No, it is where there are no double lines?

(Dr Kemp) To add to that, there have actually, in the last 12 months, been a number of additional groups formed and on the average we must remember that, going back only to 1986, or so, there were only a few, perhaps a handful, now there are 17 groups in England and Wales and two in Scotland; of course, these are outside England, but there is, in fact, one in the Orkneys and Shetland, but they were there to start with but they are not in England, which is what we are considering. There is a North East Coastal Group which goes further north from the Holderness Group; there is a Severn Levels Group, which fills in part of the Severn Estuary; there is a Dee Estuary Group, which covers the estuary; and a Cumberland and Westmorland Group, which goes

up pretty well to the west and north-western boundary. There is a little bit of the Pembrokeshire Coast which is still not covered and a little bit of the far North East, up towards the borders of Scotland, but I think we would agree that the MAFF submission is right and that this is out-of-date.

Chairman

711. Looking at the map in front of me now, the main gaps are the Severn Estuary, the Pembrokeshire Coast and then the North east, north of the Holderness Coast Protection Project. Are you saying that the gap between Holderness and Tayside Region has now been filled by a group, the Severn has been filled by a group and that only leaves the Pembrokeshire Coast, is that right?

(Mr Cowie) It leaves only a proportion of the Pembrokeshire Coast, Chairman.

(Dr Kemp) Yes, just the headland and there is a little bit, I think, right up to the north east.

(Mr Cowie) North of the Tyne is not yet covered.

712. North of the Tyne is not covered and then what on the other coast, the west coast, the entire west coast, with an exception of a little tiny bit of Pembroke, is covered?

713. The whole of the west coast is now covered, right up to the Scottish border?

(Mr Cowie) Yes, right up to the Scottish border.

714. You have very little for Scotland; is that because it does not exist or because your remit does not cover it?

(Mr Cowie) Like measles, Chairman, we are spreading onto the Scottish coastline, as well.

(Dr Kemp) I do have to add, Chairman, that in the Institution we take the United Kingdom view whereas, of course, MAFF, covering England and Wales, and so on, there are demarcations, but when we have a coastal engineering database in the Institution and when we send out all our questionnaires, in this sort of matter we covered the whole of the United Kingdom and there were two groups which we identified from the results which were in Scotland.

715. Would it be fair to say that today approximately 95 per cent, say, of the English and Welsh coastline is now covered by a regional group?

(Mr Cowie) Yes, Chairman.

Mr Cummings

716. What distance north of the Tyne does it cover?

(Mr Cowie) As far as I am aware, Chairman, I am not absolutely certain but I think it is from South Tyneside Borough to the border.

717. To Berwick?

(Mr Cowie) I think so.

718. Can you tell me why it is not covered?

(Dr Kemp) They have not decided to do it. I think these things are spontaneous and voluntary.

(Mr Waters) They are local initiatives, Chairman, some of them.

---

[1] *Not printed.*

Mr Cummings *Contd]*

719. Certainly, along that coast, Chairman, we have received evidence from the Druridge Bay Group, who are extremely concerned at activities taking place there, the removal of sand and, of course, there is also a nuclear power station, perhaps, planned there sometime in the future, and I find it absolutely incredible that no grouping has come together to look after some 85 miles of the most beautiful coastline in Great Britain?
(Mr Cowie) I would share that view but it obviously has not spontaneously happened up till now.

720. Who would provide the stimulus?
(Mr Cowie) Your local authority engineers, I would suggest.
(Mr Waters) Yes, it can come from the elected representatives or, indeed, as my colleague has said, from the technical people, who are the engineers and perhaps planners in the local authorities concerned.

721. You know of no moves afoot to try to have a grouping established in that area?
(Mr Waters) I do not.
(Dr Kemp) What the Institution has done, Sir, is right from an early point we recognised the potential of these groups, right back to their formation in 1986, and so in the last two or three years we have hosted informal meetings of representatives of the regional groups at the Institution and then we have circulated informal notes as to the problems that they raised at these meetings. This perhaps has engendered interest so that, for instance, some of the later ones, like the Kent Coastal Group, have been in touch with us and said could we indicate what the terms of reference were, and so on, and put them in touch. How we act is that we have an on-line database which enables anyone to, on the 'phone, access the data on our database and that includes the names and addresses and the constituent authorities in all the known regional groups at that time. This can be brought up on a day-to-day basis and we have just sent out an up-to-date questionnaire on a three-year basis so that if anybody up in that area wishes to contact any other group and say, "How does this work and how is it going?" they can get this instantaneously on the 'phone from the Institution, but we are not in the position of saying, "You ought to have a group".

722. Do you know if the Durham coastline is covered by a group, Mr Cowie?
(Mr Cowie) Durham is, yes.

Chairman

723. As I understand it, the initiative for the formation of one of these voluntary groupings comes from engineers working within the local authority, the maritime authority; is that right?
(Mr Waters) That, in general, is how they come about, yes.

724. The attitude of the local authority really determines whether or not a group is likely to come about?
(Mr Waters) Yes, Sir.
Chairman: So where it is missing it is because the local authorities probably have not paid that

problem the attention perhaps they might do; that sounds to me a nice election campaigning issue in the North East.
Mr Battle: Yes, but for whom, Chairman?
Chairman: Where the cap fits.

Mr Pendry

725. Can I turn to research and information. You state that there is a recommendation for the establishment of a national co-ordinating committee to develop coastal research strategy and this has been accepted. I do not know if it is clear who has accepted it and perhaps you could clarify that and who will act on it and when it will be acted upon and what role your institution has to play in all this?
(Dr Kemp) I think this is a very important question, Sir, and I have got a note in front of me so I hope it will be factually correct and will provide an accurate record of what has happened so far. I am quite sure your Committee, Chairman, is conversant with the report of the Co-ordinating Committee for Marine Science and Technology, which was set up following the House of Lords Committee on Science and Technology in 1986. In responding to the report which that Committee, the CCMST committee, submitted to Government, the Government established an Inter-Agency Committee for Marine Science and Technology, which has the acronym IACMST, which I have difficulty in remembering, but there is, therefore, a follow-on committee from the CCMST. Between the submission of the original CCMST report and the Government response to it, whether they accept it or not, the CCMST committee was still in existence and they perceived that coastal engineering research and development was an important factor which they could address by means of a working group. This is perhaps a rather long way of answering your question, but it does, I think, indicate the way in which things have developed. They set up a working group on which the Institution, amongst others, was represented and they produced recommendations in relation to coastal engineering research and development. One of the working group's recommendations was that a co-ordinating committee, or working group, should be set up under the umbrella of this new Inter-Agency Committee to further the proposals which the working group had made about research and the way it should be organised and the investment in it, and so on. An informal meeting took place after that between the then Chairman of CCMST, Sir John Mason, the ex-Director General of the Met. Office, and the President of the Institution of Civil Engineers to consider the part which the profession, as represented by the Institution, might play in any future work of this new Inter-Agency Committee. On 19 March the President, Professor Severn, of the Institution met Sir John Mason and they agreed that a meeting should be arranged between the Institution and the Chairman of this new Inter-Agency Committee, Professor Shanks, to further discuss what role the Institution could play. In the meantime, the new Inter-Agency Committee met, on 27 September last year, at which one of the Vice

Mr Pendry *Contd]*

Presidents of the Institution of Civil Engineers, Mr Dennington, was present and there it was mooted that the Institution should be invited to play a part in relation to the coastal engineering aspects of this very wide remit of this new Agency. Following that, on 17 October last year, the Institution of Civil Engineers' President did meet Professor Shanks, as Chairman of this new Committee, to discuss this proposal in more detail and it was left that the Inter-Agency Committee would make a formal proposal to the Institution as to the part it felt it could play and, I think, initially, it was felt that perhaps it might chair or invite some neutral chairmanship for such a part. To answer your question, finally, in one sentence, it is that we are awaiting a reply.

726. Are you expecting that soon?

(Dr Kemp) I believe that Mr Dobson, who is the Director General of the Institution, is at the moment trying to find out what the position is.

727. My next question I think you have partially answered, in an earlier answer to a question from the Chair, and I think you were saying there was no corporate view, but you mentioned the need for long-term monitoring of the coast and coastal defences, databases for planners and designers and a national wave recording network. How do you think each of these issues should be addressed, by whom and how funded?

(Dr Kemp) This, from an engineering standpoint, I think is probably the most vital question that we are trying to address and I wondered, having perhaps thought this might be a matter of debate which might be discussed, quoted from your White Paper on "This Common Inheritance", where they say, "We must base our policies on fact not fantasy and use the best evidence and analysis available". That, I think, is the nub of so many of the problems on the coast; whereas, especially from an engineering point, if you are designing a building you know in advance what the loads are you can sit down to design it, on the coast there are tremendous areas of ignorance and some of the fundamental areas of ignorance relate to the forces acting on the coast and notably to wave action. One will recollect that after the 1953 surge the Waverley Committee reported, and you will find, I think, no reference to wave action there and the influence that waves played because there was not any. And after the 1989–90 storm, when Towyn occurred, the Institution took the initiative of sending a questionnaire round to all maritime authorities in the United Kingdom, asking a whole variety of questions about the storm, the damage which occurred and whether, in fact, they had any wave data, and only one response, out of all those which came in, stated that they had measured any waves; that is 1953 to 1989–90. In the meantime, it would be unfair not to mention that there has been a system of wave collection over the years, that this was organised mainly through the Institute of Oceanographic Sciences, and in 1988 they rationalised that and this particular aspect fell out, so that from that point onwards there was no national plan for data collection, as such, for its own sake, to serve the United Kingdom interest. There are a number of other facets involved, as well. The Proudman Oceanographic Laboratory, for instance, has research into what is a very important factor, which is the probability of the simultaneous occurrence of a very high sea level and intense wave action as well; they need information on that. The Met. Office has a method of forecasting waves, based on a numerical model, which has a 25-kilometre grid on the coast, which, as you can imagine, spans an awful lot of authorities, so that they can attempt to forecast offshore what the waves are here and there, 25 kilometres apart, and they are greatly in need of extra information on a wave basis. Coming down to the engineering aspects, in the CCMST report and in our own report on research requirements, we have emphasised there is a great need for field studies in order to assess the impact of wave action on sediment transport, and so on. That, again, requires information on waves. MAFF have set up a committee quite recently which incorporates the wave aspect into the sea level or surge aspect which they are concerned with so far as warnings to authorities on the coast are concerned, so they need information there and so does NRA. We have a whole complex of people nationally who need information, produced on a strategic and co-ordinated basis, which serves different interests but which can be co-ordinated. Therefore, I think the answer to your question, Sir, is that it should be a national body which does it, it should be nationally co-ordinated, there should be a national database which collects this data and this data should be available to anybody who wishes it.

728. It seems to me your answer seems right and all those strands should be brought together?

(Dr Kemp) I think you mentioned monitoring, Sir. Monitoring is the other side of the coin because waves are the dominant destructive force. Monitoring beaches, in the way they behave and the way they respond, which may seem to be quite clear on a day-to-day basis but not so clear on a long-term basis, is something which also needs addressing. The National Rivers Authority are in the process of drawing up guidelines, I think, for their own monitoring process but lots of local authorities also do monitor. Again, clearly, there is a great need that any monitoring which takes place should be on a common basis so that data is interchangeable and there is a good case for being able to draw a lot of this together on a national basis and a national database basis.

Mr Field

729. Does side-scan sonar data have any part to play in this?

(Dr Kemp) I suppose it does. Do you mean in relation to wave?

730. No, I will come to the waves in a moment, in the collection of data?

(Mr Waters) On the matter of monitoring, some of the regional coastal groups are starting to carry this out so they get the information that way. Chairman, the most important thing in many ways is the tide gauges, a national system of tide gauges, a

Mr Field *Contd]*

national system of wave-rider buoys, or whatever, so that we can get the long-term trends.

Chairman

731. Which do not exist at the moment?

(Mr Waters) Tide gauges do; I was myself designing some for the Department of Transport, as it then was, many years ago, so I do not know how far that programme has got, but it is the long-term trend from which to extrapolate future design. So far as records are concerned, what we have to do at the moment is use our computer capabilities to use wind records and then to correlate those to offshore wave heights and then use computer programmes to transfer that wave condition inshore, and that obviously has problems due to scale and conditions and you cannot put all the variables into mathematical modelling. With respect to the physical action on the foreshore, obviously, as you have appreciated, the shape of the foreshore is going to govern the coastal process at any one particular point; so, yes, we do think it is important that you get records beyond the low-water point so, in that case, you have got to have some form of underwater survey, of which side-scan sonar is an important feature.

Mr Field

732. I have obviously asked my questions back-to-front, unfortunately, but just to pursue the point, there is an enormous amount of data being collected now by side-scan sonar, as a result of the licensing of exploration by the Department of Energy. It has been put to me by the British Geological Survey people that that information should be compulsorily deposited with a national databank, run by some organisation, and that currently it is all being held to the bosoms of the oil companies who procure it, although in my own case they have generously made it available on request to archaeological organisations. One wonders, if all this is being done, is it a difficult or a simple matter to ask them to draw it all together and deposit it somewhere?

(Mr Waters) Speaking off the top of my head to that one, I would have thought that it would be best if that information were concentrated with the hydrographer and in that way it got onto charts.

733. I wanted to ask you about the hydrographer, because he seems to go merrily upon his way, with great Government subsidy, but does not seem to play a part in all this very much. He commissions surveys from Trinity House and others to do his hydrographical surveys but, apart from printing his charts, does not seem to have a direct input into all this in terms of the direction it should take. For example, the side-scan survey data, he does not seem to be saying to Government, " Well, you ought to be making them have this as a public record? "

(Mr Waters) Yes, I take your point.

734. He is not a positive creature, he is reactive one rather than a proactive?

(Mr Waters) I suppose it is reactive to a point but they are an enormous source of information to the coastal engineer in carrying out planning and their

information is made available, at a cost, and it is one method, using geographical information systems, it is our one method then of determining long-term movement in offshore seabed profiles, so I agree, yes, perhaps reactive. The only problem I can see at this level of consideration of having the input of a number of surveys carried out by different bodies is whether or not they are of a scale which could be got on commercial charts; that occurs to me as a difficulty.

735. He might be a depositor for the data, might he not?

(Mr Waters) It does seem to me, off-the-cuff, that that might be very reasonable.

736. Sorry to go back to the wave thing, but MAREX do a very good wave and meteorological collection buoy, which they have sold successfully internationally. They have developed the Doppler radar for current technology in plotting, to save sending a man out in his rowing-boat and dropping buoys all over the place, and so on, and you can survey vast kilometres of the sea at a stroke, so to speak, and yet we have had the greatest difficulty in getting water companies, the NRA and Government to pick this up and actually use it and it is not very expensive technology. Do you have ideas as to why it is that MAREX have had such a struggle on this?

(Dr Kemp) I do not know the answer to the MAREX one, but I may say it does pinpoint an important point and that is the lead authority. Although we did not, Chairman, answer your question about who should take responsibility it does need somebody to grasp this nettle. Whether, in fact, the hydrographer carries and stores information of this type, whether in fact the Institute of Oceanography has wave data, and so on, these databanks can be held anywhere but they must be known about, their content must be known to all the users. What it does need is, exactly to your overview, that this information is a great potential source of information which can be used but it is not being drawn together. It is not really a very big job to have not necessarily one standard database but at least a database at the centre which indicates where this other information is available, what it holds, how it can be accessed and the people who run it are, as you suggest, cognisant of the fact that it is relevant to coastal engineering and to dredging and other matters of that kind.

Chairman

737. In all this, how do we compare with other North Sea countries, for example?

(Dr Kemp) I think that one has to recognise that if you are going to start with the Netherlands, which is the obvious place to start, they have had a national system there for nearly 200 years, the Rijkswaterstaat has been the centre of authority and one could almost say, I suppose, that the reason why they have had this is because the man in the street realises that he may be in danger and he is going to make sure that his political masters provide this service. So that the structure, I would say, in the Netherlands has been practically unchanged for 200 years, insofar as their overall central——

Chairman *Contd]*

738. So they have worked out a structure whereby all the data from various sources can be collected together and is then readily accessible?

(Dr Kemp) They do it. It is a smaller country; in a way that is a great advantage because they have one great technological university, they have one great laboratory, at Delft, which is concerned with this particular aspect, which was very largely developed when they started to consider the reclamation of the Zuider Zee into the Ijsselmeer, so that they set up a laboratory and advised the Government on that and since then, of course, it has advised them on the enclosure of all the other sea-arms which reduces their coastline considerably. I took a course there for a year in 1958 and at that time they had a very fine wave-collecting data system, so that they have 40 years', 50 years' information. You said how do they compare with the European ones. Denmark has a system which, I think, is rather similar to ours, in the sense that they have local authorities, and so on, but they do have a central authority which is responsible for coastal zone management.

Mr Field

739. Would you happen to know whether the Norwegians require their oil companies to make their wave data information available publicly?

(Dr Kemp) No, I do not know.

(Mr Waters) I do not know.

740. The North Sea oil companies must have a phenomenal stack of wave data now because of the engineering requirements of the rigs?

(Dr Kemp) Yes, I think it has been kept in-house, very largely. This has always been the feeling over the many years that I have been concerned with it, but perhaps they do release some of it.

(Mr Waters) Offshore wave data is available in the public domain.

(Mr Cowie) Nearshore is not available.

(Dr Kemp) Nearshore is where the gap is.

Mr Pendry

741. Your Institution is urging a substantial research into the physical aspects of coastal processes but are you happy that enough work is being done on the biochemical aspects of these processes?

(Dr Kemp) Clearly, it is a matter which civil engineers are cognisant of but, apart from the pollution side, have not been greatly involved with; but, nevertheless, there are interfaces which do affect sedimentation. The interaction between these biochemical processes and silts, and so on, greatly affects their deposition and the shear stresses which subsequently develop on the beds, their resistance to further movement, and this is especially so in saltings and other similar semi-tidal flats, where this activity is noted. We do know that these biochemical aspects are shortly going to be addressed in depth and in breadth; we understand that the NERC project, Land/Ocean Interaction Study, which is due to start, I think, this April, will be looking at just this problem, in other words the interaction between pollutants, whether they are natural or whether they are artificial, from whatever source, whether they are delivered into the sea by rivers, streams, from overland, or whatever, will be considered in relation to marine processes. This project, I understand, is going to be funded to the extent of £16 million, spread over three years, so it is a very considerable, wide-ranging project, with possible extensions and I am told that there will be a press launch in about a month's time which will officially explain what they are going to aim at. The NERC expects and I think has already been promised close working relationships with the NRA, with MAFF, with DOE, higher education establishments and conservation bodies as well, so that it is broad-ranging, also, in the range of authorities which will be involved in this. I cannot add very much to that because I do not know until this is launched exactly what it will do, but I am assured that this particular aspect to which you refer will be very fully covered.

(Mr Cowie) Chairman, perhaps I might add that the biochemical aspects really must be addressed at the same time. Going down to the micro level, my authority some 10 years ago spent a quarter of a million pounds on studying wave energy and sediment transportation. I am in the process of having to devise some system of trying to assess the sediment budget for the coastline; that, we believe, is influenced by sludge-dumping, which is all part of the biochemical aspects of the thing, forming a blanket over a source of materials supply. That, plus the changes in bathing water standards, yes, it has all got to be done in conjunction, Chairman, no doubt at all.

742. Are you happy that the £16 million is the kind of size which is needed for that three-year period, on your earlier point?

(Dr Kemp) I cannot really comment. I think if £16 million were put into coastal process studies, if this Committee recommended that, we would be delighted because nothing like that was spent on the Anglian one which cost, I think, £1·65 million and so £16 million on that comparative basis is very big money. I can only comment on a direct comparison between a coastal process study and this particular one on the basis of the amount of money being put in.

743. Can I ask you about some of the engineering and economic problems involved in the policy of controlled retreat. I think we would all agree that the coastal engineering techniques are developing fast, but is there a need for the training or retraining of engineers and other professionals and how would this be dealt with best, as far as your Institution is concerned?

(Mr Ford) I think controlled retreat is certainly being looked at as an option for rural coastlines, I think where the benefit/cost ratio difficulties are going to occur and where it is not possible to justify protection under the existing rules. I think there are other arguments which have been put forward for controlled retreat. For example, the need to retain a natural landscape has certainly been proposed, the need to continue to reveal geological features from

Mr Pendry *Contd]*

the point of view of the earth scientist, but in terms of the engineering problems it does need a very careful and comprehensive study, it must be, I think, within the context of an overall plan, it cannot be something which is simply allowed to happen, it must be something which is deliberate as a result of extensive studies, so that the full and long-term consequences are understood and accepted.

744. Is that taking place?

(Mr Ford) I believe it is starting to, through the process of preparation of shoreline management plans and through, eventually, the preparation of coastal zone management plans.

(Mr Waters) There is one case, of course, where it is happening in an *ad hoc* way and that is the Holderness coastline, which is soft clays which are being eroded at a fairly constant rate, that it is policy to protect, as hard points, the towns of Hornsea and Withernsea, which will, over a course of years, steadily become further obstructions out to sea, but the rest of that land being what is held to be " low value " agricultural land is being allowed to erode; so it is happening in some areas.

745. Is the technical training and the retraining taking place to cope with that?

(Mr Waters) There are two aspects of controlled retreat, though certainly that is an uncontrolled retreat where it has been decided to protect those areas and the rest is going to go as it will. The other aspect of controlled retreat has certainly been adopted in the United States and elsewhere and in some areas in the United Kingdom, whereby it is controlled. It is designed retreat in that you allow an area to be inundated, or whatever, at a controlled depth, level, what have you, in order to reproduce the habitats, wetlands, and whatever, which we have lost over the last 50 years by the provision of coastal development, hard defences and that sort of thing. However, the cost of providing those is probably commensurate with actually defending the land in the first place, so there is a difficulty in that point of view. So far as training and retraining engineers, I think probably there is enough training in hand. Mr Cowie might like to comment on that.

(Mr Cowie) I think there is no doubt that some training, retraining and simple updating are necessary. We have some very good opportunities in the country with, for instance, Liverpol University's part-time MSc course and their short courses; this is a very good topping-up process for engineers' expertise and can be replicated or expanded elsewhere. In the coastal groups we actually can and do our own in-house training; certainly in my own group and I know of two others at least, at every single meeting they have a presentation or a paper or a seminar on one particular topic of mutual interest, which is part of the whole exercise. I think there is a change in emphasis from structures to more environmentally-friendly solutions and this requires some slightly different skills and approaches, but that is being addressed by the universities, polytechnics and the engineers themselves. For a slightly broader perspective, perhaps the European initiative, the Dutch actually offer scholarships to overseas students to study their processes and everything else; larger authorities, and certainly unitary authorities, can and do need the expertise to be able to talk eyeball-to-eyeball with particular specialists and, therefore, we feel it appropriate that that is taken as part of the engineers' processing of training and retraining. Funding is a little difficulty; perhaps there is some cause for the Department of Education and Science to look upon that, but if it is a statutory duty imposed upon a local authority perhaps SSAs and grant systems will take that into account that way. There has to be a national benefit to training; if it is fragmented, to some extent it is at the moment, there tends to be a migration between particular sources of expertise to other sources of expertise, which is far from beneficial to the overall profession and the country. Perhaps a separate matter altogether, almost, is that not really enough engineers actually engineer and that is a matter of image of engineers, peculiar, perhaps, to the United Kingdom.

(Mr Waters) Within the Institution, of course, we have our seminars and conferences and there is an obligation on engineers to continue their own professional development, so within that scope people are supposed to keep themselves up-to-date, as necessary.

(Dr Kemp) I think, in the last 30 years, certainly on an undergraduate basis, that this subject has been given more and more attention; in other words, those aspects of coastal engineering which are best taught at the university, which are usually the more esoteric and mathematically-oriented ones, are taught much more frequently now and, as Mr Cowie has said, this has led to some MSc courses. I think there is a problem when it comes to encouraging consultants or engineers working in consultancies to be released to take courses which perhaps will bring them bang up-to-date on things such as mathematical modelling and advanced techniques and this is a question, I think, of (a) their being lost to the organisation, and (b) who should fund them, because it is not cheap. I mentioned that I went to Delft in 1958 for a year's course there. They have a very national view, they have a duality, in my opinion, of (a) providing a service to the world in general, by giving courses which relate to coastal engineering, river engineering, and so on, but they do give very generous scholarships, especially to those coming from under-developed countries and this, I think, is possibly regarded as an investment because you establish your reputation, you establish your expertise and you make it possible for them to study in a particular country. I do, just almost to peripherate at this, think that we have in recent years made overseas students pay quite high fees to come even on undergraduate courses, whereas at that time, when that was introduced, we had practically no one coming from Malaysia, whereas before that we know we had an excellent intake and, therefore, you have people who are cognisant of what we can offer. Coming back to the question in general, Delft have grasped this nettle and provide international courses. I think one could do it, but it is no mean effort to provide a course of this type and I wonder whether one is now into the European dimension, so far as that goes.

Mr Cummings

746. I have several questions relating to hard and soft engineering. Could you tell the Committee what proportion of the existing coastal defences could be classified as "hard" engineering structures as compared with "soft" coastal defences and to what extent does the grant-aid system create a bias in favour of hard coastal defences and would you like to see grant-aid arrangements changed in any way?

(Dr Kemp) If I could start off by addressing part of that question, Chairman, one of the problems is that hard and soft defences very often exist together, that there is a wall fronted by a soft defence and the wall is providing a backstop. So far as answering the question as to how much, I have been in touch fairly recently with the NRA who have done a coastal survey and have an awful lot of data and I have the impression, from what they said, that they could produce this information. They themselves are not clear as to how they could differentiate between hard and soft because, as I say, there are lots and lots of composite structures. The question of whether, in fact, perhaps implicit in your question, soft defences are regarded as desirable and whether local authorities actually use this type of defence, we found when we three years ago conducted our survey for our Institution database that 38 authorities said that they used beach replenishment and that 21 schemes had been implemented in the last 10 years. So that it is sometimes perceived as an uncommon method of defending. It is by no means uncommon but it has been influenced in the past by the fact that up to about a decade ago grant-aid was not available for soft defences and so that authorities that wanted to put in a soft defence would pay 100 per cent and if you put in a wall you paid 25 per cent of the cost and so, yes, there is a definite relationship there.

(Mr Cowie) Could I perhaps amplify on that by picking up what the differences are. As Pat Kemp said, traditionally, years ago, it was much more prevalent that there was a distinction between the two; now, I think, it is much more a difference between a one-off cost and a continuing cost which causes the problem. It is more easy than it was years ago to get assistance for one-off soft solutions. Traditionally, I suppose, in coastal terms, things have been allowed to drift until you have actually got a real problem which you called in to try to address and then, almost implicitly, you are faced with a hard solution. Beach nourishment, sand-trapping and that sort of thing have not actually been seen as sea defences traditionally; they were viewed as maintenance rather than structural solutions and, historically, structures were the only ones which were available for grant assistance. I think we collectively would wish to see greater credence given to what I can only describe as environmentally-friendly solutions which have little, if any, capital or one-off costs but may have a long-term even low cost per annum, but still a long-term cost which is a soft solution will be just as effective in the long term as a multi-million pound hard solution.

747. Are you suggesting that perhaps the authorities ought to be responding in a more positive manner in relation to grant-aid for the soft options?

(Mr Cowie) Yes, I think so. If I can take a particular example in my own authority, we have a system which is done by our Coast Management Officer, which regularly traps sand to be washed away by the next high tide, to be trapped, washed away, trapped, washed away, and that is an excellent coast defence but it does not come anywhere in the pecking order for expenditure.

Mr Field

748. Could I just pursue this matter, Chairman. Mr Jonathan Sayeed had an Adjournment Debate on sea defences and I raised the point in that debate, courtesy of himself, that a lot of coastal defence authorities found that it was actually better to allow the coastal defence to fall down because it was a big financial burden on them to repair it and they could get no grant assistance for repair. The Ministry assure me that that is no longer the case but I continue to get representations that it is and I wonder what your view is of that?

(Mr Cowie) I can answer for one specific case. It is possible to get grant-aid for repair in that sense.

749. So that has been an improvement in recent years?

(Mr Cowie) We have had it in my authority within the last six months, but it is not general.

(Mr Waters) I think, Chairman, also, you probably have to find some form of words which indicates that it is an emergency repair and at that point and where there is no budget set aside for it, in which case MAFF then do come up with a grant. But overall it can be a case in point with the coastal authority with a long coastline that the maintenance budget can be very large, cannot be sustained and, almost willy nilly, the level of maintenance drops until there is physical damage, at which point a capital works scheme is put into effect and that is in effect how everything operates.

Mr Cummings

750. The Committee understands that engineering companies obtain materials for beach replenishment from marine aggregate extraction. Do you know whether they ever experience problems with securing supplies?

(Dr Kemp) I understand that one of the problems is, and I think there are many problems in this area, that dredging companies, like any other company, have a forward look, they have a certain amount of capital investment in their dredgers and, therefore, as the pressure is coming on from coastal authorities who may have schemes involving soft defences where very considerable amounts of material are required and sometimes at fairly short notice, as far as long-term planning is concerned, this places an extra pressure on the actual companies themselves. There then is, because there are a limited number of licences, a pressure which, as in every other commercial organisation, results in a cost factor being introduced so that the price becomes a governing factor. On a comparative basis, I believe

Mr Cummings *Contd]*

that the aggregate in this country might cost three or four times the cost in the Netherlands and I can only presume it is partly because the Netherlands have a tremendous amount of sand that they can dredge but partly because of the supply and demand problem in this country. This is being addressed in some ways by the assumption that non-aggregate materials could be used, that is materials which are not used by the construction industry because they are not suitable in grading. The Construction Industry and Research Association has a contract at the moment to look into this, that is to study the technical problems associated with the use of materials which would not traditionally have been used and then to go on to consider their actual viability on the coast, so the search seems to be on for alternative sources. Coupled with that, I believe the Anglian Sea Defence Study incorporated in their study a geological investigation which, incidentally, as it were, identified possible additional sources of material in the nearshore area which, perhaps changing the View Procedure rules, could be used in the sense that if it were put straight from the offshore area onto the beach adjacent to it it could have a net beneficial effect.

751. Are you saying, basically, it is a question of cost or resources?
(Dr Kemp) As far as I can understand, it is that there are a certain number of licences at the moment, that dredging companies have a certain output which they have planned ahead and if the pressure is increasing for material of a particular specification, which is rather a high one at the moment for beach recharge, then this will have an effect, of course.

752. But there is no shortage of reserves?
(Dr Kemp) They say that there is, of resources, but I do not know whether that is a shortage based on the licensing areas so far, or whether that is based on any specific knowledge of the overall results of prospecting, which I think is going on all the time. Perhaps my colleagues can comment in relation to specific schemes.
(Mr Cowie) I was more concerned about the effect of mineral extraction on potential sea defences. We are very much short of information upon what the impact of extraction of minerals is. Certainly, it can be the case that there would appear to be plenty of material out there but unless you do a proper sediment audit to see what is really happening to it then you really cannot say whether extracting even half a million or many millions of tonnes is going to affect the sea defences in the longer term.

753. Who would be responsible for analysing what the reserves are off the coast?
(Mr Cowie) At the moment it is the Crown Estates. I think there is no reason why they cannot actually do it but there is a necessity, in my view, to expand the area which they look at when they are looking at any one particular spot, because you can have a closed environment where sediments are moving around between banks offshore, many, many miles offshore, and it may appear that there is a constant replenishment. Once you start interfering with that balance, before you know where you are

you can have a major problem in sea defences; so it may be a matter that there is plenty of stuff there but unless you really have the information——

754. Basically, we can quantify coal reserves, we can quantify oil reserves and gas reserves but we cannot quantify reserves of aggregate?
(Mr Cowie) Yes, we can, Chairman, but we do not; at least, not sufficiently, in my view. You talk about coal reserves, you can calculate how much stuff is there very easily. If you extract aggregates from the land you can see the size of the hole you leave, you know immediately what the problem is, but when you are talking about sediments underwater it can be many, many miles away where the real effect is seen when you extract something elsewhere.

755. Would you like to see a survey being carried out?
(Mr Cowie) Yes, certainly. In a large number of locations around the coast, Chairman, there is a very great need to be sure that dredging and sand-winning, and so on, is not causing damage, or potential damage, to sea defences.
(Mr Waters) I think we are informed, Chairman, that there is a control and that the Crown Estate in issuing licences to the dredging companies is acting on a cautious and conservative basis and that if there is, in their view or in the view of their advisers, any cause for concern I believe it is Hydraulics Research who are called upon to carry out an investigation as to the likely other effects of taking off an area of shingle. I say taking off an area of shingle because I think there has been, in the past, some conception, especially with old-style dredgers which have buckets on, that people dug holes; in fact, it is a skimming operation, it is a sweep and, therefore, it does not have the effect of digging a great big hole off somebody's beach which then fills up with the material you have put there.

756. Do you believe that marine aggregate extraction has an effect upon the coastal zone?
(Mr Waters) Evidence, such as it is, suggests not, but there is some level of unhappiness that that can be held to be a definitive view.
(Mr Cowie) I am unhappy, Chairman, on the basis of a lack of information.

757. Can I take this one a stage further, because you are responsible for an area, of course, which has various problems; I believe the Committee are visiting Sefton tomorrow. If someone applies for a licence to extract several million tonnes of material from the seabed, what redress do you have, as an authority?
(Mr Cowie) The particular area there is one where the local council is the landowner; we are the landowner and, therefore, we issue a licence to win the sand. It is then up to the authority, if there is sufficient evidence, to say "thou shalt not extract". At the present moment there is no evidence for or against that extraction damaging the sea defence. There are changes taking place out on the banks which might or might not be as a result of the sand-winning; it is that lack of evidence, lack of information, which causes me concern. It may be

Mr Cummings *Contd]*

perfectly alright, it may be from entirely different results, the closure of the Port of Preston might be part of that but, yes, it is the information which is short to say that it does not have an effect.

Chairman

758. On this question of studies and information, would a study of the effects of dredging or extraction off Worthing be validly applied to extraction or dredging outside Yarmouth, as to whether or not that should be permitted there?

(Dr Kemp) I think there are two aspects to this, Chairman. One is that if you conduct a study off Worthing and you know all the hydrodynamics, you know what the waves are, you know the depth and you know the relationship, therefore, between the velocities of the bed and the movement then, of course, this can be extrapolated, just as laboratory experiments, fairly large-scale laboratory experiments in oscillating tunnels which are almost full-scale can be extrapolated. In relation to a specific site, such as off Great Yarmouth, I would very much doubt whether an investigation would be complete if there were any suspicion that the conditions there were significantly different from the ones at Worthing. The whole question of the threshold of movement of materials goes back 50 years or more and I think the information which is being used is the summation of many factors and not just one of translation between, shall we say, Worthing and Yarmouth. That is not to say that I think that the question is understood, I think it is one of the many imponderables, because a lot of research is going on even today and has been since I was in university on the interaction between waves and currents and the way in which they move sediment. It is an ongoing research area and I suppose the best criteria are being used at the moment I would only take up the question, I think it was, Mr Cowie made that I am not completely happy that the extrapolation of these criteria to an area where there are offshore sandbanks, where there may be circulating patterns and sand may be fed from offshore, inshore and back offshore, might not be misleading as much of the individual criteria would not apply.

759. Would you say then that, in view of the lack of comprehensive information and knowledge and the need for further research, each site should be specifically investigated before extraction or dredging takes place?

(Dr Kemp) I think, unless there are very good reasons for saying we are completely happy, in the light of our information, that this is a site where perhaps the criteria are so beyond the assumptions so that the situation is so outside the criteria you would use that there is no reasonable possibility of it happening. On the other hand, there are two studies which are going on on the South Coast, one funded by SCOPAC and one funded by HR and MAFF, looking into shingle mobility and so that does indicate (a) that more information is required, or (b) that there is a need to reassure authorities that what is being done is, in fact, not harmful.

Mr Field

760. Could I just ask you if you would have qualified your reply to the Chairman's question if he had asked you if it had any bearing when that data was below the 18-metre line?

(Dr Kemp) That, I think, is what I would call a base-line, which traditionally has been used, perhaps for the last 50 years, as a depth at which in certain studies it has been shown that there is very little significant movement, but I have not been involved scientifically in this for many years and I would like to, therefore, see what a particular authority would say if I said, "What substantive information have you still got for that?", bearing in mind that today there are very different wave climates in very different areas of the coasts and "Can you extrapolate results which may have been produced in the United States to the Norfolk Broad area?"

Mr Cummings

761. What account is taken of the "intangible" benefits of coastal defence options and how are these factors quantified and could you tell me and the Committee how MAFF is responding to these issues?

(Mr Waters) There has been quite a lot of work, initiated originally by Middlesex Polytechnic, on determining abstract benefits, leisure use of cliff-top and other, which cannot readily be put into the benefit/cost calculation like the value of property or the residual life of cliff-top property, for instance. Certainly it is being used, they have produced the formulae by which one can produce how much cash can be allocated to the number of people who will set out on a Sunday afternoon to walk that specific length of promenade, or cliff-top, or whatever, and those can be put into the calculations and they will be accepted by MAFF, especially if it is a borderline case on the benefit/cost ratio and we have done so.

Mr Field

762. Do you evaluate sea defences as to whether they reduce the amount of tropical hardwood?

(Mr Waters) The tropical hardwood question has been addressed up to now individually by local authority coast protection authorities where, within the authority, there has or has not been a move by the elected representatives to say, "We do not wish to have tropical hardwoods used". I do not know of a case where MAFF have either agreed or not agreed to fund an extra expense because tropical hardwoods have or have not been used, I do not know of a case where it has been put to them in those terms.

763. Is there really any engineering alternative to teak, in terms of groynes and that type of defence?

(Mr Waters) Regrettably, if I take the tenor of your question, regrettably tropical hardwoods are by far the best to use in the coastline because of their resistance to rot and resistance to wear.

764. Would it be possible for your Institution to tell the Committee just how much, in percentage terms, of our sea defence actually employs tropical hardwood?

Mr Field *Contd]*

(Mr Waters) I think it is possible but we have not got that information with us.

765. I think it would be interesting for the Committee to know?

(Mr Waters) As to the alternatives, there are difficulties. The most popular now is to consider things like natural rock deposit. They use a lot of room, they can be resisted in amenity areas because of dangers especially to children. I find it slightly strange because the Mediterranean, for instance, uses a lot of rock and nobody complains there but there is a different attitude. One other thing which comes about if you are using rock groynes on the British coastline, you have, in some areas, quite considerable tidal movement, that means at low tide there is quite a large mass of weed-encrusted rock available to see, whereas in the Mediterranean, of course, tidal movement might be only a few inches.

766. You say that MAFF has not provided coast protection authorities with sufficient guidance on preparing coastal management plans. Do you think this guidance ought to be the responsibility of MAFF or the Department of the Environment?

(Dr Kemp) MAFF have encouraged the maritime authorities to produce coastal zone management plans; now whether it should have been MAFF or DOE I am not sure, but MAFF actually indicated, I am not sure whether by letter, I have never been able to establish whether they formally asked for this, but certainly the word was put round that this would be a desirable thing to do but, as I say, at that stage, there seems to have been no indication. The result of this has, I think, been quite noticeable in the sense that all the authorities concerned could also recognise that coastal zone management plans were extremely necessary and that they should get on doing something about it and so many authorities, either separately or together, counties and district councils, sometimes in co-ordination and sometimes not, set out along this path at different paces, some have not started at all, and it seems to me it is rather like the London Marathon, where everybody sets out while the Committee is still working out what the route should be. It would have been a great help if there had been just the minimum amount of guidance by saying that " coastal zone management should be prepared, addressing basically the following factors and that these should be carried out in co-ordination with authorities in your structure plan area ", or even something a little bit more specific. Without giving guidance on this I think there has been a tremendous amount of difficulty, where, as far as I can gauge, certain draft coastal management plans are only shoreline management plans, are confined to a narrow area of the coast, and others cover the lot and there is a feeling by the structure authorities that the strategic questions are not being addressed by those on the coast and those on the coast feel that the county authorities should not really have much to do with it because they do not pay for it.

767. I still want to press you, Dr Kemp, on DOE or MAFF?

(Dr Kemp) At the moment it is both, Sir, to be quite honest, because DOE is a planning authority

and MAFF are the coast protection authority and they both, therefore, have an influence. The coastal management cannot be prepared without——

(Mr Waters) They are the funding authority.

(Dr Kemp) The funding authority, yes.

768. If you want to tighten up the policy you will have to have a lead authority; which is it to be?

(Dr Kemp) Absolutely. I think the one which fulfils the job description accurately, which one will be prepared to draw up; but at the end of the day I do not know. Somebody has got to grasp the nettle and, as I said earlier, I think the engineers——

(Mr Waters) We do not think it is us, Sir.

769. It is so unlike an engineer to be so unpositive?

(Dr Kemp) They do not mind who it is providing they know who it is.

Chairman

770. Part of the difficulty, I suppose, that you are in is this, that you are contractors, you rely upon business from one or both of these great bodies and you do not want to say something which might upset your friends, so I do not think it is fair really to ask you publicly anyway what your feelings are about the respective merits?

(Dr Kemp) I think there is another point, Chairman. If an engineer were asked specifically to look into this I think he would want to know what the present situation was, he would want to know what MAFF do, what DOE do, what their structure was and, in fact, very importantly, what the attitude of their Minister was as to how important it was.

Mr Field

771. To press you on a point there, for example, the Department of the Environment has sponsored a geomorphological survey, of which I am sure you are aware, into predicting slope stability problems. That is a leading piece of technology, entirely under the auspices of the DOE, looking into the future and it would seem to me that there is nothing of that ilk in the MAFF confines at the present time, so it would have led me to believe that perhaps the DOE have the engineers and civil servants who would interpret that sort of data and usefully employ it in conjunction with people such as yourselves and other Institutes. You must know, in your heart of hearts, which of the Departments is more likely to have the predominant expertise within it to take this up, but I take the Chairman's point that perhaps you cannot say?

(Mr Waters) MAFF do have an engineering staff, their regional engineers are extremely helpful, we find them very good to get on with from the point of view of individual schemes. I think, Sir, if I dare, whether it was MAFF or DOE, the answer you get when you go to them for the consideration of schemes is, " I will never get it past the Treasury ".

772. The Treasury is always an excuse for everything in this House!

(Dr Kemp) Certainly, DOE or MAFF will take up a matter and will run with it, as the case that you have just described, but one has difficulty in finding out

Mr Field *Contd]*

when they are running with it and why. It does not seem to fit into this overall plan. I think once one of them is made a leader, or somebody else, another authority, given the funding, I think that their in-house expertise might have to be expanded or changed, or whatever, I think you could say. We do not know whether the in-house expertise of either of these is really directed towards doing the whole thing.

773. We did raise this earlier in our deliberations. NERC does some wonderful work in terms of sediment exploration, they spent a whole year in the North Sea exploring sediment, etc, but, of course, that all comes under the DES. It does seem to me that there is a lot of very interesting stuff going on but, as you rightly say, it is not co-ordinated?
(Dr Kemp) This is where the Inter-Agency Committee, which you will remember from a question raised earlier, is trying to draw this sort of thing together.

774. Could you say what are the effects on coastal defences of a lack of, if any, planning control for ports, harbours and land below the low-water mark?
(Mr Cowie) I think, Chairman, traditionally there were a lot more difficulties than there are at the present time. There is no doubt a port creates a discontinuity in almost every case in the coastal processes and that did cause difficulties, but nowadays I believe there are much more environmental issues rather than sea defence or coast protection matters, matters relating to control of activities on dust, pollution, traffic, rail traffic, road traffic, noise, and so on. On land below low water, it really is then mineral working related and that is back again to the knowledge of the effects of the discontinuity.

Chairman

775. Many of our nuclear power stations are located along the coast and a number of major toxic waste sites are also on the coast or at sea level. With all the talk we have about rising seas and global warming, and the rest of it, can you tell me whether sufficient work is being done to protect these sites from sea incursion which might be anticipated in the future and what still needs to be done?
(Dr Kemp) I would really start off by saying that the information we have from the Energy Board of the Institution of Civil Engineers, which is really the appropriate source of information here, is that the nuclear power stations and, in fact, others were designed with very liberal safety margins, so far as the incursion from the sea is concerned, at the time they were constructed. Secondly, they perceive that the major danger from flooding is while the power station is operational, there would be considerable problems if it were flooded while it were operational and the operational life, I think, is something like 40 years, or so, so that there is a very limited impact there of sea level rise and it would appear that if, in their opinion, there were a sea level rise then they could enhance the defences to match that. The long-distance future of it is, as far as I am concerned, well-known; once the power station has ceased to be an

operational power station it seems to me that, in their opinion, the risk is then manageable and can be contained, but without knowing the specific sites involved I think I will not be able to say any more.

776. I think, in a sense, you have answered the question because you say that when these power stations were built a safety margin was built into what was known about sea levels at the time of construction. It is only very, very recently that the question of global warming and the consequent effect on the oceans have come to our public awareness, and scientific awareness, presumably, also, and possibly at the time these safety margins were being built in they were taking the average in those days and have not taken into account an increase in the average overall. When you are talking about the life cycle of a nuclear power station one has to bear in mind that at the time it has to be decommissioned the current philosophy is to entomb it on site. Whether that entombment on site would be sufficient to deal with the kind of scenario I am discussing is something I wonder whether thought is being given to, hence my question?
(Mr Waters) We were involved at one stage and certainly when there was consideration of a number of proposed sites for nuclear power stations, exactly this sort of problem was addressed, especially with respect to the East Coast, and the design criterion was, as Pat Kemp has said, that their operational life is, in comparative terms, quite short and they then have to be entombed and it was certainly a criterion for the sites being examined that they had to be defendable or if the coastline over the 200 years, or whatever, retreated behind them that they would remain behind as islands and that the level of defence built round them would be sufficient for that to happen.

Mr Field

777. Just to pursue that point, the NRA were saying at a public meeting recently that their prediction of the mean sea level rise is still one foot, over the next 90 or 100 years. Is that something with which the Institution agrees, or do you have a greater figure than that?
(Mr Waters) No, we depend upon the IPCC.
(Dr Kemp) I think, as far as I can gather, that this inter-governmental committee have made proposals and these are being interpreted in this country differentially because I think even on the East Coast the annual rise has been set differentially between NRA and MAFF, depending upon the location of the particular defence structure in question.
(Mr Waters) There are tectonic differences across the country, in any case.

Chairman

778. The answer is really that nobody really knows the full impact of climatic changes, it depends upon your computer model what kind of answer you get and the extent and rate of sea rise is really imponderable at the moment?

Chairman *Contd]*

(Mr Waters) It is imponderable because we cannot model something on which we have not got previous information to produce.

779. The only question is that there may be, from the way scientific discussion seems to be going, a risk of which account has to be taken?

(Mr Waters) Yes, indeed.

780. Really, my question is are we taking account of it to a sufficient extent currently?

(Dr Kemp) I think the Met Office has got a very well-established unit which is looking at this and they are the front-runners, I think.

781. What about toxic waste sites, lagoons of all kinds of nasties, sitting there on the coastline, waiting for the sea to come along and wash it all out somewhere?

(Dr Kemp) I have no information on it.

(Mr Waters) It is clearly a scenario, I suppose perhaps the one case even of a long-abandoned gasworks, where the ground beneath them is full of phenols, and things; if, in fact, that area then becomes eroded one can only say that in areas like that then the retreat option is not one which can be adopted.

782. This is something which if we had a body which was responsible nationally for strategic management it would have to look at, presumably?

(Mr Waters) Yes, Chairman.

Chairman: I think that concludes our battery of questions for you and I am most grateful to you, gentlemen, for your attendance and assistance this morning.

Examination of witnesses

MR TIM COX, Borough Planning Officer, examined, MR GEOFF COWIE, Director of Engineering and Surveying, further examined, MR JOHN HOUSTON, Coast Management Officer, MR TONY SMITH, Head of Coast Protection Unit, Metropolitan Borough of Sefton, MR IAN DEANS, Joint Countryside Advisory Service, examined.

Chairman

783. May I now open this session on the inquiry by the House of Commons Environment Select Committee into coastal zone protection and planning, first of all by thanking the Sefton Metropolitan Borough Council for the assistance that they have given us and for making all the facilities available for us to come and see for ourselves some of the problems associated with coastal zone protection, both erosion and build-up where perhaps it is not required, and also for the very useful documentation with which you have supplied us. I wonder, Mr Cox, as presumably you are leading the team, you would care to introduce yourself formally for the purposes of our record and also your colleagues who are with you.

(Mr Cox) Thank you, Sir Hugh. I am, as you know, Tim Cox, the Borough Planning Officer for Sefton. On my immediate right is Geoff Cowie, who is the Director of Engineering and on his right is Tony Smith who is the Principal Engineer dealing with coastal matters. On my left is John Houston, the Coastal Management Officer for Sefton and immediately on his left is Ian Deans, who directs the Joint Countryside Advisory Service which provides specialist advice not only to Sefton but to our neighbouring boroughs of Knowsley and St Helens.

784. During the course of our inquiry we have heard a great deal about coastal zone management plans which appear to be very informal documents prepared very much on an *ad hoc* basis by various groupings of people and those local authorities interested. They do not seem to have any statutory basis. Can you tell me how you see these coastal zone management plans fitting into the statutory planning system and whether you feel they should be given a statutory status in their own right?

(Mr Cox) Yes. I think first of all I do not see any difficulty in principle in coastal zone management plans forming part of a statutory land use plan in the town and country planning sense of that word. Clearly the content of a coastal zone management plan may be wider and at one level may involve land management issues which are inappropriate to the statutory land use plan but it does seem to us fairly important that any management plan is underpinned, if I can put it that way, by some statutory force and the only device currently available to us to achieve that is the land use planning system under the Town and Country Planning Acts. I think one ought to say that there are clearly some limitations to that approach. For example, such a statutory power cannot extend, as I am sure Members of the Committee know, beyond the low water mark which is in fact the extent of the local authority's jurisdiction under planning law. I think there is also a question of whether all the types of things that a coastal management plan might actually include would actually be appropriate in a land use plan. For example, the appropriate types of coastal defence may not be appropriate in a land use sense, although the impacts of that may well be dealt with in terms of land use policies. I strongly feel that the coast should be looked at and included within the panoply of land use planning operations that we carry out and should be included in our particular case in Sefton in our Draft Unitary Development Plan and, as you will know, that actually has happened. We have put policies into that particular plan. In terms of whether the wider concept has some statutory force beyond that, I have some doubts as to whether that is actually necessary, though I think it is important that a framework is established rather wider than individual local authorities which provides the basis for individual actions by the local authority whether it be by inclusion of policies in land use plans or whether it be giving legitimacy to expenditure by local authorities upon management coast protection or other measures.

785. Much of the evidence that we have received on this matter suggests that coastal zone management plans should be dealt with not on a borough council basis but on a wider regional basis because it is difficult to say that what you do on one part of the coast will not have an effect further up or down the coast but obviously there needs to be some integration of different schemes and ideas. Now, the Sefton Coastal Management Scheme has a limited geographical area and also is concerned with a very specific range of issues. Do you see that fitting into the kind of coastal zone management plan that others have talked to us about and how do you feel that can be best achieved?

(Mr Cox) The answer to that is, yes, I certainly do see that as being the case and I would agree with you entirely that Sefton coast as a unit is of too limited an area. I think perhaps Mr Cowie might like to talk a little on that because he is very much involved with the coastal cell, as it is called, which covers the area from Llandudno to the Ribble Estuary.

(Mr Cowie) Yes, Chairman, I think I quite clearly see the Sefton Coast Management Scheme as it now exists as being one of the building blocks of a total plan. It has to be said if you talk about cells that Sefton's coast itself is a single cell in its own right, bounded by the Rivers Ribble and Mersey and therefore can be considered in that sense as a single cell. However, the extensions of that which I think are necessary and have been demonstrated by the actions we have taken locally to form coastal cells, the Liverpool Bay cell and the Ribble to Morecambe Bay could indicate that it is possible and proper to link those functions together to make sure that the operation of the elements does not respect any arbitrary boundaries which may be administrative. Certainly we go west to Llandudno and we have a very close working relationship with the northern one which is Ribble to Morecambe Bay. Quite clearly

Chairman *Contd]*

there are a whole range of issues and parts of plans that are already in existence and the Sefton Coastal Management Scheme is one very important one, as is the work that the coastal cells do to bring the whole issue to one final conclusion which is a coastal zone management plan whether that be an individual one for the Sefton which is closely linked to the other ones by consultation and negotiation or one that is arrived at by the range of authorities coming together to collectively devise one single plan.

786. You have mentioned " cells " and Sefton is a cell in its own right. People who talk of regional management plans envisage a much larger cell especially for coastal planning purposes. If there were a larger cell how would you envisage integrating your policies within the regional framework bearing in mind that a cell could cross national as well as county boundaries?
(Mr Cowie) Chairman, the Liverpool Bay cell does actually cross national boundaries because we go from Ribble right through to Llandudno. That in practical terms means that there are two different legislative systems in operation but it does not really affect the elements or how we consider them. In our UDP system we are consulting with our neighbours but the cell also allows that consultation to be broadened. The half dozen authorities that that makes up all have views upon their own elements of the work necessary and they are closely related. The Welsh situation actually has an additional element to it in that on the North Welsh coast there are two agencies, the local authorities and the county authorities. At the present time the county authorities have chosen not to join in the coastal cell work but are keeping a watching brief from the sidelines. That situation actually does not apply in the Wirral and Sefton in that we are unitary authorities looking after our own interests. I think there is no difficulty in the practical application provided there is at some later stage than the present time a means of directly involving, perhaps in the cell-axis system, members as well as officers.

Mr Jones

787. Can I follow up the answer to that particular question and ask you first of all why the counties would not participate directly but only opted for what you might describe as " observer status "?
(Mr Smith) They have been asked to be kept informed of our activities but unless there is a specific issue raised which requires their presence they have decided to take a watching brief for the time being. They have not disassociated themselves entirely from the activities of the Liverpool Bay Group.

788. This is all about strategic planning and they are the strategic planning authorities for those particular areas so is it not rather remarkable that they have chosen to opt out?
(Mr Smith) Possibly because the Liverpool Bay Group is a coastal engineers' group and for the moment at least it concentrates on coastal engineering issues.
(Mr Cowie) I think I would use the word " unfortunate ". It is difficult to say as officers what it

is but it is unfortunate, is as far as I would go. They do have a responsibility and would, as Mr Smith said, participate if they really felt it was something important in the broader spread which clearly coastal zone management plans will have. They will cover things other than coastal engineering matters. They will cover land use matters and so on so, yes, I am reasonably confident they would. They certainly would if it was a statutory issue.

789. But as you say this is an engineers' group?
(Mr Cowie) For the moment.

790. Engineers are practical people—I say that as a son of an engineer—and I can well imagine they can bring their views together and come up with some sort of an agreed solution. If these coastal units go much wider than engineering considerations, and there are lots of planning considerations involved, I am not sure how you envisage within these cells any conflicts being resolved. Earlier today I raised the point that the authorities on the north side of the River Dee were pursuing a different planning strategy from those on the Welsh side; namely a concentration on leisure on the north side and a concentration on industry on the south side, two things that could well come into conflict. There does not seem to be any mechanism for resolving that conflict at the moment and I wondered how within the enlarged cellular unit you could actually resolve genuine   differences   of   priorities   between democratically elected local authorities?
(Mr Cowie) I think there are two elements to that one. I think we looked at a location today on the banks of the River Alt where there are different pressures from the environmentalists and the engineers to solve that particular problem in a particular way. The simple way would be to construct a hard wall but we have resolved our differences in that sense by talking it right the way through and coming to a conclusion as to which are the most important elements in making up the environment. I think that sort of situation can exist even in a broadened coastal cell which would include other disciplines than engineering. I am quite sure that planners are equally capable of coming to a consensus view at the end of the day of what is the right solution for all the elements.

791. Supposing they do not, that is really the question I am putting to you. Who is to referee the system and who is to bring together any differences and make a decision as to which should prevail or what compromise should be put forward?
(Mr Cox) That point ties up with the point I would like to make which is, in my view, clearly the current framework is inadequate. That is not to say there are no good things within it and there is no good work done within in. My own view, if you take this particular area for instance, there is a lot of work going on the Mersey Estuary Plan, the work we are doing on our part of the coast and the cell work Mr Cowie has talked about, there is the Ribble Estuary Liaison Committee and the question is how do those all come together and the problem at the moment is there is no adequate framework for them to come together and there is no adequate way of

Mr Jones *Contd]*

actually dealing with the problems that you are actually talking about. The only way that can be done is by a much more formal organisation than exists which feeds advice down to individual planning authorities, local authorities, which they are required to take into account in either, or probably both, the preparation of their development plan, that is a form of regional guidance if you like, and I think also quite importantly if that sort of approach is required that the individual spending plans for works should actually fit into that framework and it should be a requirement for funding that they fit that framework. You then have two potential incentives, if I can put it that way.

792. That is a top-down system. The Institution of Civil Engineers told us only yesterday that they favoured a bottom-up system because that was the best way to bring things together. Now I can see some considerable conflict there obviously. Have you any comment to make?

(Mr Cox) Perhaps if I speak first and Mr Cowie, who speaks with several hats on, may want to add to that. I think that any workable system is really going to be a combination of both. What is coming up is an enormous amount of expertise, understanding, practical experience. I think it is yet to be proved in my view that that system is going to actually move forward fast enough in terms, if I can call it, of this strategic overview without any framework above it. I think that framework should be sufficient to give direction without actually telling everybody precisely every single thing they should do and should provide—the word I have used is—a " framework " so that things fit together but should not be prescriptive in the sense it is an agency whose authority is required for the crossing of every " t " and the dotting of every " i ". There has to be a balance drawn. I think the bottom up experience is extremely valuable. That is not in any way to be underplayed but I do think from my point of view there has to be some framework at a higher level which actually provides the incentive to the mechanism to ensure that things do actually gel together and it is not just left to the goodwill and good working relationships or whatever it may be of a group of disparate authorities or individuals.

Mr Cummings

793. Is this your personal opinion or have you indeed spoken to any planners of the county councils to which Mr Jones has referred?

(Mr Cox) I have not spoken to planners from the county councils particularly on this matter. In that sense it is a personal view based on the experience I have had over the last 10 years.

(Mr Cowie) If I could perhaps pick up a point. As engineers we believe the important element in the first instance is the practitioner on the ground. That does not have to be necessarily an engineer but it is the practitioner. They are the ones who know what is required to be solved. I think that the process should start from there. It does have to go up—yes it does— but if we have a situation which lies like a lead balloon on top I think that will stifle the necessary

work at the sharp end. I am quite happy with the situation that there will have to be some national organisation which is responsible for the whole issue at the end of the day, but the driving force in my view is at the bottom where the problems actually are seen, recognised and understood. They can then be fitted into a regional or national scene but the initial thrust must be from the practitioner.

Chairman

794. To see if I have got this right, as I understood the evidence of the Institute of Civil Engineers yesterday, they said there existed voluntary coastal zone groups covering 95 per cent of the coastline of England and Wales which they were advocating very very strongly.

(Mr Cox) Yes.

795. That they said made sense, the way it was organised, because those groupings came into existence to cover stretches of coast where there was interaction between what happened to one part of that coastline and another. You would get the Severn Estuary which would be one area where what happens on one side of the estuary then affects what takes place on the other. You would have the Wash Basin and you would have the North-east coast. It was broken up into rather large groupings, but in addition it was felt there was a need for some national body which would give a uniform direction and lay down national policy. Of course, we have had a lot of discussion as to whether the responsibility for the National Trust should lie with the Department of the Environment, the Ministry of Agriculture and Fisheries or the National Rivers Authority or whether we should create a new national body altogether responsible for the coastlines of the British Isles. At that point the Institution of Civil Engineers became rather coy and they did not want to favour one body rather than another. But now today we are getting a further breakdown, are we not, and we are saying that the regional groupings that we started with yesterday should in fact be made up of building blocks or cells. For example, we have Sefton here as a cell which could be a building block, fitting in and dovetailing with other blocks to form one of these regional groupings. That is how I understand the way the evidence is going.

(Mr Cowie) Yes.

796. This is something obviously the Committee would have to consider because it is quite clear from what we have heard that the whole system at the moment in so far as coastal protection is concerned is very fragmented and there is a lot of overlapping and no co-ordination going on. We are looking at the best way in which that can be achieved. Now, turning to the building blocks—and that is why we are here in Sefton, the cell—if I turn to your paper at paragraph 2, you explained to us that the Management Scheme really turns upon the steering group. That seems to be the decision-making body, if that is correct, of this particular cell. You give us the composition. There are elected councillors involved, a number of outside bodies such as English Nature, the Forestry Commission and of course it is fully

Chairman *Contd]*

supported by the relevant officers of the council who are involved in these matters. The impression one gets reading that is that you rely a great deal upon the good working relationship between the individual members that meet together. That of course is very important. Is it that you are particularly fortunate in the individuals that happen to come together and could you envisage situations where other cells may be riven with personality clashes or political problems that would mean that they would not work terribly well? Or to pick up something Mr Cowie said a few moments ago, that we are all beating the elements or combatting nature. Is it faced with the power of nature that you drop your human differences and are able to co-ordinate and work together? I raise that question because obviously if we are trying to create something that we hope will work properly, we want to have a profile which would minimise the usual personality clashes that can spoil the best of intentions.

(Mr Cox) Yes, I wish human beings always worked together even in extreme circumstances but we know that does not always work out. I think, yes, I have to say I think we have been fortunate in the Sefton coast. I think we are helped by a relatively simple pattern of land ownership. For example, there are three principal land owners on the coast. I think it is probably a truism that any organisation functions far better if people work well together irrespective of the framework that is set up. I do not think one could assume automatically that a voluntary system would work consistently well across all sections of the British coast. I think that would be gross optimism, I think looking at the pattern of activity round the coast, you have a vast variation from areas where there is not a perceived problem apparently to areas which are, as it were, market leaders in their fields. I think probably if I ask Mr Houston, who has been the Coastal Management Officer in Sefton since 1979, and who has carried out much of the " working together " process, perhaps he could actually comment on his experience and how far he feels that that could be replicated elsewhere. I think the question depends at what scale we are talking. Are we talking about the Sefton coast or are we talking about a much larger sub regional scale.

(Mr Houston) I think the Sefton Coastal Management Scheme has worked, as Mr Cowie mentioned, as a cell within its own right so that the main landowners were all receiving the same scientific information. There would be problems, for example, if two neighbouring landowners each had their own consultant giving conflicting information about coast erosion, for example. That would lead to conflict. We are very fortunate that the knowledge base we have in Sefton is equal to all landowners so we are working under the same guidance. To think could it work elsewhere—I think it can. There have been some experiments that have been done. I know for example in Cornwall a similar experiment was tried using the National Trust as the lead because they were the major landowner and the National Trust got into problems. They felt uneasy about acting as the lead agency even through the management of the coast and they did not like the

idea they could then lead others and dictate to others. I think if management schemes are going to work there has to be a lead from local authority level to take on this statutory function. I think that is really the basis of why the Heritage Coast Scheme has been successful—they identified something common. They might not be talking about common geomorphology, they might be talking about common visitors or landscapes or something in common along the coast. It would be interesting to see how it works where on the face of it they have nothing in common.

797. In both your answers you seek to attach a great deal of importance to land ownership as being one of the ingredients that assists or could obstruct a sensible scheme coming through. Is that right or not?

(Mr Cox) I think it is a matter of scale, is it not? What we are looking at on the Sefton coast, is the Sefton Coast Management Scheme developed as a land management exercise. That was its origin. I think if you then change that scale, as Mr Cowie was talking about, to the area from Llandudno to the Ribble then the sort of considerations we are beginning to think about are less dependent on individual landowners and more dependent on the co-operation of statutory authorities and agencies.

798. This is probably, you will know, where a national authority can sort out the differences that might arise. Is that how I understand the position?

(Mr Cowie) I would concur with that view. At some stage there may well be the necessity to iron out what may appear to be implacable differences. My experience in general terms is that there is common, I hesitate to use the word " enemy ", but there is an acceptance of a common denominator somewhere. You used the word, Chairman, about working against nature or combatting nature. I would say working with nature rather than against it but it is a common interest and in almost all the cases we are likely to be talking about there is going to be some common interest, whether it be the elements, whether it be land development or whatever. In the small number of cases there will be where there are differences that cannot be resolved by discussion or negotiation, yes, there may well have to be someone at the top who wields a big stick.

799. We have met instances in this inquiry where two local authorities side by side are pursuing diametrically opposite solutions and one is causing enormous damage to the other. Obviously that is a situation one has got to try and find an answer to.

(Mr Cox) I think our concern is reflected in the nature of that upper level authority, as it were, the concern that it should not be, if I can put it this way, a big-brother type approach but it should actually encourage the structure of and provide incentives for the activities already going on at anything from grass roots to sub regional level.

800. Can you help us where the Institute of Civil Engineers were unable to help us yesterday in that. Where would you place the ultimate responsibility, as it were, for making sure we have a co-ordinated plan right round the coastline?

Chairman *Contd]*

(*Mr Cox*) I am not an engineer and I cannot be held to the Institute's views.

(*Mr Cowie*) Neither am I!

(*Mr Cox*) I think Mr Cowie and I may well actually agree on this one. Our view is that the appropriate national institution or department is the Department of the Environment because the Department of the Environment has a view across a whole range of matters, from land use planning to the running of local authorities and so on. It seems to us that is the most appropriate department of state, if you like, to actually provide the guidance.

(*Mr Cowie*) I think I can perhaps today, Chairman——

801. Wearing a different hat?

(*Mr Cowie*) Wearing a different hat, I would certainly say that I believe that it ought to be the Department of the Environment. To explain the difference I think to some extent it is a matter of what one's experience is. By saying that I have no adverse view on either NRA or MAFF as to what performance they have done so far but so far as this authority in my view here is concerned, I believe that the breadth of responsibility that lies with the Department of Environment makes it the appropriate body I think MAFF is relatively constrained and the NRA would be constrained. As for setting up an entirely new one, I am sorry, I believe this is one that could do it and let them do it.

Mr Jones

802. Others would argue, of course, that it is precisely because of the breadth of the responsibility, that the Department of the Environment would be unsuitable because what one needs is a greater focusing and if one is coming into conflict with other plans and programmes within the Department of Environment it will not come to the top very often.

(*Mr Cowie*) I think the argument there though is that at some stage even if you go somewhere like MAFF or the NRA, they do not have the whole picture for all the various elements. I suppose at the end of the day it becomes the responsibility of the Prime Minister to decide everything.

Chairman

803. It is refreshing to find a local authority saying something kind about the Department of Environment!

(*Mr Cowie*) They are pulling their building down!

Mr Jones

804. On the planning, Mr Cox, I think it was touched on earlier on about the jurisdiction of local authorities in planning matters affecting the coast, and I wanted to ask you whether you agreed with the majority of witnesses who suggested that local authority jurisdiction should be extended out to sea?

(*Mr Cowie*) I think I would wish to start off on that one by saying, yes, we certainly would. Beyond low water is essential in our view for Sefton. If you consider mineral working and dredging and things of that nature then you have got to get below low water.

805. How far?

(*Mr Cowie*) I will come to that in a moment, Chairman. As far as the Irish Sea is concerned it is virtually a closed basin itself and almost anything that happens offshore affecting the whole of that sea, can have an affect upon our coast. The most vigorous weather systems and the prevailing sea bed drift is west to east so whatever happens out there finishes up on Sefton's coast. On how far, I think that is a bit of a variable feast throughout the whole of England and Wales and probably the same for Scotland, in that it can be as close as a couple of metres, I suppose, in somewhere where there is nothing but hard rock but anywhere else it has got to be horses for courses and we have to look at where the majority of prevailing influences will be on the authority's area. I think for Sefton perhaps we ought to have some responsibility way out for the whole of the Liverpool Bay. That is not to say we can control everything but it does have an effect upon our coastline. One of the things that happens is that there is dumping of sludge and that seems to be forming a blanket over part of our cellular resource and therefore we need to have some input into where it is dumped, if it is and to be sure that it does we need research and investigation.

(*Mr Cox*) I think it is a fairly thorny question about how far you should go out. If you think about what problems you might face in any particular point on the coast, the answer might be different in each case. Clearly, as Geoff says, the things like marine aggregates and control of moorings or water space or marine conservation issues may on particular coasts suggest that that boundary ought to be in some way a movable feast. The only way I can perhaps see this working within the current framework is that you have something equivalent to an Article IV direction which allows you to apply to the Secretary of State to extend your boundaries. You then have to make the case to him about why you actually want to do it. Clearly if that case were thought to be persuasive then the boundary might be extended. I find great difficulty saying it should be 12 miles out or three miles out because it seems to me you have almost got to say, "What is the problem you are actually trying to deal with out to sea beyond the low water mark?"

806. Some of these problems are obviously going to be well beyond our own territorial waters and whatever we did we could not extend the boundaries beyond that. The Article IV direction solution that you were proposing, is that in response to a particular situation or are you suggesting that that should be applied in the context, for example, of a coastal management plan?

(*Mr Cox*) Yes. I think there may be a number of circumstances in which you would want to apply for that sort of extension. One would be because you have done a coastal management plan. Another would be because a particular problem emerges that you need to make a response to. There could be two circumstances, one would be a preplanned circumstance and the other would be a response to a particularly urgent problem which perhaps you did not actually envisage occurring. I think you would have to in my view try to cover both sets of circumstances.

Mr Jones *Contd]*

807. Some of these circumstances might be mineral extraction. We have had some evidence from Crown Estates that the South Wight District Council has powers under Section 18 of the Coast Protection Act 1949 to control offshore dredging adjacent to its coast. Would that be a possible option?

(Mr Cowie) A couple of things on that one, Chairman. We expressed concern in our evidence that the planning powers were below the water mark, as we have already said, the prospect of offshore mineral extraction, waste disposal and dredging presents difficulties of control in what is a highly sensitive coastal area. Section 18 allows and provides to prohibit extraction of materials from any portion of the seashore within their area or lying to seaward therefrom. It is broad and not very clear but the order is subject to the approval of the Minister for the Environment. The powers exclude mineral extraction, as I read it, more than 50 feet below the surface which includes oil and gas. Sefton coast is under the influence of the Rivers Mersey and Ribble, both of which have been maintained by dredging and the construction of river training walls. Offshore dredging for navigational purposes I do not believe falls within the scope of section 18. Anyway in this location, certainly on the Mersey, it has been undertaken for so long and is so essential to the future of navigation and the viability of the port that the possibility of obtaining an order is not a serious option. Offshore dumping and construction of retaining walls, I believe, again do not fall within the scope of section 18. They are controlled by national legislation whereby a Parliamentary Bill is required for operations below high water mark, in some respects, yes, it includes most of the things that concern us.

808. Thank you. Now your Unitary Development Plan is due for approval sometime later this year, is it not?

(Mr Cox) It will go to public inquiry later this year.

809. So you have had quite a long consultation period so far but obviously some yet to come. Can you give us some idea of what the public reaction has been to what you are putting forward?

(Mr Cox) The reaction has been generally favourable. I think in saying that one has to look at who the respondents have been. The bulk of the respondents have been individuals on environmental grounds of one sort or another and to that extent you would actually expect them (perhaps that is a bit presumptuous of me) but perhaps it is not surprising they feel quite strongly this is the way forward and it is something they would strongly support. There have been a very limited number of objections in relation to development interests and again you will actually expect that probably to be the case because the general argument put forward is that these policies would be constraining which is why they were defined in that way, or some of them were defined in that way, in their operational freedom and their ability to carry on what they consider to be perfectly legitimate activities.

810. They are very site-specific objections, are they?

(Mr Cox) Yes. The Dock Company, for example, is worried about whether these policies will actually constrain its activities on the dock. That perhaps again is understandable and we now have to have discussions with them to perhaps reassure them in some cases that that is not the case. But some of the policies are restrictive, there is no doubt about it, and that is what they are there for.

Chairman

811. I would just like to pursue that for a moment. Maybe I have been partly answered by Mr Cowie a moment ago when he talked about an Act of Parliament and powers given by Acts of Parliament to various bodies. You have a very important port here. Presumably it is acting under powers given to it by a special Act of Parliament, some of them general powers within which it carries out a whole range of detailed operations. Are there occasions in which you feel that perhaps your right to apply the planning law to those intended or actual operations would be preferable to the blanket permission given by the kind of Act of Parliament that ports have?

(Mr Cox) I think the answer to that is generally yes. The experience we have had over the last four or five years when the character of the port's trade has changed significantly—you saw on our trip today probably the worst difficulty that we have had which is through the importation of large amounts of coal. There have been others also in relation to the export of scrap and its processing either for transport or for customers at the other end of the supply chain. There are two problems really in relation to the port. The first arises from their powers under the general development order which allows port authorities to carry out a range of activities, some of which are substantial in terms of what is required for the shipping of goods, the storage, holding and all those sorts of operations. That, for example, has allowed the storage of coal without the need for planning permission because it is clearly in transit and it is unloaded from a ship, it is held on the quay side or stockpiled and then transported out largely by rail. Our ability up to the Environmental Protection Act coming into force in 1990 was very limited in terms of what we could do and we have, in fact, instituted court action in relation to the dust problems. We have now reached an understanding, an undertaking which has been given to Parliament in relation to the Mersey Docks current Bill which is in the House, whereby they have agreed to carry out environmental assessments on projects which would otherwise escape the need for planning permission, by virtue of the fact that the Planning Act does not bite in these particular circumstances. I know other authorities elsewhere have had similar problems associated with the exercise of development rights by docks and harbour companies or other powers which they may possess. I think the problem, if I can put it that way, for the company is that they are very reluctant (to use an apparently vogue phrase at the moment) to have their playing fields made uneven because they see concessions, as it were, to one local authority as worsening their competitive position relative to other ports. I think dealing with it by a local process, which

Chairman Contd]

is the way we have had to do it in this particular case, is not a very satisfactory way of handling it and it is a matter of some concern to me as a planning officer to find that developments which are clearly set out in the 1985 EC Directive on Environmental Assessments and are also included in the 1988 regulations by which it is implemented in this country, actually escape the need for that type of assessment despite the obvious serious environmental problems that occur or impacts that might occur. I think that is something which on a national scale needs to be addressed. I would not necessarily say it was peculiar to the coast as such. It happens to occur on the coast in this particular case and obviously it relates to dock and harbour companies which are coastal operations but that is a matter of very real concern and has caused considerable anguish, certainly in this local area.

Mr Cummings

812. What consideration have you given in your plan to the 5 million tonnes—which could become 10 million tonnes or even 15 million tonnes—of imported coal where promises have been given that it will not be removed by rail. We have had a brief look at the site this afternoon. I have my doubts whether that amount of coal will be moved by rail. What provisions can you make in a plan where you see an area completely surrounded with houses which have residents. What do you do in your plan to accommodate this?

(Mr Cox) Well, the first obvious point is that if you are going to put provisions into the plan, you have to be able to operate some form of control which ensures those provisions can be followed. That mechanism is clearly the process of granting planning permission. If planning permission is not required there is in effect nothing you can do about it unless you actually have some voluntary arrangement with the particular company involved. I think that clearly is the problem. Our plan very clearly says in relation to any activities in the port, the environmental impacts of those activities must be top of our priority list in terms of concerns. That is an empty policy if the mechanism by which that operates just does not apply.

813. Circumventing normal planning procedures?
(Mr Cox) It effectively circumvents normal planning procedures, yes, that is absolutely right.

814. I certainly will not envy you in the future! The Committee realises that very few local planning authorities have considered coastal planning policies in their development plans. Is there anything to prevent other authorities developing similar coastal policies to those in your UDP and if so do you know of any other authorities who have embarked on this particular operation?

(Mr Cox) I think the quick answer to the question is no. There is nothing preventing any other planning authority going down the same route that we have gone down but examples of it are few and far between. I am not aware of any other district authorities certainly which have actually taken this view although other authorities such as Hampshire within their County Structure Plans have got policies relating to the coast.

815. Is this because of the diligence of the officers or the politics of the council?
(Mr Cox) I am tempted to say pass on that, Chairman!

Mr Jones

816. Can I put you an alternative—
(Mr Cox) I am sure it is a combination of both.

817. Is it because you are one of the relatively few, if any, unitary authorities with a large coastal section?

(Mr Cowie) I think I will try and answer this one, Chairman. We are not the only unitary authority with coast but you are quite right, I think we have the longest length of coastline. The unitary authority system means that the buck stops here. I think that is one of the things that has concentrated the minds. I think perhaps in the shire counties and shire districts there is a split of responsibility. It may be that there is somebody saying, "It's them or it's them." I personally view the unitary authorities, subject to some constraints, as being the right way forward.

(Mr Cox) Can I just elaborate, if I may Chairman, on what I see as constraints on that view. I would agree with Mr Cowie that unitary authorities are the way forward because they actually bring the whole range of issues together, avoiding the problems of two different bodies, as it were, rowing about what should be done. I think my only reservation on that is that the unitary authorities should be sufficiently large to have both a coherent geographic remit and to be able to develop the expertise necessary to deal with the matter involved. I think my only concern about unitary authorities would be if they turned out to be very small unitary authorities, then that might be a matter of concern.

818. What do you mean by "small"?
(Mr Cox) Can I just give an example. Before Sefton was created there were five local authorities covering this stretch of coastline. During that period very little was done about the coast in the way that happened after re-organisation in 1974. I am not pretending that was solely as a result of them being brought together. All I am saying that that those five authorities obviously would find it more difficult to come together than a single authority which then became responsible for a coherent coastline. I think if you actually ended up with stretches of coastline such as the Sefton Coast, split between two or three or four authorities, the process of providing that consistency of approach would become more difficult.

(Mr Cowie) It is not entirely a function, I believe Chairman, of numbers of bodies. It is partly a function of the size of the coastline. I think there is a case to be made that there may well be a numerical minimum for a unitary authority to be functional— that in my view is probably about 200,000. That is not part of the issue——

Chairman

819. If the unitary authority is your cell—coming back to my first group of questioning—and you have a regional authority that brings together the various cells, then perhaps we overcome your problem, Mr Cox?

(Mr Cox) Yes, I think if you look at a cell of the size we are talking about, Llandudno to the Ribble, the fact that it goes across a national boundary means that it would be inconceivable that that would be a single local authority.

820. My experience of London local government is that there is a real danger of making your local authorities too large. They are then too remote and their councillors do not know what is happening from one end to another and that causes problems of a different nature and does not make for good local government. There must be identification.

(Mr Cox) I would not disagree with you at all on that. I was looking at it in terms of coastal planning.

Mr Jones

821. That could be overcome, could it not, by arrangements between neighbouring authorities. You are talking about arrangements over a very much broader area which would also be possible over a narrower area. Supposing, for example, Sefton was partitioned between a unitary authority of Southport and Crosby and Bootle went into Liverpool. It would not cause any great problems in co-ordinating along there providing there was some form of structure.

(Mr Cox) Providing in that form of structure the priorities of each individual component actually gelled with each other.

822. That is what we covered in earlier questioning when I think we agreed the thing was going to have to be refereed by the Department of the Environment come what may.

(Mr Cowie) I think the point is perfectly well made. It is more a function, in my view anyway, that it is the unitary authority almost irrespective of size which is the crucial issue, that the buck is in one particular field. There is a case for arguing about numerical size or geographical issues, yes, and they are answerable by the cell system or whatever but it is the buck passing which I feel has to be eliminated.

Mr Cummings

823. On the question of unitary authorities I do not subscribe to Mr Jones's view that the Department of the Environment should be the referee. They made a mess of it in 1974 and there is every indication that the next round will end up in a mess as well. I do believe there has got to be some form of independent commission. If I could deal with the unitary authorities, do you believe that they will have a detrimental or a beneficial effect upon coastal planning and protection?

(Mr Cowie) I certainly think, Chairman, it would be beneficial plain and simple because the buck stops there.

824. Of course, Sefton has 300,000 odd people. You are referring to authorities of 200,000. Do you see the cell being broken down into two cells?

(Mr Cowie) No, I do not. All I was answering was an earlier point, Chairman, that there may well be a minimum size. That is not to say that Sefton is not perhaps optimum. All I am saying is there is an option for a minimum size and I do not think it is much less than 200,000, if at all. There is also a maximum.

Chairman

825. I think the way the Department's mind is working at the moment from discussions I have had is that there is not going to be any single pattern that everything has got to be 200,000 or 300,000. That is not the approach. It is what does the geography and community identity require? And some unitary authorities may be much larger than others depending upon those kind of criteria. I think that is a far more practical approach because people have to identify with their local authority and not feel it is some strange creature that has very little to do with them.

(Mr Cowie) I think, Chairman, part of the proposals, as I understand them, is that Cumbria may be one unitary authority and, yes, fine if that is 3,000 square miles and whatever the population is, so what. If that is going to work then yes it is still a unitary authority solution.

Mr Cummings

826. You suggest in your submission, paragraph 5.10, that environmental impact assessment regulations do not recognise the sensitivity of the coast. Can you give the Committee some examples and suggest how the system should be amended. Do you think local planning authorities have the expertise necessary to evaluate environmental assessments for coastal developments, some of which may involve highly complex engineering and scientific analysis?

(Mr Cox) Can I perhaps start off with the first part of that question, Chairman. I think the answer is yes we do have some reservations. The environmental assessment regulations relate to particular processes but there are also, as you are probably aware, indicative criteria about whether things might possibly be justified on environmental assessment. For example, you saw today when we stood there on the high point at Formby Point and looked southwards there are proposals in that area for a major golf course development. That could have very serious effects upon the habitat of that particular part of the coast. It is that sort of development which currently falls outside or does not form part of this indicative criteria in the environmental assessment regulation. So I think there are some areas there that could do with some improvement.

(Mr Cowie) I think, Chairman, I would pick up the point about whether authorities can evaluate environmental assessments. I think if we are talking about something like Sefton I believe that we certainly can evaluate. For instance, in my department, there are national and international experts in certain spheres and they are perfectly capable of evaluating environmental issues. I am sure

Mr Cummings *Contd]*

that Mr Cox would say the same of his staff. I do not have any difficulty in a unitary authority of the size of Sefton actually doing that, however complex they are, because we handle other matters equally complex in respect of other matters such as highway schemes and so on. I think this comes back again perhaps to the size of the organisation and the ability to have the expertise in-house. But even if they do not and perhaps if we take some of the North Wales coastal authorities they do not but they do have real access to a lot of that information and expertise by belonging to a coastal group through mutual help.

(Mr Cox) Could I just add to that, Chairman. I think that it is very important even if a local authority does not have that expertise, they should build that expertise and they should actually own the process of looking at development and seeing what its impacts are and seeing what the solutions are. They should not depend entirely on some external body to actually give them advice in that sense although in specialist cases they may well actually ask for that specialist advice elsewhere. I think, Mr Deans, who actually runs an advisory service of that sort might actually like to contribute.

(Mr Deans) Yes, Chairman. If I can add to that comment a little, I would certainly go along with what Mr Cowie has said to you, that local authorities do already have to deal with a number of very complex and detailed matters. They have a considerable amount of expertise available on matters of special interest and detailed concern. Surprisingly though, very few authorities have a scale of expertise on environmental matters of substance and detail. I think it might be useful just to indicate to you the difficulties that local authorities may have because of this lack of expertise. My service is one of only three nationally advising district level authorities on matters of environmental concern and the value which the grouping of authorities that have chosen to take advantage of this relationship have gained from this I think is invaluable. That is not to say, of course, that every authority needs to have an expert on any environmental matter within their own staff. I think that would be unreasonable and unnecessary. I think as casework and policy is becoming more attuned to environmental needs which are very wide in scale and deep in complexity, it has become increasingly important for authorities to have somebody to whom they can turn for advice, impartial advice, advice which they can have some confidence in and some ability that they can comprehend and understand, not advice which is so complex in its own right that it does nothing other than raise further questions. The question then, Chairman, is to whom do they turn for that advice? Do they sustain a level of expertise within their own service, their own staffing, or do they go outside for expertise elsewhere? I think that really is a matter for the authorities to decide what is the most expedient and efficient source to gain information on whatever issues are put forward. I do think it is important that authorities do have a base level of understanding on environmental matters across the board, not just in the planning sphere necessarily which they can call on and utilise in order to gain an insight.

827. In your submission you stress the importance of managing the " urban " coast in order to reduce pressure on wilder parts of the coast. This is in paragraph 6.5. How do you propose local authorities implement such a policy?

(Mr Cox) I think the importance of the urban coast is often under-rated. It seems to me that it is extremely important that that resource with all the investment that has taken place in infrastructure and so on is actually recycled, if I can put it that way, efficiently. It is almost an analogy to the principle of urban regeneration because what you want to minimise is the need to, as it were, swan off into undeveloped areas in order to provide facilities or opportunities for economic development. In that sense it seems to me the opposite side of the protection of the unspoiled coast is the proper use of the developed coast. I think the mechanism for that is relatively simple. It is already in place. It is the planning system which is already there. It is set up to deal with that sort of thing. The only problem that occurs is whether the resources, as is in the case of urban regeneration in inner cities, is actually sufficient to achieve that recycling. We have an example that we saw this afternoon which was the Southport sea front, a major leisure area in a fairly serious state of decline and really potentially it will have to provide future leisure facilities for people without actually trying to provide those in areas of the coast which have natural habitat value or whatever it may be. I think there ought to be adequate financial provisions to recycle the British seaside resort. At the present time the financial resources available are generally insufficient to do that. We have a major problem in Southport in getting basic sea defence works in to allow the private sector to have confidence for investment in the area. So, one area is the planning area and the other is how you actually get the financial side of the equation to work in order to achieve what we call recycling.

Mr Jones

828. I am sorry to take you back yet again for the third time to the questions of who takes the strategic overview. It is extremely important and you had some reservations about the idea of the NRA having that strategic overview, I think because of its narrower focus than the Department of the Environment. Were there any other objections to the NRA?

(Mr Cowie) I do not think so.

829. It does not matter to you that it is not an elected body; it is a quango?

(Mr Cowie) Not particularly, Chairman, no.

830. Right, OK.

(Mr Cox) Can I just add to that. It depends on how wide the focus is. Coastal zone planning, I think probably we all struggle to know precisely what the bounds of it actually are. It is probably not very fruitful to go down that particular route but if you see that as a major element of strategic planning then I would have some concern that there should be some democratic accountability for these strategic

Mr Jones *Contd]*

decisions which, as it were, at the lower level are actually going to be taken account of and implemented.

831. I think it depends on whether you are thinking largely in terms of coastal defence or coastal planning which is a much broader issue and the two do not necessarily need to be in the same hands at the end of the day.
(Mr Cox) No.

Chairman

832. There is a form of accountability because the NRA are responsible to the Secretary of State, who in turn is answerable in Parliament. Certainly the NRA is accustomed to coming before this Committee and explaining its actions as and when we have our attention drawn to matters that people are not happy with. So there is probably not a direct relationship to elected members but nevertheless the checks and balances are in place.
(Mr Cox) Perhaps, Chairman, to us it seems rather distant at a local level but much less distant to you at the national level.
(Mr Cowie) Perhaps it is a minor point but one which I see as quite serious, in that sense it does not matter who it is providing we have got some input in the decision-making process from the bottom.

Mr Jones

833. Can I just sound you out about one proposition on the coastal defence side, and that is if the NRA took the strategic view but the local authorities were the executive bodies to carry out the work?
(Mr Cowie) Yes, that is tenable but I think we really have to be taking a broader spectrum than just that narrow little bit of coast which is only half an inch wide in practical terms. I think the inter-relationship between whatever the solutions to whatever the problem has an effect so far in and so far out of that line to simply tie in one element of it into one organisation seems to me to be wrong and it is part of my original statement of the Department of the Environment having that breadth of responsibility.

834. I know but to have a lot of tiers going from local authority, we keep using the term "cell" for different sized cells but the local authority at the bottom perhaps through co-ordinating groups, if I can put it that way, at regional level up to some other level. One has to be careful not to have too many cooks.
(Mr Cowie) Yes, I am concerned at it finishing as a matrix rather than a pyramid.

835. A pyramid upside down is what worries me where everything bears down on you. Can I switch the subject then a little back to a point you touched on in your previous answer about having a financial problem of dealing with the sheer size of the sea defence works that are necessary. I take it that though you are just about able to get along, it is not really a satisfactory situation from the point of view

of maintaining your existing coastal defences. How would you like to see the changes in the financial arrangements work because that is obviously a conclusion one must draw?
(Mr Cowie) I think a couple of points. One, I believe that things like major storm events cause particular problems, as to whether they are maintenance or whether they are a capital involvement. It is a little difficult maintaining defence in the face of that. 1990, for instance, as far as Sefton was concerned we had an instance where the Belwin formula, theoretically provided us with significant assistance. The cost of the damage to our coastal structures was about a quarter of a million. We actually received a supplementary capital allocation but no direct financial assistance. I think something of that sort would benefit from re-addressing.

836. How did that quarter of a million compare with the product of a penny rate and having to pay a quarter of everything above that. That is the Belwin formula, is it not?
(Mr Cowie) Yes, that is right. The quarter of a million was less than the old penny rate but quarter of a million is still a dickens of a lot of money. So in that sense I think maintenance ought to be grant aid related. The other question is what is maintenance? I am sorry if I hark back to yesterday and what we have seen today. If we are going to maintain the sea defence as we do at Formby Point using Christmas trees and fencing and so on to trap sand and you have then got land which does not make a dramatic change in the loss of land but it helps to maintain it in that sense, is a continuing drain on expenditure. It is a little unfortunate if the only solution to get cash for it is to put in a concrete wall, one would not wish to do it. That is changing to a degree, it is quite true. The two agencies responsible for funding do begin to take a slightly more relaxed view on what is eligible. That is the second point. I think we would wish to see some change in making sure there was some funding mechanism which would enable coastal groups to maintain themselves to provide the necessary inter-relationship between the authorities and the works that are going to be necessary in any one of the authorities' areas.

Chairman

837. I would like to turn, if I may, to paragraph 4.4 of your paper where you raise the very interesting point of the Sefton Coast Data Base which has been produced as a result of the co-operation between Sefton and Liverpool University. It seems to be quite clear that you arranged that partnership because you found there was a great need to obtain a database to help you with your work. Would you say that that is the ideal way in producing that database? Other authorities do not necessarily have the kind of partnership you have been fortunate in forging and they may not have the same access to a university to be able to do so. Can you make any suggestion how other local authorities who need a database could be helped to do so?
(Mr Cox) I think we found that very beneficial.
(Mr Cowie) I would hate to use the word "ideal".

Chairman *Contd]*

I think that is a bit presumptuous. We certainly found the relationship with Liverpool University beneficial and I would recommend any other authorities who are trying to do the same thing to endeavour to find an appropriate educational establishment with whom they can jointly do the exercise because the benefits are mutual. The academic institutions range over a broader range of disciplines than research organisations or in terms of consulting engineers for that matter and are less likely to have a particular axe to grind. Co-operation with an academic organisation encourages a local authority to participate to a greater extent and perhaps you would get the experts thinking for you to some extent. I am well aware that the Anglian Coastal Group has produced its own database with the aid of consultants. Yes, it seems to have worked but I believe they would have done better if they could have got in with one of the local universities over that side of the country. We have always co-operated in national research in providing access to the database and that is a beneficial point again. We have co-operated and worked with the Institute of Civil Engineers' database but even that is no substitute for a local database. University, yes, certainly here where we have had traditionally over many years a maritime influence, we found that particularly useful and still do as the University is represented on the coastal cell. Brian O'Connor is a regular participator and we join with him in trying to feed through the education system means of training and re-training engineers. So the whole thing all ties together in my view.

838. Do you think this research and survey work taken at a local level, if it were replicated right round the country there would be a case for co-ordinating this and linking them into some national database or some national research work?
(Mr Cowie) There are certainly elements of it which would be useful if it were known what was being done in some areas. Quite clearly things that happen on the Sefton Coast may not be directly relevant to things which are happening on the Anglian coast, but nevertheless there are elements of data gathering and research which can be of use all round. Yes, some national means of putting all that information together is very useful in my view as you would know where to go to get whatever you thought you might need.
(Mr Cox) I think that comes back in a sense to the question of what sort of national guidance is needed. I think that is one area of national guidance that would be helpful to get a consistent base level approach to data collection on the way to the understanding of coastal processes, if not all round the country simultaneously, initially perhaps in key problem areas.

839. That in fact was going to be my final question to you. What special guidance, from all the experience you have had now over the few years in grappling with these problems, on coastal issues do you feel you should have from central government?
(Mr Cox) I think the first level is guidance from central government is to say that the matter of the coast is important and it is a matter we should look at in an holistic sort of way.

840. That has never been said to you?
(Mr Cox) It has now been said. It is in PPG12. At the end of the sentence it was suggested as an issue which should be looked at in a similar way to that which we have tried.

841. When was the PPG12 issued to you?
(Mr Cox) I think it was this month.

842. I wonder if that is because we are having this inquiry?
(Mr Cox) A little throw away sentence gives an example of the need to look at the coast in the overall sense. We are all agog for the draft guidance on coastal matters which we are reliably informed is about to emerge from Marsham Street at some time fairly soon. But I think guidance should set the framework for an approach which is consistent. It seems to be that the British approach to coastal matters is essentially extremely British. Some of it is brilliant and marvellous; some of it is non-existent. It depends very heavily on a whole set of voluntary co-operation, people being enthusiastic in senior positions and so on. The whole thing is a real patchwork quilt of endeavour and achievement. I think that is one level of guidance. The second level is guidance in terms of the acceptability of things like set back zones or exclusion zones; guidance in relation to how we deal with the question of sea level rises and how we respond to that; (are those things going to be supported eventually when it comes to conflict with landowners or developers or whoever it may be?); the question of national tidal power, barrages and so on—there are a whole host of things—how we approach major leisure developments on the coast outside established leisure areas. Again I would say those are the sort of things which are on my list but it has to be guidance because that guidance had to be applied to the individual circumstances of any particular piece of coastline. I think Mr Cowie might have other thoughts as well on that front.
(Mr Cowie) Yes, I can only repeat the business about it not becoming a statement of dogma, that it has got to be done in this way, but what ought to be taken into account is things like the formation of administration and funding of whatever coastal arrangements there are and guidance on the actual preparation of management plans, which factors to take into account. We think we know but they ought to be on a unified basis.

843. Really what you are saying is that you want help and advice in technical areas where you have not got the resources to obtain the information yourself. What you do not want is instruction.
(Mr Cox) That is right.
(Mr Cowie) Yes.
(Mr Cox) What we want to achieve nationally is a consistency of approach and to travel at a consistent speed so we do not have areas where there is very good practice and other areas where there is nothing at all.

Chairman *Contd]*

Chairman: Any further questions? Thank you very much, gentlemen, for your help this afternoon. We will mull over our experience here in Sefton and hopefully we will not be overtaken by a General Election and we will be in time to produce a report where you may find the advice you have given us is properly reflected. Thank you very much.

APPENDICES TO THE MINUTES OF EVIDENCE

APPENDIX 1

Memorandum by the Anglian Coastal Authorities Group

The Association of Chief Technical Officers (ACTO) has responded in detail to your request for evidence on coastal zone protection and planning and the Anglian Coastal Authorities Group (ACAG) supports this evidence, but wishes to emphasise the importance of the role of Coastal Authority Groups and the need for greater support being given to the groups by statutory means.

Because of the nature of ACAG its evidence is directed to the roles of the many authorities responsible for protection and planning in the coastal zone.

The wide variety of geology on the shorelines of the United Kingdom present difficulties in producing national criteria for the design of coastal engineering systems and coastal management plans. The formation of coastal engineering groups covering coastal cells or groups of similar coastal cells is addressing these difficulties.

ACAG was one of the first coastal groups to form and is a leader in producing coastal management plans. It is anticipated that most member authorities will have such plans within two years and some plans are already finished with coastal engineering works under serious discussion with the Ministry of Agriculture, Fisheries and Food (MAFF).

Allowing environmental considerations to be included in cost-benefit calculations and acknowledgement of long term plans in principle, with detailed approval required for individual schemes are welcome changes in MAFF thinking. These changes will allow whole coastal cells to be considered as a unit and capital works programmed to make maximum use of the residual life of existing structures.

National statistics of coastal works need to be produced in order to prioritise coastal schemes and justify bids for financing future programmes.

Finally it is most important to recognise that the ACAG formed because engineers at local level felt the need to co-operate with each other. They are taking the initiative in working with planners, conservationists and leisure users in order to produce comprehensive coastal management plans based on the engineering opportunities for working with the sea. These local plans must develop into regional plans to provide planning controls based on local needs. This " bottom up " approach is already working in practice and will be far more successful than any approach superimposed from elsewhere, providing the plans are produced within a policy framework which takes into account the environmental and leisure needs.

7 *October* 1991

APPENDIX 2

Memorandum by the Association of British Insurers

1. INTRODUCTION

The Association of British Insurers is the main trade association for United Kingdom insurance companies. In 1990 its member companies received a total of over £65 billion in premium income worldwide, amounting to over 90 per cent of the worldwide business transacted by the whole United Kingdom insurance company market. Within that total, some £14 billion was in respect of general (ie non-life) insurance for risks situated in the United Kingdom, of which the two largest categories were property damage (£5·0 billion) and motor (£4·9 billion). Looking specifically at cover for United Kingdom property, it is estimated that ABI members transact over 90 per cent of the business carried out by all United Kingdom insurers, including Lloyd's. In many cases the cover includes insurance against flood risks in respect of property loss or damage and/or business interruption.

It can therefore be seen that ABI members have a very substantial interest as property insurers in the provision and maintenance of effective United Kingdom defences against coastal and estuarine flooding. They also have an interest in less immediately obvious ways, for example as major investors in commercial and industrial property; as providers of business interruption insurance to companies whose operations might be disrupted by serious flooding; and as motor insurers, since experience of recent coastal floods such as that at Towyn in February 1990 has shown that motor vehicles submerged in salt water to any significant depth become total write-offs. Finally, insurers are concerned in their capacity as providers of third party liability cover: for example, to operators of hazardous waste sites, since the inundation of such sites as a result of coastal flooding could result in heavy claims for environmental pollution, and to owners of coastal defence works, for negligence in maintaining or extending them.

The present submission is concerned primarily to set out the Association's concerns with regard to the current system of organising and funding coastal flood protection and sea defence in the United Kingdom, but before doing so insurers' approach to underwriting flooding risks will be briefly explained.

2. INSURANCE COMPANIES' UNDERWRITING OF UNITED KINGDOM FLOODING RISKS

For **household** buildings and contents insurance, cover against flood has only been widely marketed since the early 1960s, when the older and more restrictive fire and theft insurance covers were largely superseded by wider-ranging policies, including those offering cover for both buildings and contents within the same policy, thus minimising the danger of a claim " falling into a hole " between the two. At that time insurers pledged themselves to offer flood cover to all personal and small commercial risks, with the possible exception of a very small proportion of high hazard properties.

In underwriting terms, the flood risk within household policies has in the past been regarded as of insufficient importance (in relation, for example, to fire and theft) to justify rating household flood cover on a geographical basis. Moreover, since the great East Coast flood of 1953, such household flood claims that United Kingdom insurers have had to meet have predominantly been in respect of inland flooding, from melting snow, overflowing rivers or localised heavy downpours of rain. However, the advent of postcoding enabled insurers to analyse losses for all types of peril on a geographical basis much more easily than before. For contents cover, therefore, geographical rating was generally introduced some years ago, but primarily because of the escalating cost of burglary claims, with a widely varying incidence in different parts of the country.

Recently, the high and rising cost of household subsidence claims, primarily associated with structures built on shrinkable clay soils, has caused some insurers to introduce geographically differentiated rating for house buildings insurance and others to make plans to this effect. Some insurers are now considering the introduction of geographical rating for other weather damage claims, following the heavy cost of the windstorms of October 1987 and January and February 1990.

While current systems can look at historical data, the absence of an authoritative analysis of the relative vulnerability to flooding of different coastal areas, linked to postcodes, prevents the use of predictive assessments.

As regards **commercial and industrial property** insurance, cover against flood is an optional extension to the standard fire policy, and is included in the scope of " All Risks " insurances. Where businesses have conducted their own risk evaluations and decided that the siting of their premises is such as to minimise the risk of flooding, they sometimes decide to dispense with insurance protection against flooding. For major commercial and industrial property insurance risks, the business is often rated on an individual basis, taking into account any special features, eg if a factory is sited on a known flood plain, the premium is likely to be loaded to take account of this. By these informal means there is therefore already a degree of flood hazard evaluation incorporated into this side of the business. However, it should be noted that the construction of buildings situated on known flood plains often takes account of the flooding risk, so that—up to a certain point—any given level of normally anticipated flooding there is likely to cause less damage than a flash flood.

In such cases, the properties are usually visited by a risk surveyor, and appropriate recommendations are implemented for physical protection, eg storage of vulnerable stock. The insurance policy may also be varied in order to provide for a substantial measure of self-insurance (via an excess or deductible) in the event of a flood claim.

Looking to the longer-term future, United Kingdom insurers are bound to take account of the possibility that global warming, besides giving rise to a degree of sea level rise around the United Kingdom coastline, could result in an increase in the frequency and severity of storm surges, in the worst case threatening potentially devasting inundation of coastal lowland areas.

It is with this possibility in mind that the ABI is now closely monitoring information and research being undertaken in the United Kingdom bearing on the relative vulnerability to flooding of United Kingdom coastal lowland areas; the question of the ABI commissioning its own more tightly focused research is now also being seriously considered.

3. THE ORGANISATION AND FUNDING OF COASTAL FLOOD PROTECTION AND SEA DEFENCE IN THE UNITED KINGDOM

(a) *The Size of the Coastal Flooding Problem*

The Committee is likely to have received considerable evidence on this point from bodies with much more direct experience and expertise to offer than has ABI. Nevertheless, the Association would like to stress that, because of the degree of prosperity and technological advance that has been achieved by the United Kingdom, the potential for material losses, and thus insurance claims costs, is now much greater for any given depth and extent of coastal flooding than at any time in the past. For example, were the 1953 East Coast flood to be repeated today, it would not be unrealistic to expect the resulting losses to be at least three times as large in real terms as they were then. Moreover, this is not the " worst case scenario " which it is possible to envisage. There is firm geological evidence that the east of Scotland was hit by a tsunami about 7,000 years ago, resulting from a submarine landslip off the coast of Norway, and producing waves which may have reached as high as 26 feet in the area of what is now Inverness (Dr Alastair Dawson, Coventry Polytechnic, in a paper given to a DYP Insurance and Reinsurance Research Group Conference in February 1991). In this context, it should be noted that several inland earthquakes have occurred near the west coast of Norway in recent years.

It is for this reason that the ABI has over the past year been collecting information on the effectiveness with which United Kingdom sea defences are currently organised, with a view to making representations at an appropriate moment to the responsible authorities. The remainder of this section is based upon the information obtained in this way and the conclusions which have been reached following its analysis.

(b) *How Sea Defence is Organised*

(i) *England and Wales*

The primary bodies in England and Wales responsible for sea defence are the Ministry of Agriculture, Fisheries and Food (MAFF), the National Rivers Authority (NRA), (both of which operate through the Regional Flood Defence Committees (RFDCs)), and local authorities. The NRA was created by the Water Act of 1989 and became operational in September 1989, taking over the main regulatory and water management functions from the previous Regional Water Authorities. The Authority's statutory duties, in addition to flood defence (both coastal and inland), include coastal protection (against erosion), water resources, pollution control,[1] fisheries, recreation, conservation, and navigation. Its coastal flood defence work comes under the ultimate control of MAFF, which sets broad policy in this area.

RFDCs usually consist of two members from MAFF, a number of local authority officials, and local council members. The Committees normally meet three to four times a year. There are nine RFDCs, three of which are further divided into local defence committees (in Anglia, Southern and Wessex Regions).

Although RFDCs consider all local flood protection needs, one of their functions is to prepare, and gain approval for, new sea defence schemes. MAFF engineers vet all projects (even if there is no financial consideration for the Ministry) and ensure that there is a degree of co-ordination. In addition, other interested parties affected by the intended programme of work are informed. The criterion for undertaking any flood defence or coastal protection work is that there should be a demonstrable net benefit to the community above the cost of the work.

(ii) *Scotland and Northern Ireland*

Flood defence is organised entirely differently in Scotland and Northern Ireland.

---

[1] Although it should be noted that under the latest Government proposals it is envisaged that the NRA's responsibility for controlling pollution of rivers, water supplies etc, should be taken away from it and handed over to a new Environment Agency, to be responsible for all aspects of pollution control.

In Scotland, the responsibility for flood prevention lies with the Regional councils and three Island councils which are charged with a number of responsibilities for flood under the Flood Prevention (Scotland) Act 1961 and for coast protection under the Coast Protection Act 1949. Only one of the councils (Central Regional Council) has no coastline and is therefore not concerned with coastal flooding. The councils determine what schemes need to be undertaken in the light of available resources and competing demands. For agricultural land, however, flood prevention is the responsibility of the owner. Although the Regional and Island councils are empowered to consider flood prevention needs, they have no statutory duty to do this. As with coastal protection, the councils' powers are discretionary, with primary responsibility lying with the owners of the properties concerned.

The majority of council schemes must be submitted for approval to the Secretary of State, and are then considered with respect to technical, economic and environmental criteria on the basis of net benefit to the community. For flood prevention work, following clearance, a scheme can be considered for grant aid. For coast protection work (other than in an emergency), the Minister must give clearance and also decide whether the scheme is eligible for grant aid.

There is no single co-ordinating body for flood defence in Scotland and the grants provided to councils (explained in (c)(ii) below), do not include a separate item for coastal flood defence. Information received suggests that officials do not consider coastal flooding to be the potential threat to Scotland that it is to England and Wales. This is possibly due to the favourable post-glacial uplift in Scotland, but nevertheless there are some black spots which suffer periodic floods, eg, Largs.

Northern Ireland also has no overall body responsible for the instigation or maintenance of coastal defences. The Northern Ireland Office reports that if a scheme is required that relates to tourism or harbours, the Department of Economic Development is responsible. If a road or other type of communication is involved, the responsibility lies with the Department of the Environment. However, if there is an essential scheme which falls within the interests of neither Department, the scheme will be sponsored by the Department of Agriculture. The Office also reports that since the cost of coastal work in its experience is usually much greater than the value of the land at risk, the Department of Agriculture has never yet had to undertake any coastal defence works. However, the Office has stated that a working party chaired by the Department of the Environment is looking at the wider implications of sea-level rise for the region.

The lack of any body with overall responsibility for sea and coastal defence in Scotland and Northern Ireland appears to reflect the pre-existing belief that the threat of coastal flooding is not significant in either country.

(c) *The Financing of Coastal Protection and Sea Defence*

   (i) *England and Wales*

The support for coastal flood defence work is slightly different from that for protection projects in the proportion of grant aid available per scheme.

For England and Wales figures for flood defence, both inland and coastal, are available but it has proved difficult for the ABI to break down exactly how these funds are allocated and spent. The RFDCs are funded by grants from MAFF and the Welsh Office and by levies on County Councils. Each RFDC is given a Grant Earning Ceiling (GEC) which is set each year by reference to each Committee's programme of work for the year in question and the local resources considered to be available. The RFDC may spend above its GEC but this will not attract any grant aid for that expenditure, the cost of which must be met from local funds.

The type of work which the Government (through MAFF or the Welsh Office) will directly support is limited; it does not support the maintenance of sea defences but will support agreed schemes of a "capital" nature. For maintenance work of designated coast protection authorities (of which there are 88), indirect support is provided retrospectively each year from the Department of the Environment's (DOE's) general revenue support grant to local authorities. Most protection authorities are allowed a separate element in their returns to the DOE covering all the local services they provide, including maintenance of the coast. It is not possible, however, to work out exactly what proportion of the total maintenance costs are supported in this way.

Anecdotal evidence to ABI suggests that most coast protection authorities do not experience difficulty in recouping the greater part of their total maintenance costs, so long as they meet the net benefit criteria referred to earlier.

The level of sea defence capital grants for which Committees qualify can be either 15 per cent, 35 per cent, 45 per cent or 55 per cent, depending on the flood defence requirements as indicated in their GEC; a supplement of 20 per cent is available for certain work. The stated aim is to provide more support where flood defence needs are high and local resources are low. ABI has been informed by MAFF that local authorities' flood defence work is on average 26 per cent grant aided, with a 20 per cent supplement for sea defence work.

For coast protection work the grant rate can vary between 24 per cent and 70 per cent depending on a formula which takes into account the total cost of the scheme, local resources, etc.

The 1990–91 programme of capital works submitted to MAFF for grant aid amounted to £36 million, in respect of which grant aid of £14·7 million was provided for coastal protection and £10·1 million for sea defence. (No comparable figures have yet been obtained for Welsh Office funding.) The NRA operates a Medium-Term Plan covering a five year rolling programme of intended capital schemes for flood works (both inland and coastal). The present Plan to 1995 involves projected expenditure of £1 billion (at 1989 prices). It has been reported separately that £40·4 million will be spent in 1993–94 for coastal defence; this represents a doubling in expenditure in real terms over 1989–90, but suggests that coastal defence works constitute only a small proportion of the NRA's budgeted expenditure.

In summary, given the method of supporting both capital and maintenance flood defence and coast protection work, it is clear that the majority of these funds come directly or indirectly from central Government, although a sizeable proportion of capital expenditure has to be found from local sources.

(ii) *Scotland and Northern Ireland*

As explained in (b)(ii) above, in Scotland responsibility for sea defence rests with the 12 regional and island councils, but these have no separate item for coastal defence in their budgets. Any coastal flood protection work is financed out of the councils' water and sewerage programmes, with capital works funded by borrowing which is limited by capital expenditure allocations set annually by the Secretary of State. Government grants are available: for coast protection the rate is fixed for each individual council at between 20 per cent and 80 per cent of the total cost; for flood prevention, the basic grant available is 30 per cent, but higher rates can be requested subject to the circumstances of the case. For agricultural land, the owners may apply to the Agriculture and Fisheries Department for grant assistance for flood prevention work.

Information on the funding position in Northern Ireland is not available to ABI. Similarly, ABI has been unable to discover exactly what amounts have been spent in either country.

(d) *Areas for Concern*

(i) *Lack of Information Available*

A number of people whom ABI has consulted, including academics who have specialised in this research area, have made reference to the lack of detailed empirical data on the United Kingdom coast. The NRA, for example, relies on 5m contour maps provided by the Ordnance Survey, which are of little value in estimating the relative vulnerability to coastal flooding of low-lying areas.

It has, however, been reported that the NRA Anglian Region has commissioned Sir William Halcrow and Partners, consulting engineers, to undertake a detailed study of the state of sea defences from the Humber to the Thames Estuary, as a result of which it is to be hoped that much more information will in due course be published (at least for this part of the coastline) than has hitherto been available.

(ii) *Problems of Co-ordination*

Information provided by an ABI member company on how sea defence is at present organised in the Blackpool area suggests that there are a number of problems with regard to targeting funds where they are required, in the amounts and at the speeds necessary. It is understood that similar concerns were expressed in a 1989 study by Noble Denton Weather Services, whose report questioned the way in which liaison in the north-west of England between local councils and the NRA worked in practice. (However, information received for the south-east region suggests that the situation there is much more satisfactory, particularly with regard to the effectiveness of MAFF's co-ordinating activities.) It also seems curious that there are different bodies responsible for sea defence in England and Wales, Scotland and Northern Ireland and that the Ministry responsible in England and Wales is MAFF, rather than the Department of the Environment.

On this latter point, ABI has been informed that the MAFF connection reflects its historical interest in coastal matters, but that MAFF and the Department of the Environment maintain very close links on issues relating to coastal protection and sea defence. However, an added complexity is MAFF's dual interest, since it is also responsible for approving proposals for new development of agricultural land and buildings (including those in coastal areas). In addition, the Department is responsible for considering, and in some cases approving, planning applications for commercial development in areas vulnerable to coastal flooding, and the increased flood defence requirement thereby arising implies additional funding by MAFF, which may cut across its own budgetary priorities. Moreover, in certain areas, coastal land and its defences are privately owned by, for example, British Rail; it was a BR wall which was breached causing the Towyn floods.

On the other points of concern, the need for closer co-operation has been recognised by the NRA. Its 1990–91 Corporate Plan states that a draft guideline document is being enlarged to include aspects of how the Authority liaises with local and regional bodies concerned with flooding. This is expected to feature in the NRA's " Flood Defence 2000 " strategy, now being finalised.

(iii) *Method and Amounts of Funding*

The way in which sea defence and coastal protection is financed appears to be unnecessarily complex and varies considerably between England and Wales, Scotland and Northern Ireland. Anecdotal information suggests that, in some areas at least, much of sea defence work which is in fact maintenance (and therefore ought not to qualify for supplementary Government aid) is on occasion classified differently in order to attract immediate grant aid for the work. In addition, it has been suggested that the nature of the funding system distorts spending in favour of " hard " defences (sea walls, etc) and against " soft " defences, eg artifically constructed shingle banks, which in some circumstances may be more effective, but which require regular maintenance (*The Economist,* 10.3.90, p. 132).

Sea defence work tends to be *ad hoc* and variable, with local authorities having to react to events with little advance warning. The method of financing this work should reflect this. An important concern is that, since part of the costs of sea defence is paid for through levies on local councils, reasonably high priority work may remain undone in areas where it is considered that sufficient funds are not available locally to complete the work. This problem could become more acute as the financial wealth of some local authorities continues to deteriorate.

The NRA Corporate Plan 1990–91 once again appears to have recognised the problem. It suggests that it would be helpful to increase the percentage rates granted by government for sea defence work in future years, but this is not a decision which the Authority is in a position to take.

4. CONCLUSIONS

The potential danger to the United Kingdom of flooding due to sea level rise and global warming is significant. To meet this threat, the amounts of money being spent on identifying and protecting important areas have increased considerably and will continue to do so over the next few years. In addition, the need to plan effectively at local, regional and national level is recognised and on record.

Four areas of concern have, however, been identified:—

—paucity of detailed and publicly available empirical data on the coastlines at risk,

—problems of effective liaison at a local, regional and national level and between England and Wales, Scotland and Northern Ireland, with no single authority responsible for strategic planning for the whole of the United Kingdom; this is further complicated by ownership of parts of the coast by farmers and commercial organisations, such as British Rail, who are presumably also responsible for its defence,

—the apparent overlap in responsibilities for coastal flood defence, with the NRA, MAFF, the Department of the Environment and local authorities all involved in different ways in England and Wales, and with different arrangements again in Scotland and Northern Ireland, and

—the apparently complex way in which funding for sea defence is organised, its partial dependence on limited local authority funds and the lack of detail as to the amounts allocated for coastal as opposed to inland flood protection work.

It would appear that some of these points are now being addressed, albeit in a rather piecemeal way. In particular, progress is being made in mapping some of the most vulnerable sections of coastline. The NRA is reviewing the way in which flood defence planning is organised in England and Wales. However, it has no responsibility for Scotland or Northern Ireland, where no similar mapping or planning initiatives have been undertaken or are apparently contemplated.

The main concerns which remain thus relate to:—

   (i) the paucity of publicly available information on the relative vulnerability of coastal areas to flooding,

   (ii) problems of co-ordination of policy and strategy at the United Kingdom level,

   (iii) the difficulty of ensuring, under the current system, that adequate resources are in all areas devoted to the maintenance (as opposed to the improvement) of sea defences,

   (iv) the apparent lack in Scotland and Northern Ireland of the necessary mechanism for detailed planning and its implementation provided in England and Wales by the NRA, and

   (v) even in England and Wales, the scope available under the present system of interlocking responsibilities for inter-departmental conflicts between MAFF and the Department of the Environment.

6 *November* 1991

APPENDIX 3

Memorandum by Association of Chief Technical Officers

1. INTRODUCTION

1.1 This memorandum of evidence is submitted by Waveney District Council on behalf of the Association of Chief Technical Officers (ACTO). The Association represents technical officers of non metropolitan District Councils throughout England and Wales who are the officers responsible for the management of their professional and technical departments. It is the body to whom the Association of District Councils, which serves 60 per cent of the total population, looks for its professional engineering advice.

1.2 Many of the Association's members are employed by Maritime District Councils within whose boundaries lies almost all of the length of the English and Welsh sea board. As such they are the Officers responsible for the protection of the coast through the implementation of their Council's Coastal Protection policies and strategies.

1.3 Many Chief Technical Officers head multi-disciplinary departments and are also responsible for their Authority's planning function. Those who are not, have a corporate involvement in the planning process. Whilst recognising the relevance of wider planning issues from this position of involvement and liaison in the planning process, this memorandum of evidence has been prepared from an engineering standpoint to allow the appropriate professional planning association to address matters more specific to their interests.

1.4 Following the synopsis and a general statement the evidence is grouped under the five areas identified by the Environment Committee as the principal issues of coastal zone protection and planning.

2. SYNOPSIS

2.1 Maritime District Councils are in good position to retain and expand their involvement in coastal zone protection and planning because:

(i) District Councils have a continuing and long involvement in coastal works which predates the 1949 Act. They also have powers under the Land Drainage Act 1976 to carry out sea defence works. This proven performance and valuable experience cannot be under-estimated.

(ii) District Councils are readily accessible to the public and accountable to them through the democratic process for the protection of their coastline. When inhabitants of coastal towns and villages find their homes at risk, District Councils are regarded as the primary point of contact to deal with these emergencies.

(iii) District Councils have other important responsibilities and roles within the coastal strip including land use planning, recreation, tourism, environmental protection and public safety issues. These are addressed and co-ordinated through local democratic and planning processes which involve wide and extensive consultation with all interested parties. This enables the production of comprehensive and co-ordinated plans for the protection and use of the coastal strip.

2.2 Through the initiative of District Councils, coastal groups have provided a forum for discussion and consultation on future schemes and a platform for the production of overall coastal management plans. Threats to the coastal environment have been identified and positive action is being taken. To make these groups more effective they should be established within a statutory framework. There should be strong links with one Central Government Department to allow full discussion on policies and strategies for coastal works.

2.3 The present administrative arrangements and division of responsibilities work well within a basic framework. Contrary to some uninformed opinion there is wide consultation with affected and interested parties. There are deficiencies within the present financial arrangements covering coastal works and sea defences. These need to be addressed too.

2.4 We believe that the process of change necessary to improve coastal zone planning is one of evolution. If any amendments to the present arrangements are to be introduced, the most opportune moment is at the creation of Unitary Authorities. We see these democratically elected and accountable organisations as being the future co-ordinators of coastal management. At the same time, there could be a review of the legislation to include lengths of the coastline that are excluded at present.

2.5 More guidance on the recommended responses to the likely consequences and affects of global warming and climatic change is essential. To date limited advice is available on increased sea level but more research needs to be carried out on the effects of increased storminess and wave action.

3. Position Statement

3.1 The Coast Protection Act 1949 placed the responsibility for Coast Protection on Maritime District Councils. Prior to that Act Local Authorities performed the function either under local Arts, as owners of land threatened by the sea, or as highway authorities.

3.2 Therefore, many Maritime District Councils have a continuing and long involvement with and experience of the provision, design, construction and maintenance of coast protection works particularly along their developed seafronts much of which, in its original form, predated the Act. This is a situation which the Department of the Environment failed to recognise within their recent consultation Paper on the Future of Local Government. They omitted to register this statutory responsibility of District Councils.

3.3 The responsibility for flood prevention, including sea defence works, falls to the National Rivers Authority through the Land Drainage Act 1976. However, it should be noted that Section 98 of that Act empowers District Councils to construct, improve and maintain sea defence works. Both Organisations are required to consult widely with affected and interested parties. Both have coastal defence plans and programmes mainly dealing with the Engineering aspects (see Figure. 1).

3.4 Maritime District Councils have protective works which have been located and constructed not only to provide defence against erosion, but also to prevent overtopping of defences by high tides under adverse storm and surge conditions. In this dual function "coastal works" are essential for the protection of lives and property. This is particularly the case when the coast protection works form an integral part of, or give support to, the resort's promenades, walkways, roads, paths, steps and slopes giving access to beaches and foreshore, its ornamental and amenity features, its public shelters, boating pools, water sports activities, etc. Therefore, they can be a vital element in an Authority's tourism and public protection policies, contributing to their economic well being.

3.8 Maritime District Councils are sensitive to local needs and requirements. Their performance and experience in the provision of coast protection works should not be underestimated. They have a knowledge of local conditions and problems, often of a unique nature, which it would be difficult, if not impossible, to replace. It is on occasions constrained by the problems of finance, a factor which would no doubt apply whoever carried that responsibility.

3.9 The provision and maintenance of coastal works should be carried out by a locally accountable and experienced authority.

3.10 District Councils have other important responsibilities and roles being along the coastal strip including land use planning, tourism, recreation, environmental protection and public safety issues. These are addressed and co-ordinated through the local democratic and planning processes. Their close relationship is illustrated by Figure 2, from which it can be seen that each function is closely linked to the others, requiring District Councils to develop a co-ordinated strategy.

3.11 The implementation of these processes has established a system of consultation with other affected authorities and the public on Coastal Zone matters and on future coast protection proposals. This basic framework is influenced financially by the Government through MAFF or the Welsh Office and politically by local democracy.

3.12 Coastal Groups have been established largely as a result of the initiative and support of District Councils. The membership of these groups comprise the Technical Officers of Maritime District Councils and other disciplines or interested organisations within a coastal cell(s). At present there are 14 such coastal groups along the coastline of England and Wales. (See Figure 3). The general objectives of the groups are:—

1. To provide a technical forum.

2. To create a library of records, studies, schemes and problems.

3. To identify the need for further studies and opportunities for liaison and collaboration.

4. To work towards an overall coastal management strategy.

4. The Dynamics of Coastal Change Both Human and Physical

4.1 This issue is the most diverse as the dynamics of natural coastal processes have a local, national and international dimension. Similarly the human activities have specific and general aspects.

4.2 The physical processes affecting a section of coast are influenced by the physical characteristics of the coast and its geomorphology, the near shore and off shore features and its exposure to wave attack. Successful protection of the coast is dependant upon observation and understanding of the processes at work and the selection of the appropriate coast protection mechanism or structure if required. In this respect local knowledge is of great value and a local presence enables regular monitoring to be carried out and responses made to specific events.

4.3 There are many examples of local research initiatives to investigate specific problems of wave climate, cliff stability, salt marsh erosion, on shore/off shore movement of beach material, etc. Whilst some will be of use in the local context only, others will be of general value and may involve a shared sponsorship and pooling of knowledge. National and international research is necessary on those aspects of coastal processes and meteorology which are common to all, particularly global warming, sea level rise, increased storminess, statistical storm return periods and basic design data. Each aspect depends upon the other. Local knowledge is useless without fundamental design data. The solutions to coastal defence problems are site specific and there are numerous examples of imposed " standard " structures which failed because their performance did not match the conditions. Once the coastline is unprotected then nature will quickly determine its own coastline.

4.4 Natural processes of protection are likely to play a far greater role in the future. It is extremely important that the fragile state of equilibrium is not upset by excessive extraction of deposits. These deposits feed beaches, the coast's first line of defence.

4.5 Whilst local knowledge is important, the processes influencing and acting upon the coast do not recognise administrative boundaries. This situation has been recognised and addressed by the setting up of coastal groups around the coastline. Each group is based upon a coastal cell or cells and their membership largely comprises representatives of the Maritime District Councils from the particular length or section of coast. The general objectives of the groups are to share knowledge, to consult on proposals, and to resolve conflicts through a coastal management strategy. Some groups are multi-disciplinary and include representatives from the various bodies with an interest in the coast.

4.6 The human aspects of coastal change are becoming increasingly important. The demographic changes, increasing wealth, greater leisure time, the increasing awareness of the environment have brought into focus the need to conserve flora and fauna. These have generated pressures and conflicts resulting in the need to manage the coastal zone in an integrated manner. Maritime District Councils are responding to these pressures by preparing Coastal Management Plans. There is a pressing need to plan for the various demands being placed upon a valuable but finite resource. Clearly many conflicts will need to be resolved.

4.7 Some coastal management plans have been prepared which only cover beach or cliff management. This view is far too narrow, the wider aspects of land use, recreation, tourism, conservation, landscape, and archaeology must be encompassed.

5. The Risks to Coastal Settlements and Coastal Ecosystems

5.1 Risk and its evaluation is very much a matter for local determination and reaction. Local people express their point of view via the local democratic process. Local conditions play an important part in the overall assessment of risk, ie high tides, condition of defences, mechanical failures, tidal surge effects, wind action, overtopping of structures and closeness of estuaries. Local accountable presence, with adequate local knowledge, commitment and resources are essential particularly for areas at risk pending action which might eliminate that risk.

5.2 Nationally and regionally the co-ordinating role of one government department, ie MAFF or the Welsh Office with staff experienced in and dedicated to the function is very important in evaluating schemes and approving grant aid and supplementary credit approvals. A key element of this process is the overall benefit/cost evaluation for a particular scheme. The recent decision by MAFF and the Welsh Office to accept the intangible benefits such as tourism, recreational value and environmental considerations, is welcomed, as is the Government sponsored associated research being undertaken by the Flood Hazard Research Centre of Middlesex Polytechnic. Their inclusion has enabled the construction of schemes needed to protect threatened existing coastal developments which otherwise might not have been considered as financially viable.

5.3 The nationally applied design parameters for coastal defence schemes are of critical importance. Storm return periods are being questioned in the light of the storms of the winter of 1989–90 and there is considerable uncertainty about future engineering responses to global warming and climatic change. Responses to these issues must be basis initiated centrally and their implications and recommendations dealt with on a national/regional and even/or a local basis. Whilst there is current research under way further resources need to be employed to ensure that reliable and timely advice is forthcoming to the practitioners on the ground.

5.4 Local inhabitants are extremely stressed by flooding and the potential for repeat events. In the first instance they look to the local democratic system for adequate warning systems, action during events and more particularly prevention. Experience shows that the response to coastal disasters is initially spearheaded by Maritime District Councils. This was highlighted by recent floods at Towyn and Lymington. Responding to emergencies and supporting local people, which is dependant upon local organisation, planning resources and knowledge is a primary function of local councils. In addition to the threat of natural disasters, district councils also have a primary role in respect of oil pollution of the coast, dangerous cargoes washed ashore, litter, and as far as local people are concerned a monitoring role with respect to water quality and bathing beaches. In the light of greater local accountability there is a need to maintain and strengthen these local roles.

This is particularly important with respect to water pollution and its effect upon the coastal and bathing waters of a tourist resort and its associated beaches, which may find its water quality failing EC standards. The associated publicity can have considerable adverse effects on local tourism interests and can lead to the loss of the prestigious Blue Flag Award for clean and tidy bathing beaches and water.

5.5 Risks to coastal eco-systems arise as a consequence of man's activities. Building works, recreation activities, exploitation of mineral resources and their consequential effects, including the subsequent need to carry out coastal defence works can and do result in damage to the environment. Paradoxically the construction of coast protection and sea defence works can both despoil and protect sensitive coastal sites. There is a strong recognition of the need to work with the national and local environmental agencies in order to construct acceptable engineering works. In rare situations there will be objectives on both sides which cannot be reconciled. In these instances the overview provided by the Public Enquiry will determine which case is the most important and the course of action to be pursued.

5.6 District Councils and their staff are aware of the pressures and threat to our coast and its heritage and by involving all of the diverse interests concerned endeavour to ensure that the decisions made are soundly based.

6. COAST PROTECTION SYSTEMS, ENGINEERED SEA DEFENCES AND THEIR ENVIRONMENTAL CONSEQUENCES

6.1 In the past certain coastal structures were built which were very successful from an engineering standpoint but nevertheless may now be considered to be environmentally and aesthetically unfriendly. At that time the need for these " hard defences " to perform effectively far outweighed the consideration of any environmental damage that they may have caused. There is now far greater sensitivity and understanding of the need to produce " soft " engineering solutions which work with the natural processes and also fit into the environment. " Soft " engineering solutions generally rely on using natural materials and systems to achieve a more acceptable defence.

6.2 There will always be situations where this may not be achievable. For instance, where the prevailing conditions are such that the only effective solution is a " hard " structure of concrete and steel. In these situations the appearance of the structure can be softened by imaginative use of high quality materials and the use of appropriate finishes. In any event its design must be appropriate for its location taking into consideration its function and its relationship to its surroundings.

6.3 An environmental impact analysis has become an essential part of any coastal scheme. This study examines the impact of the scheme upon the environment and its effect on nature. To be successful and acceptable the study must be prepared with the support and co-operation of the recognised environmental and conservation organisations. Through this consultation process, in conjunction with the design of the scheme, all points of view are available when it is submitted for planning approval. Scheme preparation is a long and complex exercise (See Appendix 1) which if used properly produces solutions that should suit and meet all objectives.

6.4 As has been stated before there are many situations where the existence of coastal structures is necessary to conserve sites of special environmental importance. In some areas the existence of defences has enabled the development of flora and fauna of national or international importance. There are also places where the stabilisation of coastal erosion has threatened sensitive areas, particularly for sites of geological importance. In nearly all of these circumstances it has proved possible by consultation and negotiation to arrive at solutions which are acceptable to all interests. This may include doing nothing at all.

6.5 Tropical hardwoods have traditionally been used in coastal defence works. Its continued use can only be supported if it is from managed renewable sources. In its own way this has advantages and disadvantages. For those countries supplying the favoured hardwoods there are potential benefits, but for the end user there is a likely increase in costs. Nonetheless, the need to conserve tropical rain forests is well supported amongst engineers.

6.6 The ever increasing use of rock armour as an alternative material to concrete and steel creates another environmental problem. Access to coastal sites is generally only feasible via road transport. This is both costly and environmentally unacceptable when travelling through rural areas. As a consequence it has proven more economic to obtain rock from Scandinavia, particularly for eastern region contracts.

6.7 The existing legislation provides for consultation with the organisations charged with the responsibility for safeguarding the environment and for the resolution of conflict and the consensus view is that it works reasonably well.

7. PLANNING POLICY IN COASTAL ZONES

7.1 There is an ever increasing development pressure on the coastal zone. The seaside is a highly desirable place in which to live. The consequences of development in areas of land which are threatened by flooding or coastal erosion results in the need to provide coastal defence systems. Engineers have always recognised that building on unstable land or land subject to risks from flooding or erosion is not a sensible course of action, but without appropriate legislation find difficulty in preventing it.

7.2 Advice given to refuse permission on the grounds of the site being unsuitable because of these risks has largely been considered irrelevant to the planning process. The presumption has been in favour of granting planning permission and experience has proved that when an application is refused on these grounds, subsequent appeals have resulted in the decision being overturned.

7.3 The guidance on planning needs to be strengthened so that the advice given on sound engineering grounds can be accepted and acted upon. Many authorities are now adopting policies which approve development in the coastal strip only if it meets certain criteria. There can be developer contribution towards the cost of protection; minimum floor level heights; sensitive zones in which no new development will be permitted.

7.4 Coastal engineering and development control is only one facet of the overall management of the coastal zone. The development of coastal management plans will integrate all the various land and sea use aspects with the other elements. The preparation of such plans will identify the conflicts and problems and ensure that they are managed and resolved in a satisfactory way.

8. THE ROLES OF THE MANY AUTHORITIES RESPONSIBLE FOR PROTECTION AND PLANNING IN THE COASTAL ZONE

8.1 The overall role of one Ministry, now MAFF or the Welsh Office, in overseeing the works of Maritime District Councils and the NRA works well. The Regional Engineers know their area, are accessible and very helpful and knowledgeable. The District Councils know their responsibilities and boundaries and have the mechanism to co-ordinate their works with their neighbours. Their statutory powers are exercised to ensure the essential coastal protection and sea defence works are built, renewed and maintained.

8.2 The Maritime District Councils are the primary point of contact for local people and without doubt the split in responsibility for coastal protection and sea defence work does cause confusion. The inhabitants of coastal towns and villages whose homes are at risk find that on sea defences not controlled by District Councils they have less easy access to representatives and yet for coast protection they can telephone their local Town Hall or Councillor.

8.3 The confusion created by the difference between sea defence works (constructed to prevent flooding of land by the sea) and coast protection works (which have the function of protection against erosion and may also provide protection against encroachment by the sea) extends to the administration and financial fields. To redefine all such works as coastal works to be carried out in accordance with the provisions of the Coast Protection Act 1949 would be a major contribution to the resolution of the current confusion.

8.4 A Maritime District Council is the one authority which has a direct involvement with planning matters, coast protection, leisure activities, tourism, environmental protection and public safety. Through these various activities and the provision of these services, they have established strong and direct links with other bodies who have an interest in the coast. These responsibilities and roles can only be strengthened by the establishment of the proposed unitary authorities.

8.5 It is for these reasons that the responsibilities and roles of district councils in respect of coast protection and planning should remain in their control and eventually transfer to the unitary authorities when they are created.

8.6 There are many other bodies both private and public which are able to undertake works or authorise work which are outside of the control of the normal statutory processes. These include Crown Estates, English Heritage, MOD, British Rail, port authorities and private landowners. In some instances the role of these bodies is not understood and they are perceived as a greater influence than they really have. Notwithstanding this they can and do undertake activities or by inactivity create situations which have significant implications for the coastal environment. Some regulation needs to be imposed to ensure that no organisation can carry out work which may be detrimental or damaging to the coastal environment without first undertaking the necessary consultations.

8.7 The coast, for coast protection purposes, is defined by the Coast Protection Act 1949, there are anomalies in these definitions which need to be addressed and it would be appropriate to review the boundaries of the Act. There are many large estuaries where coastal processes take place and yet these areas are not encompassed by the Coast Protection Act 1949. Similarly large areas of saltmarsh which form an important natural coast protection feature and are being eroded are outside the boundaries as defined by the Act, particularly where these features are within estuaries.

8.8 The MAFF or the Welsh Office now has the responsibility of co-ordinating strategy and controlling the funding of coastal defence in England or Wales. With their encouragement, the coastal groups are now providing an effective forum for debate on coastal management issues, which enables those participating to inform others of their roles. By meeting and participating in open discussion an appreciation of other points of view is gained and problems and conflicts can be aired and resolved.

8.9 Working within the present legislation coastal groups are addressing the production of regional plans. In some areas, working from a local level, members are producing local management plans which could be used as a model for others. (See Figures 4 and 5.) Adopting the hierarchy for the production of regional coastal management plans shown in Figure 6, the Coastal Groups can produce Coastal Management Plans covering the zone within their cell.

8.10 The potential formation of unitary authorities provides the opportunity to put the control of coastal works, coast protection, coastal zone monitoring, management and planning in one organisation. The development of coastal management plans will be an essential tool for the future control of development within the coastal zone. Controlled through one central government department, planning in the coastal zone can be developed in an effective manner.

8.11 The cost of coastal works is met by Central Government by means of grants, supplementary credit approvals and the SSA (standard spending assessment). It is a very complex system which is not widely understood and it is difficult to explain in clear terms. The present system also requires County Councils to contribute towards schemes over which they have no control and little, if any, input. It would be appropriate for the present financing arrangements to be revised.

October 1991

FIGURE 1

Inputs In District Council Coastal Protection Plan

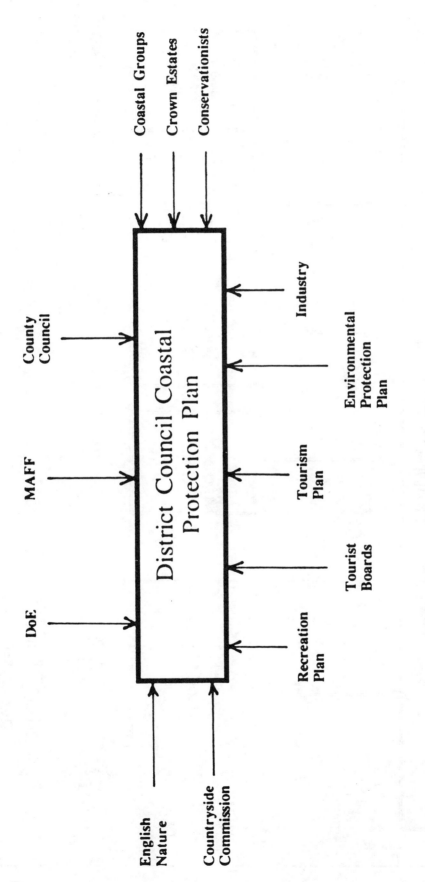

District Council Issues

FIGURE 2

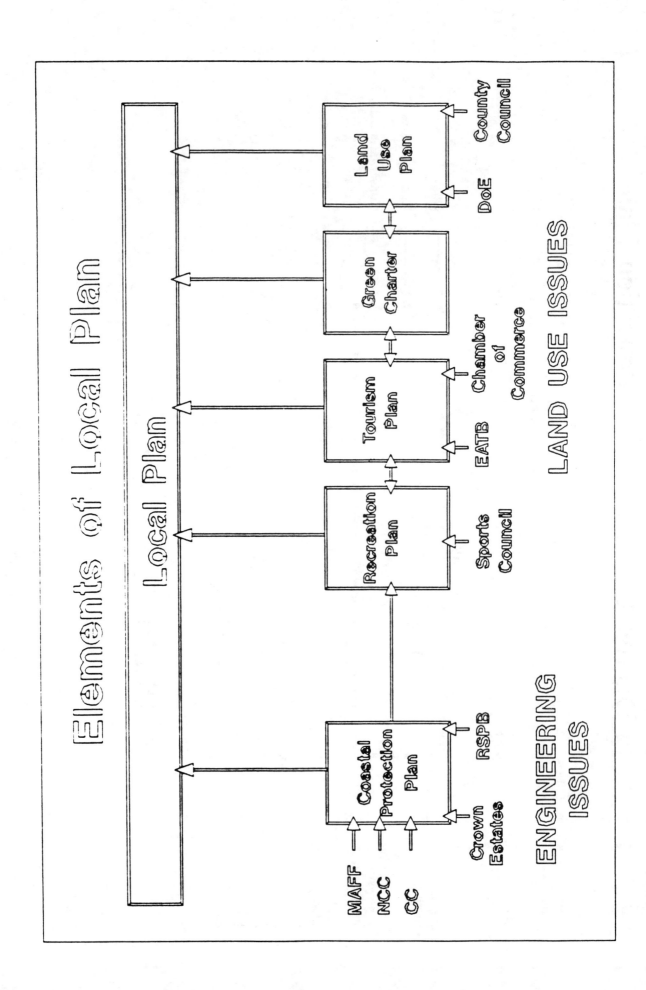

FIGURE 4

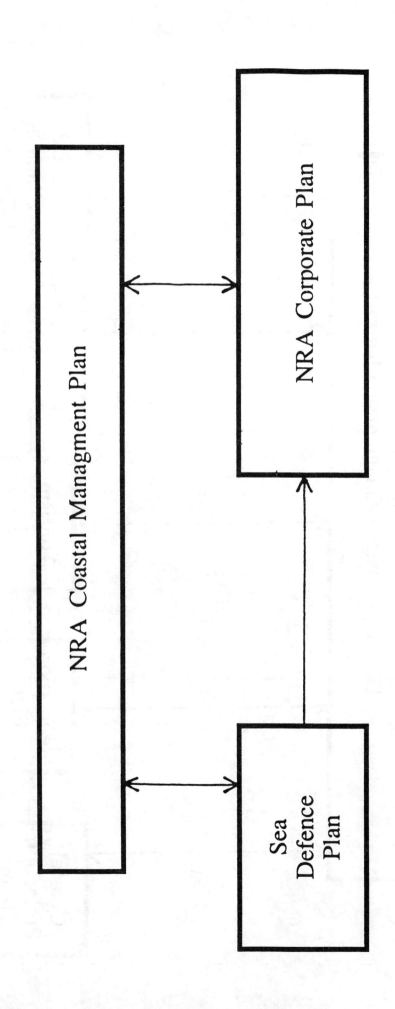

Elements of NRA Coastal Management Plan

NRA Coastal Managment Plan

NRA Corporate Plan

Sea Defence Plan

FIGURE 5

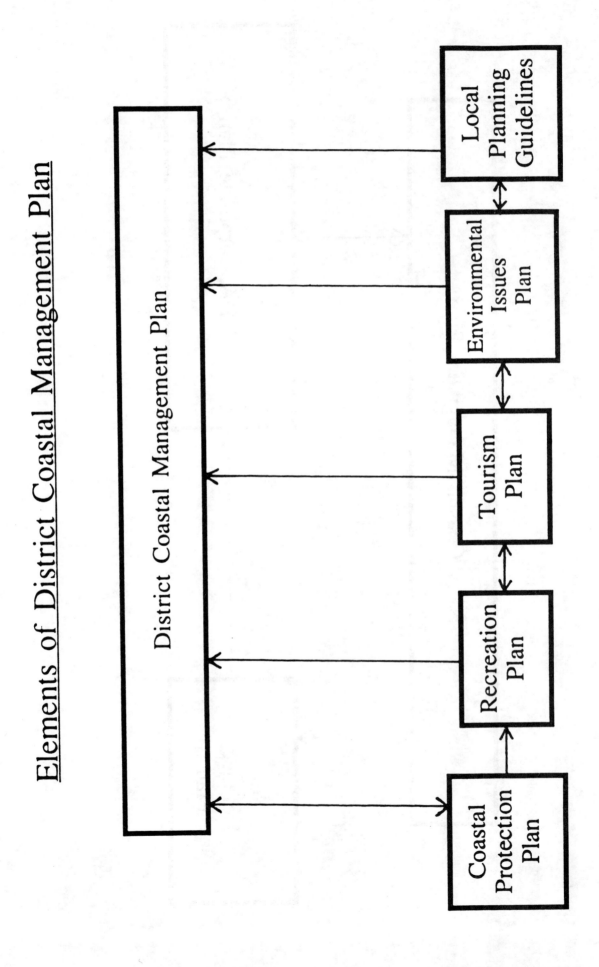

Elements of District Coastal Management Plan

FIGURE 6

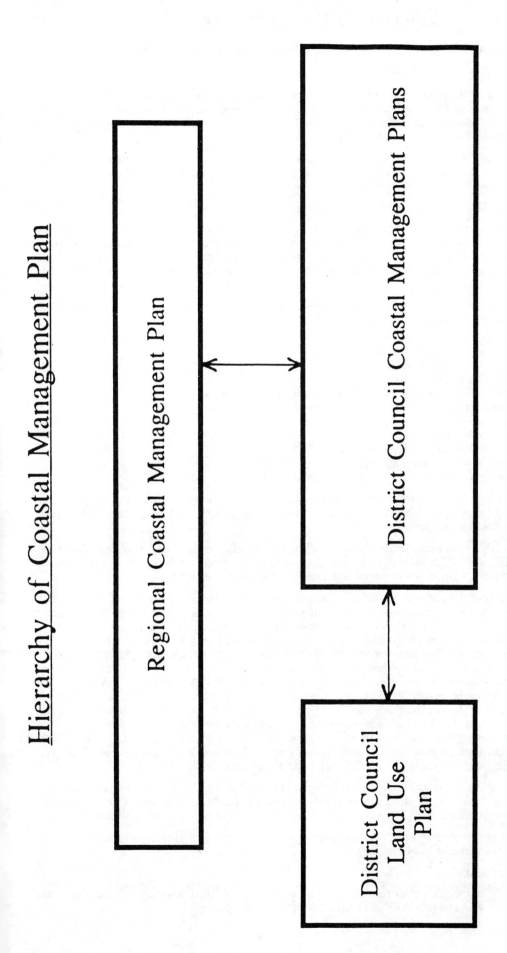

Hierarchy of Coastal Management Plan

Regional Coastal Management Plan

District Council Coastal Management Plans

District Council Land Use Plan

COAST PROTECTION CAPITAL WORKS

Additional Notes to be read in conjunction with the Scheme Preparation Flow Chart

Box No	Remarks

1.　Problem can be manifest in a number of ways—condition following inspection, public concern, collapse etc. At this stage an assessment is made of the desired location of a scheme of works in the Authority's forward programme and, if available, the policies in the Coast Protection Strategy Document should be brought to bear.

2.　MAFF are always keen to learn of problems at the earliest posible stage. There is no specific format to this—usually a fairly informal method of advising the Regional Engineer either by letter or a phone call noted on file.

3.　No comment.

4.　Dependant on the timing, this basic information is either sent to MAFF under separate cover or included in the periodic returns. As the new return format is provided on an annual basis (Form FLDD 1) and only includes anticipated expenditure for the next three years some schemes included in the Authority's capital programme may not be advised to MAFF as a matter of course by this method.

4.　The range of consultation is, to a degree, determined by the site and the type of work envisaged. This list gives typical examples:

> National Rivers Authority
> Adjacent Coastal Authorities
> Crown Commissioners
> MAFF Fisheries Division
> Department of Transport (who in turn contact Trinity House)
> Associated British Ports (" quasi " adjoining CPA)
> RSPB
> Relevant Wildlife Trust
> County Council—ref Heritage Coast
> Countryside Commission
> English Nature

At this stage there is no formal requirement to carry out consultations with these or any particular body—it is simply done to reduce the likelihood of problems at later stages when such consultation is mandatory.

5.　Although shown sequentially on the flow chart it is common for 4 and 5 to be taking place simultaneously. This stage includes site investigation work, model study site surveys and revision of estimates. Options are explored and a preferred solution is devised.

6.　With a good idea of what the engineering solution to the problem should be and the receipt of comments from conservation and environmental agencies a decision should be made whether or not to have an Environmental Impact Assessment carried out. Sea Defence works have clearly laid down procedures for dealing with this matter under Statutory Instrument 1988 No 1217 " The Land Drainage Improvement Works (Assessment of Environmental Effects/Regulations 1988 ". There are no equivalent provisions for Coast Protection Works which fall back to the general requirements of " The Town and Country Planning (Assessment of Environmental Effects) Regulations 1988 ". The general burden of responsibility for deciding the need for an ETA falls to the local planning authority who would determine whether the scheme fell into the category of a " schedule 2 " scheme—most typically because it is in a particularly sensitive or vulnerable location. They may then require Technical Department, as applicant, to comply with the regulations governing Environmental Assessment—issuing of relevant notices etc. There is a growing need to include EIAs for all major coast protection works.

7.　Planning Authority will refer scheme automatically to Parish and County Councils. Thereafter they will decide, on the circumstances of the scheme, on who to target details of the application. This could include NRA, Local residents, Countryside Commission, English Nature etc.

8.　Just to note that in the event of strong objections the Secretary of State for the Environment can call in a scheme for public inquiry.

8A.　No comment.

8B.　Essential to include a clear resolution from Council—one of the documents to accompany the Engineer's Report to MAFF is a certified copy of this resolution.

9.　No comment.

10.   Notice referring to the scheme has to be published in local press and the London Gazette. It must also be served on adjoining Coast Protection Authorities, the County Council, the local Fisheries Committee and the NRA. Twenty eight days are allowed for notice of objection to be made.

10A.   This is licence to carry out marine works—issued by MAFF.

11/11A/ No comment.
11B

11C.   MAFF need to ensure that the contract contains standard clauses allowing them access to the works and to satisfy themselves that there are no unacceptably restrictive clauses.

11D.   No comment.

12.   Engineer's Report will be accompanied by:—

      (i)   Scheme drawings
     (ii) Cost benefit analysis
    (iii) Certification of service of notices
    (iv) Copy of the notice
     (v) Copies of publications containing the notice
    (vi) Certification that the Authority owns the land—or—that consent from the landowner has been obtained
   (vii) Evidence that the scheme has planning approval
  (viii) Statement of Need
    (ix) Certified copy of the Council's resolution
     (x) Evidence that the licence for carrying out marine works has been obtained.

13.   MAFF's approval that the scheme meets the requirements of the Coast Protection Act. This consent is not sufficient to allow work to proceed if grant aid is being sought and the authority intends borrowing to cover the balance after grant aid.

14.   Having received and approved the returned tenders MAFF will issue approval to grant based on the tender sum. This is accompanied by award of Supplementary Credit Approval to cover the non grant aided element of the costs. Further consent is required from MAFF in the event that actual costs exceed the tender sum by more than 5 per cent.

15.   No comment.

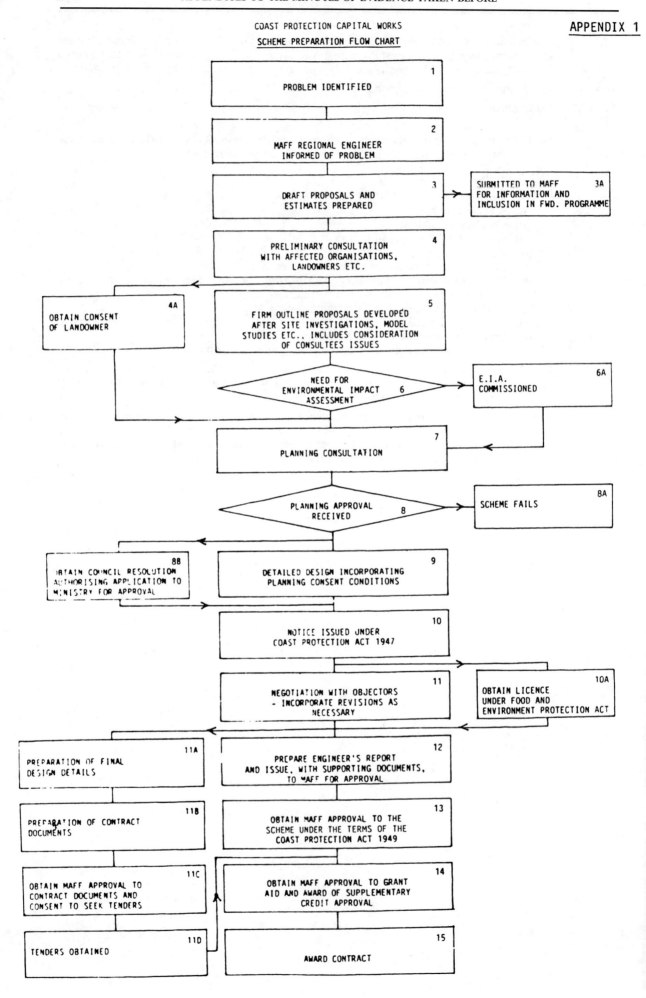

COAST PROTECTION CAPITAL WORKS
SCHEME PREPARATION FLOW CHART

APPENDIX 1

APPENDIX 4

Memorandum by: Association of County Councils, Association of District Councils, Association of Metropolitan Authorities, County Planning Officers Society, District Planning Officers Society, Metropolitan Planning Officers Society

CONTENTS

1. GENERAL

1.1 This memorandum is submitted with the support of the Association of County Councils, the Association of District Councils, the Association of Metropolitan Authorities, the County Planning Officers Society, the District Planning Officers Society, and the Metropolitan Planning Officers Society.

1.2 It has been prepared on their behalf by the National Coasts and Estuaries Advisory Group which comprises officers of local authorities who are at the forefront of coast protection and planning practice. The Group was established in 1990 to disseminate knowledge about the coastal zone and to publicise "good practice".

1.3 The joint nature of this submission demonstrates the common concern which exists over the current situation in the coastal zone. It is testament to the inadequacies of current statutory and administrative arrangements for the coastal zone, and the measure of agreement over the solutions needed.

2. LOCAL AUTHORITY RESPONSIBILITIES IN THE COASTAL ZONE

2.1 The Association of District Councils member authorities have a variety of powers and functions relating to the coastal zone. These include responsibility for development control, local plan preparation, coast protection, together with the provision of services on amenity beaches and elsewhere. The latter includes litter clearance, refuse collection, operation of tourist facilities, maintenance of parks and gardens. All the coastal districts in England and Wales are in membership of the Association.

2.2 The Association of County Councils member authorities have responsibility for strategic planning, minerals and waste planning while other county council services such as (environmental) education, waste disposal, highway planning and construction can have an important impact in the coastal zone. All coastal counties are members of the Association.

2.3 The Association of Metropolitan Authorities member councils have responsibility for the majority of both county and district council functions. Some of Britain's most important estuaries lie within the metropolitan areas.

2.4 All three types of authority have other responsibilities which impinge on the coastal zone including management of land in their ownership, environmental improvement work, control of pollution, emergency planning and countryside management.

3. DEFINITIONS

3.1 To the general public, the coast is usually epitomised as cliffs and beaches facing the open sea. However, estuaries are at least as important in terms of their variety of wildlife habitats, the development and natural threats facing them and the opportunities posed, and the issues to be addressed. The coastal zone must be defined as embracing both, as integral parts of Britain's varied coastline.

3.2 Misunderstandings sometimes occur because of confusion over the meanings of particular terms. Some people see coastal zone planning as essentially about erosion and flooding. It is our view that coastal zone planning must embrace all aspects of the coastal land and marine environment, including nature conservation, recreation and development.

3.3 The Committee has defined its Inquiry as focusing on coastal zone protection and planning. " Protection " has a very specific meaning at the coast and is considered in this submission in this technical sense. However, " protection " can also imply preserving and maintaining the status quo, while " planning " concerns future change. It is our view that there is an even greater need in the coastal zone to consider " management "—indicating regulation and control of existing uses. This is needed to achieve a balance between the requirements of navigation, waste disposal, drainage, flood prevention, commercial fisheries, recreation, mineral extraction, and (possible) energy generation.

4. PRESSURES ON THE COASTAL ZONE

4.1 The United Kingdom coastline contains some of the most beautiful landscapes and important habitats anywhere in Europe. It is nearly 25 years since the last national review of the coastline in England and Wales— " The Planning of the Coastline "—published by the Countryside Commission in 1970. Only one of its recommendations was implemented—namely the establishment of the concept of " Heritage Coasts ".

4.2 The designation of Heritage Coasts has been a success as far as the conservation of Britain's most prestigious stretches of coastline is concerned, and has led to the development of some worthwhile management initiatives. However the problems now facing the coast are of an altogether different order, requiring a more comprehensive approach.

4.3 Most of the 1,460km of Heritage Coast coastline is on rocky and sandy shores; none of the larger estuaries is substantially within an area designated as a Heritage Coast. Yet they are a major component of the coastal zone; with 9,320km of estuarine shoreline forming almost half (48 per cent) of the longest estimate of the British shoreline. Our estuaries are unrivalled in Europe for their number, size and diversity of form and together they comprise 28 per cent of the entire estuarine area of the Atlantic and North Sea coastal states—more than any other European country. According to the Estuaries Review of the former Nature Conservancy Council, estuaries are amongst the most fertile and productive ecosystems in the world, making them of great conservation importance.

4.4 In the White Paper " This Common Inheritance " published last September, the Government identified some of these broader issues. Thus, there were welcome commitments to a number of actions, including an end to the dumping of sewage sludge at sea and the implementation of North Sea Conference decisions. There was also a reference to the possible extension of Marine Consultation Areas. These and other possibilities for future action were timely and welcome.

4.5 We must, however, express our disappointment that the White Paper did not reflect an appreciation of the need for a much more strategic approach to the whole of the coastal resource or of the urgency of providing a framework through which policies for the coastal zone can be properly integrated.

4.6 The coastal zone faces a whole variety of human pressures, some of them new, all of them potentially severe. These include:

—increasing demand for recreation;

—the effects of fossil fuel extraction, processing and storage;

—exploitation of marine aggregates;

—the major growth in fish farming;

—problems of marine pollution;

—proposals for new fossil-fuel power stations;

—military development and use;

—a variety of major estuarial and coastal developments sometimes involving land claim. Examples are barrages, large recreational schemes and airport proposals.

4.7 The key, in our view, to grasping these issues and creating a strategy which properly addresses them is to treat the coastal zone as a whole: that is to say to consider the coastal land, the foreshore, and the area below low water as integral parts of a single zone, across which there are complex interactions. This integrated approach has been missing hitherto.

5. The Inadequacy of Current Arrangements

5.1 The inadequacies of current arrangements for coastal zone protection, planning and management can be summarised as follows:—

—a mismatch between legislation and present-day coastal conditions.

—continuing damage to estuaries in spite of the growing awareness of their value and of the designations intended to protect them.

—continuing pollution of the marine environment by land-based industry. Despite new powers available to the National Rivers Authority, there is gross pollution in many rivers.

—difficulties in meeting international obligations because Ramsar and SPA sites are limited to existing SSSIs, and because of Government indecision over proposals for Ramsar designation.

—narrowly defined port and harbour authority legislation and responsibilities conflicting with modern environmental and recreational requirements.

—the dual role of Crown Commissioners as landowner/developer and environmental protector of the sea bed.

—the plethora of bodies with powers and responsibilities in the coastal zone, and the absence of any statutory or defined obligation on any one authority to formulate any overall strategic view.

—lack of co-ordination between the various bodies with powers and interests in the coastal zone—both geographical, ie between local authorities, and functional, ie between local authorities, NRA, port authorities, etc.

—lack of forward planning with a resulting tendency to react piecemeal to circumstances, for example, coastal protection works and flood defence schemes.

—the wish of the Ministry of Defence to withdraw from its management responsibilities in all but those areas actively used.

—lack of knowledge about the cumulative impact of coastal activities such as off-shore mineral extraction, shellfish harvesting etc.

—the absence of a national database or co-ordination of available information on coastal and marine matters.

—the lack of progress with marine conservation in comparison with land-based conservation practice.

—the absence of geological data pertaining to the sea bed between the low water mark and 10 metre depth water.

—confusion over how to meet the as yet undefined challenge of sea-level rise or climatic change.

5.2 It is against this background that coastal local authorities have attempted to carry forward good coastal planning. A great deal of innovative work has been undertaken by these authorities and there are examples of good co-operation between the various agencies involved in the coastal zone. But much remains to be done. Some of the issues above can be tackled within the present system, by the publication of Government guidance and advice, and through Government backing for work and initiatives already underway. However, many of the issues can only be properly addressed through a comprehensive review of coastal planning and management systems, including legislation. We will now consider these in more detail.

6. The Multiplicity of Agencies in the Coastal Zone

6.1 The number of organisations with powers or responsibilities in the coastal zone is considerable: Land and fundus owners, port and harbour authorities, county and district councils, Crown Estate Commissioners, water companies, National Rivers Authority, and over 20 Government departments and agencies including Department of Transport, Department of Energy, Ministry of Agriculture, Fisheries and Food, Department of Environment/Welsh Office. (A complete list appears in an appendix at the end of this memorandum.) There is a need to review the powers and responsibilities of each with a view to possible rationalisation.

6.2 Despite this large number, no single body has the responsibility or authority to take an overview of the coastal zone or ensure co-ordination. Local authorities' powers end at the low water mark whilst the remit of other organisations tends to be functionally specific. This leads to overlap, conflict, confusion, omission, piecemeal action and a lack of longer-term planning.

6.3 An example of overlap concerns the National Rivers Authority (NRA) and harbour authorities who are able, under different pieces of legislation, to make by-laws for the same area of water. The requirements of the different bodies will not necessarily be compatible as the responsibilities are different. The harbour authority has a responsibility to maintain safe navigation within the estuary while the NRA will be concerned with the water quality. These could potentially conflict, particularly with respect to dredging requirements.

6.4 Spurn Head at the mouth of the Humber Estuary offers a good case study of the plethora of agencies with duties, powers or interests in the coastal zone. Spurn is a fragile peninsula of sand and shingle which depends upon the erosion of the boulder clay cliffs of Holderness and the silt of the Humber Estuary for its origins and continued life. Protection works constructed in this century largely by the military authorities have prevented the natural process of erosion, but now nature is gaining the upper hand. The body responsible for the protection of the coast is the Holderness Borough Council which hitherto has regarded the expenditure necessary for the continued existence of three miles of sand and shingle to be a low priority. Research is necessary to decide on a course of action. There are however nearly 30 separate bodies with an interest in the area. Who is to take a lead?

7. The Need for Co-ordination

7.1 Historically, many estuaries were utilised to delineate the boundary between local authorities. Thus the estuaries fall within the jurisdiction of several councils. The Dee Estuary, for example, falls within the areas of two county councils and six district councils. Moreover, since it also marks the boundary between England and Wales, governmental planning and environmental responsibilities are also split between the Department of the Environment and the Welsh Office. Despite the Estuary being designated as an SPA, a Ramsar site, and an SSSI, there is as yet no co-ordinated approach to strategic planning.

7.2 There are many examples of good co-ordination and consultation. Management or coastal plans have been or are being produced for the North Norfolk Coast, Poole Harbour, the Exe Estuary, the Milford Haven and the Taw/Torridge Estuary amongst others. One specific example is the Northumberland Coast Management Plan prepared by Northumberland County Council in conjunction with the other local authorities and a range of organisations which have a direct interest in the management of the Northumberland Coast. In some other areas, liaison committees or agreed arrangements exist to enable consultation to take place.

7.3 However, even where these do exist, they depend on voluntary involvement by the agencies concerned, and are essentially self-initiated in the absence of any one organisation having a statutory or defined " lead " role. And none of the management or coastal plans, nor the liaison forums, has executive authority but relies on voluntary implementation by relevant agencies.

7.4 The ability of an organisation to carry out coast protection works without consultation was illustrated when English Heritage let a contract for coast protection work around Hurst Castle on the Hampshire coast. The first indication that any work was to be undertaken came when the successful tenderer approached the New Forest District Council's Leisure Department to explore the possibility of transporting via Keyhaven Harbour, the materials needed to carry out the work! Had the District Council promoted the scheme as coast protection authority, there would have been a statutory requirement to consult adjoining landowners and other interests but the reverse is not always the case.

7.5 A second example concerns the now closed lighthouse at Spurn Head on the Humber. Trinity House constructed its last sea defence works to protect the lighthouse in the early 1980s without informing the coast protection authority or any of the other parties with interests. Less than 10 years after construction, the defence of the lighthouse was abandoned on closure of the light. No maintenance is now being undertaken.

7.6 The achievement of satisfactory planning and management on the coast is as much an administrative exercise as a technical one. Co-ordinating the activities of the many authorities and agencies involved calls for positive leadership, backed up by the appropriate experience and skills.

8. The Need for Government Backing and Support

8.1 The examples in paragraph 7.2 illustrate the coming together of local authorities in many areas, often also involving other agencies, in order to discuss and tackle issue in the coastal zone. This regional and sub-regional activity is a welcome development but is not sufficient; a clear national perspective is needed.

8.2 One of the greatest needs is quite simply that a higher priority and profile should be given to coastal planning and management at all levels. In this central government has consistently failed to give a lead. Coastal issues have traditionally not been afforded a high priority by Government, although several recent announcements suggest a change. The commissioning by the Department of the Environment of a research project into planning policy for the coast, and into earth science information in support of coastal planning is greatly welcomed, as is the Secretary of State's intention to issue a Planning Policy Guidance Note (PPG) on the coastal zone. The former should fill a number of gaps in knowledge about the coastal zone, while hopefully the latter should provide much needed advice to local authorities. It is hoped that the PPG will give official encouragement to the preparation of management plans. Welcome though the PPG will be, it will, however, only represent guidance within the existing system and will not address the more fundamental issues raised in this memorandum, which would require fresh legislation.

8.3 Current national coastal policy consists simply of national and international protective designations, such as Areas of Outstanding Natural Beauty, Heritage Coasts and Ramsar Sites. This is inadequate. Some areas, which do not qualify for such protection, nevertheless have a special appeal and role, which should be recognised. The importance of the coastal zone for recreation, particularly when near to large centres of population, should be important considerations. The Solent is an obvious example.

8.4 As well as giving a lead on general policy for the coast, central government should prepare more detailed policies on specific issues which require a national perspective. A current example is the provision of yacht moorings. The sheer number of yachts now berthed in some estuaries and harbours is causing problems. A national strategy is needed to guide additional moorings and marinas to areas where there is environmental capacity for further growth. This could be effected through the regional guidance which provides a framework for the preparation of structure and unitary plans.

8.5 Press reports in late August suggested that the Secretary of State for the Environment is considering the introduction of a national coastal policy. This would be most welcome and would provide a context for the more specific administrative changes which are needed.

9. THE CASE FOR EXTENDING STATUTORY PLANNING TO INSHORE WATERS

9.1 For historical reasons, the town and country planning powers of local authorities end at the mean low water mark, although a few extend out to sea by ancient charter or private legislation.

9.2 An example of private legislation to extend the boundary of planning powers is that proposed in association with the winning of the Wytch Farm oil reserves which lie under Poole Bay. British Petroleum proposes to construct an artificial island around 1·5 kilometres off-shore to accommodate the oil rig and related facilities. In view of the public rights to navigate and fish in the United Kingdom's territorial waters, a Private Bill is necessary to enable the island development to proceed. Such an island would be outside the jurisdiction of the planning authorities, and therefore any building or other development on the island would be outside the normal planning controls. However, BP is to seek Parliamentary agreement to the inclusion of a provision in the Act for the island to be in the County of Dorset. This will allow the local authorities to exercise their normal duties and functions in respect of the island, including responsibilities under the Town and Country Planning Act.

9.3 The number of developments below the low water mark may be relatively few, but their impact on the environment and on other users can be considerable as shown by the above example. Moreover, the attitude of mind, which is induced, of land and sea being separate entities to be considered separately is most unfortunate. The principle of statutory planning on land has been broadly accepted over the last 50 years. However, there is no equivalent in the marine environment, perhaps because " out of sight is out of mind ".

9.4 The situation in river mouths and estuaries is even more unsatisfactory. There has been no legal delineation of river mouths or estuaries; instead administrative practice relies on an extra-statutory procedure adopted by the Ordnance Survey in the nineteenth century which produces arbitrary lines. Moreover it is uncertain whether planning control is legally exercisable in these subtidal waters even within local government boundaries.

9.5 The extension of statutory planning powers to in shore waters offers an opportunity to introduce a unified system for authorising development in different parts of the coastal zone.

9.6 A proposal for a marina at Swanage illustrates the sorts of problems which can arise in the absence of such a system. The scheme was for a combined marina and housing development in Swanage Harbour. A Private Bill was required to secure permission for the marina (because of the navigational implications) whereas the housing development was covered by the town and country planning system. Permission was refused for the marina but not the housing. The aim was to market each property with its own mooring space and the project was only viable with the two elements in combination. However, there was no mechanism for considering it in this way.

9.7 Parliamentary Private Bills have been used instead of the local planning process to seek consent for some recent major damaging estuarine developments, where parts of the development affect statutory rights of access or navigation. The use of this procedure for developments on estuaries (and other habitats) of high wildlife importance causes concern, not least because such domestic legislation is exempt from the terms of the EC Environment Assessment Directive, although there are now proposals to change this.

9.8 In addition, this ties up valuable parliamentary time and prevents proper consideration of the local implications of the proposed development. The proposed Mersey Barrage will, if it goes ahead, require the lodging of a Private Bill because of its effects on navigation in the estuary. However, the other aspects of the proposal could be dealt with through the processing of a planning application under the Town and Country Planning Act. Such a " two prong" approach has been pursued in other cases, but this option has been rejected by the Mersey Barrage Company who propose to seek all the necessary approvals through a parliamentary bill. Thus the only recourse for concerned local authorities groups and individuals will be to petition Parliament—a daunting, costly and time-consuming procedure.

9.9 Current proposals for incinerator and power plant schemes on the Thames Estuary clearly illustrate the complexity that can arise with large-scale proposals in estuarine locations, and the difficulties local authorities face in assessing such proposals, or in ensuring they are properly controlled.

10. Environmental Damage Despite Protective Designations

10.1 The main legislation governing conservation in Great Britain is the Wildlife and Countryside Act 1981. This Act gave the then Nature Conservancy Council—now "English Nature" in England and the "Countryside Council for Wales" in Wales—a duty to identify and designate Sites of Special Scientific Interest (SSSI). The importance of this designation is recognised in Government circulars which indicate a presumption against damaging development, but no binding constraint. The SSSI procedures require landowners and occupiers to consult with the NCCs successors when activities that might damage the site are being considered.

10.2 However, problems with the existing SSSI system limit its effectiveness in a number of ways. Although most activities which occur on estuaries notified as SSSIs are controlled by the Wildlife and Countryside Act, some potentially damaging operations are not. In particular, those governed by common rights, authorised by the granting of planning permissions and by private Acts of Parliament specifically override the protection afforded by the SSSI designation. And in any event, the designation gives no guarantee of protection from damaging development. As a result, lasting damage to SSSIs continues to occur. According to the Nature Conservancy Council, in the three years from April 1986 to March 1989, 146 SSSIs (around 3 per cent of the total) were permanently or lastingly damaged, and overall some 564 (10 per cent) suffered some damage. Estuarine SSSIs are even more threatened; during the same period 56 estuarine SSSIs were damaged (17 per cent of estuarine sites) of which 27 (8 per cent of the estuarine total) suffered at least long-term damage.

10.3 In addition, many SSSI boundaries have been drawn along the mean low water mark. As a result such designations fail to give any protection to those areas below mean low water mark even though they may be of equal scientific value.

10.4 This also raises doubts about whether Britain can fulfil its international obligations under the Ramsar Convention and the EC Birds Directive 79/409. In April 1989, 39 areas in Britain had been designated as Ramsar sites with 115 awaiting designation. Under the EC Birds Directive, member states are required to designate suitable areas as Special Protection Areas (SPAs). The United Kingdom Government requires that before any area can be considered for Ramsar or SPA designation, it should be an SSSI. This has delayed the process. More importantly, because many SSSIs end at mean low water mark, the contiguous sub-tidal and marine areas are not afforded protection. Yet the Ramsar Convention aims to stem the progressive loss of wetlands including all parts of estuaries down to six metres below low water, whilst it is implicit that SPAs on estuaries need to cover subtidal and marine areas as well as intertidal and land habitats. Thus the United Kingdom is failing to properly protect sites of international value.

10.5 The absence of a government decision over a proposal for a Ramsar/SPA designation, means the area concerned does not have the protection this designation brings. The Mersey Estuary, for example, is facing several major development proposals—a tidal barrage, expansion of Liverpool Airport incorporating land to be claimed from the Estuary, and a new road crossing. And there have been consultants' studies on the feasibility of petro-chemical development on 300 hectares of land and 330 hectares of mud flats along the Estuary's southern shore. A proposal for designation as a Ramsar site and a Special Protection Area was submitted by the Nature Conservancy Council in 1984; seven years later, a Government decision is still awaited.

11. The Role of Port and Harbour Authorities

11.1 Port and harbour authorities have a responsibility to maintain safe navigation within their areas of jurisdiction, and to that end can make by-laws. Their narrow remit can result in an unreasonable curtailment of recreational use and can in the event cause conflict with such uses or with environmental considerations.

11.2 The role of these authorities is becoming ever more important as water recreational activities continue to grow and new types of craft are developed. Some authorities have not adapted adequately to the changes either in terms of staffing and organisation, or in their charging policies. Some give an overriding priority to commercial shipping and unreasonably restrict recreation uses. Many are hindered by out-of-date legislation which deals inadequately with conservation and recreation issues. In some cases, there needs to be better liaison between planning and harbour authorities, especially to ensure that development control and navigation policies are consistent with each other.

11.3 The exemptions from planning control which port and harbour authorities enjoy under the Town and Country Planning General Development Order are unacceptably wide-ranging and prevent local planning authorities from controlling development which can cause serious environmental problems in surrounding areas.

11.4 Harbour authorities receive mooring fees or harbour dues from craft moored within their jurisdiction and with the growth in the numbers of pleasure craft, this represents a substantial revenue in many harbours. And there are substantial capital gains to be achieved from marina and other development in many harbours. There is thus a strong financial incentive to a harbour authority to increase recreational use in its area which may be at the expense of conservation of the natural environment. The privatisation of trust ports and their increased commercial orientation will create a temptation to pursue a maximisation of such income in the absence of any requirement for the privatised ports to include environmental management and conservation in their terms of reference.

11.5 Even where a harbour authority acknowledges the importance of environmental issues, it can be constrained by legislation. A case example is the Lymington to Yarmouth ferry service operated by Wightlink who wish to replace their existing ferries with vessels some 20 metres longer and with a greater capacity for vehicles. To cater for this, dredging will be required to straighten the channel and widen the approaches. Both the District and County Councils have expressed their concern on three points. The saltmarshes in the area are rapidly eroding and a recent study identified wash from ferries as one of a number of contributing factors; larger ferries will exacerbate this problem. Increased shore traffic for the ferry, especially at peak times, will lead to congestion, while the number of small boats moored along the river give reason for concern on safety grounds. Despite all this, neither the District nor County Council can intervene. Moreover the Harbour Commissioners, who have also voiced concern, believe they cannot deny access and would need to carry out the dredging work to ensure safety.

12. THE CROWN ESTATE COMMISSIONERS

12.1 The unique position of the Crown Estate Commissioners (CEC) has been criticised from many quarters. The CEC, which owns much of the foreshore and the sea bed on behalf of the Crown, has considerable powers over the use to which the sea bed is put, but few of the responsibilities for environmental protection which should be associated with such powers. In the absence of any planning system which covers sea areas, the CEC, under pressure from the Government, has assumed a quasi-planning role by default, despite the fact that it is not a publicly accountable body. At the same time the Government views the CEC as independent and allows it to pursue its function as principal landowner in the marine environment. It is the body responsible for issuing licences for fish farming and therefore for balancing the interests of commercial fisheries and conservation of the environment. Yet it receives the income from the licences which is an incentive to give greater weight to the fishing interest!

12.2 It is in effect both judge and jury on many developments in the coastal zone. In response to intense public pressure the Commissioners have recently adopted a stance which appears more sympathetic to the environment and less commercially aggressive. Nevertheless, we believe that their role should be re-examined, having particular regard to their lack of accountability and the interpretation of their financial objectives.

13. COAST DEFENCES

13.1 Coast protection is the prevention of the erosion of the coast, and sea defence is the protection of land from flooding by the sea. Maritime district councils are responsible for coast protection works under the provisions of the Coast Protection Act 1949. The NRA is responsible for sea defence works under the provisions of the Land Drainage Act 1976. Although the two responsibilities are clearly defined there are areas where sea defence and coast protection need to be carried out together, and the two functions can overlap. Despite the split in responsibilities, discussions between the two organisations resolve any particular difficulties when they arise.

13.2 However, the split in responsibilities may inhibit a wider strategic view and there may be a case for bringing both under the purview of a single organisation. The two relevant Acts are now respectively 43 and 25 years old, while the future of the NRA is itself under review by the Government, in the context of the mooted creation of an Environmental Protection Agency.

13.3 The responsibility at a national level for administration of the two activities lies with the Ministry of Agriculture, Fisheries and Food whose regional engineers liaise with the scheme promoters and give technical approval for the work. The Ministry has the responsibility to withold approval if the proposals would have a detrimental effect upon adjoining sections of coast. The system of grants and supplementary credit approvals is also administered by MAFF and to qualify the scheme must have a positive benefit cost. The widening of the definition of benefits to include intangible aspects such as tourism, recreation and amenity is welcomed as this enables environmentally sensitive areas to be protected when previously works of this type were not financially viable.

13.4 Whilst these arrangements adequately cover works by district councils and the NRA, problems can still arise when landowners undertake works, as described in Section 7.

14. SEA BORNE LITTER

14.1 Despite international attempts to legislate on the dumping of waste of all kinds (eg MARPOL Convention), little effective action is taken on enforcement. Apart from major incidents of oil spillage and wrecked toxic cargoes, there is a continuing polluting process which ends up on the beaches, estuaries and river banks of our coast. Oil, chemicals and ordinary domestic refuse (from ships) are served up on each tide.

14.2 The Environmental Protection Act 1990 requires local authorities and landowners to keep public access land clear of litter. To comply with the Act fully on coastal parks and beaches would require a daily sweep of (i) all areas above high water mark to collect all the wind blown litter and (ii) of the area between low water mark and high water mark to prevent the litter there from blowing above high water mark. The Crown Estate Commissioners as landowners of the tidal zone have been exempt from responsibility to clear their land.

14.3 One council has estimated that the additional cost of meeting the full requirement of this part of the Environmental Protection Act in just one coastal park as £60,000 pa.

15. MARINE DREDGING OF AGGREGATES

15.1 The extraction of aggregates from off-shore deposits occurs in several locations around the coast. Rightly or wrongly, some of the problems of coastal erosion has been blamed on this activity. It is an issue which raises serious procedural questions of control, particularly with respect to the need for an Environmental Assessment, and who is responsible for requiring and evaluating an Environmental Statement in respect of a specific proposal. Moreover there is concern about the cumulative and long term impacts of aggregate extraction which emphasises the need for wider-ranging research on this important subject. A recent example which raises these issues is the proposal for Helwick Bank off the West Glamorgan coast.

15.2 Paradoxically, there is an increasing demand for shingle to be used in coast defence schemes, largely in the form of beach replenishment. This is an environmentally attractive "soft" engineering solution to many coast protection or sea defence problems. However, the solution of one problem which is environmentally acceptable should not be achieved by the creation of another environmental problem. There are moves to investigate the use of inshore shingle deposits for coast defence purposes where there is a significant balance of benefit in environmental terms.

16. MARINE ARCHAEOLOGY

16.1 Nowhere is the confusion of marine responsibilities at Government level more apparent than in archaeology. Little has happened since the 1989 report of the Joint Nautical Archaeology Policy Committee—"Heritage at Sea—Proposals for the better protection of archaeological sites underwater". Among the actions required are:—

(i) New legislation for the protection of underwater archaeological sites and artefacts.

(ii) Allocation of resources to establish a national and local record, similar to the land based Sites and Monuments Record, (some preliminary work in Hampshire and the Isle of Wight has started but funds are pitiably small).

(iii) Archaeological impact assessment to be carried out by developers of the sea bed (eg oil and gas, cables, aggregates).

(iv) Establishment of a marine heritage protection agency with survey, protection, training and educational responsibilities.

17. COMMERCIAL FISHING

17.1 New methods of harvesting shellfish in estuaries and inshore waters by mechanical or hydraulic dredging are destructive and may cause long-term damage to the ecology of coastal waters where they are practised. Common rights of fisheries within coastal waters present severe difficulties in controlling such activities.

18. GLOBAL WARMING, SEA LEVEL RISE AND CLIMATIC CHANGE

18.1 If, as is generally predicted, global warming continues, one likely consequence is a rising sea level. This would particularly affect the low coastlines of South-East England, the shallow and intertidal parts of estuaries and the low-lying land surrounding them. Average sea-level rise worldwide is currently estimated to have been 10–20cm over the last 100 years. A continuing rise is predicted: 20cm by 2030 and a 65cm rise by the end of the 21st century. There is, however, an absence of Government guidance to local authorities on how to anticipate and respond to this prediction.

18.2 There is also uncertainty about the likely effects of climatic change. In the short term, the effect of increased storminess is likely to be more severe than the effect of sea level rise. This will be due to the greater chance of storm surges coinciding with high tides. This issue is raised, for example, in the Strategic Planning Guidance for Wales (Interim Report on the Emerging Strategic Issues). These are matters which require urgent discussion at all levels of government and which require the development of a long-term view on coastal management.

19. THE NEED FOR MORE RESEARCH

19.1 The marine environment is not particularly well researched, while information which has been collected is neither collated or referenced in one place, nor necessarily shared or co-ordinated between agencies. The research commissioned by the Department of the Environment into earth science information in support of coastal planning is a significant step towards remedying these inadequacies. It will review the sources, location, accessibility and level of detail of existing information, and identify other essential and desirable information

19.2 One particular knowledge " gap " is geological information. Surveys by the British Geological Survey (BGS) cover United Kingdom land, and the sea bed beyond 10 metre water depth. Whilst some localised survey work of the sea bed between the 10 metre depth and the foreshore has been carried out in a few areas, this knowledge is generally non-existent. The BGS has no funding to undertake such survey work and thus fill this knowledge gap. Information from such surveys would greatly assist, *inter alia*, in the design of more " environmentally friendly " coastal protection works, by identifying the location of suitable deposits which could be dredged for beach replenishment.

20. CONCLUSIONS AND THE WAY AHEAD

20.1 Coast protection, planning and management suffer from inadequacies in current legislation, divided responsibilities and a general lack of co-ordination.

20.2 There is no comprehensive planning legislation framework for the marine environment to parallel that covering land in the form of the Town and Country Planning Act. That which does exist is often out-dated and is related to individual functions or agencies, creating a recipe for conflict and fragmentation of effort.

20.3 There is a plethora of bodies involved in coast protection, planning and management—landowners, port and harbour authorities, county and district councils, Crown Estate Commissioners, water companies, National Rivers Authority and over 20 Government departments and agencies. This means that proper co-ordination is essential. There are examples of good co-operation between agencies but this is by no means universal. In some areas, a forum has been established to promote a better understanding of the issues and a wider co-ordination. But this is necessarily a voluntary co-operation without any statutory back-up. In many areas there is nothing.

20.4 There is an absence of strategic thinking on coasts and estuaries. Priorities, policies and by-laws are set by individual agencies in relation to their own statutory functions and often limited geographical interest, without sufficient reference to adjoining areas or bodies, or to environmental and other factors.

20.5 On top of this, there is a lack of knowledge in important topic areas and a lack of guidance and advice from Government.

20.6 To remedy these deficiencies we would argue for the following:—

(i) a national planning perspective for the coastal zone, together with Government encouragement, guidance and support for the development of a strategy for each coast and estuary. This could be effected through regional guidance prepared by the Secretary of State as a framework for the preparation of structure and unitary plans.

(ii) the establishment of a standing forum—embracing all disciplines—for each stretch of coast or estuary, to ensure consultation and co-ordination, plus the designation of a " lead agency " to convene it. Each would be charged with responsibility for taking a strategic and longer term view of coastal protection, planning and management within its area. Each would discharge this responsibility by preparing any non-statutory/management plan, by giving guidance to authorities preparing structure and local plans, and by advising on the implementation of plans and major development proposals. Government should encourage and take any necessary steps to ensure the convening of a forum for each stretch of coast or estuary.

(iii) Government encouragement to the preparation of a non-statutory plan where needed for a coast or estuary, which in many cases could take the form of a management plan. Such plans would cover a wider range of issues than is possible in statutory plans and would be particularly appropriate where an estuary or stretch of coast lies within the jurisdiction of more than one local authority.

(iv) the extension of statutory planning powers to below the low water mark to ensure that all development and changes of use of land and water within the coastal zone are brought within the planning system. The new seaward limit of planning powers would best be fixed after detailed study, so an arbitrary mileage limit is not offered here. Indeed the appropriate distance will vary from area to area. So the ideal approach might be to provide the Secretary of State with an enabling power to define the precise boundary in each area after consultation with the relevant standing forum. In parallel with this extension of statutory planning control, there should be a review of the permitted development rights of harbour and port authorities.

(v) a review of responsibilities in the coastal zone with a view to rationalisation. Specifically this should include the role of port and harbour authorities and the position of the Crown Estate Commissioners.

(vi) a speeding-up of the process of designating Ramsar sites and Special Protection Areas, together with an extension of existing designations, where necessary to ensure full coverage of all areas qualifying for such protection whether land or water.

(vii) additional research and dissemination of knowledge pertaining to the coastal zone and the marine environment in particular. Specifically, this research should cover the likely effects of global warming and climatic change. The surveying of the geology of the sea bed and its underlying strata in the shallow waters between the foreshore and 10 metre depth would also be of great value. A national database on the marine environment is needed to facilitate access to relevant information.

October 1991

Appendix

ORGANISATIONS WITH REMITS IN THE COASTAL ZONE IN ENGLAND AND WALES

GOVERNMENT DEPARTMENTS

Department of Education and Science (research)
Department of Employment (health and safety)
Department of Energy (offshore oil and gas, and tidal and wave power promotion)
Department of Environment (pollution issues—aggregate extraction, designation of SPA/Ramsar sites)
Department of Trade and Industry (shipbuilding policy, telecommunications, marine shipping management)
Department of Transport
Foreign and Commonwealth Office (marine law co-ordination)
Her Majesty's Customs and Excise (ship registration)
Home Office (pleasure boat licensing and beach safety)
Ministry of Agriculture, Fisheries and Food
Ministry of Defence (survey, ownership, bombing/firing ranges)
Welsh Office

OTHER PUBLIC BODIES

Civil Aviation Authority (hovercraft)
Countryside Commission (Heritage Coast, Areas of Outstanding Natural Beauty)
Health and Safety Executive (implementation of health and safety policy)
Health and Safety Commission (shipboard safety)
Local Authorities (planning by-laws, emergency planning etc)
National Rivers Authority (fisheries, pollution, coastal defence)
Natural Environment Research Council (marine research)
English Nature and Countryside Council for Wales (Sites of Special Scientific Interest and Marine Nature Reserves, research, advice to DOE)
Sports Council (water borne recreation, grant aid)
Tourist Boards (tourism, grant aid)
Trinity House

OTHERS

British Telecommunications plc (submarine cables)
Crown Estate Commissioners (foreshore and sea bed ownership)
Harbour Authorities and Commissioners (by-laws, shipping management, dredging)
Hydraulics Research (coastal protection research)
Landowners, managers and leaseholders
Sea Fisheries Committees (fisheries management)
Water companies

APPENDIX 5

Memorandum by the Association of Sea Fisheries Committees of England and Wales

The Association of Sea Fisheries Committees wishes to submit the enclosed evidence for consideration by the Environment Select Committee. I am sure that you will be aware of the interest generated by various bodies in England and Wales this year on the subject of, Coastal Zone—" Management ", " Protection " and " Planning ". All of which has taken place without direct involvement by Sea Fisheries Committees.

You will see from the Evidence that SFCs have a wealth of knowledge and experience in coastal management of the fisheries and shellfisheries and all their Fishery Officers are fully trained. It would appear from the various reports of the seminars and meetings that have been held throughout the year, that little thought has been given to the in-shore fishermen and their livelihood. I am not sure if this consideration is within the terms of reference of the Select Committee but if you require any further information please do not hesitate to contact me.

COASTAL ZONE PROTECTION AND PLANNING

1. The management of the marine environment, especially the part that forms estuarine and coastal waters on the United Kingdom has been the subject of both seminars and discussion papers during the course of 1991. Much of the information has originated from conservation interests who have formed working groups to address the issue as perceived by them with the exception of English Nature (The Nature Conservancy Council), other regulatory bodies tasked with the management of coastal and estuarine sea fisheries were not always consulted. This is clearly unsatisfactory.

2. The Association of Sea Fisheries Committees of England and Wales, would like to draw the attention of the Environment Select Committee to the role and status of Sea Fisheries Committees around the coast of England and Wales and to their concerns about and interest in the coastal zone below high water mark.

3. Sea Fisheries Committees are statutory bodies set up originally circa 1890 to manage and develop the sea fisheries around the coast. They operate currently under the primary legislation of the Sea Fisheries Regulation Act 1966. Membership of the Committees is drawn from appointments by the maritime County and Metropolitan Borough Councils, who fund the Committees, and appointees of the Minister of Agriculture, Fisheries and Food and the Secretary of State for Wales, with the addition of a National Rivers Authority appointee. The Minister/Secretary of State's appointees include active commercial fishermen among their numbers.

4. Sea Fisheries Committees manage the fisheries by byelaws (controlling such things as fish and mesh of nets sizes). Their Fishery Officers enforce the byelaws and relevant parts of National and EC fisheries legislation. The Committees own and operate both inshore and offshore patrol vessels and as well as operating to seaward have specific estuarine interest particularly in relation to shellfish fisheries. The Sea Fisheries Committees role has changed over the years and they now have a responsibility for the protection of migratory fish. Their area of jurisdiction seaward of high water mark is presently in the process of being extended from the three mile limit to the six mile limit.

5. Sea Fisheries Committees have strong contacts with the coastal community and all official bodies who operate within their area of work, such as the Ministry of Agriculture, Fisheries and Food, The Welsh Office, National Rivers Authority, HM Coastguard, Police, Customs and Excise, Harbour Authorities etc etc, as well as bodies such as Countryside Council for Wales/English Nature, National Trust and National Parks.

6. There is a wealth of experience and knowledge held by each Committee on matters related to its own District. Although not always having specific powers the Committees are usually consulted on the fishery implications of pipeline discharges and major harbour or coast protection works, and have strong interests in water quality, dredging for marine aggregates and dumping at sea. Over the last decade Committees have also become increasingly involved in environmental conservation matters as well as their primary concern in sea fish conservation.

7. The Sea Fisheries Committees are supportive of the needs for sensible forms of coastal zone management policy that reduce the scope for conflict between competing users of the same environment. Indeed part of the Committees role for over a 100 years has been to balance the exploitation and conservation of coastal fish and shellfish resources.

8. Over the last decade, with the upsurge of wider interest in the marine environment there has been tendency for both government agencies and environmental non-government agencies to overlook the wealth of experience the Sea Fisheries Committees have in the equitable " conservation " of biological marine resources. Inadequate prior consultation by bodies with less experience of maritime matters has frequently resulted in quite unnecessary misunderstandings and friction. Such problems have more often arisen from national initiatives than at the local level, where good working relationships have usually developed between the local officers of the various agencies.

9. A particular legislative point that needs to be rectified is that Sea Fisheries Committees at present are only allowed to make byelaws that can strictly be interpreted as being for fisheries management purposes. If there are needs for fishery operations to be regulated so as to be sensitive to broader environmental concerns, then the Committees would seek to have their role expanded to cover such matters. The Sea Fisheries Committees long experience of sensitively enforcing complex byelaws on the shore and at sea should not be lightly disregarded.

10. The Association and its Member Sea Fisheries Committees would be pleased to give further information to the Select Committee if so required.

9 *October* 1991

APPENDIX 6

Memorandum by the British Geological Survey[1]

Additional information from the BRITISH GEOLOGICAL SURVEY, based on a seminar given by Mr R S Arthurton, Manager of BGS's Coastal Geology Group, to the Local Authorities National Advisory Group on Coasts and Estuaries in December 1991 entitled **Aspects of Geoscience in Coastal Zone Planning.**

INTRODUCTION

The coastal zone is of enormous importance throughout the world. It is the region where much of the world's population lives or works, or use for their recreation. However, it is also the region where we derive many of our living and non-living resources and increasingly it is where, intentionally or unintentionally, we dump much of our waste. The coastal zone has special importance to a crowded island nation such as Britain, but despite this importance the coastal zone is in geoscience terms perhaps the most poorly understood and most imperfectly documented part of Britain. It is the view of the British Geological Survey (BGS) that this results in an inadequate knowledge base to support effective Coastal Zone Protection and Planning.

For the purposes of this contribution from BGS the **Coastal Zone is defined to include:**

* **on the landward side**, all land at risk from the effects of marine processes both directly, eg by flooding, and indirectly, eg by coastal landslip.

* **on the seaward side**, all sea bed within the 18 metre bathymetric contour. This contour serves as an **arbitrary** limit and has been chosen in order to correspond to the minimum water depth currently stipulated by the Crown Estate for the licensed dredging of aggregates from the sea bed.

The coastal zone is a very dynamic system, where the physical conditions, such as tides or sediment type, can have a profound effect on the biological and biochemical conditions. Human activity has affected the coastal zone for thousands of years in some areas, but the rate and magnitude of those changes have increased enormously in recent years. One of the tasks of the geoscientist is to document the nature and extent of natural and anthropogenic change in the coastal zone, and to help to develop options for managing that change. The geologist has a particular role in providing the temporal framework for change. If planners are concerned with making provision for the 50-year event then it is necessary to have a record which goes back 500 years. If the planner is concerned with the 500-year event then the record needs to extend back 5,000 years. The geologist provides that record. Therefore an essential element in any consideration of coastal zone protection and planning must be the evidence of past coastal change, for the past is likely to be a key to the future and planning for that future.

The geologist also has a fundamental role to play in the documentation of non-living resources. The entire coastal zone in many ways can be regarded as a non-renewable resource. However, within that zone there are particular non-living resources such as sand and gravel, and calcium carbonate, which are increasingly being extracted. Also, important resources such as oil and gas are extracted from below the coastal zone. The extraction of these resources is impacting on the coastal zone to a varying but increasing degree. It is therefore essential that those coastal resources are fully assessed and the likely impact of their extraction is evaluated.

PROTECTION AND PLANNING ISSUES

The major issues of coastal protection and planning to which geoscience is relevant include:

* Risks to human life, settlement, property, agriculture and ecosystems

* Exploitation of the coastal zone resources

* Conservation and amenity.

The risks include:

* the catastrophic flooding of low-lying coastal tracts by the failure or overtopping of sea and tidal defences, either natural or engineered

* the recession of coasts by progressive marine erosion, and the consequent failure and mass-movement of coastal cliffs by landslip and mudflow

* the disruption of navigation channels and port facilities by sandbanks and siltation

* the contamination of marine sediments in the coastal zone and in estuaries giving rise to a threat to human health and natural ecosystems

* the salinisation of coastal aquifers.

[1] Submitted as supplementary evidence to that of the Natural Environment Research Council (*see* Appendix 16).

The main areas of exploitation are:

* the dredging of sea bed materials for commercial aggregates, or for fill or beach replenishment

* the use of coastal sites to provide intake and discharge facilities for cooling water and waste; also terminals and pumping facilities for oil and gas

* the reclamation of low-lying coastal land and estuaries, particularly for industrial development; also the construction of barrages across estuaries

* the development of coastal sites for recreational purposes both onshore and offshore

* the use of coastal water and estuaries for fisheries, including mariculture.

Conservation and amenity interests include:

* the conservation of coastal landforms and cliff scenery, in particular the Heritage Coasts

* the conservation of wildlife habitats, marine and terrestrial

* the conservation of Man's coastal heritage, including coastal archaeology

* the maintenance of beaches for recreation and protection.

ROLE OF THE BRITISH GEOLOGICAL SURVEY

A large number of organisations including universities, consulting organisations and public bodies are involved in geoscientific studies of the coastal zone. However, for the most part they do not carry out strategic long-term studies. The BGS has carried out geological mapping and other geoscientific studies of the landmass of the United Kingdom since 1835. An offshore Regional Mapping Programme, largely funded by the Department of Energy, commenced in 1966 and since then a range of geological maps covering the majority of the United Kingdom designated shelf area has been produced. The offshore surveys have largely been constrained by operational considerations, except in certain estuarine areas, to areas of shelf with water depths greater than 10 metres. The bulk of the offshore survey work has been carried out by the Marine Operations Group of the BGS based at Edinburgh, assisted by other BGS scientists.

In the shallow nearshore areas data acquisition has been underway on an opportunistic basis for a number of years but with no integrated strategy. The net result of this is that in many areas the data are sparse. In 1990 the BGS set up a **Coastal Geology Group**, based at its headquarters at Keyworth, to investigate and improve the geoscientific knowledge of the nearshore and particularly to integrate knowledge of the sea bed and sub-sea bed geology with that of the adjoining land. The Group aims to provide *inter alia* a national inventory of the non-living resources of the sea bed, and to interpret the long-term record of nearshore erosion and sedimentation against which modern coastal processes may be assessed.

Marine sand and gravel resources

The Coastal Geology Group of the BGS carry out desk studies for the Crown Estate on the marine sand and gravel resources around the United Kingdom, and with the expertise of the Marine Operations Group have executed surveys of sand and gravel resources in three areas off eastern and southern England, the work being largely funded by the Crown Estate and the Department of the Environment. None of these surveys, nor any of the desk studies, has extended inshore of the 18-metre bathymetric contour, and thus knowledge of nearshore marine sand and gravel resources remains sparse.

Coastal Geoscience Data

People and organisations concerned with management and planning issues in the coastal zone or the development of policies for the coastal zone must have access to a comprehensive geoscientific database. This is currently unavailable.

Geoscientific data relating to most of the subtidal part of the coastal zone are sparse compared to the amount available for the United Kingdom designated shelf area further offshore. This reflects the general dearth of data acquisition in the coastal zone. However, the British Geological Survey is aware of the existence of considerable amounts of reliable data relating to this zone in a wide variety of formats that are dispersed through many organisations without an easy means of access by potential outside users. In many cases such data become lost or are destroyed.

The British Geological Survey wishes to fulfil its national role in respect of its database by identifying and incorporating as much existing information as possible into its database and processing the data to a standard format that its public and private sector customers need. The task of acquiring data would be facilitated if there were a statutory obligation on the part of those who produce such data, (particularly that produced partly or wholly at public expense), to deposit a copy with the British Geological Survey, in much the same way that certain types of borehole data are presently lodged with the BGS.

Geological Survey of the Coastal Zone

Using existing data where these are available, and carrying out new surveys where they are not, the BGS believes that there is an urgent need to geologically map the coastal zone. The scope of the coastal mapping programme includes investigation of the 3-dimensional distribution of the sediments and rocks that make up the coastal zone; an inventory of non-living resources particularly those of the sea bed; geotechnical assessment of vulnerable coasts addressing particularly the problems of consolidation, foundation conditions, erosion and mass movement; sediment dynamics; sediment geochemistry; and interpretation of the long-term history of coastal change particularly in the context of changing sea level.

The proposed coastal survey is of a strategic nature, and while its results would be of considerable application in specific site projects, its main value would be in the provision of geoscientific data and data interpretation to the planner, engineer and coastal manager at regional and national levels. The proposed survey would provide baseline information against which future change in the coastal zone, whether physical, chemical or biological, may be measured by subsequent monitoring. The survey results and those of subsequent monitoring will facilitate realistic forecasting of future change.

The British Geological Survey attaches importance to the effective dissemination of the results of its survey programmes. For the coastal zone, information from the resulting databases would be made available in a variety of forms including thematic computer-generated maps and plots at a scale required by the user. This provides a dynamic Geographic Information System for the coastal zone.

APPENDIX 7

Submission by the University of Liverpool Centre for Marine and Coastal Studies

I. OBJECTIVES

Coastal zone protection and planning, taken together constitute coastal zone management which is a subject being examined by the European Community and many groups within the United Kingdom. Management is concerned with the achievement of objectives so it is important to understand what objectives are being pursued by any protection and planning measures which may be taken.

In this paper it is assumed that these objectives are as stated in 1–4 below.

Coastal protection objectives:—

1. To avoid damage to human investment in the coastal zone by the action of waves, rain, wind, currents, tides and raised sea level.

   The protection of human life is not included here because human life is only put at risk if it has already been decided to protect an investment (in say a village on low ground behind an embankment). If the investment is abandoned there need be no risk to life.

Coastal zone planning objectives:—

2. To enable the many legitimate human activities which take place in the coastal zone to occur without adverse effects on each other.

3. To avoid long-term changes of the kind which reduce options for future generations.

4. To ensure the continued viability of coastal communities, particularly those in remote areas where coastal or marine activities are essential to the economy.

II. THREATS TO ACHIEVEMENT OF THE OBJECTIVES

1. *Threats to Human Investments*

The sea is a powerful force which can change natural features by erosion of headlands and cliffs, and can destroy the human constructions such as sea-walls intended to protect them. Land previously claimed from the sea and defended by sea-walls is frequently flooded when high tides coincide with onshore storms. These problems will be exacerbated by the rise in sea level which is expected to occur over the next half century.

Preventing damage by the sea is very expensive if conventional methods such as the construction of concrete walls is used. In many cases these methods prove to be futile because the walls are undermined by erosion which itself can be caused by attempts to defend land elsewhere. Other methods which depend on " soft " constructions such as sand-dunes or salt-marshes require a wide zone and operate by gradually adjusting to the pressures of the sea. They are therefore unsuitable for defending coastal towns and industries which are built directly along the coast.

Human investment in coastal and estuarine towns is so large that it is not practical to abandon these; they must be defended. All the available resources will need to be concentrated on them and there will be little available for defending cheaper farm land, especially if by doing this, erosion in the heavily built up areas is increased. The current surplus of farm land which is being taken out of use with set-aside schemes strengthens this view.

Possible Solutions:—

In the long term abandon any commitment to defend agricultural land against encroachment by the sea. Owners of threatened land would adjust to the fact that their occupancy of it is equivalent to a leasehold with a varying time to run, depending on their distance from the sea. Land prices would gradually adjust to this situation. This policy may need to be applied to some villages so that they eventually fall into the sea as has happened in the past.

In the short term it may be necessary to provide compensation for ceasing to defend land at public expense.

No planning permission should be given for housing or industry on land which would be at risk during the lifetime of the investment.

Soft defences with a buffer zone behind, within which there should be no long term investment, should be adopted wherever it is not of vital importance to defend an area.

2. *Adverse Effects of Activities on Each Other*

Special problems arise between activities involving the sea which do not normally occur on land.

One problem is the common ownership of the resource which does not encourage responsible exploitation to ensure its long term value. For example, the sea can provide a continuous supply of food and can absorb a limited amount of waste of an appropriate kind, provided it is required to do these to a limited extent. Common ownership usually results in an unsustainable over-use of a resource.

Multiple use is another problem and this arises from the common ownership of the sea's resources. Unlike the situation on land many activities take place in the same area. Bathing, recreational boating, fishing, waste disposal, transport and mineral recovery can all take place in the same piece of sea. Frequently these activities physically interfere with each other and are often incompatible.

A third problem is the invisibility of the damage which may be done. Waste disposal does not always have visible effects and some of the most damaging wastes with long term effects are not only invisible but are very difficult to detect. Damage to the sea bed by trawling or sand and gravel extraction or even anchoring by pleasure boats is quite invisible to those doing the damage and to those trying to prevent it.

The final problem which will be mentioned is the dispersing and diluting effects of the sea on any wastes discarded. This means that the problems may occur many miles from the point of discharge and, because the concentrations of any toxic wastes will be low, these effects may not be noticed until years later and be hard to link to the cause. An example from the Irish Sea illustrates this point. Toxic heavy metals are discharged by industry into the sewers, then processed in treatment plants and discharged into rivers or through sea outfalls directly into the sea. As a result of years of this practice fish contain high levels of these metals. In the worst case, in the River Mersey, fish are regarded as unfit for human consumption.

In addition to the special problems mentioned above the very heavy development pressure on the coast from tourism, recreation, housing, industry and ports increases the problems of planning these developments in a rational and compatible way.

Possible Solutions:—

Improved, possibly formalised, liaison between users and conservers of the environment and legislators.

A rationalisation of responsibilities between the large number of organisations responsible for the regulation of marine and coastal activities. This might include the extension of local authority control of developments below the low tide level.

Strategic plans for the use of the coastal zone would help to avoid conflict between activities and over-development of any one activity. These plans could be at local, regional and national level depending on the activity concerned, ie marina development would be local or regional while port development should fit into a national strategy.

The heavy pressure on the coastal zone could be reduced by reserving the zone to those activities and developments which of necessity must take place there. Approval would not then be given to other developments in the coastal zone.

A very high priority should be given to the elimination of the discharge into sewers of toxic heavy metals and other persistent toxic materials which can bio-accumulate.

3. *The Preservation of Options for the Future*

Threats to future options mainly arise from two sources:—

The destruction of important wildlife habitats and rare species which could result in a reduction of bio-diversity.

Man-made structures which are almost impossible to remove after use but which interfere with other activities. For example, sea bed debris from past oil extraction which interfere with trawling.

Estuaries and soft coastlines are the most likely habitats to be lost by man's activities. They are important for wildlife, especially estuaries with their important fish, invertebrate and bird populations. Estuaries are subject to pressure from land claim for waste disposal, agriculture, industry and ports. Soft coasts such as sand dunes are under pressure from bathers, golfers, motor cyclists etc. The threat to these habitats will be increased by any rise in sea level. For example, in estuaries the important intertidal mud-flats will be squeezed between the rising sea and any existing sea-walls.

Man-made structures as massive as those associated with ports are probably impossible to remove even when the ports go out of use, as has happened in a number of areas as ships have increased in size. The impossibility of removal is partly a result of their sheer size but also because industries and towns have developed alongside the ports. This loss of estuarine habitat is therefore likely to be permanent. Offshore structures and their associated sea bed equipment are less of a problem but are still expensive to remove completely if an uncluttered sea floor is desired.

Possible Solutions:—

> The solutions suggested for the protection of investment would go a long way towards protecting wildlife habitats by allowing land claimed in the past for agriculture to revert to intertidal habitat. They would also allow natural processes to occur which would preserve the balance between different but interrelated habitats such as saltmarsh, sandbanks and mudflats.

> The problem of long-lived man-made structures would be best solved by requiring the method of decommissioning and removal to be determined at the time permission for the development is given. The costs of removal would need to be set aside at an early stage to avoid these having to be met out of public funds in the event of the developer going into liquidation.

> In the case of existing port constructions, their conversion to other coast related activities, while not restoring the original habitat, would reduce pressure elsewhere on the coast. For example, conversion of a disused dock into a marina would reduce the need for marinas elsewhere whilst its conversion to a waste site and then office accommodation would not.

4. *Threats to the Viability of Coastal Communities*

The coastal communities most at risk are those in remote areas which rely on a limited range of economic activities. Many factors affect their viability such as transport and communications, but these are not specifically in the realm of coastal planning. The two important coastal activities which can be lost are port related activities and fishing. The first of these can be lost due to changes in ship design which result in a port being unable to accept newer, larger ships, or changes in the pattern of trade. Fishing opportunities can be lost by the closure of traditional fishing grounds (eg Iceland) or by competition with fishing vessels from other areas which result in the local fishing grounds being over-exploited so that yields fall.

Possible Solutions:—

> The alternative use of redundant commercial ports for recreational purposes could be encouraged and helped by development grants.

> Although the overall fishing controls are determined by the EC through its Common Fishing Policy the United Kingdom does have control on the distribution of its own quotas. Within the limits available local fishing vessels should be given priority for fishing within the inshore zone (12 miles?).

> Where it is impossible to ensure the viability of the local fishing industry at its current level decommissioning of surplus vessels, as allowed by the EC, could be used to provide capital for new developments. There seems to be a marked contrast between the treatment of fishermen who find they cannot make a living and the treatment of farmers. Fishermen are essentially left to their own devices, whereas farmers are helped by intervention payments, Highland Livestock Compensation Allowances, the Set-aside scheme, the Countryside Premium scheme, etc.

III. Other Factors Which Can Effect the Success of Coastal Zone Planning

The topics discussed briefly above are those which occur on the coast which directly affect the success of coastal zone planning in achieving its objectives and which therefore should be brought into a framework of coastal management. There are, however, a number of other factors which whilst not directly involved with the coast can have a serious effect on the success of any policy for managing the coast. Whilst it is not expected that the present enquiry will address these problems it is necessary to be aware that any recommendations for the coast could be frustrated by decisions on these other factors.

Some relevant factors are:—

> Transport policy

> The success of a local or regional decision to develop tourism or a port will depend on good communications with the more populated areas of the country. Ports in particular need good rail or motorway links into the national transport network.

> Energy policy

> Sea-level rise will be closely linked to CO_2 emissions. Coastal zone planning problems will therefore be reduced if these emissions can be reduced by energy conservation or the use of energy sources not emitting CO_2 (renewable sources or nuclear).

> Common Fisheries Policy

> The United Kingdom government could use its influence to press the case for special protection for communities dependent on their local fishing grounds.

D F Shaw
8 *October* 1991

APPENDIX 8

Memorandum by the Centre for Marine Law and Policy
University of Wales College of Cardiff

INTRODUCTION

The Centre for Marine Law and Policy of the University of Wales College of Cardiff was established to bring together staff in different disciplines so as to promote research and teaching in this field. Current membership is made up of staff from Cardiff Law School and the Department of Maritime Studies and International Transport whose collective expertise includes legal, scientific and managerial skills. For the purpose of producing this document we have been joined by a colleague in the Department of Extra Mural Studies who has a particular interest in marine conservation. The existence of the Centre has facilitated collaborative studies in a wide range of issues at both national and international levels covering such topics as port policy, sea bed management and environmental protection. Staff in each department have research interests in coastal zone management and are also involved in teaching various aspects of the subject at undergraduate and postgraduate levels. While many of these activities could be carried out independently, placing them in the context of the Centre makes it easier to achieve a balanced, broad-based perspective. We believe that such an approach is essential not only for studying coastal zone management but also for carrying out that management.

The most important point that we should like to make, therefore, is that there is an urgent need for an holistic approach to be used in the development of policies for the coastal zone and that management of coastal activities must be integrated in accordance with these policies. We do not, therefore, think that the issues of coastal zone protection and planning should be examined in isolation and recommend instead that these issues be looked at in the broader context of coastal zone management.

THE COASTAL ZONE

There is no universally accepted definition of the coastal zone. Taken at its broadest it can include the hinterland of rivers draining into estuaries and extend seawards to the extent of national jurisdiction. At its narrowest it may be no more then the coastal fringe of land where the vegetation is under maritime influence together with the intertidal zone and shallow sublittoral. The best way of resolving these differences of opinion might be to set limits on the maximum extent of the coastal zone, preferably on biological criteria, and then adopt a pragmatic approach whereby narrower limits are used according to the needs of management for particular purposes. Something of this type has been applied in the United States. The Coastal Zone Management Act 1972 avoids fixed boundaries but allows States to take action in relation to coastal waters out to territorial limits and inland to the extent necessary to control shorelands.

The important point, however, is to recognise that the coastal zone is not merely a composite of a terrestrial zone and a marine zone whose management problems can be solved in a composite way but that the whole is more than the sum of its constituent parts with special features and problems of its own deserving of special treatment. One of the reasons for the current problems in management of this zone is the tendency to treat it as no more than a boundary rather than as a unit in its own right. On the other hand, it must not be forgotten that there are very real differences between the terrestrial and marine components of the zone. There would be no advantage in moving from a two-part system of land and sea to a three-part one of land, coastal zone and sea which failed to recognise the inter-relationships of all three.

THE IMPORTANCE OF THE COASTAL ZONE

Worldwide, over 70 per cent of people now live at or near the coast and over 60 per cent of cities with populations of over 2·5 million are located there. The coastal zone is attractive to man for several reasons all of which have influenced the pattern of coastal developments. These include:—

—Transport. Shipping led first to coastal settlements at natural harbours and, later, to the development of man-made facilities.

—Industry. Industry is attracted to the coast for transport reasons and/or because of the need for an abundant supply of water.

—Residential use. People live in the coastal zone because of employment needs and because they like it. Modern transport patterns enable people to live at the sea and work some way inland.

—Recreation. Many coastal towns owe their existence to Georgian and Victorian fashions for bathing. Modern requirements are for marinas, holiday complexes and leisure parks.

—Exploitation of natural resources including fish and oil and the use of the sea bed for mariculture.

—Conservation. The coastal zone is scenically attractive with a rich and varied wildlife. The desire to preserve its natural heritage and to enjoy it by visiting is on the increase.

—Cost and Availability. Coastal land, even with the costs of reclamation, is economically attractive, increasingly so as the availability of other land declines.

MANAGEMENT NEEDS

The threat of sea level rise has focused attention on the need to plan for coastal protection. We do not feel competent to comment on the various predictions made but believe that the real damage, if any, will come from extreme storm conditions and will not be that closely related to the actual change in sea level, if any. There is clearly a need to plan for an increased likelihood of flooding but we do not think this should be done in isolation of other issues of coastal zone management.

The risks to human settlements are obvious although difficult to predict without any degree of certainty as to the level of change. There are, of course, risks to the environment as well but these can be viewed merely as part of a natural dynamic process. The coastline around the United Kingdom is very varied. It is also variable because it is moulded by a dynamic system of erosion and deposition in which sediments eroded from one place build new coastlines through accretion elsewhere. This means that man's interference with the natural processes at one site may have profound effects on coastlines many miles away.

Although the coastline is constantly changing the rate of change is not constant. The rate is determined largely by climatic conditions and there is good evidence to suggest a long term cycle in the rate of change. Most of the sand dune systems around the United Kingdom, for example, were built up during the 17th century during a period of violent storms. Although they have developed since then, there have been no changes in basic distribution of the dune systems.

The coastal environment is at much greater risk from man's interference with these natural processes than from sea level rise itself. Hard defences, for example, interrupt the natural process of erosion and may lead to beach starvation elsewhere on the coast. Of course, these consequences can apply whatever the motive for man's intervention and the coast has changed as much through the needs of industy, housing and recreation as through coastal protection. Soft sediments, in particular, have suffered. Recent reports on estuaries by the Royal Society for the Protection of Birds and the Nature Conservancy Council illustrate the environmental consequences of these changes. Physical change to the coast, however, is only one of the manmade risks that threaten the coastal environment. Pollution and over-exploitation of resources pose equally important threats.

The impact of these changes is great because the coastal zone is of great natural importance. The shallow coastal waters are highly productive and have a high biodiversity. Much of the coastline, especially where cliffs have made it unsuitable for exploitation, is of great scenic beauty and conservation importance. The nature conservation of much of the coastal zone has been recognised as being of international importance. Many stretches of the coastline are under some type of conservation designation and, although the record offshore is less good, there are many sites worthy of designation.

The coastal zone is thus an area with a multitude of habitat types and more than its fair share of people. The management problems are how to accommodate the desires of these people to carry out their livelihood with the needs to protect a fragile environment, all within the framework of physical changes of an unspecified magnitude.

FAILURES OF THE PRESENT SYSTEM

An appreciation of the need to protect the environment is still very new and it is not surprising that there are failures in the present system. These are exacerbated, however, by the failure to recognise the coastal zone as a unit. Instead, the coast is viewed as a boundary not only between land and sea but also between different administrative units. Natural features, such as estuaries, frequently form the boundary between neighbouring jurisdictions. These boundaries are inappropriate from a biological viewpoint and provide a poor basis for environmental management.

Similar problems apply to the law. Planning legislation provides for the control of development across all sectors to varying degrees but does not apply below low water mark and, apart from this, most legislation is framed in sectoral terms. There are over 80 Acts dealing with the regulation of activities taking place in the coastal zone.

At present management is based on sectors of interest with a broad division between things that happen on the sea and things that happen on the land. Furthermore, the units of management vary from sector to sector and tend to be too small. This leads to a lack of integration with the following results:

—The scale of physical change in coastlines is much greater than the unit of management so that coastal protection etc in one geographical unit may affect the coastline in another unit outside the control.

—The management of one activity may affect another sector of interest. Aggregate dredging, for example, affects fishing, navigation, pollution control, coastal protection, conservation, etc.

—Similar activities may be subject to different management regimes because they fall on different sides of the land/sea boundary. The construction of a pier or a marina complex, for example, has many features in common with the building of a bridge or a hotel complex but below water mark planning law does not apply. Similarly, a national nature reserve may adjoin a marine nature reserve but the protection afforded by one type of designation may not be used to protect wildlife in the other.

—Activities that combine elements of terrestrial and marine activities may be inadequately controlled because of gaps in the law and/or in the administrative framework. The Scottish fish farming industry is a prime example of this.

REQUIREMENTS

There is a need to develop a planning policy for the coastal zone as a whole together with the means of implementing it. At present, planning policy stops at low water mark but development does not. The private bill procedure does not provide a satisfactory substitute because it does not operate under any general policy framework. It would not be enough simply to extend terrestrial planning policies. There are many activities, such as fishing, taking place within the coastal zone that have profound effects on the environment but are not subject to any planning policy. Similar problems have arisen through the failure to integrate agriculture policy within land use planning policies. The main characteristic of the coastal zone is its varied nature, geographically, biologically, and socially. The key to controlling the many and varied activities going on there without detriment to the environment is to develop a planning policy that incorporates spatial and temporal zoning. Development of such a comprehensive policy is a long-term objective. In the meantime, the following actions could be taken.

—A thorough review of the relevant law to pinpoint those areas where there are no adequate legal controls.

—A review of the geographical jurisdiction of administrative authorities and the geographical limits on the extent of particular laws imposed by statute with a view to removing anomalies. One specific example here might be to consider extending planning controls beyond low water mark. Many activities that take place on and immediately beyond low water would, apart from their location, fit within the definition of operational development under planning legislation. It has been argued that local authorities are not equipped to deal with the peculiar issues involved in development control at sea. For some activities, such as offshore oil rigs, this may be the case but not for what are essentially terrestrial operations. Navigation interests could be drawn into the planning process by imposing a requirement on local authorities to consult with the appropriate bodies.

—A review of the role of the Crown Estate Commissioners as managers of the marine estate and not just in relation to fish farming. In the absence of any, more suitable, controlling body they have taken on the role of a planning body for offshore activities requiring licences or leases from them and have a statutory responsibility for environmental assessment of fish farms. They do not have the features usually associated with a regulatory body, namely, a clearly defined remit and public accountability. Their constitution, under the Crown Estate Act 1961, makes it impossible for them to become an environmental protection agency and their commercial vested interest is at odds with principles of natural justice.

—The publication of national guidance for local authorities. Local authorities, such as Hampshire County Council, have produced plans for the management of waters within their coastal zones but these remain empty policy without the appropriate powers. Initiatives at regional level, such as the Standing Conference on Problems Associated with the Coastline (SCOPAC) while very much to be welcomed, suffer in the same way. Some form of guidance from central government would give more weight to these local initiatives and would be a good first step towards the goal of integrated coastal zone management. The Government's proposals to modify the Scottish concept of Marine Consultation Areas and extend their use to England and Wales could be linked in with this guidance.

There is much to be said for proceeding along a non-statutory route. The legal complexities are so great that it would be unwise to wait until these have been fully analysed before proceeding. Certainly, much can be done by improving communications and developing informal co-operation between the many user groups and regulators involved in coastal zone activities. At the end of the day, however, the issues are too complex and developments proceeding at too rapid a pace for this approach to be entirely successful if for no other reason than that regulatory bodies can only operate within their remits leaving some sectors of activity outside of the system. At the very least, legislation needs up-dating. Ideally, some form of Coastal Zone Management Act is needed in order to assign responsibilities for all activities and to lay down clear pathways for co-ordination and integration.

There is a further question as to which arm of Government should be responsible for producing guidance and promoting legislation. All Government departments are involved in regulating activities in the coastal zone with the Department of the Environment, the Ministry of Agriculture, Fisheries and Food and the Department of Transport having the greatest involvement. Whatever approach is adopted, it must ensure proper integration and avoid departmental rivalries. One method might be to set up a specialist Coastal Zone Unit, perhaps within one of these Departments. There have been suggestions that the responsibilities of the proposed Environmental Protection Agency might encompass coastal zone management. This would have the advantage of promoting the environmental issues of coastal zone management but would require a considerable widening of the combined remit of Her Majesty's Inspectorate of Pollution and the National Rivers Authority if these do, as predicted, merge to form the Agency. Coastal Zone Management is far more than pollution control.

APPENDIX 9

Memorandum by Sir David Scott, Civil and Marine Ltd

Harbour maintenance dredging currently accounts for the removal from the sea bed annually of approximately three times the amount of material taken by dredging for sand and gravel over the same period. The total amount of material removed from harbours in England and Wales is currently some 80 million tonnes per annum, whereas the total for sand and gravel dredging is about 26 million tonnes.

Completely different regulations govern the practice of "Harbour Maintenance Dredging" and "Sand and Gravel Dredging". The former is carried out largely at the discretion of the Harbour Authorities, whereas sand and gravel dredging is scrupulously regulated by the Department of the Environment and the Crown Estate, who will not consider an application for a licence unless and until Hydraulics Research Ltd, who are very experienced in this matter, have given their view that there will be no adverse effects on the coastline.

Because of its immediate proximity to the coast, harbour maintenance dredging is likely to have an effect on coast erosion, particularly in an area such as the Isle of Wight, where extensive dredging of this type is carried out in the ports of Portsmouth, Southampton, Poole, Cowes and their approaches.

The products of harbour maintenance dredging are dumped well out to sea, permission for this being granted by the Ministry of Agriculture, Fisheries and Food. I appreciate that some of this material may be contaminated, but by no means all of it. For instance, clean sand often arises from the dredging of fairways. If, instead of being dumped at sea, it could be returned, in part or whole, to certain suitable beaches or mud flats, the coast erosion problem would be ameliorated.

On another matter, I feel sure that the Committee must be aware of the damage currently being done to the fisheries industry by the widely prevalent practice of beam trawling. This covers vast areas of the sea bed, hundreds of times greater than the area touched by sand and gravel dredgers, and yet there are apparently no restrictions placed upon it.

27 *February* 1992

APPENDIX 10

Letter from DG XI, Commission of the European Communities

In reply to your letter dated 18 December 1991, I am very pleased to inform you that the Commission is preparing a comprehensive strategy on integrated management and planning of the Community coastal zones which is now at its final stages of internal discussions.

Also of note, the Dutch Presidency recently proposed a Council Resolution on the same lines, that were adopted on 13 December 1991. Furthermore, it is expected that the Portuguese Presidency should give a special support to this file in the first half of this year.

Please find enclosed a copy of an article by our Director General from the Council of Europe review (Natureuropa 11/91), as well as a copy of the recent Council Resolution (in French)[1].

Apologising for the rather poor translation of the English version of Natureuropa (original in French), and the lack of an English version of the Resolution, I am at your disposal for further information on this subject.

Olivier Bommelaer
16 *January* 1992

---

[1] *Not printed.*

APPENDIX 11

Memorandum by the Flood Hazard Research Centre, Middlesex Polytechnic

INTRODUCTION

The views summarised herein are those of the Flood Hazard Research Centre (FHRC) at Middlesex Polytechnic. The FHRC is an independent research and consultancy unit which was established some 20 years ago. It has a staff of 20 and has undertaken research on many aspects of sea defence, coast protection, flood defence, water quality, environmental appraisal, and planning. These contracts have been undertaken for a wide variety of organisations including the Ministry of Agriculture, Fisheries and Food; the Department of the Environment; the Transport and Road Research Laboratory; Friends of the Earth; the World Bank; the OECD; the WHO; and United Nations agencies.

Although the Centre is currently engaged in work in Bangladesh, India, Hong Kong and Iran, the majority of our research experience has been derived in the context of the United Kingdom. Most recently, we have been involved in benefit-cost-analyses (BCA) of coast protection schemes at Hurst Spit on the Solent, and Herne Bay in Kent. A Manual on the application of benefit-cost analysis to the assessment of coast protection and sea defence schemes, prepared under contract from the Ministry of Agriculture, Fisheries and Food, will be published next year.

1. COASTAL ZONE MANAGEMENT

1.1 At present, the management and planning of the coastal zone is fractured in three ways: between sea defence and coast protection; into relatively small coast protection and maritime authorities, whose boundaries are not related to any physical boundary features or processes; and between land and sea.

1.2 In practice, the differentiation between the sea defence and coast protection function of a particular scheme is not usually clear cut (and has required court cases in the past to determine) but responsibility is differentiated in an unhelpful way.

1.3 The designation of the coast protection authority is largely for historical reasons, and, in most cases, at the District Council level. In planning terms, this is arguably too low a level since coast protection should be part of an overall coastal zone management policy and coast protection works frequently impose externalities outside of that coast protection authority's boundaries.

1.4 The development of regional groups such as SCOPAC is therefore to be welcomed and supported, but they are inadequate on their own. A geographically wider approach will be necessary if the retreat or abandon options to sea level rise are adopted. Indeed, a fragmented local approach to coast protection is likely to lead to pressure for maintaining current protection levels where there is no economic rationale for so doing. The present fragmentation has also encouraged the neglect of the externalities resulting from the decisions of individual organisations: those costs which are as a result imposed upon other organisations.

1.5 Integration of hazard management with other aspects of development remains difficult; as with flood plains, there are cases where development is permitted which then generates pressure for a protection scheme to be undertaken. No perfect solution is likely to be possible; a blank prohibition of development is not desirable since the policy objective is to make the most efficient use of coastal zone land (ie maximise the increase in outputs net of any increase in required inputs).

1.6 As with flood risks, there is evidence that the property market is imperfect: property is bought and developed by purchasers who are unaware that there is a risk of flooding, erosion or ground instability. Legal searches are apparently inadequate to pick up these risks and, in some cases, sellers have been known to be economical with the truth (eg one person we interviewed at Swalecliffe was told by the seller of the house that the reason the garden was rather run down was " heavy rain ": the house had been flooded to a depth of two to three feet six months earlier). These market failures inevitably create strong local political pressures for the affected properties to be protected.

1.7 In many areas, the risk of erosion has increased because protection of areas updrift has reduced the supply of sediment reaching areas downdrift. Other actions, such as the construction of long sea outfalls, harbour works or the removal of material, can have the same affects. Consideration might be therefore be given to zoning some areas of sea shore as sources of sediment material where maintenance of erosion at these sites is important in order to provide sediment to protect sites downdrift.

1.8 The free market economist will argue that the person undertaking such works bear the costs of any such increase in erosion downdrift. A minimum requirement is that those seeking to undertake specified works must bear the costs of the necessary model studies to demonstrate that there will be no significant affects downdrift.

1.9 Maritime authorities or regional management units should have their jurisdictions extended seaward of their current shoreline boundaries. Structure plans need to encompass both the land area and the associated sea areas.

1.10 Consideration might be given to a more coherent and effective system of zoning uses of the sea and sea bed, particularly where these uses interact with those on the land and where potential uses are in conflict.

2. CLIMATIC CHANGE

2.1 The main generic options for adapting to sea level rise are: protect, retreat or abandon. Abandonment or retreat both require to be managed if planning and investment blight, and the resulting short term inefficiency losses, are to be avoided. The first danger is that investment will be made in areas which will be abandoned in the absence of knowledge that abandonment will occur. Equally, knowledge of future abandonment is likely, in the absence of some planned process of abandonment, to result in premature abandonment through withdrawal of investment. Or, in some cases, people who will be trapped in property which they wish to leave but for which they can find no buyer.

2.2 As part of this management process, the structure plan process requires consideration of the implications of climatic change in general and sea level rise in particular.

2.3 That a number of nuclear power stations (eg Dungeness, Hinkley Point) are located on sites which are already subject to coastal erosion raises issues either as to future decommissioning policy of nuclear reactors or as to the commitment to coast protection for these areas over the medium to long term.

2.4 On current real values, generally the protection of agricultural land is not economic. However, considerations of the " option value " of protecting high quality agricultural land against irreversible loss to the sea may justify protection. A case may be made for such an " option value " if it is considered that there is a risk of scarcities in agricultural products at a future date, possibly as a result of the wider effects of climatic change. An upper bound to this " option value " can be estimated (Penning-Rowsell et al 1992).

3. ENVIRONMENTAL ECONOMICS

3.1 We welcome the publication of " Policy Appraisal and the Environment " by the Department of the Environment. In particular, that it emphasises analysis rather than techniques, and that the function of analysis to aid the clarification of the issues involved in a choice, rather than as a substitute for thought. Equally, we are pleased that it stresses that economic values are not the complete definition of value. Overall, we find it an excellent introduction to the purposes and capabilities of economic analysis.

3.2 In general, we recommend that economic analysis be used critically. With any form of analysis, the decision maker needs to remember that violation of the assumptions underlying the analysis will seriously damage the user's credibility. In particular, we consider that neo-classical economics strains its assumptions when applied to environmental goods (Green and Tunstall 1991b). We would, for these reasons, strongly recommend against the use of a neo-classical framework to value environmental losses and that great care be taken to value non-use values in general (Newsome and Green 1991).

For example, it has been shown (Bateman et al 1991) that a large difference between a willingness to pay for an increase in the availability of an environmental good, and the willingness to accept compensation for an equal decrease, is consistent with neo-classical theory. Secondly, we do not as yet understand the motivations that result in non-use values (Green and Tunstall 1991a), nor whether these motivations can be included into a neo-classical framework (Bateman et al 1991).

3.3 We note that, notwithstanding the Communique of the Dublin Heads Government Summit, there is apparently no agreement between Government Departments as to whether HM Government has adopted a policy of sustainable development. If it has, then the development of some criteria to test whether policies, programmes and projects satisfy that objective is urgently required. At present, we and others are recommending such criteria on an *ad hoc* basis (Penning-Rowsell et al 1992). Some uniformity and consistency between Departments would be desirable.

4. BENEFIT-COST ANALYSIS

4.1 In general, coast protection is only likely to be economically efficient to protect the following: buildings which are within a few years of being lost; sites of significant recreational value; and sites of significant environmental or heritage value. The protection of agricultural land will, after the artificial price inflations of land prices caused by the Common Agricultural Policy are removed, seldom be justified except on the basis of its option value, as discussed above.

4.2 Encouraged by institutional shortcomings, a general omission in the past from the BCAs of coast protection schemes have been the externalities of protection: any reductions in sediment supply or movement resulting from the proposed scheme must be valued as a disbenefit of the scheme and included in the BCA of that scheme (Penning-Rowsell et al 1992).

4.3 We welcome the principle of the prioritisation studies undertaken by the National Rivers Authority and others although we are unable to comment in detail because we have not seen the studies themselves.

4.4 Pending the publication of the National Audit Office report, we will not comment in detail on the quality of the benefit-cost analyses undertaken on coast protection and sea defence schemes. A comparison of the methods and techniques employed in the benefit-cost analyses undertaken by the different government departments would be interesting.

4.5 A benefit-cost analysis can be no better than the options compared. For this reason, we concur with English Nature's argument that the best environmental option should be included in the comparison of scheme options (English Nature 1991).

4.6 However, the inclusion of the environmental gains and losses resulting from scheme options into benefit-cost analysis is problematic at present. In part, this is for the reasons outlined above as to the limitations of including such affects within a neo-classical framework. But, in addition, there is often no scheme option, including the baseline "do nothing" option, which does not result in some environmental disbenefits. For example, at Hengistbury Head in Dorset, without a scheme an important archaeological site and environmentally important mudflats would have been lost. Conversely, reduction of erosion has adverse effects on an important geological site.

4.7 For this reason, we have proposed a process solution to the choice of scheme options for coast protection and sea defence schemes (Coker and Richards 1991; Penning-Rowsell et al 1992). However, this does not resolve the problem of how much it is worth spending to preserve an important environmental site which would otherwise be lost through coastal erosion. To date, the protection of such sites has been justified by the quantifiable economic benefits resulting from protection to other land uses (eg Hengistbury Head, Aldeburgh). This is unlikely to be the case for all sites potentially at risk.

5. WATER POLLUTION

5.1 We would wish to see the development of water quality indices for estuaries and coastal waters related to the uses of those waters. Such systems have been developed for rivers (House 1980). The present systems of classifying the water quality of coastal waters are too coarse for management purposes.

5.2 The distinction between pollutants for which the sea has a natural capacity to render harmless, or assimilate, and those which are cumulative is essential.

5.3 We would wish to see consideration of the use of economic instruments to control the discharge of materials for which an assimilative capacity exists.

5.4 The economic costs of pollution of coastal waters are not established (Newsome and Green 1991); research would clarify some of these. However, we regard it as misleading to define the problem of controlling pollution as the search for some economic optimum. Instead, it is unlikely that there will at any point be sufficient information as to the benefits and costs associated with different pollution standards to determine such a hypothetical optimum. The issue is thus one of devising the most efficient method of moving in the right direction and searching for the current best level. In particular, it is undesirable to treat present least cost solutions as fixed: the desire should be to induce the development of better (and lower cost) technologies.

5.5 It must also be recognised that the treatment of wastewater, and other potential pollutants, is highly capital intensive. Consequently, dischargers cannot respond rapidly to changes in required discharge standards. Improvements in discharge standards come largely from increased investment and technological innovation.

5.6 To the economist, pricing systems have the virtue of maximising the scope of people or organisations in the market to respond in that manner which results in the lowest costs. Equally, economists have a belief that only competition will cause companies to forego making excess profits. Therefore, economists tend to have a preference for "marketlike" approaches to control pollution in place of regulatory instruments.

5.7 Two strategies which we would, therefore, like to see considered are: first, the setting of charges (over and above the simple recharging of the administrative charges of setting and monitoring discharge consents) for discharges. Secondly, the use of marketable discharge consents using deflating discharge "bubbles". In the first case, we would wish to see a medium term (eg 10–15 years) racheting system of price rises built into the system. Since the optimum level of discharges will not be known in advance, the prices cannot be so set to induce this level. Instead, the purpose should be to induce innovation by indicating that discharge costs will rise, provide a reasonable investment planning framework to dischargers, and to search towards an optimum which will itself change over time.

For marketable permits, a "bubble" approach, a fixed limit on the total polluting load to a given receiving area, should be considered. Again, in order to induce innovation, this "bubble" should be subject to a specified rate of decrease over a medium term horizon. By decreasing the available supply of permits, the market price should rise so inducing a search for lower cost treatment methods.

Neither approach is universally applicable as the transaction costs involved (defining the equivalence of loads of different pollutants, monitoring the discharges) will only be relatively low when the capacity of the receiving waters is large relative to the loads. There is a real risk that the apparent gains in efficiency of economic instruments over regulatory approaches will be lost because the added transaction and administrative costs of creating such marketlike solutions will absorb all of the theoretically expected gains. Claims that the administrative costs of such marketlike solutions will be no greater than those of regulation are false (Green 1990).

Whilst there is probably the greatest scope for the use of such economic instruments to control atmospheric emissions, there may be some scope for their use in relation to coastal waters. This should be explored.

6. ENVIRONMENTAL ASSESSMENT

6.1 Environmental Assessments are required for projects " which would be likely to have significant effects on the environment by virtue of factors such as its nature, size or location ". This is somewhat imprecise.

6.2 We note that the average quality of Environmental Assessments completed in the United Kingdom has been found to be poor and welcome the actions by the Department of the Environment and others to raise the standards. However, that, at present, an Environment Assessment cannot be rejected as failing to meet the minimum standard required undermines the principles of the legislation.

6.3 Compared to several countries in Western Europe, and particularly in comparison to North America and Australia, the scope for public participation is presently restricted to rather minor forms of consultation. In today's society, the population affected by any proposal is highly educated and will include experts upon one or more of the issues raised by the proposed scheme. In this context, early public participation is essential, since without it the proposing authority inevitably loses credibility; once lost, this credibility is not easily regained.

We note that on a number of occasions, the coast protection authority concerned has instructed its consultants that any scheme proposed must, as a result of the consultative process, be such as not to give rise to objections leading to a Public Inquiry. That is, a solution was required which was agreed by all parties as being the best of the possible alternatives. In our opinion, such consultative processes have both resulted in better schemes and have resulted in lower costs. Notably, the now extensive use of " soft " protection has been partly driven by this process.

However, it has to be said that the enthusiasm for a consultative approach was partly driven by the experience of losing Public Inquiries; in some cases, several times.

6.4 The public can, however, only participate effectively if information can be made available to them in usable form. This problem has yet to be solved.

7. FLOOD RISKS

7.1 Any repetition of floods of the severity of those of 1953 are likely to cause severe problems in resourcing and managing the emergency response. The armed forces are considerably smaller than they were then and privatisation has meant that the direct labour forces and other resources (eg schools) controlled by local authorities are also much smaller. It is important to ensure that the availability of private contractors for such emergencies is clearly specified in their contracts. Similarly requirements apply, for example, to Hospital Trusts: an emergency may require the clearing of beds and the direction of casualties to the particular hospitals. Privatisation can also mean that the existing problems of incompatible radio equipment, and hence command and control of the resources deployed, are made worse.

7.2 For the obvious reasons, areas that are at risk from coastal flooding contain large numbers of CIMAH sites. Indeed, in the first Quantitative Risk Analysis of Canvey Island, the greatest risk of a major release of toxic or flammable gases arose from potential flooding. In such areas in particular, an integrated, rehearsed emergency plan including the flood risk is crucial.

7.3 In general, we would thus wish that a statutory duty be placed upon local authorities, under a new Civil Protection Act, to establish an integrated emergency management structure for civil emergencies covering not only sea and river flooding, and dam failure, but also other potential hazards.

7.4 When designing a sea defence scheme, or indeed, any engineering structure attention should be paid to that system's manageability under failure. A system which can rapidly change from operating in reasonable safety to complete failure creates difficult management problems. Any effective response must then usually be a response to a prediction of failure rather than to an actual failure.

Such a potentially rapid transition to failure is characteristic of sea walls which protect land lower than sea level. An effective warning (as opposed to a forecasting) system and evacuation system is essential in such cases. An effective warning system is one where those at risk are warned before the event and can respond effectively to the threat. In our experience, whilst the flood forecasting systems are often excellent, few flood warning systems are effective (Parker and Neal 1990).

The public typically need to be informed of the nature of the risk and what to do in the event of an emergency if the response is to be effective. We have in mind something akin to the "quatre heure" criteria adopted in France in relation to dam failure.

In this country, effective emergency planning is hamstrung by the misplaced fear that people will panic. So there is reluctance to tell them anything unless the threat is immediate. In reality, few cases of panic have been reported in any disaster. However, people often report that they panicked: when their self-reports are analyzed in detail, a statement by an individual that s/he panicked is associated with statements that they were frightened and did not know what to do (Green *et al* 1990). Consequently, it is probable that current policy creates the very conditions it is intended to avoid.

Equally, the current strategy rests upon the implausible assumption that once people have been told what to do, they will do what they are told and nothing else.

7.5 Our research reveals that whilst most development in coastal flood risk zones is piecemeal, the flood loss potential is rising rapidly. Decisions to develop land are complex and, on economic efficiency grounds, at least some development is likely to be justifiable particularly where the alternatives threaten areas of high environmental value. However, we believe that structure plans which focus residential development in coastal flood risk settlements are highly problematic. In some areas, we have found planning departments consider that their job is to promote economic development and that floods are nothing to do with them.

7.6 Integrating considerations of risk into the planning process requires some form of hazard mapping. Whilst some countries (eg France) have mandated that all communities prepare such maps, we have some concerns about the time and cost required to prepare detailed maps. However, indicative, as opposed to "definitive", maps are required as the basis of planning guidelines. In this regard, a problem with the preparation of such maps covering sea level rise is that the Ordnance Survey does not at present provide any contours under the five metre contour. This policy is under review and the problem may be resolved by developments in the use of Geographical Information Systems (GISs).

7.7 Other than from dam failure, sea flooding presents the highest risk of death from any form of flooding—once a sea wall is breached or overtopped, the area behind rapidly fills with water. Towyn was typical of sea floods although larger in scale than most recent such floods. From the accounts of victims of coastal flooding it is clear that we have been lucky in recent years to have avoided deaths.

However, provided that people do not try to wade through the flood waters, the depths of water are not usually sufficiently great to threaten residents in two-storey buildings. On the other hand, residents of single storey dwellings may face several hours being trapped in several feet of water and in near freezing temperatures. The residents of single storey houses are often elderly or infirm and consequently particularly vulnerable. Evacuation planning, either self-evacuation or aided evacuation, is a necessity in such conditions.

7.8 We welcome the introduction of both a "one-stop" bureau approach and trained counselling at Towyn; the first flood in the United Kingdom to our knowledge where this has been undertaken. It is known that to the flood victims, the stress of the event, the affect upon their health and worry about flooding in the future are all far more serious consequences of flooding than the damage to their home and its replaceable contents (Green and Penning-Rowsell 1989). Indeed, we find typically that some 70 per cent of victims of such floods report that it has affected their health. Flood victims also typically report that a number of people's lives were shortened.

What is unknown at present is both how effective are such interventions and which are the most effective form of interventions. Present limits on central government funding for emergencies can impose severe restrictions on such interventions: a local authority which has already had to respond to one crisis can be under severe financial pressure if it then has to respond to a second within the same financial year.

We are concerned that such interventions are being developed through professional, outside help rather than by reinforcing the community's ability to help itself. We have two reasons for such concern. The first is that this approach is necessarily limited to the scale of event which can be handled in this way before the supply of trained personnel is overwhelmed by the demand. The second is that communities have been shown to have enormous powers of self-help which, if properly supported, can aid the recovery of the whole community.

October 1991

REFERENCES

Bateman I, Green C H, Tunstall S M and Turner K 1991 *The Contingent Valuation Method applied to the Environmental Effects of New Roads,* Report to the Transport and Road Research Laboratory, Enfield: Flood Hazard Research Centre/University of East Anglia.

Coker A C and Richards A 1992 *Ecological Evaluation and Economic Valuation,* London: Belhaven (in press).

English Nature 1991 *Submission to the Ackers Committee,* Peterborough: English Nature.

Green C H 1990 " Economics in a time of failure ", paper given at the *European Environmental Risk Management Conference,* Kiev, Ukraine.

Green C H, Parker D J and Penning-Rowsell E C 1990 " Lessons for hazard management for United Kingdom floods ", *Disaster Management 3*(2), 63–73.

Green C H and Penning-Rowsell E C 1989 " Flooding and the Quantification of ' intangibles ' ", *Journal of the Institution of Water and Environmental Management 3*(1), 27–30.

Green C H and Tunstall S M 1991a " Is the economic valuation of environmental resources possible? ", *Journal of Environmental Management 33,* 123–141.

Green C H and Tunstall S M 1991b " Theoretical Bridges and Empirical Trains: Economics and the Environment ", paper given at the *Regional Studies Association Conference,* Oxford.

House M A 1980 " Water Quality Indices: An Additional Management Tool? ", *Progress in Water Technology 13,* 413–423.

Newsome D N and Green C H 1991 *The Economics of Improvements to the Water Environment,* Report to the National Rivers Authority, Reading: CNS Scientific and Engineering Services.

Parker D J and Neal J 1990 " Evaluating the performance of flood warning systems ", in Handmer J and Penning-Rowsell E C (eds) *Hazards and the Communication of Risk,* Aldershot: Gower.

Penning-Rowsell *et al* 1992 *The benefit assessment of coast protection and sea defence schemes,* London: Belhaven (in press).

APPENDIX 12

Memorandum by Hampshire County Council

Hampshire County Council has directed particular attention to the coast since the reorganisation of local government. It organised a series of Solent Sailing Conferences and has produced a number of publications about the coast, the most recent of which, *A Strategy for Hampshire's Coast*, was published in June 1991. It has set up a Coastal Conservation Panel, which meets regularly under the chairmanship of the Leader of the Council. In pursuit of its conservation objectives, the Council has a policy of purchasing coastal land when opportunities arise and now owns 3,800 acres on the coast. It also has an annual budget for coastal environmental improvements.

1. The County Council is concerned that coastal *management,* as well as protection and planning, should be considered by the committee. In its view the integration of planning and management is the most urgent need in areas under pressure. Many of the activities which affect the coastal environment, including the operations of certain statutory authorities, such as harbour authorities, are outside planning control. Sympathetic and effective management is of the essence.

2. It is important to recognise the diversity of the coastline in terms of the problems, priorities and pressures. There is unlikely to be a uniform solution, or priority, which can be applied. It would also be wasteful of resources to impose new controls and administrative structures in areas where these are not needed.

3. It is equally important to recognise that the achievement of satisfactory planning and management on the coast is as much, if not more, an administrative exercise as a technical one. The main need may be to co-ordinate the activities of the many authorities and agencies involved. This calls for positive leadership, backed up by the appropriate experience and skills. Central government should recognise this need and encourage initiatives which emerge, where appropriate.

4. Misunderstandings sometimes occur because of confusion over the meaning of terms. To some people coastal zone planning and management is essentially about erosion and flooding. In the County Council's view it should embrace all aspects of the coastal environment, including nature conservation and recreation. In many areas a fresh approach is called for, and it may be necessary to resist the temptation to broaden the scope of existing arrangements, which have been set up to deal with one particular subject. It is important, too, that a proper balance is struck between conservation and other interests.

5. One of the greatest needs is quite simply that a higher priority should be given to coastal planning and management at all levels. In this central government has consistently failed to give a lead. The insistence in recent speeches by ministers that the main thrust should be through the completion of statutory district-wide local plans, though understandable, is particularly disappointing. The experience in Hampshire has been that the coast does not feature prominently in statutory local plans. It is, of course, essential that statutory plans contain firm policies for the coast, but they can only go so far. Their scope is tightly constrained. In many areas a wide-ranging non-statutory plan is called for and central government should give guidance and encouragement to their preparation.

6. Coastal zone planning and management must consider the relationships between activities on land and water. In this context, the termination of local authority jurisdiction at low water mark over much of the coastline is not helpful. Although the number of developments below the low water mark may be relatively few, the attitude of mind which is induced, of land and sea being separate entities to be considered separately, is most unfortunate.

7. The exclusion of most of the inshore waters from local authority jurisdiction has also hindered proper research and consideration of such issues as fish farming, maritime archaeology and the nature conservation value of the sea bed.

8. The unique position of the Crown Estates Commissioners has been criticised from many quarters. Intense public pressure has been brought to bear in a number of areas in which they have jursidiction for them to adopt an approach more sympathetic to the environment and less commercially aggressive. The County Council believes that their role should be re-examined, having particular regard to their lack of accountability, the interpretation of their financial objectives, and the uniquely " public " nature of much of the Crown Estate on the coast.

9. National coastal planning policy should not rely simply on the existing national and international protective designations, such as Areas of Outstanding Natural Beauty, Heritage Coasts and Ramsar Sites. Some areas, which do not qualify for such protection, nevertheless have a special appeal and role, which should be recognised by central government. The importance of the area for recreation and its location in relation to large centres of population should be important considerations. The Solent is an obvious example.

10. There are an unfortunate number of examples where relatively recent development has been located in areas at risk from flooding or erosion. Some National Rivers Authority regional offices are working with planning authorities to prevent this in future, but many statutory plans still do not include adequate policies on this subject. Central government should require risk areas to be defined and suitable planning policies to be drawn up. It must also uphold such policies at appeals.

11. The distinction between coast protection, district council responsibility, and sea defence, which is the responsibility of the National Rivers Authority, is unrealistic, inefficient and is not conducive to prudent forward planning. On some parts of the coast, the distinction between the two codes is blurred. On others, there has been a lack of liaison between neighbouring authorities to ensure that remedial works do not exacerbate problems elsewhere. The fragmentation of responsibilities and resources has hindered the development of expertise, integrated research and a pro-active approach. The Ministry of Agriculture, Fisheries and Food has not noticeably taken a lead, except in one or two areas. The funding of works at strategic locations sometimes imposes a heavy financial burden on the District and County Council directly involved. The use of more natural environmentally friendly methods of coast protection, such as beach replenishment, may be deterred because the maintenance which these inevitably require does not qualify for grant aid. Well over half of the Hampshire coastline is not covered by the Coast Protection Act with the result that landowners are left to patch and mend as best they can. In short, the responsibilities, legislation and funding of coast protection and sea defences are in need of urgent review with a view to giving overall responsibility for both aspects to one properly resourced authority. The prospect of rising sea level makes such a review all the more pressing.

12. The role of harbour authorities is becoming ever more important as water recreation activities continue to grow and new types of craft are developed. Some authorities have not adequately adapted to the changes either in terms of staffing and organisation, or in their charging policies. Many are hindered by out-of-date and restricted legislation which may not be geared at all to conservation and recreation issues. In some cases, there needs to be better liaison between planning and harbour authorities, especially to ensure that development control and navigation policies are consistent with each other. The exemptions from planning control which harbour authorities enjoy are unnecessarily and unacceptably wide-ranging.

13. As well as giving a lead on general policy for the coast, central government should prepare more detailed policies on specific issues which require a national perspective. A current example in Hampshire is the provision of yacht moorings. The sheer number of yachts berthed in the Solent is now causing problems and the County Council considers that there should be a national strategy, which encourages the provision of moorings elsewhere, to spread the load.

14. A relatively detailed, but none the less important, issue is the classification of small boatyards in the Use Classes Order. For this purpose they are bracketed with large shipyards and general industry. Significant changes, which would be extremely damaging in many cases, can, therefore, take place outside planning control. A new Use Class for predominantly recreational boatyards is required.

24 *September* 1991

APPENDIX 13

Letter by Hampshire County Council

COASTAL ZONE PROTECTION AND PLANNING

Further to your recent telephone conversation, I confirm that the County Council wishes to add an item to the Memorandum of Evidence, which it has already submitted to the Environment Committee.

The County Council is convinced that public ownership of coastal land is by far the most effective way of conserving the character of the coast in areas under pressure. For many years it has had a policy of acquiring sites as opportunities arise. At the present time the Council owns over 3,800 acres on the coast. The sites range from country parks, of which there are three on the coast, to quiet areas of marshland. All are managed with conservation as a main objective.

The Council strongly urges the Committee to consider carefully the advantages of adopting such a policy on a national basis, backed up by appropriate legislation and resources.

If the Committee wishes to explore this concept further, I can provide more details of the County Council's coastal land holdings. I enclose a copy of a note outlining the work of France's Conservatoire du Littoral, an existing example which would be worth investigating.

21 *October* 1991

CONSERVATOIRE DE L'ESPACE LITTORAL ET DES RIVAGES LACUSTRES

(French State Coastal Conservancy)

GOAL

Enacted in 1975 by the French Government, the *Conservatoire du Littoral* is responsible for:

—the acquisition and the protection in perpetuity of the most sensitive and endangered coastal land

   —along the coast of metropolitan France and overseas counties,

   —around the 12 lakes larger than 2,500 acres,

—the management by local governments of the land acquired in order to open it to public access and insure the protection of natural values.

MEANS

—Annual investment budget, fiscal year 1990:

   ● State funding: about 80 million francs ($15 millions),

   ● Other funding: local government, EEC, donations, sponsorship: about 50 million francs ($9 millions)

—Total investment on 1976–1990 period: about 1,2 billion francs in 1990 currency ($220 millions).

—Staff: 40.

LAND ACQUISITION

—Number of sites: 270 in 300 townships (out of a total of 1,000 townships).

—Acreage: 90,000 acres:

of which:	scrub	35%	moor	11%
	wood and forest	18%	grassland	5%
	wetlands	7%	cultivated land	2%
	dunes	12%		

—Coastline protected:
—Metropolitan territories		222 miles
—Overseas territories (Guadeloupe, Guyane, Martinique, Réunion)		21 miles
● Lagoons		36 miles
● Lakes		10 miles
	Total:	289 miles

MANAGEMENT

Under the responsibility of local governments: townships, counties, local authorities on the basis of agreements with the *Conservatoire du Littoral*. Local governments often call on conservation groups and forested areas are managed by the *Office National des Forêts*.

SOME EXAMPLES OF REMARKABLE SITES

—Famous landing sites in Normandy, such as Utah Beach, Omaha Beach, Pointe du Hoc, batteries of Merville . . . with historic and physical rehabilitation.

—Milliau Island in Brittany where are located 225 different plants species on 60 acres.

—Marshes of Yves and of Brouage on the Atlantic Coast managed by the *Ligue pour la Protection des oiseaux* (similar to the US Audubon Society or British RSPB).

—Camargue area: Eight estates totalling 4,250 acres as a contribution to the protection of migratory birds.

—Rayol Botanical Garden on the Riviera.

—Agriate Massif in Northern Corsica protecting 12,500 acres and 15 miles stretch of coast.

EUROPEAN TWINNING PROGRAMME EUROSITE NETWORK

Nine out of the 33 nature coastal sites selected for twinning arrangements by the European Community belong to the *Conservatoire du Littoral.*

Platier d'Oye	— Titchwell Marsh, GB	—	Wexford Wildfowl Reserve, IR
	Marais et Vasières Charentais		
	— Het Zwanenwater, NL	—	Elmley Marsh, GB
Baie de Canche	— Dunas de St Jacinto, P	—	Lucio des Cangrejo (Marismas del Guadalquivir), E
	La Palissade (Carmargue)		
	— The Raven Nature Reserve, IR		
Domaine de Certes	— Ria Formosa, P	—	Het Zwin, B
	Domaine du Marquenterre		
Baie d'Audierne	— S'Albufera de Mallorca, E	—	Parco Nazionale del Circeo, I
	Massif des Agriates		
Estuaire de l'Orne	— Beaulieu Estuary (North Solent SSSI), GB.		

ORGANISATION

—Administrative supervision: Department of the Environment.

—Board of Directors:
    chairman: Mr Guy Lengagne, Congressman and former Minister of Sea and Fisheries

      13 public officials   5 national elected officials
        4 officials from conservation groups   12 local elected officials

—7 *ad hoc* Regional Coastal Councils
—Director General: Mr Patrice Becquet
—6 Regional Offices.

DECISION-MAKING

for each acquisition:

—consultation of the township and the *ad hoc* Regional Coastal Council concerned,

—decision by the Board of Directors,

—implementation: Head office and Regional office.

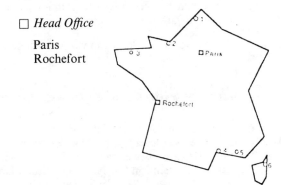

☐ *Head Office*

Paris
Rochefort

○ *Regional Offices*

1 Boulogne-s/-Mer
2 Caen
3 St-Brieuc
4 Montpellier
5 Aix-en-Provence
6 Bastia

APPENDIX 14

Memorandum by the Institution of Water and Environmental Management

1. The Institution is grateful for the opportunity to submit a memorandum of evidence in connection with the enquiry on coastal zone protection and planning.

2. COASTAL ZONE DEFINITION

First and foremost, the Institution feels that the extent of the coastal zone must be carefully defined. Land and water interact at the coastal zone. The area of interaction will vary from site to site. Coastal zone protection and planning should be all-encompassing, including not only engineered coastal defence systems (shoreline management) but also wider coastal management issues such as planning and all aspects of natural, physical and human environments. The latter will include flooding, erosion, pollution, fisheries, migrant and domiciled birdlife, unique geological or physical features and historical buildings, etc, many of which are inter-connected.

3. PLANNING POLICY IN COASTAL AREAS

Planning response to applications for development in areas which are sensitive to physical change due to flooding or erosion should give more consideration to the long-term and/or indirect consequences of permitting that development. Possible consequences of development include an interruption in natural coastal processes and/or an increase in future expenditure on coastal defence systems. Wherever it is technically viable, soft defences should be considered in order to provide a defence solution which is sympathetic to the natural physical system/ecosystem. Delays in both the permitting process and the appeal stage should nevertheless be avoided at all costs.

4. THE ROLES OF THE MANY AUTHORITIES RESPONSIBLE FOR PROTECTION AND PLANNING IN THE COASTAL ZONE

A widescale integrated approach to protection and planning in the coastal zone is important. Therefore, it is necessary that all bodies, both statutory and non-statutory, with legitimate interests in the coastal zone must be involved in protection and planning. Such organisations will usually be nationally recognised, and may include the relevant planning authority, local authority, National Trust, Royal Society for the Protection of Birds, National Rivers Authority (NRA), Ministry of Agriculture, Fisheries and Food, Countryside Commission, English Heritage and local interest groups. Local participation in the preparation of coastal zone management initiatives is crucial. Local coastal cell liaison groups should similarly be encouraged to have a significant input. Finally, it should be remembered that coastal processes are not delineated along municipal boundaries. The coast should therefore be treated as a continuous series of components.

5. Sea defence, coastal protection and land drainage should be the responsibility of one agency. This should be the NRA unless the Government sets up its proposed Environment Agency in which case it should be transferred to that Agency with all of the other NRA responsibilities. Environmental protection should also be co-ordinated, again in the NRA or, if it is set up, the Environment Agency. Environmental protection should also be co-ordinated, possibly as the responsibility of the prospective Environment Agency. Planning should remain under the remit of the local authority, while having regard to other interested agencies, especially the NRA which should be overall supervisor on all coastal protection matters. Some members believe that the NRA should have the power of veto over high risk developments. Finally, MAFF should have responsibility for policy direction and for facilitating projects of national importance by way of increased grants. In this way, the status of coastal zone planning and management will be raised.

6. In the short-term, interested parties in the coastal zone should be encouraged to communicate via coastal-cell type groups to help resolve conflicts and build bridges towards future policy.

7. Finally, planning policy with respect to pollution and ecosystems should acknowledge an inherent, international or cross-boundary perspective.

8. THE RISKS TO COASTAL SETTLEMENTS AND COASTAL ECOSYSTEMS FROM, FOR EXAMPLE, FLOODING, COASTAL EROSION, LANDSLIDES AND POLLUTION

The degradation and loss of coastal ecosystems due to land claim for industrial and/or agricultural development is likely to be exacerbated by global warming-induced sea level rise as these ecosystems are " squeezed " between mean low water and hard defence structures. Experience overseas, notably in the United States, indicates that it may be possible to recreate some coastal ecosystems on low-lying agricultural land. The creation of a fully functioning natural system may, however, take several decades. In view of continuing development pressures and the possible long term effects of pollution compounding the sea level rise problem, it may be prudent to instigate experiments to investigate the practical viability of recreating coastal ecosystems in Britain.

7 October 1991

APPENDIX 15

Memorandum by the Joint Nature Conservation Committee

THE JOINT NATURE CONSERVATION COMMITTEE

The Joint Nature Conservation Committee (JNCC) was established by the Environmental Protection Act 1990, to be responsible for the provision of advice and research into nature conservation at United Kingdom and international levels. It is jointly funded by English Nature, the Countryside Council for Wales and the Nature Conservancy Council for Scotland. The Committee is composed of the Chairman and one other member from each of the country Councils, three independent members and representatives from the Countryside Commission and conservation interests in Northern Ireland. The Committee is supported by specialist staff.

This submission aims to provide a United Kingdom perspective on the issues which are important when considering coastal planning and managment, in relation to the conservation of the wildlife resources, including earth science interests. It has taken account of the views of English Nature and the Conservation Service, Northern Ireland.

BACKGROUND

The coastline of Great Britain is extensive, at high water when all the tidal inlets and larger islands are included, it measures in excess of 18,600 kms. Despite the exploitation which has taken place historically and in more recent times, there is still a high proportion of natural and semi-natural habitats and ecosystems with considerable wildlife interest. For the purposes of this submission the coastal zone is taken to include the rivers and tidal inlets, the terrestrial coast (sea cliffs, sand dunes, saltmarshes, shingle structures and coastal grazing marsh), intertidal shores (including sandy beaches, mudflats and rocky shores) and the marine zone to the three mile limit, together with the plants and animals which live there.

The zone is dynamic and the geomorphological processes can operate over a wide area. For example, river-borne sediment may originate in the uplands, be deposited in estuaries to form tidal flats, or moved offshore by the tides to be redeposited as a sandy beach. This may in its turn act as a precursor to the development of a sand dune. Eroding cliffs also supply sediment which can be transported along the coastline for some distance before acting as a sediment source for features such as sand dunes and shingle barriers. Many of the animals which inhabit this zone also range over a wide area. Migrating shore birds require a series of feeding stations, including many tidal estuaries. Some fish live in the ocean but breed in the shallow nearshore waters, or even upstream in the rivers. Marine mammals may also travel long distances feeding in one location, breeding in another.

Man's impact on these systems can come from sources remote from the area affected. Pollution originating from the land is transported by the rivers into the sea; land claim can affect the tidal regime and sedimentary processes in estuaries; coast protection may influence sea defence structures elsewhere along the coast.

The scale of the coastal geomorphological processes which operate in this highly dynamic zone and the wide-ranging nature of man's impact on it are fundamental to the consideration of the nature conservation requirements.

THE COASTAL RESOURCE

The coastal wildlife resource includes a number of habitats and species concentrations unique to the zone. The marine and terrestrial components each have their own physical and biological characteristics, which are distinct. In addition it provides an interface between the land and the sea where the transitions which occur between the two are often biologically rich. Many terrestrial habitats also occur here, some showing a truly maritime influence, others are more typical of inland areas.

The importance of the whole coastal/marine zone for nature conservation, cannot be over emphasised. In the marine environment the rocky coasts and sea bed of the south-west England contrast with the cold Scottish northern isles. The former lie within the warm-temperate lusitanean region, the latter include arctic elements of the north-east Atlantic flora and fauna. Along the coastal fringe there is a similar north-south differentiation, though superimposed on this is the effect of the prevailing westerly winds, which subject the west coast to a greater maritime influence than the east. Further variation is introduced by the underlying topography and geological strata.

The marine communities around Great Britain are the most diverse of any European Atlantic state's border, reflecting both the range of geological and geomorphological features and the climatic variability. The marine algal flora (seaweeds; excluding microscopic species) includes about 840 species and the fauna about 7,000 species including 33 mammals, 188 birds and 300 fish. Marine species currently protected under Schedule 5 of the Wildlife and Countryside Act 1981, include all species of whales, dolphins, porpoises and turtles as well as the walrus. Only one species of marine fish is included (the allis shad), but there are five marine invertebrates (two anemones, a worm, a sea mat and a shrimp) and a plant restricted to saline lagoons.

Table 1 below, gives an indication of the size of the main coastal habitats in Great Britain. The information is derived from measurements taken from the Ordnance Survey 1:50000 scale maps and existing habitat information.

TABLE 1

Coastal habitats in Great Britain

Habitat Length (km)	England	Scotland	Wales	Total
Sedimentary shore	3,067	2,511	858	6,436
Shingle shore	541	2,993	100	3,634
Rocky shore	576	5,675	463	6,714
Saltmarsh	1,424	360	183	1,967
Sand dune	325	894	149	1,368
Cliff > 20m	876	1,391	439	2,708
Artificial embankment	1,764	226	180	2,170
Habitat Area (sq km)				
Seabed LWM–3 miles	—	—	—	68,285
Marine inlets and foreshore	—	—	—	4,331
Tidal flats	1,509	536	308	2,353
Saltmarsh	325	61	67	453
Sand dune	93	400	71	564
Shingle (vegetated)	34	7	1	42

The coastal fringe is equally rich. Our estuaries, for example, which are composed of a mosaic of important habitats, lie at the junction of two major bird migration routes, one coming from arctic Canada, Greenland and Iceland, the other from central Siberia, Scandinavia and northern Europe. Many species require a number of sites to provide feeding stations en route as well as winter feeding areas and British estuaries support 4 per cent of the north-west European population of wildfowl and 15 per cent of the east Atlantic flyway populations of waders. Their importance is reflected in the 30 or so coastal/marine bird species which are included on Schedule 1 of the Wildlife and Countryside Act 1981.

Despite the obvious importance of the bird populations they are a manifestation of a complex and highly productive ecosystems. Sand dunes, saltmarshes and shingle structures are important habitats in their own right and support a characteristic and often rich flora and fauna which includes species specially adapted to the rigours of these highly dynamic and sometimes stressed environments.

Sea cliffs are particularly varied, reflecting the combination of solid rock types and climatic factors which influence the vegetation. Highly exposed sites may be rich in rare plants, particularly where calcareous rocks are present. Vertical cliffs may support large colonies of nesting sea birds including many species which spend the majority of their life at sea. A large proportion of the European population occur on Britain's cliffs, and for seven species their numbers exceed 50 per cent of the world population.

PROTECTED STATUS

The significance of the zone in wildlife terms is reflected in the number of Sites of Special Scientific Interest (SSSI), National Nature Reserves and the extent of reserves owned or managed by non Governmental organisations like the Royal Society for the Protection of Birds (RSPB) and the National Trust. There are approximately 800 SSSIs on the coast and these are made up of examples of all the major natural and semi-natural habitat types occurring in Great Britain. Most of these sites include intertidal shores where many of the important communities of marine plants and animals occur. Parts of 111 of 155 estuaries, approximately 120 sand dunes and 10 major shingle sites, and a high proportion of the existing saltmarsh habitat are notified as SSSIs. Of the National Nature Reserve series 15 per cent of the 234 reserves are coastal and there are two Marine Nature Reserves.

The RSPB have many coastal nature reserves and the National Trust are by far the largest land owner of coastal properties, with their Enterprise Neptune having acquired many kilometres, mainly of coastal cliffs.

Outside protected sites the long stretches of relatively undeveloped coastlines still survive. Some of these are included in Areas of Outstanding Natural Beauty. There are also many Heritage Coasts, defined by the Countryside Commission.

The United Kingdom Government is party to a number of international wildlife conventions. These require the listing and safeguarding of Wetlands of International Importance under the " Ramsar " Convention. Classification of Special Protection Areas by the United Kingdom Government is a statutory responsibility under the EEC Directive 79/409 on the Conservation of Wild Birds. Many of the sites covered by these designations lie on the coast.

MAN'S IMPACT ON THE COASTAL ZONE

Given the extent of statutory designations and nature reserves, both public and private, it might appear that the coastal zone is well protected from adverse impacts. However, this is far from the case. Historically man has exploited the coast for agriculture, industry and recreation and the marine environment for fishing, port and marina developments among other activities. In so doing he has destroyed or damaged a great deal of the natural habitat. The recently published Estuaries Review (Davidson *et al* 1991) provides an assessment of the current status of the wildlife on the estuaries around our coast. Historically some 25 per cent of the estuarine habitats have been destroyed though a process of piecemeal and cumulative land claim. This process continues and the results suggest that at present rates at least match the historical rate of between 0·2 and 0·7 per year (over the last 200 years). The current spate of proposals for amenity and tidal barrages would further stress several internationally important sites if built.

Although comparable studies have not been carried out in the same detail for other coastal and marine areas it is clear that damage is no less severe. Sand dunes have been affected by pine plantations, sand extraction, building development for housing, industry and recreational use including golf courses. Saltmarshes have been enclosed for agriculture and saltmarshes and tidal flats have been claimed for industry, marinas and port development. Shingle structures have suffered equally with gravel extraction forming the main loss. Coastal grazing marshes, themselves the product of man's use of land have been ploughed, replacing the rich plant associations in the traditionally managed drainage dykes and pastures with intensive arable land. Coastal cliffs tops have been "improved" for agriculture and siting of caravan and car parks have destroyed habitats.

River systems have been canalised and increasing amounts of water abstracted from them, resulting in greatly reduced water flows. This in turn has affected the discharge of sediment to estuarine and coastal systems. Industrial discharges have created problems from pollution, some of which are only just beginning to be recognised for the environmental damage that they do. For example there is a possible link between the presence of toxic compounds such as organochlorine insecticides and polychlorinated biphenols present in coastal waters, and the recent outbreak of phocine distemper virus in 1988 when over 18,000 common seals died.

Marine systems are affected not only from the knock on effects of land based development and pollution, but also from activities taking place on or under the sea. Amongst these some of the more important impacts are associated with over fishing, aggregate, capital and maintenance dredging, cultivation of non native and native fish and shellfish stocks and oil pollution.

All these activities have taken their toll on the natural environment of the coast. The impact of land based development of the coastal margin, including the upper levels of the tidal zone, has been to squeeze the habitats into an ever narrower zone. This, not only destroys the wildlife habitat, but also reduces its effectiveness to adapt to the ever changing conditions which exist there. The way in which this has affected the wildlife resource has different manifestations around the country and these are considered below.

SOME EXAMPLES FROM AROUND THE UNITED KINGDOM

England

In lowland England particularly around the south and south-east coasts the enclosure of saltmarsh and other tidal lands has helped to create high quality agricultural land. At the same time the development of housing and industry has taken place, both adjacent to estuarine land, and also on vulnerable cliff tops. Having established claim to this land society is loath to give it up and a great deal of money and effort has been put into defending the land from flooding and erosion. Today in excess of 50 per cent of the south and south-eastern shore line is protected by some form of artificial coastal structure.

In this same area sea level is rising due to the sinking of the land as a result of adjustments still taking place following the end of the last glaciation. The effect of this is to exaggerate the coastal squeeze, as sea level rise causes a narrowing of the shore between high and low water. The way this situation is dealt with has important for coastal defence and in its turn land use policy.

A number of other important issues are more significant in England than elsewhere in the United Kingdom. These include: land claim; sediment extraction both from the shore and in inshore waters and water quality.

Scotland

A different set of problems occur in Scotland. Here coastal defence is much less of an issue, because there is less enclosed land, a higher proportion of the natural shoreline is resistant to erosion and also because relative sea level is static or falling in most areas. The presence of major populations of breeding birds on the cliffs and the importance of the coastal waters for feeding results in there being a potential conflict with the oil and fishing industries. Fish-farming which now occurs in most sea lochs, particularly on the west coast has contributed to the degradation of formerly pristine marine habitats. Agricultural use, most signficantly over-grazing on sand dunes, combined with recreational pressures has caused the break up of the surface vegetation and erosion.

Wales

Sea defence is an important issue in a number of locations and amenity barrages pose a significant threat at a number of estuaries. Otherwise many of the issues of concern are common to those of England.

Northern Ireland

Recreational use is a major theme here as is the question of sand extraction from the beaches. The multiple use of the internationally important Strangford Lough provides an interesting example of a site where reconciliation of the needs of conservation with those of man is urgent particularly in view of the proposal to designate it as a Marine Nature Reserve.

The European Dimension

The European Commission (EC) with the support of others has initiated a review of coastal zone management. A workshop (jointly sponsored by the Department of the Environment (DOE), National Trust, Countryside Commission and the EC amongst others) met earlier this year. The communication that was issued stressed the need for a more integrated approach to management which took as its starting point the protection of the remaining areas of natural and semi-natural coastlines. At the same time it was recognised that development had to be sustainable and that this needed to be considered in the wider coastal/marine context.

Past initiatives have focused mainly on the need to deal with pollution issues which cross state boundaries. However, it is increasingly clear that the coastline itself and the inshore coastal waters are suffering damaging impacts through land based development, and other activities such as marine dredging which affects both the marine and coastal environments.

North Sea and Irish Sea

Both of these areas are the subject of international co-operation in recognition of the problems which cross country boundaries. The ministerial conferences, which have become a feature of the North Sea collaboration, focused initially on control of pollution at sea. The importance of the coastal margin is increasingly being taken into account and the production of a " Directory of the North Sea Coastal Margin ", which is due to be published in draft form by the JNCC shortly, will provide a wider view of the importance of the North Sea coastline of Great Britain. It is envisaged that this could form a model for other North Sea states to follow. This in part complements the Irish Sea Study report which was completed last year. This report will itself be taken into account in the production of a West Coast Directory which is currently being contemplated. Both of these initiatives are being developed in collaboration with the marine branch of the DOE.

Resource Data and Impact Analysis

The development of a coherent policy for the coastal zone requires an adequate data base. This must include information not only on the resource, but also on the factors which affect the coast. A start has been made on collating data on the coastal resource and the figures given in the table above provide a first assessment. More recently the " Coastwatch " project, which is being completed by the JNCC, surveyed the whole coastline of Great Britain and provides information on the extent of all major coastal habitats and the occurrence of damaging activities.

At a more detailed level the Nature Conservancy Council carried out the Estuaries Review and the information on the estuary resource of Great Britain and man's impact upon it provides the basis for future policy consideration. The JNCC will extend and develop this approach in the context of a wider review of the entire coast. This will run in parallel with the Marine Nature Conservation Review, which is a fundamental survey and evaluation of the marine nature conservation resource. These data together with other sources of information like that being collected by the JNCC's Seabirds Team in Aberdeen, have already been used to provide the comprehensive inventories included in the Directory of the North Sea Coastal Margin.

The recently completed " Atlas of coastal sites of nature conservation importance sensitive to oil pollution " which was produced by the Nature Conservancy Council (NCC) in collaboration with the Marine Pollution Control Unit of the Department of Transport and British Petroleum, provides an important example of the way baseline data can be used to consider coastal zone management issues. These co-ordinated data gathering exercises are of considerable importance and funds are required to continue and develop these activities.

Conclusions

It is relatively easy to catalogue the destruction and impact of man's activities in the coastal zone. Despite all that has been outlined above the coastal zone still represents an area which has many attributes of importance for wildlife conservation. This applies particularly to the marine environment, though even on the coastal margin where man's activities have been most obvious, extensive features of nature conservation interest still occur. In considering the options for the future planning and management of this resource there are two important principles:

1. The coastal zone represents a complex sequence of habitats which include the rivers which flow into the sea, the coastal margin, including transitions between the land and the sea, and the coastal waters which surround the land. The nature and scale of the geomorphological processes which operate there and the distances over which animals may travel reinforce the interrelated nature of the zone both within sites and between sites.

2. The cumulative and far-reaching impact of some of man's activities, suggest that a piecemeal approach to management is not adequate if we are to protect the best areas of coast at the same time as providing for their sustainable use. The table included as Annex 1, taken from the Irish Sea Study Group Report, provides a possible framework for considering conservation management on the coast.

Recognising the special needs of the coastal zone in terms of protecting it from the continuing degradation brought about by the variety of human activities which degrade coastal waters and destroy natural habitat is the first stage in determining the action which is needed.

RECOMMENDATIONS

To date activities in the coastal zone have largely taken place on a piecemeal basis with little regard for the cumulative nature of their impact nor of the knock-on effects across habitat, county or country boundaries. Regulatory control has similarly been unco-ordinated resulting in decisions which often compromise one or more of the other interests.

1. The development of an adequate data base to identify the most important and sensitive zones on the coast and the relationship between these and adjoining areas, is an important prerequisite in the formulation of any strategy designed to conserve the natural features. These data bases will also be important in providing information for European initiatives including the European Environmental Agency when this is established.

Resource Inventories. The rapid completion of resource inventories and the establishment of data bases for the United Kingdom coastal and marine zone is a high priority.

2. Monitoring of activities and changes in the, particularly at the coastal margin will be needed if the implications of the predicted climate change and the associated sea level rise, are to be assessed, particularly in relation to major development proposals.

Monitoring. It will be important to establish a system of coastal zone monitoring sites for the repeated measurement of change in natural and semi-natural systems and the factors causing or likely to cause change.

3. Research is a vital element in determining the impact of man's activities on the natural coastal zone.

Research. Research in the coastal zone should be more precisely directed towards elucidating processes and impacts underlying change, defining limits of acceptable change and setting standards for environmental quality.

4. One of the keys to developing sustainable strategies for managing the coastal zone is to establish mechanisms to facilitate co-ordination and communication. These should be based on the characteristics of the zone together with the nature of man's use. These groups could operate at the site level as already occurs in an *ad hoc* way for some estuaries and on Heritage Coasts, or by subject as in the case of the sea defence and coast protection initiatives. Annex 2 provides a review of some of the current initiatives.

Co-ordinated Action. Given the need for integration in the management of the coastal zone, a system of regional coastal zone planning (and management) bodies should be set up. Their area of responsibility should extend to the limits of maritime influence on land and to three miles offshore. These bodies should be charged with the responsibility producing coastal zone strategies for their area.

5. Site protection using land based legislation can achieve nature conservation within the areas covered so long as there is a willingness to control damaging activities. However, the artificial nature of the low water mark boundary for the delimitation of interest and the failure to secure more than two Marine Nature Reserves is an impediment to adequate control.

SSSI Limits. The question of extending the seaward limit of SSSI designation, and with it the limit of responsibility of the planning authority must be addressed.

6. In some areas, notably where development and associated sea level rise has squeezed the coastal zone, policies are required to restore coastal habitat. In areas where the opportunities for landward movement are restricted by a sea wall or other man made obstruction, consideration should be given to the development of options for retreat. At the present time there are methods for grant aiding the building or repair of a coastal defence but none to compensate those who may loose their land or property.

Compensation. Consideration should be given to the establishment of a mechanism to allow compensation to be paid where it is more cost effective to allow natural forces to reassert themselves. This might involve loss of land or property by erosion or the recreation of natural habitats in set back areas.

Prepared by Dr Pat Doody, Coastal Conservation Branch, with the collaboration of the following JNCC specialists:

Dr Keith Hiscock, Marine Nature Conservation Review
Dr Don Jefferies, Vertebrates and Terrestrial Pollution Branch
Dr Roger Mitchell, Marine Conservation Branch
Dr Chris Newbold, Fresh Water and Pollution Branch
Mr David Stroud, Ornithology and Landscape Ecology Branch
Mr Mark Tasker, Seabirds Team.

REFERENCE

Davidson, *et al* 1991. "Nature conservation and estuaries in Great Britain." Peterborough, Nature Conservancy Council.

11 *October* 1991

Annex 1

THE COMPONENTS OF AN INTEGRATED FRAMEWORK FOR MARINE CONSERVATION MANAGEMENT

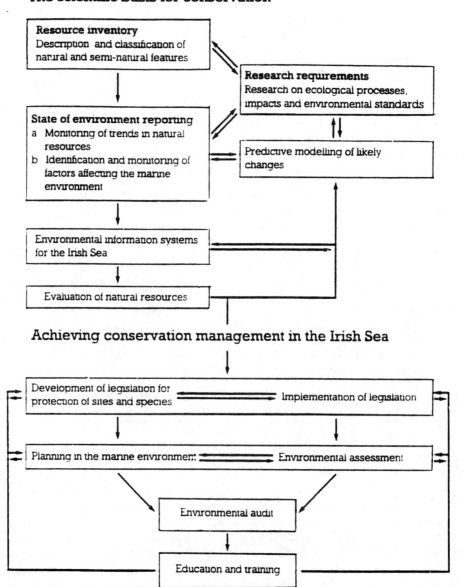

Annex 2

COASTAL ZONE MANAGEMENT IN THE UNITED KINGDOM

A review of the current situation

by

Dr J P Doody, JNCC

POLICY STATEMENTS

Much has been said and written about the need for a coastal zone mangement strategy for the coastline of the United Kingdom. Foremost amongst these is the report of the Marine Conservation Society. "A future for our coast?" (Gubbay, 1990). Other important documents include:

"Planning and management of the coastal heritage, AGENDA FOR ACTION". Produced following a symposium held in Southport in October 1989, hosted jointly by Sefton Metropolitan Borough Council and the North-West Branch of the Royal Town Planning Institute.

The NCC's Estuaries Review, "Nature Conservation and Estuaries in Great Britain" (Davidson, *et al* 1991) and the RSPB's "Turning the Tide" (RSPB, 1990). Both of which deal specifically with the conservation of estuaries.

In addition there have been two recent conferences concerned with similar issues. The Royal Society of Arts, "The future of Britain's estuaries" held on 13/14th February, and the Britain Tidy Group's recent "A future for our coasts". Both identified the need for some form of national review of policy in the wider coastal environment.

PRACTICAL INITIATIVES

In addition to the United Kingdom's involvement in the North Sea Task Force and the Irish Sea Study Group, there are a number of practical initiatives which in their own way try to fulfil some of the requirements of coastal zone management. At the Local Authority (LA) level Hampshire County Council have prepared a wide range of management plans and policy statements since 1972 and they are currently involved in the preparation of a coastal strategy, embracing all the issues affecting the coastal environment. In South-East England, particularly sea defence and coast protection are major considerations. The National Rivers Authority (NRA) now have an important role and the initiative of the Anglia Region to develop a strategic sea defence strategy based on a detailed study of the whole coastline is important. At a more local level liaison groups such as the Standing Conference On Problems Associated with the Coastline which covers the south coast of England, have been set up to co-ordinate activities between coast protection authorities in recognition of the implications of protection of one part of the coast for adjacent areas. There are also a number of groups concerned with the conservation of some of the larger estuaries, such as the Dee, the Mersey and Humber.

Each of these represents a practical expression of the need to integrate activities in the coastal zone. However, other than the Hampshire initiative, which attempts to look at the whole coast, they have a relatively restricted remit (pollution, coast protection) and none of them are intended to address the national policy issues.

CONCLUSIONS

The evidence from the United Kingdom concerning coastal zone management suggests that there are a number of important developments which address some of the requirements of integrated management in the zone. However, these largely act at a local level, in isolation and are unco-ordinated. There are two primary areas where action is suggested:

1. Co-ordination of, and rationalisation of current initiatives at a national level. In this context the following have a role to play.

Heritage Coast Forum. Largely concerned with the reconciliation of recreation and landscape conservation on designated areas of Heritage Coast in England and Wales. Discussions to extend the remit of the forum to the whole coastline of the United Kingdom are currently being undertaken.

Marine Forum. The marine forum serves as a point of contact between a large number of organisations who have an interest in the marine environment. These range from conservation bodies like the Marine Conservation Society and the RSPB to Government departments and industry.

United Kingdom Branch of the European Union for Coastal Conservation. This group was recently established as part of the development of a European-wide initiative concerned with nature conservation on the coastlines of Europe. The United Kingdom group will provide a means of communication between site conservation managers and research workers, and those concerned with conservation in the coastal zone at a more strategic level. It will use the existing Coastal Research and Management Group newsletter as a basis for communication.

Eurocoast. This group, also has a proposal to establish a United Kingdom committee. In this case, though the precise aims are still being worked out, it is concerned more with the technical aspects of coastal management rather than those of nature conservation.

Joint Nature Conservation Committee. Through the development of a communication network and collection and collation of coastal data, the JNCC provides an important opportunity to facilitate the identification of coastal zone management issues and policy options.

2. A review of current policies for land management, pollution control, planning and sea defence, amongst other activities, as they affect the coastal zone represent a second major area requiring a national overview. Some of the areas which should be covered include:

Developing Policy Initiatives Exclusive to the Coastal Zone. These should recognise the interrelated nature of coastal processes and man's influence on them. No specific proposals are identified here since the documents referred to above already provide a check list of issues.

Extend Planning Controls Below Low Water Mark. This question, along with the many others which have been highlighted in the various documents referred to above, should form the basis of a thorough review of the planning control system as it operates on the coast.

Establishment of a Regional Locus for Co-ordinating Activities. This need not involve establishing new or even replacement groups but should build on those which exist already. Thus in East Anglia the issue of sea defence and coast protection are major considerations in coastal management. The current NRA initiative could be built on to develop a comprehensive policy covering nature conservation, planning, recreation and pollution control. Elsewhere, for example, on Heritage Coasts in England the Countryside Commission could provide the forum for co-ordinating activities. At a site level for estuaries at least, development of the estuary conservation groups with the appropriate LA or conservation agency taking the lead, as is the case with the Dee estuary conservation group (Cheshire County Council) or the Wash (English Nature).

15 *July* 1991

APPENDIX 16

Memorandum by the Natural Environment Research Council

SUMMARY OF NERC MEMORANDUM OF EVIDENCE

1. The coastal zone (land margin, estuaries and inner continental shelf waters) is, a particularly dynamic and interactive system with very high productivity.

2. It is heavily used by man for a variety of purposes. The nature of the coastal environment means that planning decisions or protective measures taken in one coastal area can be markedly influenced by events elsewhere.

3. The direct physical, chemical and biological impact of human activities is already substantial; expected changes associated with changes in climate will add their own impacts. The potential effects of sea level rise and predictions of increased storminess are obvious. Less immediately apparent are effects such as the potential for increased frequency of the intrusion of marine derived saline groundwater into coastal aquifers that serve as reservoirs for freshwater extraction.

4. Effective policies for coastal zone protection and decisions on planning require:

(i) *A sound base of knowledge of the underlying geological and geotechnical properties of the coastal zone.*

—The design and siting of engineered structures for protection against floods and coastal recession, and safety in siting of industrial and other facilities on the coast require information on the nature and stability of the sediments and hard rock that form the foundations to such structures. *There has been no integrated strategy to acquire geological data in shallow nearshore areas. As a result, the necessary data for safe and environmentally sound development remains sparse for many areas.*

(ii) *An ability to predict the impact of both natural and anthropogenic induced changes on the coastal zone.*

—Physical change is dependent on the interactions of the rock and sediments with the underlying geological strata, the current and wave climate, tides, surges and winds, and man's activities such as coastal engineering works, the exploitation of non-living resources (eg sand and gravel extraction) and human habitation and recreation. Chemical change stems from direct or indirect inputs of inorganic nutrients, organic matter and various pollutants. Biological change mainly arises from the impact of natural and anthropogenic physical and chemical changes on ecological systems, communities and organisms, from the exploitation of living resources and from recreational activities. *The capability for prediction of the scale and rate of such changes and their impacts is at present inadequate since the basic processes causing change in the highly complex and interactive coastal zone region are still poorly understood in quantitative terms.*

5. Those responsible for coastal zone planning and management need ready access to the best environmental information available. At present such information as is available is held by a wide variety of organisations, is of varying quality and stored in a variety of formats. *The current and future information needs would best be served if existing and newly acquired data on the coastal zone were brought together within well structured and accessible databases.* Such an information system should preferably build on existing major databases, such as those already held by NERC.

6. Many organisations are involved in research on the coastal zone. NERC's own basic research interests will contribute to the knowledge base for policy formulation and management. DOE, MAFF, NRA, the HEIs and many others, including the conservation bodies, also support or are involved in research in the coastal zone. Some co-ordination of effort would be desirable if gaps and duplication are to be avoided. *The development of a strategic plan aimed specifically at meeting the R and D needs for coastal zone protection and planning, building on reviews already in hand such as that on R and D in relation to flood and coastal defences would also be desirable.*

NATURAL ENVIRONMENT RESEARCH COUNCIL

Memorandum of Evidence to the House of Commons Environment Committee Inquiry into Coastal Zone Protection and Planning

INTRODUCTION

1. The coastal zone as defined in this memorandum, covers the land margin, estuaries and inner continental shelf waters. It represents a complex boundary of varying width between the land and the open ocean and is continuous with the river systems.

2. The Natural Environment Research Council (NERC), with its mission to advance underst. ing of the natural environment and the processes of environmental change and to predict future change, has supported a wide range of basic and strategic research and survey in the coastal zone over many years, both in its own Institutes and Higher Education Institutions (HEIs). It has also undertaken strategic and applied research on commission from public and private sector bodies.

3. Over two years ago NERC identified a critical need for information on particular aspects of the coastal zone environment. One resulting initiative was the establishment in 1990 of a new focus for geological studies in the coastal zone: the Coastal Geology Group within the British Geological Survey (BGS). This Group will undertake research and survey work of a strategic nature aimed at improving the geoscientific knowledge of the nearshore, and particularly the integration of knowledge of the sea bed and sub-sea geology with that of the adjoining land. The main application of the results will be in the provision of geoscientific data and data interpretation to the planner, engineer and coastal manager at regional and national levels. They will, however, also be of considerable application in projects at specific sites. The initial phase of development of the Coastal Geology Group has been supported by NERC from the Science Budget. External funding is now being sought for the further development of the Group's work in the medium term.

4. Another initiative, which draws on NERC's unique ability to promote an integrated approach to research across the whole spectrum of the natural environmental sciences, is the development of a new Community Research Project on land-ocean interactions, involving not only NERC's own research Institutes but also the wider United Kingdom scientific research community. This multidisciplinary project, entitled the Land-Ocean Interaction Study (LOIS), will elucidate the basic processes that determine the transports, transformations and fates of biogeochemically important elements and energy between the land, ocean and atmosphere at the land-ocean boundary, and their effects on the properties and health of coastal ecosystems. Its overall aim is to deliver a new capability to develop fully coupled land-ocean prognostic models of the impacts of environmental change in the coastal zone.

5. The following sections summarise comments by NERC supported researchers on scientific aspects of the Committee's inquiry. The views expressed draw on past and present experience of research and survey of the coastal zone; some selected examples of such research are cited. The memorandum identifies the need for a sound base of scientific understanding as an essential contribution to the development of effective policies for coastal zone protection and decisions on planning. It highlights areas where strategic research, survey and monitoring is needed and also comments on the development of coastal zone information systems to meet the needs of those with policy and management responsibilities.

THE DYNAMICS OF COASTAL CHANGE

6. The coastal zone is a region of intense physical, geological and geochemical activity, with biological diversity and productivity amongst the highest in the world. It is heavily used by man for residential, agricultural, commercial, transportation, waste disposal, recreational and military purposes, as well as for fishing, mariculture and the extraction of energy and mineral resources. More than 50 per cent of the world's human population lives on coastal plains (Ray 1989) and, with this proportion increasing, it is here that much of the global investment into agriculture and industrial development is taking place (Holligan and Reiners 1991).

7. Three types of coastal change may occur independently or simultaneously.

Physical changes are dependent on the interactions of the rocks and sediments with the underlying geological strata, and with currents and waves at elevations determined by climate (sea level rise), tides, surges and winds. Man's activities such as coastal engineering works, the exploitation of non-living resources (eg sand and gravel extraction) and developments associated with human habitation and recreation, will also result in direct or indirect physical changes in the coastal zone, sometimes at some distance from the original impact.

Chemical changes stem from direct or indirect inputs to the coastal zone of inorganic nutrients, organic matter and various pollutants.

Biological changes arise mainly from the impact of physical and chemical changes on ecological systems, communities and organisms, the exploitation of living resources, the introduction of exotic species (including pathogenic micro-organisms) and the loss of biodiversity.

Effective coastal zone protection and planning requires an understanding of the contributions and interactions of these three types of change, in scale, space and time.

8. The lack of a focused strategic approach to geological survey of the coastal zone, has resulted in a sparseness of geological and geoscientific data for the nearshore region. Surprisingly little is known of the geology of many vulnerable coasts. It is this deficiency that the Coastal Geology Group (paragraph 3) aims to address through a strategic programme including investigation of the three-dimensional distribution of sediments and rocks that make up the coastal zone; an assessment of non-living resources, particularly those of the sea bed; geotechnical assessment of vulnerable coasts, addressing particularly the problems of consolidation, foundation conditions, erosion and mass movement; sediment dynamics; sediment geochemistry; and interpretation of the long-term history of coastal change deposition, especially in the context of changing sea level. Apart from its immediate application to coastal protection and development, the results from such a programme will also provide the baseline geological and geoscientific information against which future change in the coastal zone, whether physical, chemical or biological, may be measured.

9. Ability to predict the scale and impacts of change is severely restricted by the at present very rudimentary knowledge of the fluxes of energy and materials into and out of the coastal zone, the rates of key physical and biogeochemical processes, and the effects of contaminants on biogeochemical transformations. The environmental systems involved are complex and difficult to observe and the instrument and technique developments and multiprocessor computers needed to acquire data and model processes occurring in such complex systems have only recently become available. A focused interdisciplinary approach, such as that planned by NERC under its Land Ocean Interaction Study (paragraph 4), is needed to help unravel those problems and provide a baseline for more realistic predictions of long-term changes.

RISKS TO COASTAL SETTLEMENTS AND COASTAL ECOSYSTEMS

Flooding and Coastal Erosion

10. The failure or overtopping of sea and tidal river defences, either natural or engineered, may result in catastrophic flooding of low-lying coastal tracts. The recession of coasts by progressive marine erosion can lead to consequent failure and mass-movement of coastal cliffs by landslip and mudflow. The risk associated with such events needs to be assessed in the context of the underlying geology and the geotechnical properties of the sediments in the coastal zone. A knowledge of the long-term history of the physical regime, and of erosion and marine sedimentation in the region is also required.

11. There is much yet to be done to understand and model the energetic surf zone, the dynamics of beach facies, sediment erosion and transport. Even basic geological data in shallow nearshore areas is sadly deficient. The formation of the BGS Coastal Geology Group will provide a new focus for such data acquisition and research.

Vertical Land Movement and Sea Level Rise

12. Estimates of the impact of climate change on future sea levels based upon models of atmosphere and ocean temperatures, ocean thermal expansion and glacier ice melt, suggest a factor of two to a factor of four increase in the rate of sea level rise by the end of the next century (30–65cm per century). Such predicted rises in mean sea level, and other physical changes associated with a changing climate, will present problems for coastal defences, estuaries and land drainage. The scale of the problem will depend on the interaction with coastal level changes already occurring due to land movement. That part of Great Britain north of a line connecting the Severn and the Wash is, for example, rising by approximately 1·5 mm per year in response to the removal of the pressure from the ice sheets which covered that area in the last Ice Age, whereas south-eastern England is subsiding by approximately 2 mm per year due to geological processes occurring in the North Sea. (It was in recognition of forthcoming problems due to the latter that the Thames Barrage was constructed to protect Greater London from flooding.)

13. The Proudman Oceanographic Laboratory (POL) is collaborating with United Kingdom and European Institutes to understand present sea level changes and predict future change. To date around the United Kingdom, the average rate of sea-level change has been comparable with the global average of 15 ± 5cm per century. However, there are large spatial variations of relative sea level change due to the different vertical land movements referred to above. POL is collaborating with the University of Nottingham in the use of advanced space geodetic techniques for the independent measurement of these vertical land movements. This should provide better spatial predictions of future sea level change and its impacts.

14. If present predictions on sea level rise are realised, there are likely to be considerable losses of coastal habitats, particularly saltmarshes and sand dunes. The Institute of Terrestrial Ecology has made some predictions for the Essex coast and adjoining areas and desk studies are being made to see how these changes can best be accommodated. It seems likely that positive management, including strategic retreats, may be better than accepting an increasing risk of flooding. Where sea defences may not be maintainable, one option could be to convert some agricultural land back to saltmarsh in order to replace lost protective zones. This would have the added advantage of replacing lost wildlife habitat. Such a process would require careful planning and management based on a sound scientific knowledge of the characteristics of the area under consideration.

Episodic Events

15. Episodic events, such as floods, tide surges, storms and the bulk release of pollutants, have major immediate environmental impacts on the coastal zone. For example, a single flash flood in a river can mobilise many years of accumulated pollutants in the bed sediments. The mechanisms by which these events affect key environmental processes, their long term effects on coastal ecosystems and the wider implications of a greater frequency of such episodic events (for example, associated with climate change) have received little attention. Instruments need to be developed which can be deployed long term in the field and triggered to record appropriate environmental information when episodic events occur.

16. A reduction in the interval between major flooding events caused by storms will be one of the principal effects of rising mean sea level. The intensity and the frequency of storms may also be exacerbated by climate change. Wave data analysed at the Institute of Oceanographic Sciences Deacon Laboratory (IOSDL) suggest that the North Atlantic is now rougher than it was a few decades ago. The meteorological data cannot explain

this result and, so far, this increased storminess has not yet shown up in storm surge statistics at the coast. Proudman Oceanographic Laboratory (POL) scientists are developing an interactive numerical model to predict tides, storm surges and surface waves in the seas round Great Britain. This model, which requires advanced supercomputing resources, will improve surge forecasts for the south and west coasts, where interactions in shallow bays are significant. It will also be used to "hindcast" the relationship of wave and surge storminess and should help understand the IOSDL observations noted above.

Salinisation of the Coastal Water Table

17. Coastal aquifers that serve as reservoirs for freshwater abstraction can be subject to periodic intrusion of marine derived saline groundwater. This is likely to increase in frequency if sea level rises. The Groundwater and Geotechnical Surveys Division of the British Geological Survey is concerned with this problem, amongst other subjects. Related surface water studies at the Institute of Hydrology (IH) will evaluate the role of land-drainage pumping stations in the management of the water table under the influence of climate change and sea level rise. IH also plans to investigate the previously unmeasured water flow from saltmarshes and the beds of estuaries at low tide, a factor critical to the hydrological balance.

Major Habitat Change

18. Natural coastal wetlands (such as saltmarsh), dune systems and mudflats play a crucial role in maintaining the physical stability of the coastal zone. Major changes can occur in these habitats and can impact both in terms of coastal protection and erosion and release of pollutants.

19. The perennial grass, *Spartina anglica* is, for example, dying back in south coast saltmarshes while spreading in the north. The growth and status of *Spartina* is of direct relevance to coastal mudflat accretion and erosion. Very large volumes of sediment and organic material are trapped by its growth. In those areas where, for reasons that are not clear, *Spartina* is dying, these materials are now being eroded with accompanying changes in nutrient flux and pollutant release. As the species moves north (a spread which could be enhanced by global warming), it may be expected to have a profound effect on coastal protection and saltmarsh and mudflat ecology. The Institute of Terrestrial Ecology has been commissioned to investigate these changes in Great Britain and is also collaborating with other EC countries experiencing similar changes.

Pollution

20. The United Kingdom's traditional use of the rivers and coastal waters for waste disposal has led to many environmental problems in the coastal zone and resulted in changes in the structure of communities and loss of biodiversity. Notwithstanding the Government's recent decision to halt most, if not all, coastal disposal of several types of waste, including sewage, the contamination of coastal and estuarine waters and sediments by discharges and waste disposal represents a problem that will continue into the future. NERC submitted evidence to the Committee's earlier investigations into the "Pollution of Beaches" and the "Pollution of Rivers and Estuaries" which examined many aspects of these problems.

21. Coastal waters and estuaries contain the breeding and feeding areas of many commercially important fish and shellfish stocks, as well as a rich variety of non-commercial species. In many cases these have been adversely affected by increasing pollution and disturbance. In some areas, genetically distinct stocks of some species have been completely lost.

22. Changes in the structure of benthic (bottom dwelling) or pelagic (surface or middle waters) communities have also occurred as a result of eutrophication. This process of over-enrichment of a water body by nutrients, often derived from sewage or agricultural run-off, leads to an over-growth of algae and a reduction of oxygen levels. The fundamental causes of eutrophication in coastal waters are not, however, fully understood. Recent research at the Plymouth Marine Laboratory (PML) and elsewhere indicates that the relationship between nutrients and plant growth is not simple. A proper understanding of the scientific basis for algal bloom development in coastal waters is essential if the right decisions on protection against eutrophication are to be made.

23. Degrees of pollution stress will vary on local and regional scales and depend on the assimilative capacity of particular habitats. Research at PML is developing methods for evaluating the degree of stress being felt by various components of coastal ecosystems. This work is leading to the identification of so-called "bio-markers" of the health of the ecosystem. The application of such techniques in the North Sea has already shown a measurable effect of pollution on the health of fish at levels below those overtly causing damage. These results indicate that biological impacts can be greater than previously thought and lead to renewed concerns about the long term sustainability of living marine resources and marine biodiversity.

24. Information available on the health of coastal ecosystems should grow significantly in future years as biological indicators of stress are applied in routine monitoring programmes. Assessment of risks from pollution does, however, require more data on the multiplicity of organic compounds discharged at relatively low levels as a result of industrial activity (organic micropollutants). In contrast to polluting metals, virtually nothing is known about these compounds, some of which may have deleterious effects on living organisms even in very low concentrations.

25. Predictions of sea level rise have implications for the release of pollutants from coastal and estuarine sediments and salt marshes. Geochemical processes in coastal waters and estuaries cause the deposition of heavy metals and other industrial waste products leading to their accumulation in anoxic muds. The muds of many United Kingdom estuaries and some coasts contain large accumulations of contaminants which are the products of two centuries of industrial activity. Changes in currents and tidal activity resulting from a rise in sea-level, or increased storm frequency, could cause erosion and remobilisation of some of these contaminated muds. On a small scale, the likely impact of such remobilisation of mud is seen in the results of dumping Mersey dredge spoils in the Irish Sea. The muds, when reoxygenated, release their contaminant load back into solution, from which they are taken up and concentrated via the food chain into commercial fish stocks. Heavy metal concentrations in some of these fish approach permissible limits for their sale.

26. In some areas, sediments have accumulated radioactive contaminants. Studies in West Cumbria by the Institute of Terrestrial Ecology and other Institutes have shown the capacity of sediment transport to cause the accumulation and concentration of low levels of radioactive waste into estuaries, for biological concentration along food chains and for both physical and biological transfer to terrestrial ecosystems. If the modified hydrographic regime caused by sea level rise resulted in the remobilisation of sediments rich in radioactive contaminants, the environmental health implications would be even more serious. The only recourse may be to dredge out the contaminated muds, *if* environmentally safe repositories can be found.

Exploitation of Non-living Resources

27. Non-living resources such as sand, gravel and calcium carbonate, are increasingly being extracted from the costal zone. Important resources such as oil and gas are also extracted from beneath the sea bed in the coastal zone. The extraction of these resources impacts on the coastal zone to a varying but increasing degree.

28. The possible extraction of sea bed materials by dredging is a particularly important area for consideration. The extraction of these materials may affect the stability of beach deposits not only locally but regionally. Information on the three-dimensional distribution of the different types of geological materials (sediments and rocks) at the sea bed is needed to develop an inventory of resources both on a site and reserve scale, and especially at the strategic level to provide information on a regional or national basis applicable over the long term. The planner also requires information on the status of sea bed sediments, whether these are inactive " residual " sediments from earlier depositional conditions or whether they form an integral and active part of the contemporary sediment transport regime.

Exploitation of Living Resources

29. The high diversity, biological productivity and accessibility of the waters of the coastal zone has resulted in their living resources being increasingly exploited. In many areas of the world, they are at risk of over-exploitation. This can lead, in the worst cases, to extinctions. There are many instances, even in United Kingdom waters, of the loss of genetically distinct stocks (eg herring populations) with a concomitant loss of biodiversity. Maximising efficient use of and improving information on remaining stocks and their needs is essential to the production of good stock management policies and will assist in building up over-exploited stocks to viable levels. One long-term sensitive measure of change in fish-stocks (including non-commercial species) is provided by the numbers and breeding success of estuarine and sea-birds. Long-term research on these populations, combined with other disciplines, will provide an integrated assessment of the state of coastal fish stocks.

ENVIRONMENTAL ASPECTS OF COASTAL ENGINEERING AND DEVELOPMENT

Costal Engineering

30. Coastal engineering works have the potential to result in important consequences for local and, in some cases, regional hydraulic regimes and thus to bring about major changes in sediment transport, deposition and erosion. The geoscientific and biological consequences of such construction need careful appraisal at the feasibility stage, if environmental damage is to be avoided or minimised. There are, for example, many uncertainties about the lifetimes and risks associated with " soft " coastal protection options (eg beach replenishment), which require careful combined environmental/engineering study. More work is also needed on interactions along and across coasts between soft and hard engineered defences (eg sea walls, groynes) and adjacent unprotected natural coastline.

31. If hard engineered protection is the chosen option, then detailed geological and geotechnical studies of the foundations are essential. If soft engineered protection is preferred, then the designers should have access to information on sea bed materials both in the coastal zone and further offshore that might be suitable for replenishment work, and an appraisal of the impact of extracting such materials for this purpose. At present, the detailed survey work to provide much of this information has not been undertaken in many areas.

32. Biological techniques, such as the use of the perennial grass, *Spartina anglica,* and other species to trap sediments and provide a " natural defence " of protective aprons of salt marsh for sea walls, broaden the options available to the planner. Extensive planting of *Spartina* took place in the 1920s, 30s and 40s with the

aim of stabilising mudflats and thus increasing coastal protection. Such biological techniques will have their own environmental impacts. For example, *Spartina* has now become the dominant species in approximately one quarter of saltmarshes in Great Britain as a result of this extensive planting and also through natural spread. It is regarded with mixed feelings by conservation bodies because it is seen as a threat to the diversity of saltmarshes and there is some, as yet unconfirmed, evidence that it may lead to a decline in the number of some wading birds.

Siting of Coastal Developments

33. The proposed use of coastal sites for intake and discharge facilities and associated industrial construction needs to be appraised in the light of geological information on the likelihood of coastal accretion or recession that might put facilities at risk. There is also a need to understand the processes of sediment erosion, transport and deposition both in the vicinity of the site in question and regionally in order to be aware of the potential threat of contamination sources from that site. For existing facilities, coastal recession trends should be monitored as should the level of contaminants in both local and regional sea bed and estuarine sediments.

34. Navigation channels and port facilities may be disrupted by the movement of sandbanks and siltation. The monitoring of coastal sediment erosion, transport and deposition should be a priority to the coastal manager concerned with the maintenance of navigation channels and deep water ports. The results need to be viewed in the context of the long term history of sandbank growth and coastal siltation.

Recreation

35. Those concerned with the development of coastal sites for recreational purposes should be mindful of the need to preserve important sea defence landforms such as saltmarsh and sand dunes in their natural state, so that the active processes of deposition and erosion that are responsible for the creation and maintenance of such landforms are not impaired. Where artificially replenished beaches serve a recreational function as well as a means of coastal protection, consideration needs to be given to the suitability of the material used for replenishment.

36. Hard coasts (cliffs and similar features), although apparently only at risk from normal erosion, can actually constitute relatively fragile habitats which are highly susceptible to public pressures from walking (erosion of footpaths), climbing and other recreational activities. The plant communities which make up ledge vegetation are, for example, particularly vulnerable on climbing cliffs, where " gardening " (the clearing of ledges by climbers) has a disastrous effect. A classification of hard coasts in terms of importance of plant communities, suitability for climbing, sensitivity to climbing or footpath disturbance would be a valuable asset for their management. Hard coasts are also important habitats for seabirds. The increasing interest in rock climbing and other leisure activities, such as sailing, in the vicinity of cliffs has led to levels of disturbance which have produced breeding failure in some seabird populations. Puffins may have disappeared from coastal areas in the south of England, partly because of such human disturbance.

Conservation of Wildlife

37. Any development of remaining coastal wetlands, lagoons and dunes should take account of wildlife conservation as a priority. Causes of deleterious effects on wildlife, including impacts on threatened species such as the otter, are not always obvious. For example, on the West Coast of Scotland otters live in the sea. Work at the Institute of Terrestrial Ecology has, however, shown that a major factor affecting their survival, is access to small freshwater pools and streams. This has important implications for land-drainage policy along coasts.

38. Conservation organisations have given particularly high priority to the protection of shorebirds. A significant proportion of the north-west European and, in some cases, the world shore bird population, depends at some point in their annual cycle on the food supplies that British estuaries provide. Some 20 species spend at least part of their non-breeding season (August–May) on British estuaries. Approximately 85 per cent of the coastal mud and sand flats have been designated Sites of Special Scientific Interest and international agreements have been signed by the Government giving some protection to these birds.

39. The fear is that man-made changes in estuaries might affect the numbers of shore birds both in particular estuaries and in the world as a whole. Changes in sediment deposition and erosion resulting from major construction works may affect their feeding grounds on saltmarshes and mudflats and their food may be contaminated by pollution. The Institute of Terrestrial Ecology is involved in detailed research about the impact of human activities, such as the building of barrages to provide freshwater reservoirs or power, the reclamation of inter-tidal land, the shellfish industry, recreational disturbance, and inflows of pollutants on bird populations. The research is designed to provide a quantitative understanding that will enable better predictions to be made on how changes in the environment of the estuary will affect the food supply in mudflats, the chances of survival of the over-wintering shorebirds and thus the effects on their local and global numbers.

Conservation of Coastal Landforms and Coastal Heritage

40. Less public attention is directed at the conservation of coastal landforms for their own sake. The aesthetic qualities of coastal landscapes, whether of cliffs or beaches, spits and nesses are the products of energy transmitted by the sea and dissipated on the coasts. An understanding of the physical processes involved and of the rocks and sediments that those processes act upon should form an important part of the knowledge-base that is required for conservation policy-making in this area.

41. The use of coastal sites for construction, and the extraction of sea bed materials from the nearshore for soft engineered coastal protection may threaten archaeological sites in the coastal zone. There is a need to carry out careful geophysical surveys where such sites are known or are believed to exist.

ORGANISATION OF COASTAL ZONE MANAGEMENT AND INFORMATION NEEDS

Administrative Arrangements

42. The existing administrative arrangements for the management of the coastal zone are complex and mitigate against the development and implementation of an integrated management policy at the national, regional and local level. Responsibility for ownership, protection, use and conservation of the coastline is very variable and frequently highly localised.

43. Political boundaries are rarely correlated with natural features and ecosystems associated with particular sections of a coast rarely operate independently of others. Therefore planning decisions or protective measures in one area can markedly influence events elsewhere.

44. At present planning regimes do not properly exist below the tidal zone. If an integrated management policy is to be developed, such controls need to be established and to incorporate a requirement for environmental impact assessments, using methodologies that improve as greater ecological insight becomes available. Such assessments are now a standard means of addressing planning for new developments on land.

Information Systems

45. DOE, MAFF, NRA, the Crown Estates, County, District and Local Authorities, landowners and the Nature Conservation agencies are all concerned with policy formulation and management in the coastal zone, and require good environmental science information to underpin their actions. Increasingly such information is needed to ensure compliance with EC regulations or to inform Government for the development of the United Kingdom input to EC policy formulation. However, the information needed is unavailable in many cases. Where it is available, it is often held by a wide variety of organisations, is of varying quality and completeness and is stored in a variety of formats.

46. In order to serve the current and future information needs of the community, facilities should be provided to bring both existing and newly acquired data on the coastal zone together within well structured and accessible databases. Ideally these databases, which might build on existing major databases, for example those already held by NERC (Annex 1), would be interlinked or combined to form a multi-disciplinary coastal zone information system. This should be widely accessible, in standard machine readable and conventional formats, to people and organisations concerned with protection and planning issues in the coastal zone. The nature of surveying for commercial commodities, such as offshore gravel, means that there will be data in this category held on a site specific confidential basis. This is the case with some of the information already stored in the British Geological Survey's marine geoscience database, currently the leading United Kingdom information source in this field.

47. The task of acquiring data for the database(s) from the wide variety of organisations that currently hold it would be facilitated if there were a statutory obligation on those producing data (particularly that derived partly or wholly at public expense) to deposit a copy with the database(s), in much the same way as borehole data is currently lodged with BGS.

48. The database(s) would be of a strategic nature and, while the information contained in it/them could be of considerable application in site specific projects, the main value would be in the provision of data to planners, scientists, conservationists, engineers and coastal managers at regional and national levels. The data would provide baseline information against which future change, whether physical, chemical or biological, in the coastal zone may be measured by subsequent monitoring. This would facilitate realistic forecasting of future change.

Research Co-ordination

49. This memorandum has identified a number of areas where research and survey data are needed. The improved understanding of the coastal zone resulting from NERC's own basic research interests in the coastal zone will contribute to the knowledge base for policy formulation and management. DOE, MAFF, NRA, the HEIs and many others, including the conservation bodies, also support or are involved in research in the coastal zone.

50. With such a large number of organisations contributing to the research base, some co-ordination would seem desirable if gaps and duplication are to be avoided. Research and development needed in relation to flood and coastal defences is already the subject of review and this should provide better co-ordination of effort in this particular aspect of coastal zone research. Similar reviews may be appropriate in other areas.

8 *November* 1991

REFERENCES

Holligan P M and W A Reiners (1991) " Predicting the response of the coastal ocean to global change ". In the press.

Ray G C (1989) " Sustainable use of the coastal ocean " pp 71–87 in " The changing global environment: perspectives on human involvement " editors D B Botkin, M F Caswell, J E Estes and A A Orio. Academic Press, London.

Annex 1

 MAJOR NERC DATABASES HOLDING INFORMATION ON OR RELEVANT TO THE COASTAL ZONE

(i) *Oceanographic Science*. With financial support from NERC, MAFF, the Scottish Office and NCC, the British Oceanographic Data Centre (located at the Proudman Oceanographic Laboratory) has created a digital Marine Atlas for United Kingdom waters, which includes, *inter alia,* much coastal zone data in a format which is readily accessible to any personal computer user. It serves as a specialised geographical information system and includes inventories to direct other users to other machine readable sources of oceanographic data such as tides, currents and waves around the United Kingdom coast.

(ii) *Marine Geosciences*. Held by the British Geological Survey, the data are derived from BGS's offshore mapping programme of the United Kingdom designated shelf area, some estuaries and other information acquired opportunistically as part of research programmes. BGS also has geochemical information from its recent surveys of 28 elements and radionuclides in the coastal zone.

(iii) *Groundwater Archive*. This database, also at BGS, combines the Well Record Archive, the Groundwater Chemistry Database and the Aquifer Properties Database. Of relevance to the coastal zone is the information that it contains on saline intrusion into groundwater and on the physical influence of marine tides on groundwater levels in coastal boreholes.

(iv) *Flood Event Archive*. This database is held at the Institute of Hydrology and contains detailed information on individual flood events across the United Kingdom. It has been used to examine the interaction of river floods and high tides for coastal flood protection schemes.

(v) *Water Archive*. This is the national archive of surface water flow data, derived statistics and associated rainfall data. With the sister Groundwater Archive at BGS (iii), it provides the major information source on freshwater resources. It is held at the Institute of Hydrology.

(vi) *Environmental Information Centre*. This data centre, at NERC's Institute of Terrestrial Ecology, has been expanded from the Biological Records Centre and now contains other survey data from ecological research.

NERC also holds additional data on:

(i) *Marine Pollution*. In the field of marine and estuarine pollution, NERC recognised the problem of collecting information from scattered sources in 1970, when the Marine Pollution Information Centre, housed at the Plymouth Marine Laboratory was established. However, this is largely a document collection and considerable resources would be required to convert appropriate data into a machine readable form.

(ii) *Remotely Sensed Data*. Aircraft and satellite remote sensed data of all environments are held at NERC and the University of Dundee.

(iii) *Land Classification of Coasts*. NERC's Institute of Terrestrial Ecology has made preliminary studies on the land classification of coasts, as part of its land use research programme. This will be incorporated in its United Kingdom wide Land Classification System.

APPENDIX 17

Memorandum by the Norfolk and Suffolk Broads Authority and the Centre for Social and Economic Research on the Global Environment, University of East Anglia

EXECUTIVE SUMMARY

The coastal zone is an integrated area of offshore water and sediment, beach, dunes and saltmarsh, cliff and foreshore, and land adjacent to the sea. At present its protection is the responsibility of a number of agencies. A particular problem is the fragmentation of effort between the sea defence and land drainage elements, administered by the Ministry of Agriculture, Fisheries and Food, and the coastal protection, including development control, elements, the remit of the Department of the Environment. The latter is essentially the compass of the Committee's enquiry. This submission seeks to retain the wider perspective. (paragraphs 1 and 2)

The main themes for future coastal management are:

(i) The prospect of sea level rise places a need to examine the future of planning and protection in a very long term perspective. This means that capital financing should be considered in 50–100 year time-scales, and that it should be programmed over decades, and not determined by annual public spending rounds. (paragraphs 1, 4, 14)

(ii) Coastal protection requires working with nature as well as forestalling natural hazard. This means a combination of hard engineering of solid defence structures, and soft engineering of beach nourishment, dune manipulation, saltmarsh accretion, and deliberate cliff erosion. (paragraphs 4–6)

To achieve such an ambitious and novel approach will require co-ordination between the sea defence, land drainage and coastal protection interests as well as the on- and off-shore planning control arrangements. That co-ordination will have to be of an unusually comprehensive kind. Present arrangements of financing and management do not allow for such effective co-operation. (paragraphs 25–28)

(iii) The economic justification of the new coastal protection strategies will involve, among other things, estimates of how society evaluates and cherishes the nature conservation and amenity interests of coastal areas. This means placing money equivalent estimates on the vital functions of natural processes that serve to maintain an integral coast at least cost to society. Such an approach is now being evaluated by the Broads Authority and the Anglian Region of the NRA as part of its investigation of flood protection for the Broads. (paragraphs 14–18)

(iv) The financing of coastal protection needs to be separated from the formulae which presently govern local authority spending programmes and capital budgets (paragraph 20). There is a legitimate local authority responsible in this area (paragraph 21). This responsibility should be reinforced through more direct arrangements for financing, and through new coastal management administrative structures on which local authorities should play an important role (paragraphs 25–27).

The Broads Authority has a special interest in this issue (paragraphs 16–17). The executive area of the Authority is particularly vulnerable to the likelihood of tidal surge. Seawater intrusion is already damaging the fen and marsh dyke vegetation which the Authority is statutorily responsible to protect. Thus the Authority, with the NRA Anglian Region, is looking at a number of options to protect the area from flooding by sea and by river. The Authority believes that the success of this approach will test the need for comprehensive financing and management of the coastal zone.

This submission therefore recommends:

(i) The creation of regionally based coastal management authorities incorporating the present mix of responsible agencies in a more structured form (paragraph 27).

(ii) The extension of planning controls across the whole of the zone, including landwards and seawards of the actual coast (paragraphs 7, 12), guided by scientifically supported coastal hazard maps (paragraphs 10 and 11).

(iii) The separation of financing for the local authority contribution from the annual pattern of local authority spending plans and rate support grant formulae (paragraph 21).

(iv) The recognition that coastal surgery, which involves deliberate erosion of cliffs and accretion of beaches, may give rise to claims for compensation, and that such claims should be met, where agreed, as part of the coastal defence budget in total (paragraphs 12, 28).

(v) greater utilisation of extended cost benefit analysis and environmental impact assessment in the appraisal of projects, ensuring that all benefits and costs are accounted for (paragraphs 13, 15, 18). Such an approach is especially important when considering a more integrated coastal management strategy involving soft engineering and coastal surgery (paragraph 18).

INTRODUCTION

1. Potential climate change and associated rises in sea level of up to 40 cms over the next 100 years ensure that there is now a need seriously to reconsider the present and future commitment to coastal protection in the United Kingdom. There is a need for a much more integrated approach to coastal protection, including

making natural processes of coastal change share some of the burden. Offshore bars and cliffs provide sediment that absorbs wave energy and retains beaches. Saltmarshes soak up nutrients, buffer the force of waves and provide a habitat for wildlife. Current management and financial arrangements are not suited to meet this challenge. If present arrangements remain unchanged there is no realistic hope for a cost effective, publicly supported and environmentally robust coastal protection programme for Britain in the 21st century. These issues are of particular relevance to East Anglia, where it is anticipated that investments of up to £300 million will be required in the next 20 years to maintain effective levels of defence.

2. All of the major interests in the Broads area and the functions of the Broads Authority depend on the security of the flood defence system. This is because most of the marshland lies below sea level, the conservation interest depends on maintaining a freshwater system, and water-borne recreation and tourism depends on secure river banks and a well–maintained channel.

3. These challenges will not be met unless there are changes in approaches on three levels:

(i) A more integrated approach to planning in the coastal zone, including the expansion of planning controls to the area below low water mark.

(ii) The adoption of project appraisal techniques to incorporate a wider range of environmental impacts, for example by the inclusion of environmental elements in cost benefit analysis.

(iii) Re-organisation of financing of coastal protection and sea defence works.

COASTAL ZONE MANAGEMENT

4. Modern approaches to coastal protection involve three inter-related components:

(i) hard engineering involving walls, groynes and barriers;

(ii) soft engineering of reconstituted dunes, shaped beaches, offshore bars and safeguarded saltings;

(iii) coastal surgery of deliberately eroding headlands, tidal washlands and new saltmarshes.

5. It is now generally acknowledged that natural processes, if properly understood and harnessed, are an essential and cost-effective component of flood alleviation. Deliberate erosion feeds offshore sediment and beaches to buffer vulnerable coasts against the force of waves, and thus provides recreational amenity. Tidal washlands and reedbeds not only absorb high water: they also sequester nutrients and toxic wastes. A flood alleviation programme can provide new environmental gains for society in the form of wildlife habitat and recreational resource.

6. The concept of coastal zone management, which combines land use controls and decisions about the particular measures taken to defend (or to leave to nature) the actual coastline has become a matter of important debate. This has the advantage of recognising the coastal zone as one of natural hazard where appropriate non-structural measures can be taken alongside the more structural mitigation of hazard. The coastal zone management approach is also far more long term than the essentially short-term approach which relies in the main on hard engineering structures.

PLANNING IN THE COASTAL ZONE

7. In a co-ordinated policy, development control should be part of a wider strategy for coastal zone management. Planning permission should not be granted in areas at risk from erosion and should be refused or made conditional upon specified building measures in flood hazard zones. In addition formal planning controls should be extended to the offshore. Appeals against refusal of planning permission should be considered in the context of this " working with nature " approach.

8. Problems of coastal erosion are recognised in planning policy guidance. For example, PPG 14, states that:

" . . . coastal authorities may wish to consider the introduction of a presumption against built development in areas of coastal landslides or rapid coastal erosion. Consideration may also need to be given to the possible use of Article 4 Directions under the General Development Order 1988 in some circumstances to control permitted development. The fact that stabilisation works may, by their size and location, involve the need for environmental assessment may also need to be recognised in the criteria established in the Local Plan ".

9. The basis for development control as part of a coastal zone management strategy therefore already exists, but to date these controls are relatively weak, and there is no statutory responsibility for presumption against development in such areas. If there were this could initiate " planning blight " and subsequent claims for compensation.

10. Extension of planning controls, and implementation of presumption against development requires defining areas where risk is sufficient to justify planning controls. This in turn necessitates determining " acceptable risk ", and delineating flood risk areas, or erosion risk zones, with detail and accuracy. Special flood hazard maps will need to be drawn up by the NRA, and the DOE will need to provide scientific advice to guide the positioning of setback lines.

11. *There is therefore a need for definition of areas at risk, and for more clearcut responsibility,* so that the issue of compensation, where proven, becomes properly incorporated in coastal protection. One of the first British Structure Plans to propose such policies is that of Norfolk.

Policy FR.3 states:

" There will be a presumption against new building in areas likely to be affected by marine erosion *within the expected lifetime of the development."*

Policy FR.2:

" In order to minimalise the effects of tidal flooding, there will be a presumption against development on land to the seaward side of sea defences including the siting of holiday chalets and caravans. On the landward side of sea defences and behind embanked watercourses, there will be a presumption against development in areas liable to flood unless the standard of defence is appropriate to the development proposed."

12. The establishment of erosion risk zones could give rise to the additional problem of blighting such areas. A facility to provide compensation would be beneficial when carrying out comprehensive feasibility studies. Funding should be available through the house defence budget (paragraph 28).

13. For a more integrated approach to be effective in the coastal zone, it is necessary for planning controls to be extended to beyond the present limit of low water mark. This would also allow planning authorities to have control over offshore mining of aggregates, a process which may have detrimental effects in terms of increasing erosion and changing patterns of offshore drift, and cliff and beach coastal accretion. All applications for offshore development should require a formal environmental impact assessment subject to thorough public scrutiny.

PROJECT APPRAISAL AND EVALUATION TECHNIQUES

14. Coastal defence works can have a major impact on other parts of the coast and the local environment. Britain's coast encompasses areas of significant recreational and amenity value, as well as of national and in some parts international nature conservation value. This underlines the importance of taking a broader, more strategic view in the evaluation of options. Scheme appraisal methodology must be extended so that it can serve as an enabling tool for a more strategic management of the coast. This requires an extended cost benefit analysis approach supported by comprehensive environmental impact assessment.

15. Whilst important improvements to the project appraisal process have been made in recent years there is still a need to ensure full consideration of the following:

(i) Evaluation of alternatives, especially a " do nothing " option which may be used as a baseline against which other options can be judged.

(ii) Identification and full valuation (including indirect effects) of the range of assets likely to be at risk from flooding or erosion.

(iii) Given the levels of uncertainty prevailing in coastal defence schemes, sensitivity analysis, that is calculations of the implications for costs and benefits of major changes in the relevant variables, is essential to understand the accuracy and meaning of CBA results. For example, there may be a lack of sensitivity analysis of extreme water levels, erosion damage, and amenity benefits.

(iv) The use of 6 per cent test discount rate will militate against investments designed to protect against flooding and erosion beyond 50 years.

16. In the Broads, as one of the National Park family, the importance of the area for conservation, agriculture and tourism, has demanded that a full environmental perspective be taken by the NRA in preparing an overall flood protection strategy. The Broads Authority has requested that the NRA take equal account of the environmental consequences of each engineering option alongside the economic considerations.

17. A study currently under way may show tidal barrier and complementary bankraising to be the most cost effective approach. However, one or more washlands may bring such large environmental and recreational benefits that they could well be included, despite higher capital costs. For this reason, the research into flood protection for the Broads has included a full investigation of the feasibility of all possible options and combinations of options, rather than allowing an early filtering out of possible solutions which are less promising in economic terms.

18. This suggests that the basis for conventional cost benefit evaluation is not well suited to the comprehensive and environment-orientated approach to coastal management which is now necessary. An extended version of cost-benefit analysis is required. *There is a need too for a comprehensive review of the financing formulae and of environmental appraisal of all aspects of coastal management, together with an examination of an appropriate management structure capable of being responsive to such reforms.*

FINANCING OF COASTAL PROTECTION

19. At present the distinction between flood defence and coastal protection and the different responsibilities entailed with these functions are blurred. The basis of financing is not entirely related to real need and in any case is spent from two quite different department remits. The management regime involves a mixture of local government and a statutory agency with contrasting responsibilities and interpretations of coastal defences. *The current management and financing arrangements are actually operating in such a way as to be counterproductive to an ecologically integrated coastal protection strategy operating over decades, not years*. The present arrangements reduce the scope for programmed investment on a staged and sequential basis.

20. Four inter-related issues influence the financing of coastal protection and flood defence and affect management of the coastal zone.

(i) The NRA is not permitted to borrow for its long term programmes, it can only spend via its current income, based on MAFF grant and local authority precepts. The NRA may borrow in an emergency, but this is for "fingers in the dyke" and even then requires ministerial authorisation. Limitation on borrowing means that major, long-term commitments to coastal protection are stifled. If the NRA is to pursue its commitment to furthering the wellbeing of nature, and to cost effective use of public money, this borrowing restriction should be relaxed, subject to appropriate democratic safeguards.

(ii) MAFF grant aid is based on financing formulae which are not made explicit. There is no obvious way in which local authorities, who are partners in coastal financing, can be sure in advance how much grant aid they will receive. This makes their own spending plans somewhat uncertain.

(iii) The county council precept for tidal defence expenditure is included as part of the annual budget for the council and is repaid according to a Standard Spending Assessment (SSA) formula set by DOE. This SSA is not set in relation to actual spending, but as part of a nominal calculation on the basis of generalised formulae. Because SSA nationally are usually less than actual spending requirements, there is inevitably a shortfall in available cash. A county council, facing rate capping because of total expenditure which is not related to need, will be forced to reduce its precept. *This holds down the rest of coastal defence expenditure, and strangles the flow of money without reference to true coastal defence requirements*. The Anglian Region of the NRA lost over £600,000 of their 1990–91 budget because of an unwillingness on the part of Norfolk County Council to go beyond a 4·8 per cent rise in precept. In 1991–92, if the Council sticks at 6 per cent, then the shortfall will be nearly £0·5 million in cumulative deficit. By 1996 the backlog of needed coastal protection could exceed £3 million—all this before any major expenditure is contemplated for safeguarding the Broads.

(iv) The district authorities, responsible for inshore coastal protection, are in a similar position. They may borrow to meet their needs and also receive a grant both from the county councils and DOE to assist them to borrow. But they do not get back the full value of their commitment, and in any case the payback is a year delayed and is also based on SSA calculations.

21. *There is a need to disengage the local authority expenditure process from the local authority contribution to coastal management*. There is a justification for some local authority contribution, though on a much more modest scale than at present bearing in mind the national significance. In any case the local authority is the democratically elected voice of local opinion that should be accountable for the final stage of any coastal defence strategy.

22. What the basis of grant should be from the local authority is a matter for careful consideration. At present it averages around 40 per cent. Certainly it should be reduced by half, and for sea defences by three quarters. Thus levels of central government grant should be of the order of 80 per cent for coastal protection and 90 per cent for sea defences.

23. The NRA East Anglia capital programme expenditure for 1990–91 was £6 million. The NRA estimate that to maintain all the defences within the catchment target standards in Norfolk and Suffolk LFDC over the next 10 years, the minimum average annual investment needs to be £13 million.

24. In the Broads area there is inadequate investment to protect the current 1 in 5 year standard; this is the case to control bank erosion, regardless of the need for bank improvement works. The programme has been slowed by the difficulty in obtaining the local authority share of the costs in recent years. Essential erosion protection schemes, such as the North Breydon Scheme, have been shelved or delayed, placing land and property under increasing risk.

FUTURE COASTAL MANAGEMENT AUTHORITIES

25. A new coastal management structure is needed, linking the onshore to the offshore in planning, management and financing, to ensure that planning considerations are fully part of coastal protection and flood defence.

26. The kind of authority that might be considered would be regional body, based on existing NRA regions, and connected to local authorities by non-statutory links but delegated powers.

27. The structure of the first phase of the Broads Authority provides a possible model. This consisted of a quasi-legal amalgam of local authorities (both county council and district council) plus statutory agencies and representation via non-governmental organisations. The equivalent for a coastal management authority would be county and district councils, the NRA (which could service the authority) plus the Crown Estate Commissioners, English Nature, the Countryside Commission, the internal drainage boards, local landowners, including nature conservation interests and other non-governmental organisations with a legitimate interest in the coast. Planning authorities at county and district level would delegate powers and local authority element would be in a position to help determine the final contribution to the total budget.

28. If the processes of erosion and accretion are integrated in a coastal defence strategy then compensation should come from a general flood alleviation budget, not from the local authority planning allocation.

29. It is submitted that without these changes the future of the British coast is in jeopardy. It is by no means impossible to overcome these impediments, and there is still time to do so.

December 1991

APPENDIX 18

Memorandum by Norfolk County Council's Director of Planning and Property

1. THE APPROACH

1.1 The Norfolk coast extends to 140 kilometres. It is subject to a variety of pressures, including risk of flooding, coastal erosion, demand for recreation, threats of marine pollution, problems of inshore fisheries, extraction of marine aggregates, conservation of habitats.

1.2 The County Council is extremely conscious of these pressures and in particular is very sensitive to the risks:

— of marine inundation (as well as fluvial flooding), which will be exacerbated by the likely rise in sea levels

— of coastal erosion, which is increasingly threatening individual buildings and even settlements.

1.3 The County includes large areas of land—in the Broads, Fens and along the north coast and the Wash—at risk from sea flooding. In Broadland alone, over 21,000 hectares of marshland are below surge tidal level and the area is protected by 210 kilometres of tidal embankments. Much of the Norfolk coastline is also subject to coastal erosion.

1.4 Awareness in Norfolk of the problems faced by the zones, and of the need to evolve appropriate forward planning policies, has been heightened by the research evidence available at the University of East Anglia[1]. Access to that knowledge and expertise has done much to facilitate a progressive attitude towards coastal management in Norfolk.

1.5 Despite the excellent relationship between the County Council and the University and the intelligence available to us, the formulation and implementation of a coherent coastal strategy is frustrated by the fragmentation of responsibility for the coastal margin. The need for co-ordination has been clearly set out in the evidence to the Committee of the Local Authority Associations[2], and is well exemplified in Norfolk[3].

1.6 We have therefore sought to utilise the potential of the statutory planning system to address some of the long-term development implications, though even here the scope of such plans is too limited to encompass the full range of relevant issues.

2. THE POLICIES

2.1 Our prime concern was to face up to the problems of planning for development against a background of rising levels, increased risk of flooding and continued coastal erosion. In view of the current evidence about global warming, we were quite clear that the question was not whether sea levels would rise, but by how much and where were the most vulnerable areas; similarly the erosion question was not whether coasts will recede but by how much and where. Although we could find no precedents in Britain among the plethora of policies in development plans to guide us, there were examples of measures adopted in North America, Europe and Australasia to draw on.

Coastal Erosion

2.2 To protect the eroding coast of Norfolk (currently at 1m per year in north-east Norfolk) was not only too costly to be realistic but would also be environmentally unacceptable and would cause damage to beaches elsewhere by depriving them of material. Hence the strategic starting point was to accept the inevitability of some erosion and to reflect that in planning policy.

2.3 The Draft Structure Plan[4] therefore included the following policy FP.3

" There will be a presumption against development in areas likely to be affected by marine erosion within 75 years."

2.4 This policy was modified in the finally submitted Structure Plan[5] to replace " 75 years " with " expected lifetime of the development " and to refer to " building " rather than " development ". The appropriate extract from the Norfolk Structure Plan is attached (Appendix A), giving the rationale for the policy.

2.5 The draft " development setback " policy was a variable one, ie the faster the coast eroded, the further away the setback line would be from the shore. Seventy-five years seemed an appropriate planning horizon, so that the setback line for a part of the Norfolk coast eroding at 1m per year would be at least 75m from the coast. However, following consultations, especially with the National Rivers Authority, we felt that a policy in which the setback line was related to the lifetime of the development was both more flexible and more defensible in terms of the development control process. The adopted policy was thus:[5]

" There will be a presumption against new building in areas likely to be affected by marine erosion within the expected lifetime of the development."

2.6 This policy statement is significant in two respects. It is the first time in the United Kingdom that a structure plan policy has been justified solely on the grounds of erosion hazard; traditionally coastal development has been controlled on the basis of landscape and amenity criteria. Secondly, it is the first structure plan to include a specific development setback policy; it should be emulated more widely (11 United States coastal states now have formal setback policies).

Flood Defence

2.7 The draft Structure Plan, in response to our concern to reduce the amount of development at risk from flooding, stated that (Policy FP.2):

"*In areas liable to flood, or which would be affected by a major incursion into the sea, there will be a presumption against major development. Where development is permitted, minimum floor levels will be agreed in consultation with the National Rivers Authority*".

2.8 Following consultation, the finally adopted policy in the Structure Plan submitted to the Secretary of State for the Environment is as follows (FR.2)

"*In order to minimise the effects of tidal flooding, there will be a presumption against development on land to the seaward side of sea defences including the siting of temporary holiday chalets and caravans. On the landward side of sea defences and behind embanked watercourses, there will be a presumption against development in areas liable to flood unless the standard of defence is appropriate to the development proposed*".

2.9 There are two elements to the above policy. The first is a clear presumption against development in flood risk areas; the second allows for the possibility of new development provided that it is so designed as to mitigate the potential for damage to property or risk to life. This could, for example, include raised floors, or access to second floor rooms, or even special loft and ladder arrangements. In this context, " minimum living surface" requirements are apparently already operating as part of building controls in America and Australia. If this kind of policy is to be implemented such requirements will need to be assessed in Britain.

3. Conclusions

3.1 The only official advice on these issues is that contained in Ciruclar 17/82[6], asking planners to " particularly bear in mind " flood hazard; we are unaware of any planning advice on erosion.

3.2 What is needed is clear advice on the role of setback lines and flood defence in the development plan and control process, together with support from the Department of the Environment at planning appeals. In this respect, the Secretary of State's proposed modifications to the proposed Norfolk Structure Plan are awaited with interest.

3.3 Such planning policies need to be set within the context of a more co-ordinated approach to the management of the coast.

17 *February* 1992

Notes

(¹) Particular thanks are due to Professor Keith Clayton and Mr Robert Kay of the School of Environmental Sciences, University of East Anglia and staff of the National Rivers Authority (Eastern Region).

(²) Memorandum of Evidence to the Committee submitted by ACC, ADC, AMA, CPOS, DPOS, MPOS, October 1991.

(³) See letter from Director of Planning and Property, Norfolk County Council, to the Committee Secretariat, dated 17 September 1991.

(⁴) Norfolk to 2006—Consultation Draft Structure Plan Review, Norfolk County Council, July 1990.

(⁵) Norfolk Structure Plan—Submitted Memorandum, Norfolk County Council, February 1991.

(⁶) Development in Flood Risk Areas—Liaison Between Planning Authorities and Water Authorities, Joint Circular 17/82, MAFF/DOE, 1982.

Annex A

EXTRACT FROM NORFOLK STRUCTURE PLAN 1991

15. Flood Risk Areas

15.1 *Policy FR.1* in areas at risk from fluvial flooding as defined by the National Rivers Authority there will be a presumption against new development or the intensification of existing development. These areas will include defined washlands, natural floodplains and other areas adjacent to rivers to which access is required for maintenance purposes.

15.2 Fresh water flooding can occur in the natural floodplains of water courses due to exceptional weather. The risks of such flooding can be increased by greater run off from urban development and constructed channels. As it is not possible to remove all risk of flooding, it is more realistic to ensure that life and property are protected by locating development outside the areas of risk.

15.3 *Policy FR.2* in order to minimise the effects of tidal flooding, there will be a presumption against development on land to the seaward side of sea defences including the siting of temporary holiday chalets and caravans. On the landward side of sea defences and behind embanked watercourses, there will be a presumption against development in areas liable to flood unless the standard of defence is appropriate to the development proposed.

15.4 There is a risk of sea flooding in Norfolk which would increase if a rise in sea levels were to take place. There are large areas behind the main sea defences which could be flooded in certain circumstances but it is undesirable to prohibit all development in those areas which are not fully protected from flooding. However, it is reasonable to minimise flood risk to the individual by ensuring that developments are constructed so as to minimise flood risk and potential damage. Where sea defences have been constructed, it is essential that they are accessible for maintenance purposes. Areas likely to be affected by marine flooding will be defined in local plans after consultation with the National Rivers Authority.

15.5 *Policy FR.3* there will be a presumption against new building in areas likely to be affected by marine erosion within the expected lifetime of the development.

15.6 Marine erosion has been taking place for centuries along many stretches of the Norfolk coastline. In north-east Norfolk, in particular, the rate of erosion is estimated to be about one metre per year. With the possibility of a rising sea level caused by global warming, it is unlikely that those areas at risk will ever be satisfactorily defended. In principle, therefore, new building in such areas will be resisted so that future occupants are not threatened with the destruction of their property by marine erosion. In practice, this entails defining in local plans a broad timescale for areas of risk and ensuring that planning permission is not given for buildings expected to survive into the period of risk assessed for the relevant location.

APPENDIX 19

Memorandum by the Standing Conference on Problems Associated with the Coastline (SCOPAC)

CONTENTS

1. THE ORIGINS OF SCOPAC

The seeds of SCOPAC were sown at a two day conference entitled " Coastal Problems " organised by the Isle of Wight County Council in April 1985.

Speakers included those acknowledged to be experts in the subject at national and international level and those attending were left with three overriding impressions viz

(a) The complexity of both the natural processes at work along the coastline and also of the many issues and conflicting interests affecting the Coastal Zone.

(b) The multitude of authorities (Government and Local), agencies and other bodies with responsibilities and interests in the shoreline area.

(c) The degree of isolation in which these authorities were obliged to take decisions which so frequently had " knock-on " effects on the interest of others.

It was resolved that something should be done to address these difficulties and in October 1986 the inaugural meeting of SCOPAC was held.

There is a very close correlation between the membership of SCOPAC and the Authorities and other bodies who were represented at the 1985 Conference and the precise length of coastline covered by SCOPAC (ie the central south coast between Weymouth and Worthing) is thus more a matter of chance than design.

Membership comprises three County Councils, 16 District Councils, three Harbour Authorities, National Rivers Authority, English Nature, Southern Sea Fisheries and the Solent Protection Society.

2. AIMS AND OBJECTIVES OF SCOPAC

These have been stated as follows:

(a) To ensure a fully co-ordinated approach to all coastal engineering works and related matters between neighbouring authorities on the central section of the south coast of England.

(b) To eliminate the risk of coastal engineering work carried out by one authority adversely affecting the coastline of a neighbouring authority.

(c) To provide a forum for the exchange of information on the success or failure of specific types of coastal engineering projects and on future proposals.

(d) To establish a close liaison with Government and other bodies concerned with coastal engineering and related matters.

(e) To identify aspects of overall coastal management where further research work is required and to promote such research.

Pursuit of these objectives over five years and the knowledge gained over that period has lead to the summarisation of SCOPAC's aims in the form of the " Three Rs " as follows:

Resources	The raising of the general awareness of the importance of the coastline in the national sense and the consequent need for a greater proportion of the nation's resources to be devoted thereto.
Research	The need for further research to be carried out in the complex natural processes at work along the shoreline and (most importantly) for this ongoing programme of research to be fully co-ordinated rather than fragmented as has been the case in the past.
Rationalisation	The need for statutory and other responsibilities of Government Departments, Local Authorities together with other bodies and interest groups to be examined with a view to improving co-ordination and/or simplification of treatment of coastal issues which it is recognised will always be complex in nature.

3. How SCOPAC Works

SCOPAC is unique among the 14 coastal groups which have been set up within the United Kingdom over the last five years, in having a membership of elected Local Government Members.

The Conference is made up of one elected member from each of the constituent authorities.

Conference meets approximately three times per year (apart from special events) in order to formulate policy, hear presentations from appropriate bodies and guide the work of the Officer's Working Group.

The Officer's Working Group is similarly made up of one officer from each member authority (mostly engineers but with some planners).

The Group meets about five times per year (in addition to meetings of the full Conference) in order to progress the work of SCOPAC. Specific topics are dealt with by Sub Groups formed for the purpose.

Full Conference and/or the Officer's Working Group have heard presentations from the following:

> Department of the Environment (Minerals Division)
> British Geological Survey
> Crown Estate Office
> Hydraulics Research Limited
> BMT Ceemaid (Hythe)
> Southampton University
> Portsmouth Polytechnic
> British Petroleum (re island in Poole Bay)
> Nature Conservancy Council
> United Marine Aggregates Limited

4. Issues Addressed by SCOPAC

(a) The Natural Transport of Sediment } Top Priority
(b) Dredging from the Marine Environment

(c) The range of Government, Local Authorities and other bodies involved in coastal issues and the associated legislation.

(d) The financing of Coastal Engineering Works (as between Central and Local Government, NRA, County and District Councils).

(e) Global Warming—practical implications for coastal engineering works of sea level rise and climatic change.

(f) Alternative materials to hardwoods for use in coastal engineering works.

(g) Coast management Plans—preparation, format, issues covered etc.

(h) Radio-activity on beaches.

(i) Seaweed.

The work on items (a) to (d) is now sufficiently far advanced for SCOPAC to have established firm views about requirements for the future and a summary of that on items (a) and (b) is given in the following section.

5. SCOPAC's Sediment Transport Study

Phase I   Bibliographic Study into all earlier research into the natural transport of beach material and other sediment relevant to (though not confined to) the central south coast of the United Kingdom.

Undertaken by Portsmouth Polytechnic (1987–89) to produce a database of nearly 3,000 items of reference.

Phase II   Analysis of material within the Phase I database to produce a comprehensive report on sediment movement and plans showing likely sediment paths identified.

Undertaken by Portsmouth Polytechnic (1989–91)—nine sediment " circulation cells " identified between Worthing and Weymouth.

Phase III   (projected)
Identification and execution of physical research targeted at gaps in knowledge of sediment transport to facilitate completion of a sediment chart for SCOPAC coastline.

Agreement has recently been reached between the Crown Estate Office and SCOPAC for joint funding of a contract to be undertaken by Hydraulics Research Ltd and Southampton University aimed at assessing the " mobility " of the seabed in a large area east and south of the Isle of Wight (ie an area of extensive existing and potential marine dredging activity).

Likely duration of contract 1991–93.

6. SCOPAC EVENTS

1989

22 February   *Visit to Westminster* to make a presentation to Members of Parliament pressing the case for coastal issues to be given a higher priority and for a review of the administrative arrangements relating thereto.

Arrangements within the House made by Sir David Price (MP for Eastleigh) who chaired the meeting.

1990

23 January   Visits to problem sites on the SCOPAC coast to which Members of Parliament and MEPs were invited.

Pennington, Hurst Spit and Barton Cliffs

Attended by four local Members of Parliament. Mr Ian Bruce (MP for Dorset South) subsequently asked a question in the House of Sir Geoffrey Howe about a unified approach to Sea Defences etc and Mr Michael Colvin (MP for Romsey and Waterside) tabled a series of Parliamentary Questions to several Ministers.

25 May   *The Isle of Wight*

Attended by two local Members of Parliament and two MEPs.

28 September   *Selsey Bill and Pagham Area*

Attended by Baroness Trumpington, Minister of State.

1991

20 June   Visit to Hydraulics Research Ltd, Wallingford, to view model of Hurst Spit and other similar research projects.

24 October   Sponsorship of one day conference entitled " Coastal Instability and Development Planning ".

Opening/keynote speech given by Mr Ian Bruce MP for Dorset South—attended by 119 delegates.

7. SCOPAC SUBMISSIONS TO THE INQUIRY

(a) SCOPAC very much welcomes the fact that the whole subject of coast protection and planning the coastal zone (which are integrally linked) is being reviewed. This is in line with the SCOPAC view that the general awareness about the importance of coastal problems and the level of national resources being devoted thereto need to be raised.

(b) SCOPAC believes that significant changes in the current administrative arrangements for dealing with coastal matters are required. This is because of the increasing pressures to which the Coastal Zone is becoming subjected, many involving direct conflict of interest such that the arrangements which may have been satisfactory in the past have become unwieldy. These changes are required at both central and local government levels.

(c) SCOPAC is of the view that these necessary administrative changes should be considered in conjunction with the other changes in local government which are currently in prospect. Management of the coastal zone including dealing with emergencies that may arise is properly a part of a wide range of services required by local communities from which it should not be separated.

(d) SCOPAC believes that a major factor which contributes to the present difficulty in managing the coastal zone is the multiplicity of authorities, agents, other bodies and interests that are involved. In formulating change therefore, the guiding principle should be a reduction in the number of separate decision-making fora while at the same time providing a mechanism for protecting these diverse interests.

(e) SCOPAC believes very strongly in the importance of the natural boundaries that are established by coastal processes. The case for working with nature rather than against it in the protection of the coast and related matters is a very strong one. It follows that new administrative " boundaries " should take account of these natural processes. If this is accepted these natural boundaries will need to be identified (as in the SCOPAC Sediment Transport Study).

(f) SCOPAC believes that the present boundary of planning legislation at low water mark needs to be revised. The advent of marina development, the construction of offshore islands and similar proposals make the concept of the nearshore area being a " no-man's-land " anachronistic.

(g) SCOPAC believes strongly in the value of local knowledge in the matter of coast protection. It submits therefore that the administrative changes for which it calls should not be such as to risk the elimination of past experience and local " know-how " as an input into the design of coast protection works and related matters.

(h) SCOPAC believes the public perception of the responsibilities of the various bodies associated with the coastline to be important. This is an argument for change in itself and it becomes particularly important during times of emergency (eg flooding) when ill defined or ill understood responsibilities not only represent a potential for inefficient response but become a source of much frustration and criticism adding to the general trauma.

(i) It is equally the SCOPAC view that coastal engineering/management agencies of a size enabling them to utilise local knowledge and exhibit responsiveness need to be well co-ordinated by a body with responsibility for determining the overall strategy for a relatively long length of coastline. In this respect the involvement of more than one Government Department with major responsibilities for the coast is regarded as less than helpful.

(j) SCOPAC believes that its work over the past five years places it in an excellent position to submit evidence to the inquiry having been formed specifically to deal with the problems which are perceived by practitioners who are actually working in the field. The cross-boundary nature of SCOPAC membership is another valuable attribute and we find the fact that there are now 13 other coastal groups around the coastline of the United Kingdom with very similar objectives particularly encouraging.

The Inquiry being conducted by the Committee is in fact into the factors which lay behind the formation of SCOPAC and the other coastal groups which we consider fulfil a valuable function pending the full implementation of the Committee's recommendations.

Autumn 1991

APPENDIX 20

Supplementary Memorandum by the Countryside Commission

Thank you for your letter of 29 January 1992.

The Commission's views on the two questions you put are as follows:

1. COASTAL ZONE MANAGEMENT PLANS

We would welcome the concept of the Regional Coastal Zone Management Plan, as proposed, for example, by the RSPB. These would be non-statutory documents. They would bring together all the relevant agencies in preparing a " Plan " which is jointly agreed and which would form a management and co-ordination tool for the coast thus preventing duality of action and helping to resolve conflicts of interest beween agencies.

The precedent for such a type of plan is the Rural (Countryside) Strategy now being prepared by a number of County Councils to achieve similar co-ordination of policies and action programmes in the countryside as a whole. The Commission, English Nature and the Rural Development Commission are preparing an advisory publication on Rural Strategies and I enclose a copy of a draft which may be of value to the Committee. We were pleased to note that the Government has commended the approach of rural strategies in its Planning Policy Guidance note 7 " The Countryside and the Rural Economy ", paragraph 1.14, which was published in January.

Heritage Coast management plans are also non-statutory plans. They include the relevant statutory planning policies from approved structure and local plans but also management and other policies for the heritage coast area. A key feature of such plans is that they are prepared through a process of public consultation and that they are plans *for the area,* rather than for the individual local authorities, to which all interests are encouraged to support. There is no reason why they cannot exist as component parts of the wider scale plans now envisaged.

Statutory development plans

It is accepted that the present town and country planning legislation provides for the preparation of joint structure, unitary and local plans. However, the Commission is concerned that such plans will only be prepared if there is a voluntary agreement between planning authorities to do so. This is an inadequate basis on which to secure effective plans for the coastal zone and for this reason the Commission advocated the retention of subject plans for coastal areas when consulted by Government on its proposals for the reform of the development plans system. This recommendation was rejected on the grounds of the need for simplicity in the new system which would essentially comprise only structure, unitary and local plans.

In its evidence to the Committee, the Commission has argued that planning for the coastal zone should be related to " coastal units "—lengths of coast which it would be sensible to plan together. For such units the Commission would advocate:

—the provision of general planning advice through Regional Planning Guidance;

—provision of strategic policies through individual structure plans for which there would be a requirement for joint working between authorities to ensure a consistent approach;

—the detailed application of structure plan policies through local plans where there should again be a requirement for joint working between authorities to ensure a consistent approach.

If you require further clarification on these points, or on any other matters raised in the Commission's evidence, please let me know.

3 *February* 1992

Draft 11 December 1991

Rural Strategies: Advice from the Countryside Commission, English Nature and the Rural Development Commission

1. INTRODUCTION

New approaches are required to help guide the rapid pace of change affecting much of the countryside. Changes in population patterns, agriculture and land-use, development needs and public expectations of the countryside are already affecting the nature of the rural environment and the well-being of rural communities. It is important that these changes are quantified, positive opportunities identified and built upon, and adverse effects managed to reduce their impact.

The Environment White Paper " This Common Inheritance " (1990) has advocated environmentally-led approaches to change in the countryside and has emphasised the need for integration of policies for conservation, and social and economic development. The Countryside Commission, English Nature and the Rural Development Commission consider that *rural strategies* offer a dynamic and comprehensive mechanism for helping to achieve such integration and have developed the following guidelines on how to prepare and use them.

2. RURAL STRATEGIES

Rural strategies are non-statutory documents which outline policies and recommendations for action in a specific rural area, usually a county. To be successful they need to bring together in partnership a range of organisations with responsibilities, interests and a commitment in rural areas, led by a single public body, usually the strategic planning authority.

The aim of rural strategies should be to identify, balance and co-ordinate policies and actions required to:

—meet the social and economic needs of rural communities;

—secure environmental objectives such as nature conservation and landscape enhancement;

—meet the aspirations of society for access to and enjoyment of the countryside.

Rural strategies should highlight where policies and activities interrelate, and opportunities for collaboration and integration over a range of topics to bring about mutual benefits. In particular, strategies should provide a framework through which to achieve a balance between maintaining and enhancing the environmental qualities of the area and social and economic development to meet the needs of local people and wider society.

The principal roles of rural strategies are to:

—present a " vision " of the future local countryside, outlining the objectives and characteristics that will be most important to achieve;

—increase knowledge of the existing state of rural areas;

—highlight the main trends and issues that will affect the opportunities for achieving the strategy's goals;

—establish a programme to achieve agreed targets, identifying the financial and resource commitments required;

—identify the main organisations with responsibility for implementing the programme, promoting partnerships between these in a way that increases the effectiveness of the resources available;

—provide a framework for monitoring change, reviewing progress and ensuring adaptability to new circumstances and opportunities;

—provide a vehicle for public involvement in setting the policies and programmes of organisations involved in rural areas.

3. THE CONTEXT FOR RURAL STRATEGIES

A number of influences are setting a new " agenda " for rural policies and initiatives. These include:

(a) *commitments to sustainable development*—" This Common Inheritance " outlines the Government's commitment to the principles of sustainable development and environmental stewardship. It recognises that " to achieve sustainable development requires the full integration of environmental considerations into economic decisions ";

(b) *the changing rural economy*—the need to continue to strengthen and diversify the rural economy, particularly in the light of the continuing decline of employment in agriculture and other primary industries. The Environment White Paper recognises the importance of activities such as light manufacturing, service industries and tourism to the Government's aim of ensuring that the rural economy continues to prosper so that it can make its own contribution to the quality of the environment;

(c) *the changing needs of rural communities*—there is increased awareness of the need for rural communities to be able to provide for the varied requirements of people in a wide range of circumstances, and for people in rural areas to have reasonable and affordable access to services;

(d) *public interest in countryside conservation issues and enjoyment of the countryside*—public awareness of environmental issues is increasing, and demands for countryside recreation, sport and tourism are growing.

Rural strategies can provide a flexible and positive means of responding to these issues. However, it is important that rural strategies are not developed in isolation, but complement other planning and policy frameworks, including:

(i) *regional planning guidance*. Regional planning guidance will provide strategic directions for rural areas;

(ii) *development plans*. While recognising land use policies within structure and local plans, rural strategies can complement and influence such plans by providing a wider perspective on rural issues;

(iii) *local management plans*. Local management plans, such as National Park plans, AONB plans and coastal management plans and at the smaller scale parish and village appraisals, will become more important as the integration of rural support measures proceeds. A clear strategic framework will become essential.

(iv) *sectoral programmmes*. Rural strategies can provide a framework for sectoral programmes (eg forestry strategies, nature conservation strategies, recreation strategies, rural development programmes and tourism plans) and help to ensure that these are fully integrated;

(v) *local authorities corporate plans, budgets and service plans*. Rural strategies can assist the co-ordination within local authorities, and between partner organisations, of corporate policies, funding, and service plans for rural areas.

4. PREPARING A RURAL STRATEGY

The processes of preparing a rural strategy can, in itself, achieve beneficial results. The involvement and collaboration of a wide range of organisations—brought about under the guidance of the lead authority—can strengthen contact networks in rural communities, foster mutual understanding of rural issues and opportunities, as well as promoting commitment to joint action to achieve strategy recommendations.

Arrangements to progress the collection of information, provide guidance, and enable discussion on ways forward include:

(a) *Steering committees*—comprising representatives of the lead authority and other main organisations responsible for compiling the strategy and ensuring its implementation;

(b) *Topic groups*—working groups of experts on specific issues, and representatives from organisations with topic responsibilities, formed to prepare topic reports and progress strategy recommendations;

(c) *Countryside conferences and workshops*—conferences, seminars and workshops enable a wide audience to discuss rural trends and issues, and contribute to strategy recommendations and proposals. These may be held prior to, or during, the preparation of a strategy, at the consultative draft stage, and to " launch " a strategy.

It is important to have " grass roots " involvement of rural communities and individuals. Ways in which this may be achieved include, for example, mobile " rural strategy" exhibitions, parish council meetings, community fora, and local media campaigns.

5. CONTENTS OF RURAL STRATEGIES

Each strategy should have its own "identity" and reflect local characteristics and priorities. However, common features emerge from existing published strategies, and we set out some of these below to assist those preparing a strategy for the first time or reviewing an existing strategy.

An *introductory section* will include:

○ a *shared vision for the future;*

○ the *purpose of the strategy;*

○ the *organisations involved;*

○ the *status of the document* and *relationship to statutory plans;*

○ *consultation procedures.*

The *"core" of the strategy* will comprise:

○ a *statement of the present countryside profile;*

○ *topic statements* highlighting issues and opportunities, and recommending policies and programmes.

The section on the *present countryside profile* should include key information on:

○ social, economic and environmental characteristics;

○ national and regional influences;

○ principal trends, conflicts and opportunities.

Individual topic areas may include:

○ *rural economy and employment*—employment structure, primary sector activities, industrial land, manufacturing, economic diversification opportunities—including in the service sector, enterprise assistance, and training and advisory services;

○ *rural housing*—settlement patterns, housing needs, land supply, housing investment programmes, and " affordable " housing initiatives;

○ *rural services*—transport policies, health and social services, education, shops and post offices;

○ *environment*—natural and cultural heritage resources, conservation schemes, environmental improvement measures, environmental protection policies, environmental interpretation, countryside management, and conservation advisory services;

○ *countryside access, recreation and tourism*—rights of way and access networks, provision for recreation and sport, rural tourism initiatives, and visitor information services.

An implementation and review section will provide a framework through which policies and proposals may be translated into action " on the ground ". It may refer to:

○ *responsibilities*—organisations which can progress strategy programmes, and principal sources of finance and other support;

○ *action programme*—drawing together the individual programmes, projects and priorities agreed by the participating organisations to forward the aims of the strategy, annual costs, target timescales, and implementation responsibilities;

○ *monitoring and review*—procedures for monitoring progress and reviewing the strategy and action programme.

6. PRESENTATION OF RURAL STRATEGIES

First and foremost, strategies should be easily understood by a diverse audience, including elected members and staff of public bodies, and members of rural communities and countryside interest groups. It is suggested that:

○ strategies should follow a logical structure—progressing from key facts and issues, through policies, to proposed actions;

○ they should be concise and avoid technical jargon;

○ illustrations can help readers to comprehend complex data and patterns of provision;

○ action programmes should enable partner organisations to identify their specific interests and implementation roles;

○ an economical format can permit wide circulation and facilitate inexpensive reprints following revisions to the strategy;

○ publication of a summary leaflet or poster (eg for libraries, community notice boards) can encourage local involvement and " feed-back " on proposals.

7. A PROGRAMME FOR ACTION

The programme is a package of measures the partner organisations agree to pursue to help achieve the aims of the strategy and is the key to " making things happen ". It should cover at least a three-year period and be reviewed annually. The programme should take account of other programmes, such as rural development programmes, already operating in the area.

Effective programmes should:

○ outline existing and proposed initiatives to be undertaken by partnership organisations which are consistent with the strategy;

○ identify intended timescales, estimated costs and potential sources of funding for projects;

○ promote the role of the strategy group in influencing public, voluntary and private service providers;

○ indicate organisations responsible for implementing initiatives.

The extent to which a rural strategy engenders visible action " on the ground ", and change in rural communities, will provide a test of its success. Also, it should help build confidence amongst rural communities in the commitment of the partner organisations.

Rural strategies may be progressed through:

○ *direct action*—partner organisations can provide enhanced rural services or undertake practical projects themselves: these can include *enterprise support and training schemes*—such as establishing " one-stop " rural business advisory services, and sectoral and marketing initiatives (eg tourism projects, farm produce marketing initiatives, rural skills training); and *community and voluntary action initiatives*—including " self-help " community care schemes, village hall improvements and voluntary conservation projects.

○ *implementation through the planning system or by agreements*—including, for example, local plan proposals, planning agreements, and access and management agreements;

○ *demonstration projects* by partner organisations—to encourage action by others (eg woodland management schemes, habitat restoration and rural retailing initiatives);

Rural development programmes, nature conservation strategies, rural tourism development action programmes and countryside management services have demonstrated how well prepared strategies—backed by local authorities and community support—can achieve substantial results. More locally, parish appraisals, the appointment of community development officers and the establishment of Local Nature Reserves have achieved a variety of environmental, economic and community objectives. Key means of securing progress include:

○ *leadership*—often provided by the appointment of a co-ordinating officer, who can ensure progress by strategy partners;

○ *financial commitment*—where appropriate the action programme can provide an informal contract to which partner organisations can " sign up ", thereby confirming their commitment to expenditure on initiatives outlined in the strategy;

○ *a framework for action*—as provided by the action programme.

○ *publishing reviews of progress.*

8. REVIEWING STRATEGIES

Rural strategies should be kept up-to-date and should respond to evolving needs and opportunities. This involves monitoring and reviewing strategies at three levels:

(a) *rural strategy*—ideally, strategy documents should be reviewed every three years; or, at least, within a five-year period. Such reviews allow the aims, policies and programmes within the strategy to be amended in the light of changing circumstances;

(b) *action programmes*—monitoring is vital to ensure the momentum of programmes. A formal annual review of progress is desirable;

(c) *projects and initiatives*—each partner organisation should monitor progress of the activities for which it has responsibility.

The review process can provide opportunities for publicising successful initiatives and fostering support for programmes associated with the strategy. Publication of an annual report is recommended.

APPENDIX 21

Supplementary Memorandum by the Department of Energy

1. This Memorandum relates to the planning requirements for generating stations in the coastal zone, and Government policy in this area. The memorandum deals in particular with the issue of the impact of the renewables on the coastal zone given that the Select Committee has shown a particular interest in this area. It should be noted that although the impact of conventional power stations on the coastal zone is very much greater than the renewables at present, there is little prior experience of the development of renewables in the coastal zone, and certain renewables could have significant effects on coastal management.

BACKGROUND

2. The coast provides siting opportunities for electricity generation, both for power stations requiring water for cooling, and for the development of renewable energy sources, such as wind, wave, and tidal energy.

3. The generation of electricity has environmental impacts on air, water, land, and public amenity. These result from gaseous emissions and particulates; discharges of liquid wastes to river and sea; disposal of solid wastes to sea and land tips; ecological impacts; and visual and noise impacts. New transmission and distribution overhead lines may also have a significant visual and amenity impact, particularly when located in scenic areas.

4. The Secretary of State for Energy requires applications made to him under section 36 of the Electricity Act 19 89, for the construction and operation of a generating station to be accompanied by an environmental impact statement in accordance with the Electricity Generating stations and overhead lines (assessment of environmental effects) regulations 1990.

5. Primary responsibility for managing and mitigating the impacts of power stations, within the context set by statute and regulation, rests with the developer concerned. The design and enforcement of control strategies for gaseous, liquid and solid waste is a matter for Her Majesty's Inspectorate of Pollution (HMIP), the National Rivers Authority (NRA), the Ministry of Agriculture, Fisheries and Food (MAFF) and the developer. The developers are also subject to the statutory requirements or environmental assessments for major new development proposals likely to have significant effects on the environment.

6. A wide range of renewable forms of energy would have an impact on the coastal environment if developed. In particular, wave and tidal energy sources, which specifically exploit energy from the sea, are inherently dependent upon coastal locations.

7. There are also some good on-shore wind sites in the coastal zone. A substantial off-shore wind resource exists around the United Kingdom but since this resource is unlikely to be exploited closer than 2km to the coast, some would argue that it lies outside of the coastal zone. Other renewables such as biofuels, and waste combustion, do not need to be located in coastal regions.

8. The following sections of the Memorandum detail: the environmental impact of power stations on the coastal zone; and the planning procedures or power stations, including the requirements for environmental statements.

ENVIRONMENTAL IMPACT OF CONVENTIONAL POWER STATIONS ON THE COASTAL ZONE

9. The environmental impacts of conventional coastal power stations in relation to terrestrial and visual impacts are relatively well defined. Two issues, unique to coastal power stations, are: the impact of cooling water discharges on the marine ecology; and the effects of cooling water intakes on fisheries. Ancillary considerations, such as the impact of dredging for intake pipes, may need to be considered. On these issues, all recent applications for coastal power stations have involved full consultation with the relevant bodies such as NRA, MAFF and the relevant conservancy council, and production of an Environmental Statement.

10. After extensive work, thermal discharges, depending upon location, are generally now not thought to cause significant impacts on the environment. Work on mitigating the impact of water intakes on fisheries is continuing, both in terms of intake design, and the development and evaluation of means of deterring fish from regions close to intakes.

ENVIRONMENTAL IMPACT OF THE RENEWABLES ON THE COASTAL ZONE

11. Renewables, although attractive from the point of view of reducing emissions of SO_2, NOx and CO_2 when compared to conventional coal-fired and other fossil fuel power stations, have environmental impacts of their own which need to be considered. The impacts will vary depending upon the site and technology concerned, but may include: ecological implications; effects on land usage; noise; and an impact on visual amenity.

12. It should be noted that the renewables with the greatest potential impact on the coastal zone, off-shore wind (if considered as being in the coastal zone), wave, and tidal, are still largely at the research and demonstration phase. Research on these renewables will increasingly assess their environmental impact and means of mitigation as they approach the market place.

13. The issues particularly related to the development and exploitation of the different renewables in the coastal zone, are detailed in the Appendix.

PLANNING PROCEDURES—POWER STATIONS GREATER THAN 50MW, AND OVERHEAD LINE CONSENTS

14. The Secretary of State for Energy is responsible for granting planning consent for new power stations over 50MW, and overhead lines in England and Wales, under Sections 36 and 37, respectively, of the Electricity Act 1989. The same planning procedures apply to all power stations greater than 50MW, whether or not they fall in the coastal zone. These procedures are outlined below.

15. When considering a section 36 or 37 application for planning consent for a new generating station or overhead line, the Secretary of State is required under Schedule 9 of the 1989 Act to have regard to: the desirability of preserving natural beauty; of conserving flora, fauna and geological or physiographical features of special interest; and of protecting sites of buildings and objects of architectural, historic, or archaeological interest. The Secretary of State is also required to have regard to the extent to which the applicant has complied with its duty to mitigate any effect which the proposals would have on the natural beauty of the countryside, or on any such flora, fauna, features, sites, buildings or objects.

16. Section 38 and Schedule 9 of the Electricity Act 1989 places an obligation on all licensed electricity companies to have regard to the preservation of amenity and to prepare, and from time to time to modify, a statement on how they deal with amenity conservation. In preparing these statements the companies are required to consult with the Countryside Commission, the Nature Conservancy Council for England (English Nature), the Countryside Council for Wales, and the Historic Buildings and Monuments Commission for England, the Historic Buildings Council for Wales.

17. The Secretary of State for Energy expects all applicants for power stations to have an Environmental Statement. Consultees include: the relevant local planning authorities; the Countryside Commission; English Nature or the Countryside Commission for Wales (as appropriate); HMIP for section 36 applications; and English Heritage for Section 37 applications.

18. Before reaching any decision, the Secretary of State will take into account the views of those parties, together with representations received from other interested parties. In granting his consent, the Secretary of State may attach conditions which take into account the views of the relevant planning authorities and other representations.

19. The Secretary of State is obliged to call a public inquiry, under Schedule 8 of the Electricity Act 1989, if there are objections from the relevant Local Authority. The Secretary of State has discretionary powers to hold a public inquiry on the basis of representations from other bodies.

20. When calling an inquiry, the Secretary of State will give the inspector a statement of matters which appear to him to be relevant to his consideration of the application in question, in the light of the views of regulatory authorities and other representations. These might well include such matters as the visual impact of the proposed development, landscaping proposals, ecological impacts, transport proposals, effluent discharges etc.

21. The statement of matters does not preclude the inspector hearing evidence and reporting on other matters, if he so wishes. The Secretary of State will take fully into account the inspector's recommendation before reaching his decision on the application.

PLANNING PROCEDURES—POWER STATIONS LESS THAN 50MW

22. The responsibility for giving planning consent for new power stations under 50 MW rests with the relevant planning authority, under the Town and Country Planning Act, 1990. The planning authority should ask for an Environmental Statement as part of the planning process. A generator can appeal to the Secretary of State for the Environment against the decision of the planning authority.

23. Although the larger proposed tidal energy barrages (eg Mersey and Severn), are well in excess of the 50MW planning threshold, as could be arrays of off-shore wind and wave devices, most renewable projects are substantially below 50MW and are therefore covered by the provisions of the Town and Country Planning Act 1990. Because of a lack of previous experience with most renewables, until recently there was little guidance for local planning authorities on how to deal with the specific needs and impacts of the renewables.

24. In the first anniversary report of the Environment White Paper " This Common Inheritance ", a commitment was given to clarify the position of renewables in the planning process in England and Wales. Accordingly, the Department of the Environment and the Welsh Office, in consultation with the Department of Energy and others, are conducting a consultation exercise on a draft Planning Policy Guidance (PPG) note on renewable energy.

25. The draft PPG explains Government policy on planning for renewable energy, and shows local authorities how they can include renewable energy policies in development plans. The PPG notes the environmental implications of renewable energy installations, outlines the relevant environmental protection legislation, and explains when an environmental assessment may be required. Considerations which apply in designated areas are also set out.

26. The PPG incorporates specific guidance on wind energy in an Annex. Annexes for other renewable technologies will be produced when they are required, and where appropriate coastal issues will be addressed.

27. As part of the development of planning guidance for the renewables, the Department of Energy has initiated a series of pilot studies to assist Local Authorities to assess the renewable energy potential in their area, and the associated planning and environmental issues.

CONCLUSION

28. The memorandum has identified the potential environmental impact of power stations in England and Wales on the coastal zone, and the associated planning requirements. The memorandum and attached appendix makes particular reference to the impact of the renewables on the coastal zone.

ANNEX

RENEWABLE ENERGY AND THE COASTAL ZONE

1. The development of renewable energy sources within the coastal zone is most likely to include the technologies of wind, tidal and wave. The development of biomass technologies using, for example, seaweed do not appear to be realistic in the near future. Other renewables technologies, such as waste combustion, are not limited to coastal applications.

2. Environmental considerations are an important factor in relation to the siting of all renewable schemes. Each of the technologies with the most direct impact on the coastal zone are outlined below.

ON-SHORE WIND

3. On-shore wind has already emerged as one of the more promising renewable energy sources of electricity generation in the United Kingdom. The coastal zone includes only a small proportion of suitable sites in the United Kingdom.

4. The major environmental impact of land-based wind turbines is visual intrusiveness, as turbines are often sited on high exposed land. Noise and electromagnetic interference can also be significant siting issues. A few test wind turbines have been sited on or near the coast eg Richborough Kent, but to date only two wind farms, at Haverigg in Cumbria, and Blyth in Northumberland, have been proposed for coastal sites. Both these schemes have avoided sensitive areas such as national parks, coastal heritage sites, and areas of outstanding natural beauty.

5. On coastal sites, the impact on birds, especially waders and wildfowl, was considered a potential issue. However, based on the experience in Denmark and the Netherlands, the effects of wind turbines on bird life is likely to be minimal. To ensure that this is the case, monitoring of the possible impact on birds is being organised for the proposed Haverigg wind farm.

OFF-SHORE WIND

6. One reason for developing wind turbines off-shore is because of their reduced environmental impact compared to land based wind turbines. Off-shore wind devices are unlikely to be located less than 2km from the coast, and as such are seen by some as being beyond the coastal zone.

7. The technology for off-shore development is essentially the same as for on-shore development, but the more vigorous climate and the relative inaccessibility of the machines place more stringent requirements on the initial design, and subsequent reliability, of wind machines. Off-shore installations are expected to have costs at least 50 per cent more than on-shore installations.

8. The visual aspects of off-shore wind energy may be important, for example in areas where tourism is a significant economic factor. This could have an influence on the way in which off-shore turbines might eventually be developed.

9. Even though the development of an off-shore wind farm is a number of years away, MAFF have already given some consideration to issues such as the licensing of off-shore wind turbines, and the impact on the coast where the cables would need to come ashore. This latter area should not cause great difficulties in environmental terms as cables for communication and electricity have been laid successfully in the coastal zone with minimum intrusion.

TIDAL ENERGY BARRAGES

10. A number of potential tidal energy sites have been identified in the United Kingdom. Site specific studies have been, or are being, supported by the Department of Energy on tidal energy barrages on the Mersey, Severn, Wyre, Loughor, Conwy and Duddon estuaries.

11. The studies are aimed at reducing uncertainties on costs, performance and regional and environmental issues, associated with tidal energy barrages, to the point where it will be possible to take decisions on whether or not to plan for their construction.

12. A key area of each of these studies has been to carry out a preliminary assessment of the environmental impact of the barrages on the estuaries, and to liaise with relevant environmental interest groups. These site specific studies are underpinned by generic environmental and engineering studies.

13. Tidal energy would have major impacts on the coastal environment, both on and off-shore. Impacts on the shore would be both direct on the coastal strip and have implications for regional development. A tidal energy barrage would change the nature of the estuary and have implications for other forms of local development.

14. In studying the ecological changes to the estuary the Department of Energy carries out research to examine the various issues, and both consults and collaborates with other Government Departments (eg MAFF, DOE) and conservation organisations (eg English Nature, the Royal Society for the Protection of Birds (RSPB), the National Rivers Authority (NRA)).

15. Local Authorities and relevant conservation organisations are consulted on site specific projects, in order that all the regional and environmental implications of tidal energy are fully addressed. After the last major phase of work on the Severn between 1987 and 1989, there was a full consultation process with local authorities and environmental organisations, the results of which have been published. On-going regional and environmental studies on the Severn include an examination of the issues relating to estuary management.

16. Currently any proposal for a tidal energy barrage would need to go through a parliamentary bill procedure, due to their implications on navigation and fisheries. The Transport and Works Bill will specifically allow tidal barrages, not of national significance, to be dealt with by a Parliamentary Order.

WAVE ENERGY DEVICES

17. Wave energy devices are classified as either shoreline, near-shore or off-shore devices according to their location. Only the west and north coast wave resource is regarded as being suitable for energy generation.

18. Wave energy technology is at an early stage of development. Shoreline devices are the furthest advanced, with, in the United Kingdom, a 75kW prototype device undergoing trials on the island of Islay.

19. Shoreline devices need to utilise a limited area of the coast, and these areas may need to be engineered (eg through removal of rock to form a suitable gully) to maximise their performance. The siting of such devices needs to be on exposed rocky coasts below the high water mark, and such land may be regarded as environmentally and visually valuable. Access roads, and the need for distribution cables, could also produce visually intrusive forms of development, although careful restoration of access roads and laying of cables underground can minimise their impact.

20. Many proposals for wave devices use air turbines, driven by oscillating columns of air caused by the motion of waves on the device. These air turbines can give rise to noise, the extent of which is site and design specific. Part of the Department's current phase of work on onshore wave power will address volume, frequency distribution, and methods of mitigation, of turbine noise.

21. The economics of shoreline devices is very site specific, and the potential for energy generation is much less than that off-shore. Shoreline devices are expected to have a greater environmental impact on the coast than off-shore devices.

22. No full scale near-shore or off-shore wave devices have yet been constructed. Off-shore wave devices may be located beyond the coastal zone. The environmental effects of off-shore devices are thought to include: the development of areas of sheltered water, and changes in tidal currents, with consequent effects on the shoreline ecology; an impact on visual amenity (eg land based transmission lines); and some effect on marine life. The precise details of these effects will be site specific and will depend upon the design of the device.

23. Near-shore devices will have a mix of the environmental impacts of both shoreline and off-shore devices, depending on their distance from the shore.

9 *March* 1992

APPENDIX 22

Supplementary Memorandum by the Department of the Environment

I and my colleagues in this Department and the Ministry of Agriculture, Fisheries and Food (MAFF) gave oral evidence to the Select Committee on 4 November. We undertook then to give further information on a number of items, and I attach the information which is now available. Paragraph references are to the transcript of the evidence given on 4 November.

Annex A: The attached note prepared by MAFF gives further information on coastal groups in England and Wales and co-ordination across the English-Welsh border (paragraph 13).

Annex B: NRA expenditure on coast defence—note by MAFF (paragraph 19).

Annex C: Harbour authorities and the environment. The attached note prepared by the Department of Transport provides additional factual information on the issues discussed (paragraphs 48 to 50).

Annex D: Contingency planning for coastal pollution—the attached note prepared by the Departments of Transport and the Environment responds to the queries raised by the Committee (paragraphs 54 to 56).

Annexes E, F and G: Allowances for sea level rise—intergovernmental panel on climate change predictions and government guidance to local authorities. Additional information has been provided by MAFF as requested (paragraphs 57 to 65).

The Private Bill procedure was discussed at paragraphs 51 to 53. We thought that the Committee would find it useful to have a note on the relationship between the Private Bill procedure and the new procedures proposed under the Transport and Works Bill, and the role of the Crown Estate Commissioners. This note has been delayed because of work on the Bill itself, but we hope to be able to let you have it shortly.

We are not yet in a position to supply you with interim copies of the consultants' report reviewing coastal planning and management policy and responsibility for the United Kingdom (see paragraph 23), nor to send you a draft Planning Policy Guidance note on coastal planning (see paragraph 30). I will arrange for these to be sent to you once they become available.

8 *January* 1992

Annex A

MAFF/WELSH OFFICE COASTAL DEFENCE FORUM

1. There are currently 15 Coastal Groups in England and Wales. Another Group is under consideration for the Cumbrian coast. At that stage some 98 per cent of the coastline of England and Wales will be covered by Coastal Groups. All are regarded as Members of the Coastal Defence Forum and a list is given below.

2. Two Groups currently have responsibilities which cross the English/Welsh border—the Llandudno-Mersey Estuary Coastal Group and the Tidal Dee User Group. The former concentrates on coastal defence matters in Liverpool Bay; the latter with representation from both sides of the Dee estuary covers in addition navigation, land use, and conservation matters relating to the estuary.

3. The Department of the Environment has a research study in progress covering land use planning and related matters along the Severn Estuary. Local authorities, the National Rivers Authority, MAFF and Welsh Offices are represented on the Steering Committee for the study. Both MAFF and the Welsh Office are encouraging the formation of a Coastal Group for the Severn Estuary, building on the contacts established through the Steering Committee.

LIST OF MEMBERS

Ministry of Agriculture, Fisheries and Food, Flood Defence Division (in Chair)

Welsh Office

National Rivers Authority

River Ribble-Morecambe Bay Coastal Group

Llandudno-Mersey Estuary Coastal Group

Tidal Dee User Group

Ynys Enlli to Llandudno Coastal Group

Cardigan Bay Group

Carmarthen Bay Coastal Engineering Study Group

Swansea Bay Coastal Group

Devon Coast Protection Advisory Group

Cornwall Countryside Coast Protection Group

Standing Conference on Problems Associated with the Coastline

East Sussex Coastal Liaison Group

Kent Coastal Group

Anglian Coast Authorities Group

Holderness Coast Protection Project

North East Coastal Authorities Group

Annex B

HOUSE OF COMMONS SELECT COMMITTEE ON THE ENVIRONMENT COASTAL ZONE PROTECTION AND PLANNING

NRA EXPENDITURE ON COASTAL DEFENCE

The Committee requested information about NRA expenditure on coastal defence; this includes sea and tidal defence works. Expenditure, including capital works and maintenance in 1990–91 was £63·5 million of which £18·5 million was government grant.

This was about one third of NRA expenditure on flood defence, which in turn was just over half of the total expenditure of the authority.

Annex C

HARBOUR AUTHORITIES AND THE ENVIRONMENT

PLANNING

The jurisdiction of the Local Planning Authority normally extends to low water mark. A harbour authority is entitled to treat certain landside developments as permitted developments for which no further planning application is needed under the General Development Order 1988. These developments are restricted to (a) developments which are specifically designated in a local Act or in a harbour revision or empowerment order; and (b) developments on operational land for the purposes of shipping or in connection with the movement of passengers or goods. It is, however, open to the Local Planning Authority under Article 4 of the GDO to seek a direction from the Secretary of State for the Environment to withdraw a specific permitted development right so that the normal planning application process would apply. This ensures that where, in exceptional cases, there is a threat to amenity or to the environment, environmental assessments can be required.

Applications for statutory authorisation for harbour works are subject to the environmental assessment regime. Similarly, proposals for works below low water mark (ie outside the planning system) which are within harbour authorities' existing powers but which are not specifically authorised by any Act or Order, can be called in and an environmental assessment required. The attitude of harbour authorities to environmental issues when planning the development of harbour facilities is being increasingly conditioned by their recognition of the need to take into account the requirements of the regulations dealing with the environmental assessment of harbour works which came into force in 1988 and 1989.

OPERATIONAL MATTERS

Harbour authorities are, of course, subject to general environmental requirements for example in connection with the disposal of dredged sediment, the prevention of pollution from ships and the handling of dangerous goods. These requirements stem from international agreements and Community law as well as from United Kingdom law (eg IMO Conventions and the Council Directive on Environmental Assessment). Examples of United Kingdom law which has particular application in ports are the Food and Environmental Protection Act 1985 and the Environmental Protection Act 1990. Harbours must by the nature of their business develop expertise capable of dealing with a wide range of environmental issues to comply with these statutory requirements and to implement them within the port area.

MOORINGS

Depending on their local powers, many harbour authorities are able to grant licences for moorings within their area, although in some cases the right to grant moorings may depend on the owner of the harbour bed (not necessarily the harbour authority). The provision of additional moorings reflects the demand for moorings from recreational boat users. The amount of additional revenue depends essentially on the level of charges imposed. Harbour authorities would of course need to ensure that the number of licences issued was not excessive for the orderly and safe management of the harbour as a whole.

Annex D

CONTINGENCY PLANNING FOR COASTAL POLLUTION

The Marine Pollution Control Unit (MPCU) of the Department of Transport is responsible for responding to oil or chemical spills from ships at sea. A National Contingency Plan is maintained for dealing with major oil or chemical spillages together with the resources to support it. Coastal local authorities have accepted non-statutory responsibility for beach cleaning in their areas with MPCU support in major incidents. Local Authorities are encouraged to submit their coastal emergency contingency plans to the MPCU for comment.

Offshore operators in United Kingdom waters are required to produce oil spill contingency plans, supported by identified resources, to the satisfaction of the Secretary of State for Energy, before they are licensed to commence operations. The task of reviewing and endorsing these plans and setting appropriate standards is carried out by MPCU on behalf of DEn, although formal approval is always given by the latter.

The National Rivers Authority (NRA) has statutory responsibility for the response to pollution incidents in rivers, estuaries and in United Kingdom waters up to three miles from the coast. In order to avoid confusion over their respective roles the MPCU and the NRA have agreed that in major ship pollution incidents at sea and in coastal waters, the MPCU will continue to take the lead in dealing with the problem, consulting the NRA as appropriate. The agreement is to be formalised shortly in a memorandum of understanding.

Harbour and Port authorities are best placed to respond to pollution caused by ships trading into, or installations, in their areas, but, with the exception of such places as Sullom Voe and Flotta, have no statutory duty to do so. MPCU encourages Port and Harbour authorities to draw up local contingency plans and to submit these plans for comment. This accords with the recently concluded International Convention on Oil Pollution Preparedness Response and Co-operation which the United Kingdom strongly supports. The Convention provides for such plans to be made mandatory and co-ordinated with the national plan—this would require legislation. If a spill is beyond the capability of the port or harbour authority, MPCU stands ready to assist and, if it is from a ship, take overall control of the response.

Each year the MPCU organises and runs a major exercise in a selected area of the United Kingdom to test the National Contingency Plan. During each exercise equipment is deployed and local authorities in the area are invited to take an active part to test their own Contingency Plans. Other Government Departments, the NRA, local harbour authorities and oil terminal operators also participate. This year the exercise was held in Lancashire and involved mobilising MPCU stockpiles at Burnett and Burtonwood and sending a team to the Preston Joint Response Centre.

The MPCU also runs residential courses to instruct local authority personnel in how to draw up an Oil Spill Contingency Plan. Separate exercises are organised to provide " hands on " experience in the deployment and use of the equipment held in the MPCU stockpiles and in 1991 these have been held at Colwyn Bay in Wales, in Dorset and Kent. Local authorities and industry conduct their own exercises and the MPCU participates as time and staff availability permits.

Annex E

COASTAL ZONE PROTECTION AND PLANNING

ALLOWANCES FOR SEA LEVEL RISE

INTERGOVERNMENTAL PANEL ON CLIMATE CHANGE PREDICTIONS AND GOVERNMENT GUIDANCE TO AUTHORITIES

INTERGOVERNMENTAL PANEL ON CLIMATE CHANGE PREDICTIONS

1. The Intergovernmental Panel on Climate Change (IPCC) has developed a number of scenarios to aid understanding of the implications of climate change. The " Business as Usual " scenario assumes that few or no steps are taken to limit greenhouse gas emissions. Under this scenario, the equivalent of a doubling of pre-industrial CO_2 levels occurs by around the year 2025.

2. Predictions based on this scenario suggest that the increase in global mean temperatures will be about 1°C above the present value by 2025 and 3°C before the end of the next century. IPCC suggest that there could be a global rise in sea levels of about 20 cm by 2030 and 65 cm by 2100; however, there will be significant local variations. These figures are best estimates, falling within the extremes of high and low estimates produced by the models, which indicate about 30 and 10 cms respectively by 2030.

3. Predictions based on other scenarios are significantly lower and suggest a global sea level rise of approx 15 cm (2030) and in the range 30 to 50 cms (2100).[1]

GOVERNMENT GUIDANCE TO AUTHORITIES

4. The MAFF Strategy agreed in 1989 is at Annex F.

5. The IPCC best estimate for global sea level rise and predictions for earth crustal movement in Great Britain have been combined to provide the best current basis of allowances for the design of coastal defences—they are set out in Annex G. This guidance has recently been issued to authorities.

Annex F

MINISTRY OF AGRICULTURE, FISHERIES AND FOOD

Conference of River and Coastal Engineers, Loughborough, 11–13 July 1989

The Greenhouse Effect—Consequences for Rivers and Coastal Zones

EXECUTIVE SUMMARY

The scientific background to the greenhouse effect was reviewed and there was general agreement on the ranges for predicted global climate changes, particularly temperature, though regional effects cannot yet be defined.

The impacts of these changes for rivers and coastal zones will be the rise in sea level, expected to be in the range of 15 cm to 30 cm by 2030, possible increases in storm severity, and changes in weather patterns which could result in greater frequencies of inland flooding and droughts.

In response to these impacts the strategy should be to:

1. Refurbish defences to reduce existing risk;

2. Continue research into river and coastal processes;

3. Monitor trends in climate, sea level, waves, beaches and saline intrusion;

4. Utilise current predictions of sea level in a review of existing standards; and

5. Keep policy and " best practice " under review as understanding develops of sea level trends, waves, surges and weather patterns.

Flood Defence Division
7 *August* 1989

1. INTRODUCTION

The Annual Conference of River and Coastal Engineers is organised by MAFF and was attended by over 200 leading practitioners in the field from Water Authorities, Central and Local Government, Drainage Boards, Universities, research organisations and Consultants. As part of this year's conference, a round table discussion on the greenhouse effect took place on Wednesday 12 July. This was chaired by Professor Wolf, Professor Emeritus at City University, and led by the following panel of experts who each presented papers in their specialised fields

Dr G J Jenkins	Meteorological Office
Dr R A Warrick	Climatic Research Unit, University of East Anglia
Dr A J Apling	Department of the Environment
Dr D T Pugh	Deacon Lab, Institute of Oceanographic Sciences
Mr I R Whittle	NRA (formerly Chief Engineer, MAFF)
Mr I H Townend	Sir William Halcrow and Partners
Mr F M Law	Institute of Hydrology

---

[1] This information is taken from Climate Change, The IPCC Scientific Assessment (1990).

2. SCIENCE

The greenhouse effect may become one of the greatest issues of our age. There is a need to be clear about the current state of our knowledge and to work towards a better understanding of the many areas of uncertainty which exist.

Incoming short wave radiation from the sun reaches earth unattenuated. Outgoing radiation from the earth is of a longer wavelength and certain gases in the atmosphere absorb part of that radiation and re-emit it in all directions. The net effect is a warming of the earth's surface and lower atmosphere. This effect has become known as the greenhouse effect. Without the present level of natural " greenhouse " gases average surface air temperatures would be some 38 degrees C less than they are today and thus they are essential for habitation in its current form. However the atmospheric concentration of greenhouse gases is increasing due to man's activities, with potentially large increases in global mean temperature. Records show an increase of about 0·5 degrees C in the last century in observed global mean surface air temperature. This warming cannot yet be attributed solely to the greenhouse effect since factors such as volcanic eruptions, solar variations, ocean current changes and the like also influence climate in the medium and long term. It has not at present proved possible to separate out these influences unambiguously.

The gases which contribute most to global warming are carbon dioxide, methane and chlorofluorocarbons (CFCs). Nitrous oxide, ozone and other trace gases are also know to contribute.

Concentrations of carbon dioxide have increased by 25 per cent since the mid-18th century, due largely to the burning of fossil fuels. Carbon dioxide accounts for approximately half the total change in global warming over the last decade. Over the same time period atmospheric carbon dioxide concentrations increased at the rate of 0·4 per cent–0·5 per cent per annum.

Methane concentrations have approximately doubled in the last 200 years and are currently rising at approximately one per cent of current levels per annum. This methane probably comes largely from agriculture and fossil fuel extraction.

CFCs do not occur naturally. CFC concentrations have recently been rising at approximately six per cent per annum and have a long atmospheric life (100 years). Although concentrations of methane and CFCs are much lower than that of carbon dioxide, molecule for molecule these gases are more powerful in their greenhouse effect, methane by 27 times and some CFCs by more than 10,000 times. These differences have led to the concept of "equivalent" carbon dioxide concentrations (ie the combined radiative effect of all greenhouse gases expressed in carbon dioxide equivalence). The best estimate is that a doubling of carbon dioxide equivalent will occur around the year 2030.

A doubling of carbon dioxide is used in General Circulation Models (GCMs) to predict effects on equilibrium global climate. The United Kingdom Meteorological Office run one of the five leading GCMs in the world which have been used for such experiments. The technique is to run the model until equilibrium climatic conditions are achieved for both control and doubled levels of carbon dioxide concentration and then compute the changes in climatic variables. Good agreement between the five models has been found at the global scale but there is significant variation in regional predictions.

In summary, the "best guess" consequences derived from these GCMs and other climate models are as follows:

(i) Global mean surface air temperatures will rise between 1·5 and 2·5 degrees C by 2050 with further increases to follow.

(ii) Global climate will change.

(iii) Global evaporation and precipitation will increase.

(iv) Global sea levels will rise.

(v) Regional and local effects are much more difficult to predict and may in some local areas be in directions which are the reverse of global trends.

3. IMPACTS

There is general agreement that the rise in global air temperature will lead to a rise in global sea levels due to thermal expansion of the oceans and to the melting of land based ice, though polar ice may increase due to increased precipitation at the poles. Best estimates indicate a range of increase of between 15 cm and 30 cm by 2030 and between 25 cm and 40 cm by 2050. These should be considered in the light of increases of some 10 cm to 15 cm in the last 100 years.

As important as the global rise in sea level will be regional effects of changing ocean currents, tides and the frequency and magnitude of storms and surges. This is an area of much greater uncertainty as the global oceanic and meteorological models are not yet able to provide adequate predictions at a regional scale.

The best that can be achieved at present is to make broad assumptions of the changes likely to occur and to evaluate their regional and local effects. In the design of coastal defences it is generally the extremes of sea level and waves which are critical. Reference was made to one United Kingdom site for which it is estimated that levels, which at present might occur only once in 100 years, could occur every five years on average by 2050.

Local impacts will be highly dependent on the nature of the coastal zone. Some natural interfaces such as salt marshes and coastal wetlands can naturally adapt to change by build up of sediment and by migration inland provided that the rate of change is gradual and the process is not constrained, particularly by man made structures.

Other subsidiary effects include the need for increased pumping from low level areas and the increased intrusion of saline water into estuaries and coastal aquifers with consequences for agriculture and water supplies.

Inland changes in weather patterns will alter the character of river systems and could well lead to increased frequencies of both floods and droughts though prediction once again awaits more detailed regional models of the meteorological system.

4. RESPONSE STRATEGY

4.1 *International action*

The Intergovernmental Panel on Climate Change (IPCC) established by the United Nations Environmental Programme (UNEP) and the World Meteorological Organisation (WMO) is undertaking a review of the science for improving predictions at global and regional levels, evaluating the environmental and socio-economic impacts and formulating response strategies. The United Kingdom is undertaking the first of these reviews of the science under the chairmanship of the Director General of the Met Office. An interim report will be made to the World Climate Conference in 1990.

Programmes studying the role of oceans in climate change and the monitoring of earth crustal movements are in hand.

Procedures for controlling emissions are being discussed and the United Kingdom supports an Umbrella Convention leading to binding Protocols.

4.2 *Local (United Kingdom) actions*

The most pressing needs are to develop a response strategy to ensure that appropriate actions are taken and public money is used to best effect.

A number of existing defences are reaching the ends of their effective lives and a first priority should be to bring these to acceptable standards relative to current conditions so that there is a firm base for any necessary improvements in the future.

On the research side it is vital to continue and to increase the accuracy of the monitoring of sea levels and climate to provide indications of future change. In conjunction with this it is necessary to study other influences so that the effects of long term climate change can be identified. It is also necessary to identify earth crustal movements to obtain absolute sea level changes.

Other monitoring needs at a more local scale include the recording of changes in the coastal zone of levels, sediment transport, wave climate, ecosystems, saline intrusion etc. which will provide the basis for policy and design decisions.

Research should continue into river and coastal processes to assist in formulating effective strategies for coping with predicted changes as these become more clearly defined.

The design life of most sea defence works in relation to anticipated rates of sea level rise allows a flexible approach to be adopted. Design should be based on current predictions but should incorporate flexibility to allow future modifications when better predictions become available. In all cases the widest range of options should be considered and these may include set back, retreat or advance from the present defence line.

There is a case for the continuing review and development of national policies including legislation in the light of future research results. This is particularly applicable to the development of appropriate links between planning and coastal management authorities within coastal cells.

Flood Defence Division

7 *August* 1989

Annex G

CLIMATE CHANGE

RESPONSE TO IMPACT ON COASTAL DEFENCE: ALLOWANCES FOR SEA LEVEL RISE

In July 1989 the Ministry of Agriculture, Fisheries and Food promulgated a strategy to respond to the impact of the Greenhouse Effect. This strategy embraced the use of current predictions of sea level rise in reviewing existing standards and best practice for coastal defences with consideration being given to incorporating flexibility to allow for future modification when better predictions become available.

The Intergovernmental Panel on Climate Change (IPCC) have since reported, offering predictions of global sea level rise for various scenarios. The " IPCC Best Estimate trend " is accepted as the most appropriate at this time.

Predictions are also available of earth crustal movements in Great Britain.

The combination of these predictions has been accepted as forming the best basis of allowances (shown below) for the design of coastal defences but consideration needs to be given in the designs to adapting defences to allow for any future change in predictions where defences have an effective life beyond 2030. In designing schemes to this response strategy authorities should ensure that their proposal is cost effective.

For ease of reference the allowances are based on NRA regions and are:

NRA Region	Allowance
Anglian, Thames, Southern	6 mm/year
North West, Northumbria	4 mm/year
Remainder	5 mm/year

Flood Defence Division
Ministry of Agriculture,
Fisheries and Food

November 1991

APPENDIX 23

Supplementary Memorandum by the Department of the Environment and the Department of Transport

THE CIRCUMSTANCES IN WHICH HARBOUR AUTHORITIES ARE SUBJECT TO PLANNING PERMISSION

Most harbour authorities require planning permission for their development proposals, but may benefit from permitted development rights under the Town and Country Planning General Development Order 1988 (the GDO). These rights confer a general grant of planning permission, making a specific planning application unnecessary. Harbour authorities may have such rights under part 11 of Schedule 2 of the GDO when carrying out works authorised by a local Act or by any order made under section 14 or 16 of the Harbours Act—but any major development requires the prior approval of the local planning authority to the detailed plans and specifications. Dock, pier, harbour and some other undertakings, if they are statutory undertakers may also have rights under part 17B of Schedule 2 to the GDO to develop their operational land in certain circumstances.

Development which is not permitted under the GDO will require planning permission in the usual way. Where the harbour authority is also a local authority it will be able to deem itself permission for development under regulation 4 or 5 of the Town and Country Planning General Regulations 1976.

A few authorities are also Crown bodies, eg dockyard ports such as Plymouth. Like Government departments they do not require planning permission for their development proposals. But all Crown bodies have agreed to follow the procedures in Part IV of DOE Circular 18/84 (WO 37/84) unless considerations of national security are involved. These procedures mirror those of an ordinary planning application except that the local planning authority does not grant or refuse permission but indicates whether or not it objects to the proposal. Unresolved objections are referred to the Secretary of State for the Environment to decide whether planning clearance should be given, if necessary following a public local inquiry.

The jurisdiction of the planning system generally only extends as far as the low water mark and it cannot therefore control development beyond this point.

2 *February* 1992

APPENDIX 24

Supplementary Memorandum by English Nature

Several witnesses have criticised the lack of a co-ordinated database for the coast and the lack of a co-ordinated research programme to cover the wide range of specialisms and interests in the coastal zone. How do you think these two issues should be addressed, and by whom?

This is a very complicated issue.

Over the years a vast amount of information has been gathered covering the physical and biological properties of the coast and its use by man. One of the simplest but most daunting tasks is to find out what information exists and how it can be obtained.

Firstly, there is a need for an inventory of research covering the relevant scientific fields which can achieve this major but basic need.

Secondly, there is a need to bring this together in some form of linked system. This has started insofar as the British Oceanographic Data Centre (NERC) is developing a United Kingdom Digital Marine Atlas (UKDMAP) in collaboration with NERC, MAFF, SOAFD, JNCC and NRA. This is a simple system which enables the user to call up charts and descriptive text which provides a summary of the information entered. It is possible to compare different sets of mapped material by overlaying them. In this way, someone trying to find out what is known about a particular area of the coast can call up all the information available in the system. UKDMAP is an inexpensive system and will be widely available. However it is a collation system not an analytical one; it cannot analyse relationships between various pieces of research.

To carry out such analysis will require a sophisticated system (eg a powerful Geographical Information System) and a lot of software development. Because of cost such a system would be available to only a limited number of users. However, the basic problem is that because much previous research has been carried out in an unco-ordinated way, the information needed for sophisticated analyses often simply is not available.

A great deal of information is available for the Severn Estuary, but even here the ability to predict the consequences of the effect of a major development proposal has proved limited, and many uncertainties exist. Moreover, information obtained for one locality often has little predictive value elsewhere. That said modelling of physical processes, for example the work of Hydraulic Research for MAFF on sediment cells, is potentially of crucial importance. A GIS based on this work, at a scale capable of accepting detailed sediment process information as it becomes available, or is updated, might for instance provide an essential central resource for Regional Coastal Groupings.

We feel that responsibility for co-ordinating coastal data and research lies with Government. However, the matter will need considerable further examination before a preferred option can be identified.

Research on the biochemical and ecological processes of the marine environment does not seem to be as well developed as research on the physical processes affecting the coast. How much work is being done to link the ecological and physical processes in the coastal zone and how is this knowledge being used?

There is a considerable amount of information available on biochemical and ecological parameters, as well as physical parameters in the coastal zone; however, the establishment of relationships between these has proved difficult. This is because the biological responses to given chemical and physical parameters are very complex. This is true in controlled laboratory conditions and even more so under natural conditions.

Where ecological responses are well defined, for example in relation to phosphates and ammonia in freshwater, predictions can be made with some confidence. However, when the responses of estuarine and coastal invertebrate animal groups to the toxicity of various pollutants was compared it was found that the toxicity of the various substances within the same animal group varied, as did the reaction of the various groups to the same pollutant. If the responses of coastal organisms to such parameters can be defined, even if only in a general or precautionary way, it will help to enable Government to set appropriate Water Quality Objectives for estuaries and coastal waters under the new statutory water quality classification.

To return to the example of the Severn Estuary, referred to in the response to the previous question this estuary has been subject to intensive scientific study as part of the pre-feasibility work undertaken prior to any proposal to construct a tidal energy barrage across the estuary. This work has covered a wide range of biological, chemical and physical parameters with the express purpose of predicting the consequences of the construction and management of the barrage. Nonetheless, considerable uncertainty remains on such basic issues, from the nature conservation viewpoint, as to the impact of the barrage on the internationally important wader populations which currently utilise the estuary.

In summary, therefore, while chemical, physical and biological data can be correlated in a very simple way, for example using the UKDMAP system referred to above, more complex correlations of biological and other processes, while they have been attempted, have proved difficult to achieve because of the inherent complexity.

APPENDIX 25

Supplementary Memorandum by the Ministry of Agriculture, Fisheries and Food

APPLICATION OF ENVIRONMENTAL ASSESSMENT PROCEDURES TO COAST PROTECTION WORKS

1. When MAFF officials gave evidence on 15 January, the Committee requested further evidence on whether coast protection works are covered by the Town and Country Planning (Assessment of Environmental Effects) Regulations 1988. This arose from evidence provided by the Royal Town Planning Institute (paragraph 146 *et seq*). This note has been prepared in consultation with officials in the Department of the Environment responsible for the Regulations in question.

2. The 1988 EA Planning Regulations implement the requirements of European Community Directive 85/337 on the assessment of the effects of certain public and private projects on the environment in so far as the Directive applies to projects which require planning permission. Schedule 1 to the Regulations, which reflects Annex I to the Directive, lists types of project for which environmental assessment is required in every case; Schedule 2, based on Annex II to the Directive, lists types of projects for which environmental assessment is required if the particular development proposed is likely to have significant environmental effects. Coast protection works require planning permission but, unlike flood relief works, are not listed in either Annex I or II to the Directive, nor consequently in the implementing Regulations, so environmental assessment cannot be required.

3. The Department of the Environment is considering whether to propose that coast protection works, ie works for the protection of land from erosion or encroachment by the sea, should be added to Schedule 2 to the 1988 EA Planning Regulations using the new power in section 15 of the Planning and Compensation Act 1991. Environmental assessment would then be required for such works where they were likely to have significant effects on the environment. The Government has promised to consult with those concerned before making regulations under section 15 of the 1991 Act.

4. Even where environmental assessment is not required for development proposals, environmental effects will be among the considerations taken into account by the planning authority in deciding whether to grant planning permission. Also, under the Coast Protection Act 1949 all capital works must be approved by the relevant Minister irrespective of whether grant aid is being provided. Proposals must be advertised and approval is not given unless English Nature (or the Countryside Commission for Wales) have confirmed they have no objection to the proposals. This acts as a separate safeguard for the environment.

5. Flood defence works, which are listed in Annex II to the Directive, require environmental assessment where the proposed works are likely to have significant environmental effects (see paragraph 123 of the government memorandum). New works are subject to normal development control procedures and the 1988 EA Planning Regulations apply. Improvements to existing works do not require an express grant of planning permission, but are subject to the environmental assessment procedures set out in the Land Drainage Improvement Works (Assessment of Environmental Effects) Regulations 1988.

Prepared jointly by
Flood Defence Division
Ministry of Agriculture, Fisheries and Food
and Department of Environment

February 1992

APPENDIX 26

Supplementary Memorandum by the National Trust

When presenting oral evidence on behalf of the National Trust recently my colleagues and I were asked to elaborate upon our concerns about the impacts of unregulated development in ports and harbours. I am writing to provide the evidence requested. In summary, the Trust's concerns relate to the nature and type of the activities which port and harbour authorities can undertake and the inadequacy of the controls over them.

THE THREAT

The land reclamation and dredging work carried out for Felixstowe Docks in order to expand its activities is often quoted as an example of the kind of damage which can arise from uncontrolled development by port and harbour authorities. In this case, the damage was to the Orwell Estuary SSSI and resulted from powers gained by the port authority under private legislation. We consequently welcome the fact that such legislation would now be subject to an environmental assessment, either under standing orders of the House of Commons and the House of Lords respectively, or as a result of an application for a harbour order under the provisions of the Transport and Works Bill.

Though these provisions will cover capital dredging schemes, however, maintenance dredging under the existing powers of port and harbour authorities will still not be controlled. This can be damaging in a number of ways including its potential for destroying sunken archaeological treasures and features of geological importance. It can also affect the harbour and sea bed and adjacent shorelines in ways which have not been fully quantified.

Harbour development works can have similar effects below the water line while on land they can give rise to buildings and associated infrastructure, such as roads, out of scale and character with the surrounding landscape as well as impinging directly upon wildlife habitats. Development for leisure purposes can also add to recreational pressures which, as well as harming the local environment, can reduce visitor enjoyment of the area in question and endanger personal safety.

Under the Transport and Works Bill, moreover, the purposes for which harbour empowerment and revision orders may be made are to be extended to include marina development, the development of surplus land and the diversion of footpaths and bridleways. The provisions of the Ports Act 1991, meanwhile, will give a number of trust ports the option to enter the private sector with all the benefits which that implies in terms of access to additional funds and the consequent ability to expand their activities. Indeed, the Government's aims under this Act were to grant access to share capital for port development; to allow development of surplus land; to facilitate diversification of port businesses and to lay greater emphasis on profitability. It is thus all the more important that there are effective controls over such activities and that they are adequately enforced.

THE CONTROLS

The majority of activities related to ports and harbours are either uncontrolled by planning legislation or constitute permitted development under the General Development Order. Instead, works undertaken by harbour authorities are governed by the Harbours Act 1964. Under this Act, application can be made to the Secretary of State for Transport for approval of Harbour Revision and Empowerment Orders which are then used to regulate maintenance and improvement works to harbours.

The Secretary of State also determines whether and, if so, to what extent a proposed order should be subject to environmental assessment (EA) under the terms of European Commission Directive 85/337/EEC. He will decide, for example, the parts of the order to which EA should apply and what information is required to aid the assessment. These provisions are covered by the Harbour Works (assessment of environmental effects) Regulations 1988 and 1989 respectively. Guidance on the implementation of EA under these provisions has been issued by the Marine Directorate at the Department of Transport, and we enclose a copy for reference.

There are a number of elements of the above procedure which concern us.

(a) The enclosed guidance note makes no mention of the environmental effects of maintenance dredging. A licence from the Ministry of Agriculture is required for the disposal of the material resulting, but not for the regulation of the dredging itself.

(b) Though Revision and Empowerment Orders as a whole are subject to EA, no adequate assessment is conducted of individual projects proposed or implied within the orders the effects of which could vary according to the intensity of their operation or their location.

(c) In this context there is no requirement upon port and harbour authorities to conduct baseline surveys of the ecological or landscape characteristics and sensitivities of the areas under their jurisdiction in order to guage more adequately the effects mentioned above.

(d) Nor do the EA Regulations compel harbour authorities generally to exercise their duties in an environmentally beneficial manner. They therefore do not have to consider the environmental implications of their proposals, or more acceptable alternatives to them, prior to the drafting of works orders.

(e) Under the 1964 Harbours Act, moreover, it appears that the Secretary of State for Transport has no powers to amend harbour *revision* orders on nature conservation grounds alone. This is a concern which was highlighted by the Trust, in conjunction with the RSPB, in a recent brief to MPs considering the Transport and Works Bill in Committee. The Government has acknowledged this point and has promised to amend the relevant wording of the 1964 Act to make it clear that it *is,* indeed, possible to incorporate environmental criteria into harbour revision orders. It is not yet clear to us, however, whether the provisions of the Bill, as currently worded, would have this effect.

For these reasons, the Trust believes that it is time to place environmental duties upon port and harbour authorities in line with those which now apply to other industries. Such duties as we propose have had broad application to ministers, Government Departments and public bodies since the Countryside Act 1968 and similar duties are contained in the Wildlife and Countryside Act 1981.

The opportunity has been taken in more recent legislation to apply them to specific Government Departments and industries as well, such as the Forestry Commission and the Ministry of Agriculture in 1985 and 1986 respectively; the electricity generating companies and the water companies, both in 1989; the coal industry in 1990; and the British Waterways Board in 1991. As we stated in our earlier evidence to the Committee, such duties as we propose above are vital to the protection of the coastal resource under the control of harbour authorities as part of a national strategy for the protection of the coastal zone.

THE CROWN ESTATE

I would like to take this opportunity to raise one other issue arising from the oral evidence we presented recently. During that presentation, I suggested that the Crown Estate Commissioners were not subject to land use planning controls. I have since been assured by representatives of the Crown Estate that, while this may be the case, the Commissioners ensure that planning permission is sought for all developments proposed for Crown Esatate *land* (as opposed to the sea bed to which planning controls do not apply) and that the normal planning procedures are observed thereafter. The Trust accepts that this is the case and we would be grateful if the Committee could consider our comments in that light.

May I, finally, ask you to pass on the thanks of the Trust to the Committee for inviting us to provide evidence to them on this subject. We very much appreciate the opportunity and look forward to the publication of the Committee's report.

4 *February* 1992

Instructions to Mar Gen D3 Staff

EC ENVIRONMENTAL ASSESSMENT

1. As part of the measures to implement the EC directive on assessment of projects with significant impact on the environment, all staff dealing with works in tidal waters will be required to notify certain environmentally sensitive applications to Ports 1, which will take any further action that may be required. This instruction applies to applications made, and notices of applications given by harbour authorities, on and after 16 March 1989.

2. To be notifiable, works will need to satisfy all the following criteria:

(a) they must be within a harbour authority's jurisdiction;

(b) they must be entirely (not partially) below low-water mark and (thus) not subject to planning control;

(c) they have to be carried out or licensed by the harbour authority under general works powers—not under its local Act or harbour empowerment, revision, or provisional order.

3. Types of case which need not be notified, because they involve works on land and/or are subject to other environmental controls, are:

(i) powerlines

(ii) outfalls

(iii) fish farms and shellfish line arrays

(iv) oil and gas production platforms.

4. In addition, the following need not be notified, since they are unlikely to have environmental significance:

(1) individual moorings for small craft

(2) pontoons for berthing small number of small craft

(3) temporary rigs for boreholes

(4) temporary works such as caissons and cofferdams

(5) oil and gas structures, temporarily placed.

5. To enable item 2a above to be determined, all submissions to the Professional Adviser at the consultation stage must include the question:

(1) Is the work within the jurisdiction of a Harbour Authority?

6. If the answer is "yes" and the case appears to meet the other criteria, the member of staff in possession of the application should submit it to his/her Executive Officer with a minute suggesting reference to Ports Division. The Executive Officer, if in agreement, will submit the case to the HEO for a decision whether it should be referred. If so decided, the AA will copy to Ports 1, retaining the file for Mar Gen D3 action, and note the worksheet that Ports Division has been informed.

Marine Directorate
Department of Transport
March 1989

The Harbour Works (Assessment of Environmental Effects) Regulations

ADMINISTRATIVE ARRANGEMENTS

THE DIRECTIVE

1. EC directive 85/337 requires member States, from 3 July 1988, to set up procedures which ensure that environmental assessments will be undertaken for certain types of project. It is concerned with projects which are likely to have significant effects on the environment, excluding any which are specifically authorised by national legislation. Projects (Annex I) which must be subjected to assessment include "trading ports". Among those which are to be assessed if their characteristics so require (Annex II) are all other harbours, including fishing harbours, and modifications to Annex I projects.

2. The directive describes the information needed for an assessment, makes provision for consultation, and says that it may be implemented either within existing consent procedures or by means established to comply with it.

WHAT KIND OF PROJECTS ARE SIGNIFICANT?

3. The directive leaves it to member States to set the criteria or thresholds for those Annex II projects which are to be subject to assessment. Practice may be expected to evolve, but the directive is considered to target:

—major developments

—developments likely to have significant environmental effects on people, wildlife, landscape, the heritage, etc, by virtue of their nature, size, location, etc.

There is bound to be uncertainty in some cases whether a project is likely to have significant environmental effects.

THE TWO SETS OF REGULATIONS

4. The regulations (1988/1336) which came into force on 3 August 1988 provide for environmental assessment, as appropriate, within the procedure for making harbour empowerment and revision orders under schedule 3 of the Harbours Act 1964. It is for the Secretary of State (or Minister) to determine whether an order which has been applied for relates to a project within Annex I or II of the directive. (Because all Scottish harbour orders go through special parliamentary procedure, and are thus specifically authorised by national legislation, cf paragraph 1 above, these regulations apply only to England and Wales.)

5. The further regulations (424/89) which came into force on 16 March 1989 deal with a residual category of harbour works, in any part of Great Britain, which are not subject to any other kind of approval. These are works proposed to be executed by, or under licence from, harbour authorities using general works powers in local Acts or orders, and which are below low-water mark and outside the local planning authority's jurisdiction. They establish a self-contained consent procedure to give effect to the directive in these cases. The appropriate Department will set this in motion when the Department of Transport informs it of a project, falling within these regulations, which has been notified to it under Part II (protection of navigation) of the Coast Protection Act 1949.

IDENTIFYING SIGNIFICANT PROJECTS

6. Within the Department of Transport, guidelines (attached) have been drawn up for sifting prima facie significant projects from hundreds of Coast Protection Act notifications. Cases which may be ignored comprise:

—projects outside the present regulations;

—projects to which other provisions apply, eg the Environmental Assessment (Salmon Farming in Marine Waters) Regulations 1988;

—projects of kinds which in practice are most unlikely to be of environmental significance.

ACTION CASES

7. On receipt of a copy notification, Ports branch 1 will as appropriate either forward the papers to Ports Office Scotland (which will if appropriate send it on to the Scottish Office), MAFF, or the Welsh Office; or, in the case of a non-fishery harbour in England or Wales, itself consider the project.

8. When a project appears to fall within the regulations (unless it is clearly of minimal significance) the appropriate Department will forthwith notify the developer and, if necessary, require further information. After considering this it will tell the developer, and where appropriate the harbour authority, whether the project is to be made subject to environmental assessment (regulation 4), ie it is an Annex I project, or an Annex II project which is likely to have significant environmental effects. This has the effect of stopping the project until the procedure is completed and consent given.

ASSESSMENT PROCEDURE

9. Regulations 5–8 set out the environmental assessment procedure. The Government has announced that the " bodies with environmental responsibilities " which are to be supplied with copies of information on environmental effects and to be consulted (regulation 7) are the local planning authorities, the Countryside Commission, the Nature Conservancy, and Her Majesty's Inspectorate of Pollution.

10. Regulation 9 sets out the considerations relevant to a decision, the forms it may take, and how it is to be made known.

PROJECTS NOT DULY NOTIFIED

11. Regulation 9 provides powers to intervene in the event of environmentally significant harbour works being carried out without prior notification, or in disregard of a requirement to obtain consent.

12. Regulation 10 provides power to serve a notice on a non-complying developer, and for the appropriate Minister to remove the works if the developer does not do so.

APPENDIX 27

Supplementary Memorandum by the Royal Town Planning Institute

Further to the presentation of oral evidence on behalf of the Institute to the Committee on Wednesday 20 November, further elaboration was sought as to whether or not an environmental assessment is required for coastal protection works. It was intimated by Chris Shepley that it was probably covered by Annex B of the European Commission Directive, ie a requirement for environmental assessment in respect of coastal protection works would be discretionary rather than mandatory.

The position is perceived to be as follows:—

A. Department of the Environment Circular 15/88: "Town and Country Planning (Assessment of Environmental Effects) Regulations 1988" sets out at Appendix A a schedule of projects subject to control under the Planning Acts. Schedule 2 equates to Annex II of the EC Directive (paragraph 9 of the Circular). Projects of the types listed in Schedule 2 are to be subject to assessment when member states consider that their characteristics so require, ie where there are likely to be significant effects on the environment. Thus, whereas projects listed in Annex I to the Directive are mandatory in respect of the requirement for environmental assessment, those listed in Annex II, ie Schedule 2, are discretionary.

B. Schedule 2 of the Circular indicates that there may be a requirement for an environmental assessment in respect of new drainage and flood defence works (Agriculture (paragraph 4)). Thus whilst the precise description "coastal protection works" is not used, it is reasonable to assume that the reference to flood defence works also includes coastal protection works bearing in mind that one of the main purposes of coastal protection is to prevent flooding. The position is confused because the Regulations do not repeat in all cases the contents of Schedule 2 of the Circular.

C. Schedule 2 of the Town and Country Planning (Assessment of Environmental Effects) Regulations 1988 (No 1199) sets out at Schedule 2 details of those instances where it is within the discretion of the local planning authority to require an environmental assessment. As stated above Schedule 2 of the Regulations does not necessarily repeat precisely the wording in Schedule 2 of the Circular. Indeed, under paragraph 1, Agriculture sub-paragraph (f), the reference is now to "the reclamation of land from the sea". Further at paragraph 10 Infrastructure Projects, sub-paragraph (e) refers to "canalisation or flood relief works".

D. It is arguable that under those two descriptions in Schedule 2 of the Regulations, the local planning authority could require an environmental assessment. However, as will be readily realised it is not specific.

The Institute would reiterate its view that an environmental assessment should be required for coastal protection works, and that this should be achieved either by clarifying the meaning of the two types of development under Schedule 2 described above or alternatively by, at an appropriate stage, amending the Regulations to make it quite clear that such development may, if it has significant environmental impact, require an environmental assessment.

To summarise it is a matter for consideration that the present Regulations also enable the local planning authority to require and environmental assessment in respect of coastal protection works. But it will be readily seen by the Environment Committee that it is a matter of interpretation since Schedule 2 is not explicit on this point.

9 *December* 1991

APPENDIX 28

Further Supplementary Memorandum by the Department of the Environment and the Department of Transport

THE RELATIONSHIP BETWEEN THE PRIVATE BILL PROCEDURE AND THE NEW PROCEDURES PROPOSED UNDER THE TRANSPORT AND WORKS BILL

INTRODUCTION

1. The Transport and Works Bill implements the proposals in the Consultation Document " Private Bills and New Procedures" (Cm 1110) which gave the Government's response on the report of the Joint Committee on Private Bill Procedure. It provides for a new Ministerial Order making procedure for railway, tramway, inland waterway, tidal barrage and other works schemes, which will avoid the need for private Bills. It also widens the scope of Harbour Empowerment Orders to cover the recreational use of sea-going ships. This will enable marinas to be authorised by order, either as an extension to a harbour or in a location where no harbour has previously existed. The Bill has completed its Committee Stage in the House of Commons, and if enacted before the Election, it is hoped that some of its provisions will be implemented by January 1993.

ORDER MAKING PROCEDURE

2. The new order making procedure, set out in part 1 of the Bill will apply to railways, tramways, inland waterways and works which interfere with rights of navigation outside harbours within the territorial waters of England and Wales. It will be based on the procedure which already applies to new road schemes and to general developments requiring planning permission such as airports and power stations. Applications for orders will be made to the relevant Department who will consider any objections received and arrange for a local public inquiry to be held in appropriate cases. The inquiry inspector will make a report and recommendation to the Secretary of State who will determine the order. By adopting an approach similar to that widely used for development proposals and with which the public are familiar, it is hoped that it will be readily understood and accepted.

3. Rules will be made covering the form of the draft order, the consultation and notification requirements and the timetabling of the pre-inquiry procedures. Anyone will be able to object to a draft order and to present evidence to an inquiry. There will be no *locus standi* requirements as with private Bills so the system should be more accessible to the public.

4. If he wishes, the applicant would be able to obtain all of the necessary authorisations and consents under a single application. The order making procedure provides for deemed planning permission under section 90 of the Town and Country Planning Act 1990 and deemed hazardous substances consent (under the Planning (Hazardous Substances) Act 1990) to be granted as part of the order for the works. Certain other consents and licences, such as those relating to listed buildings and Part II of the Food and Environment Protection Act 1985 will be granted separately by the relevant Secretary of State or Minister but will be considered concurrently with the works proposals at a single local inquiry.

5. For example, where an applicant for an order requires a licence under part II of FEPA, this would be sought with the works order, and the licence application would be passed to the Ministry of Agriculture, Fisheries and Food for processing. Any objections to the granting of the licence would be considered at the same public inquiry as for the works proposals. The Inspector's report on the licence application would be passed to the Minister of Agriculture for determination. Certain provisions in FEPA, such as the right to make representations to the licensing authority relating to the refusal of a licence, contained in Schedule 3 of the Act will be disapplied because the applicant would have had the opportunity already to state his case at the inquiry.

AMENDMENTS TO HARBOURS ACT 1964

6. Part 3 of the Bill deals with certain amendments to the Harbours Act 1964 to meet proposals made by the Joint Committee and in the Consultation Document. These amendments would widen the scope of the Act by removing the limitation on the use of Harbour Empowerment orders for recreational harbours, such as marinas. This would enable existing harbours to be partially or wholly converted to marinas without the need for private legislation. An HEO might also be an appropriate mechanism for obtaining authority to construct a marina in an estuary or along the coast if the developer wishes the marina designated as a harbour. The Bill also provides for the option currently available to promoters of seeking powers through the private Bill procedure, to be removed.

7. The Bill does not change the general procedures relating to the handling of Harbour Orders set out in schedule 3 of the Act. This means that the sort of co-ordinated approach to approvals for works, planning consents and licences set out in part 1 of the Bill for the new classes of Ministerial orders will not apply. For example, where a harbour authority wants to obtain planning permission for new housing as part of a harbour development they would still seek this separately from the harbour order. However, with these wider powers it should be possible under existing procedures in the Harbours Act to hold a concurrent inquiry into the housing and remaining harbour development proposals if this would be desirable.

IMPLICATIONS FOR WORKS IN COASTAL WATERS

8. For the new Ministerial orders system, clause 3 of the Bill provides for the Secretary of State to make orders in regard to inland waterway projects and works which interfere with rights of navigation in territorial waters in England and Wales. Clause 4 enables the Secretary of State to prescribe the nature of the works which interfere with navigation and which would be dealt with in orders under Clause 3. The intention is to prescribe those categories of works that might otherwise have to be authorised by private Bills.

9. Works in territorial waters which currently need only section 34 consent under the Coast Protection Act or can be authorised under the Harbours Act would not be subject to the new order making system. This should confine the application of clause 3 of the Transport and Works Bill to major non-harbour works in territorial waters; for example tidal barrages and artificial islands (such as the proposal, now abandoned, to construct an island for oil exploration purposes in Poole Bay).

10. If a project involved part construction on dry land and part in the sea to the extent that it interfered with navigation—a land reclamation scheme for example—it would be possible to obtain authority to the whole scheme in a single application made under clause 3 of the Bill. This offers rather more flexibility than orders under the Harbours Act where the works must be for a harbour purpose.

11. The Bill also makes provision (in Clause 7) for the Secretary of State to make an order on his own volition covering transport systems used for defence purposes or to deal with abandoned or neglected works in the interests of safety. This power could be used to secure the removal of abandoned structures in the sea up to the seaward limits of the territorial sea of England and Wales where no statutory provision is already available.

CROWN LAND

12. Clause 25 of the Bill provides a similar protection to the compulsory acquisition of Crown land as contained in section 296 of the Town and Country Planning Act. It ensures that an order made under the new procedure does not prejudicially affect Crown land. In effect it prevents the exercise of any powers in an order for the compulsory acquisition of Crown land or interest in land. If such land is required for, say, a railway project, it could only be acquired by agreement with the appropriate Crown authority.

ENVIRONMENTAL ASSESSMENT

13. The rules relating to applications for orders made under clauses 1 and 3 will make clear that all orders for works will be subject to environmental assessment in accordance with the relevant European Community Directive (85/337/EEC). Except for very minor works proposals, the draft order will need to be accompanied by an environmental statement. If the applicant believes that a statement is not required he will need to seek the Secretary of State's endorsement of that view before submitting the draft order, otherwise it would not be accepted. A similar procedure applies to private Bills under new Standing Order 27A. A project involving both landward and seaward construction works would be treated as a single scheme for the purposes of deciding whether an environmental statement was needed.

14. The requirements for environmental statements in regard to harbour orders in England and Wales are already laid down in regulations (SI 1988/1336) which put into effect the EC Directive. The Harbour Works (Assessment of Environmental Effects) (No 2) Regulations (1989) (SI 1989/424) applied the requirements of the Council Directive to harbour works below low water mark which do not require planning permission and are not authorised by or under any enactment.

Transport and Works Authorization Division
Department of Transport
6 *February* 1992

APPENDIX 29

Visit to HR Wallingford

NOTE BY THE CLERK

PARTICIPANTS

Sir Hugh Rossi (Chairman)
Mr Steve Priestley (Clerk)

BACKGROUND

HR (Hydraulics Research) Wallingford was formed in 1982 to take over and extend the work of the DOE's Hydraulics Research Station which had been set up many years before to look into problems of natural water flows and associated sediment transport.

Dr Stephen Huntington, Director of Operations, opened the discussion. Since 1982 HR has operated as a non-profit distributing company and has altered the balance between consultancy work and research, partly in order to maintain its income in a commercial environment.

CAPITAL DREDGING

Dr Alan Brampton, HR's Coastal Processes Section Manager, described the company's role in relation to offshore dredging for the construction industry as " to comment on the effect of dredging on the coast ". The work on which such a commentary can be based is increasing in complexity as local authorities and other commissioning bodies require higher grade assessment of applications to dredge.

HR has long felt that the Crown Estate Commissioners should deal with applications for dredging more openly and great progress has been made in this respect since 1985. Dredging companies, on the other hand, spend a great deal of money on prospecting and guard the results of such work jealously. Although there are moves towards requiring fuller environmental impact assessment of dredging applications, HR is not at present asked to assess the likely effects of an operation on, for example, fishing grounds although it has the expertise necessary. However, it was pointed out that the dredging industry wants " clean " aggregate (that is, not contaminated by plant or animal life). Sand on the seabed has often been deposited following coastal erosion and may have little value as a marine habitat. Nevertheless, the current state of knowledge of the ecological value of some seabed environments—such as sand—is incomplete.

The DOE's Minerals Planning Directorate operates a presumption in favour of granting permission and is regarded by HR as being too remote from local concerns. There are also regional differences in the attitude to dredging operations. Local factors, including political and economic pressures, may determine that attitude. Local elected representatives and pressure groups tend to look for a scapegoat to blame for coastal erosion. Frequent assertions by them to the effect that offshore dredging for construction material must be responsible for much coastal erosion are, in Dr Brampton's view, not proven, and completely unfounded.

REGIONAL PLANNING BODIES

HR is concerned that if local authorities are given planning control over the seabed, a combination of lack of expertise and the NIMBY attitude will result in the taking of bad decisions. HR supports the creation of regional strategic planning bodies, for example five for the English coast, in place of the existing informal groupings. MAFF has commissioned HR to produce coastal cell maps in order to assist regional groupings to rationalise their structure along geomorphological, rather than administrative, lines.

The NRA's regional structure does not provide a good fit with that favoured by HR, though it would represent an improvement on the present *ad hoc* arrangements. It was thought that regional planning authorities would be more suitable for considering applications to dredge and for determining policy on sea defence, than for actually deciding upon, and carrying out particular schemes.

THE 18 METRE LINE

Asked about the validity of the 18 metre depth line beyond which it was assumed that dredging operations would not affect the coastline, Dr Brampton explained the historical basis for the assumption. Research had been carried out off the Worthing coast in 1969–71 at depths of 9, 12, 15 and 18 metres. The data for 15m could not be gathered due to the presence of wartime munitions. At the time it was considered that movement may have been taking places at 9m and 12m depth, but not at the 18m depth. Recent review of the data even puts the evidence for significant movement at 9m or 12m in doubt. The 18m line thus may well represent an over-provision by a factor of two. It was never intended that this arbitrary limit should attain the status of a cardinal rule, but in the absence of any other data that was what had occurred. More recent research has broadly confirmed the interpretation of the Worthing experiment, and a further project is about to commence. In HR's view, each site should be assessed on its merits, and the 18m limit is " dead ".

MAINTENANCE DREDGING

According to Dr Brampton, the environmental effects of maintenance dredging are greater than those of offshore dredging for the construction industry. Removal of sand or shingle from a navigation channel dredged through a beach can interrupt natural sediment movement processes and cause coastal erosion. The dumping of silt from, for example, estuarial shipping channels, can have an unwelcome effect on the local marine environment. Such operations are generally carried out by a port under their own powers, although the location of spoil dumping grounds has to be agreed by MAFF (Marine Environmental Protection Division) and accepted by the Crown Estate.

MISCELLANEOUS

Asked about the possible erosion of coastal landfill sites, Dr Brampton said that a colliery spoil and domestic waste disposal site near Workington was subject to such erosion, but the local authority was not able to prevent it. In general, local authorities are felt by HR to be too reactive and in particular they do not research the likely consequences of their *inactions*.

HR is active in researching the consequences of global warming in terms of a possible rise in sea levels. There is, however, a general lack of research into the effects on wave formation and very little wave recording.

In recent times, HR has begun to diversify into areas of environmental studies.

The visit continued with a tour of one of the buildings housing scale models of various estuaries, rivers and coastlines, as well as test facilities. Discussions continued over lunch with the Managing Director of HR Wallingford, Dr T J Weare.

ISBN 0-10-290692-0

9 780102 906929

Printed in the UK by HMSO
Dd 0202906 C7 1475/3 4/92 3214999 19452